GEOLOGY

Illustration for cover:

Designed by Frédéric Camponova; image of the Earth from NOAA (http://www.ngdc.noaa.gov/mgg/fliers/95mgg01. html); photo of a tunnel boring machine from Wirth, with permission from Wirth TBM GmbH, Erkelens, Germany.

GEOLOGY
BASICS FOR ENGINEERS

Aurèle Parriaux

Federal Institute of Technology Lausanne
(Ecole polytechnique fédérale de Lausanne EPFL)
Lausanne, Switzerland

√CD

English translation by Patricia Bobeck,
Geotechnical Translations,
Austin, TX, USA

CRC Press
Taylor & Francis Group
Boca Raton   London   New York   Leiden

CRC Press is an imprint of the
Taylor & Francis Group, an **informa** business

A BALKEMA BOOK

CRC Press/Balkema is an imprint of the Taylor & Francis Group, an informa business

© 2009 Taylor & Francis Group, London, UK

Typeset by Macmillan Publishing Solutions, Chennai, India
Printed and bound in in Hungary by Uniprint International (a member of the Giethoorn Media-group), Székesfehévár.

Published by: CRC Press/Balkema
                  P.O. Box 447, 2300 AK Leiden, The Netherlands
                  e-mail: Pub.NL@taylorandfrancis.com
                  www.crcpress.com – www.taylorandfrancis.co.uk – www.balkema.nl

ISBN: 978-0-415-46165-8

Dedicated to my parents Alice and André Parriaux who have spent their entire lives teaching and who have stimulated me to undertake my university education, a necessary (though not the only) condition for this book to become a reality. To my family, my children Sibylle, Gabriel, Emmanuel, Nicolas and Jean-Marie who give me happiness and strength to lead the writing of this book to successful conclusion.

About the author

Aurèle Parriaux studied geology at the Swiss Federal Institute of Technology (EPFL) in Lausanne, Switzerland. He obtained his Ph.D. in hydrogeology and followed several postgraduate courses in hydrogeology, operational hydrology and geotechnics. He acquired a wide experience in engineering geology in the fields of motorway construction, geological hazards, underground water and geomaterials prospecting as well as the management of natural resources.

In 1991, he was appointed full Professor of Engineering Geology at EPFL and presently he is head of the Engineering and Environmental Geology Laboratory (GEOLEP) at the same institute. He leads a research team of about twenty people specializing in the fields of geological hazards and underground resources.

Professor Parriaux has significant teaching responsibilities. He teaches geology to students in 'Civil Engineering' and 'Environmental Sciences and Engineering'. Moreover, he teaches 'Engineering Geology' at the Universities of Lausanne and Geneva.

Parallel to his research and teaching, Aurèle Parriaux carries out expert appraisals in various fields of engineering and environmental geology. In particular, the recent appraisal of the compatibility between construction of tunnels and protection of groundwater resources.

Since the creation of the new School of Architecture, Civil and Environmental Engineering (ENAC) at the Swiss Federal Institute of Technology in Lausanne, he participates in the teaching related to territory and landscape into which he brings the geological and geomorphologic component.

Aurèle Parriaux is active in several international organizations. He was chairman of the Swiss Hydrogeological Society for six years. From 2001 to 2006 he was Director of the Civil Engineering Section of the Swiss Federal Institute of Technology of Lausanne.

In 2006, he published his book "Géologie: bases pour l'ingénieur". The second edition of this successful textbook will be published in 2009. In competition with 105 scientific books, "Géologie : bases pour l'ingénieur" received the Roberval Prize in 2007. The publisher CRC Press/Balkema, member of the Taylor & Francis Group decided to publish an English translation of the book (Geology: Basics for Engineers, 2009).

In December 2008, Prof. Parriaux was nominated Chevalier of the Order of Academic Palms by the Prime Minister of the Republic of France.

Contents

13   Weathering

14   Geology's Role in the Major Issues Facing Society

Foreword

This book is the new and improved second edition of the original text published in 2006. It is a product of my experience teaching geology at the Federal Institute of Technology Lausanne (EPFL). Since 1991, I have taught the basics of geology to civil and environmental engineering students. This is challenging because of difficult pedagogical conditions due to the heavy course load of the first year of the curriculum, the various career plans of the students, and an enrollment of more than 100 students. It was necessary early in the process to create course materials that corresponded to the particular needs of engineers, without delving too much into engineering geology per se, which is the subject of a course for more advanced students. Thus my staff and I drafted the first version of a course guide for internal use. It was the basis for a course that included extensive student participation in numerous exercises to promote mastery through problem solving. This internal document caught the attention of the management of the Presses Polytechniques et Universitaires Romandes (PPUR) who asked me if I would turn it into a "real book" to be offered to a larger audience than the EPFL. The favorable publishing conditions offered by the Presses Polytechniques et Universitaires Romandes, the EPFL, and the Natural, Architectural and Constructed Environment (ENAC) School won me over and together we created a true partnership. I thank my colleagues Manfred A. Hirt, author of the Treatise on Civil Engineering, Mr. Olivier Babel, Director of PPUR, and my colleague Laurent Vulliet, Dean of the ENAC faculty.

To create the first edition, the Environmental and Engineering Geology Laboratory (GEOLEP), under my direction, labored for more than a year to produce a manuscript worthy of inclusion in the prestigious PPUR collection. The project involved the adaptation of the internal course guide into a central part of the EPFL academic environment while at the same time maintaining its original pedagogical concept. The work group included Laurie Cortesi, my primarily collaborator on this project to whom I express my most sincere gratitude. The book owes much to her. She participated actively in the review of the text and brought her personal contribution to it, in particular, her scientific contributions enriched the chapters on mineralogy and magmatism. She also coordinated the revision of a multitude of photographs and figures that embellish the text and improved them significantly by turning them into color illustrations. She very capably managed the flow of data between the publisher and our group. Her scientific and human qualities were vital to the success of the process.

The GEOLEP editing group included three other people who contributed with their geologic and computer science abilities to create the animations that appear on the accompanying DVD: Séverine Bilgot, Grégory Grosjean, and Laurent Tacher. Tamara Lema very capably and insightfully tracked down, detected, and identified the inevitable misprints in the manuscript. Several student assistants also collaborated on the project Giuseppe: Lea Herzig, Aude Dessauge, and Alexandre Monnin. I owe gratitude to all of them, and to Giuseppe Franciosi who was my principal collaborator during the internal document period and who contributed immensely to its preparation. I also express my thanks to my colleagues François Frey and Peter Egger who kindly reread some passages. My thanks also go to PPUR and their openness to the particularities of the very diverse and rich discipline that is geology.

The preparation of the second edition in French has been done in parallel with the translation of the book into English, published by CRC Press/Balkema, member of the Taylor & Francis Group. My gratitude goes primarily to Patricia Bobeck, Geologist and translator, who made this translation using her excellent skills. This is an important step in which my laboratory has also been involved. I thank my assistant Diana Marques who helped me with the editing in collaboration with Mike Bensimon, Pascal Turberg, Julien Zigliani, Pascal Blunier, Yuri Gramigna, Clémentine Schurmann and J.L. Huesser.

The second edition includes updated versions of the thirteen chapters that made up the first edition. After the definition of the subject matter and a historical account of the scientific discoveries that have led to modern geology (Chapter 1), the Earth is described in its double context of space and time (Chapters 2 and 3). The physics of the globe is presented in its close relationship with engineering (Chapter 4). What follows is a big jump in scale to describe the basics of geological materials through a discussion of mineralogy (Chapter 5). We are then in a position to discuss the various types of rocks and unconsolidated sediments that make up the Earth. However, before talking about them, we systematically discuss the processes that have created them. First, we discuss the genesis of magmatic rocks (Chapter 6). Prior to taking up the sedimentary rocks, it is essential to study the hydrosphere (Chapter 7), which is intimately involved in the phenomena that generate sediments, whether on continents (Chapter 8) or in the oceans (Chapter 9). The processes that transform sediments into rock are at the heart of the chapter on diagenesis, followed by the principal characteristics of sedimentary rocks and their technical properties (Chapter 10). Metamorphism leads us to the final group of rocks (Chapter 11). With the geological substrate created in all its diversity, it is time to deform it through tectonic processes (Chapter 12). To this mechanical degradation, we add the issue of physico-chemical processes, which also have a geological component: the alteration of rocks (Chapter 13). Finally, a new chapter describes the important role of geology in the resolution of important socio-economic problems that we face today and will face in the future (Chapter 14).

[...] This broad scope of material could not be completely covered in a single book. This is why we have introduced "To Find Out More" in the form of book-shape icons in the margins of the text; they are numbered for bibliographic reference to books and articles that allow the reader to learn more if he wishes. As a rule, no references are made to internet sites because they can rapidly become obsolete; the reader can access them more conveniently by using a search engine. Important geological terms are printed in bold italics generally the first time they are used. They also appear in the index.

The participatory nature of the course for which the book was originally written has been preserved by guided practice through problem solving. The list of twenty problems presented to the reader in the first edition has been expanded in the second edition. Solutions for all the problems are on the DVD that accompanies the book. The DVD also has animated demonstrations of the subject matter, referenced by a disk-shaped icon in the margin. New animations have been added to those that appear in the first edition. They are the work of Séverine Bilgot and were translated by Andrea Mason.

My gratitude also goes to the numerous individuals and institutions who have graciously given permission for their documents to appear as illustrations in the book, and I am happy that the book is a vehicle for their contributions. Many thanks also go to Marcel Arnould, professor at the School of Mines in Paris, who has honored me by writing the preface.

Aurèle Parriaux
Lausanne, April 2009

Preface

Professor Aurèle Parriaux asked me to write a preface for his book *Geology Basics for Engineers*. I must disclose that the Presses Polytechniques et Universitaires Romandes sent the initial manuscript of this book to me for my opinion, with the greatest discretion and without revealing my name to the author. It was like a blind tasting for a wine-lover. I recognized a great vintage.

This book fills a gap in the French language. I am sure it will also be translated into numerous other languages. It is complete yet concise. It is current with the latest knowledge. It is highly educational. Its pedagogical value can be seen in the exercises throughout the chapters and their solutions on the DVD that accompanies the book. The DVD also contains animations, a novelty in the field. The book contains an abundance of illustrations with many original photographs that embody the spirit of observation that is at the heart of the science of geology. The clear text promotes perfect understanding and makes it easy to assimilate knowledge.

The book is first of all an excellent treatise on geology. It can also be used as a reference work. Starting with the introduction, the author explains why geological competence is today a necessary component of the engineer's toolkit, and he refers to practical applications throughout the book. He combines the fundamental presentation with applications and illustrates the two harmoniously in a highly original cross-fertilized manner.

Geology Basics for Engineers is a timely book. We all know the importance of exploration for and exploitation of mineral resources, oil, gas, water and non-metallic minerals for construction and engineering. We also understand the importance of geology for infrastructures. The development of international relations makes it possible for us to build tunnels across mountains and seaways, and each of these is an exceptional construction project. One example of an alpine tunnel is the Lötschberg tunnel in Switzerland, a masterpiece of precise geologic knowledge and coordination that resulted in the construction project finishing on time and under budget.

In the same way, environmental protection and society's growing demand for safety leads to a multiplication and a sort of democratization of geologic studies. For example, for waste storage projects, concerned citizens now demand guarantees of protection from pollution, which are very hard to prove. Citizens demand similar guarantees in the areas of protection from natural hazards and in development, urbanization, and land use.

Another growing demand calls for using the historical approach to geology, a discipline that blithely deals with millions of years, to forecasting the future. These forecasts may involve widely different scales. Scientists discuss a maximum of a million years for the underground storage of radioactive wastes with long-lived actinides down to 300 years for the storage of low and medium activity and short duration wastes. Some concrete construction projects, such as the tunnel under the English Channel, the bridge over the Tage in Lisbon, and earthen projects for high-speed trains require guarantees of up to a century without significant maintenance. We exist on the human scale. These are new challenges. We are likewise challenged by other problems of intercontinental, even planetary, scale such as the evolution of the climate and its potential control, which are magnificently exposed in the book.

For its research into the origin of the world, the evolution of the planet, the earth's changing geography, the appearance – and sometimes the disappearance – of living species, and the reconstruction of the genealogical tree of humanity, geology is a powerful cultural tool. We must also learn how to use it in the service of our fellow citizens.

As a spokesman for the readers of *Geology Basics for Engineers*, I thank and warmly congratulate Professor Aurèle Parriaux for putting his altruistic philosophy to use and sparing no effort in sharing his knowledge to serve science and progress.

I add a word of congratulations and thanks to the Presses Polytechniques and Universitaires Romandes for the quality of this book.

Marcel Arnould
*Honorary Professor at the Ecole des
mines de Paris and the Ecole nationale
des ponts de chaussées, Paris.
Honorary President of the International
Association of Engineering Geology
and the Environment.*

Addendum for the Second Edition

The success of Professor Aurèle Parriaux's book and sellout of the first edition was expected because there had been no similar textbook in the French language. The scope of its success is due to its remarkable scientific and pedagogical quality. The book received the Roberval Higher Education Prize in 2007, an award that promotes the diffusion of scientific and technological knowledge in the French language. Professor Parriaux's book was the winner of an intense competition among more than one hundred candidates representing the entire spectrum of scientific knowledge.

Motivated by a goal of service, Aurele Parriaux has decided to go beyond a mere reprinting the book. He has prepared a second edition that contains numerous improvements to the text, new animations, and new exercises. He has also written an additional chapter devoted to "Geology's Role in the Major Issues Facing Society." Within the broad scope required for addressing these problems, the author has maintained an engineering point of view. For example, the pages dedicated to pollution and underground remediation constitute a veritable instructional manual for an engineer. With thanks to Aurèle Parriaux and the publisher. I'm looking forward to a third edition.

Marcel Arnould

1   The Geology-Engineering Partnership

The engineer's job is to design and build structures that interact continuously with the soil and subsurface. The goal of this book is to give engineers an academic understanding of Earth science and the basics needed to conduct engineering activity appropriate to the various environments that exist on our planet. The objective is to provide:

- The knowledge and ability to analyze the geological processes that create, deform, and weather rocks;
- An understanding of the principal types of rocks and unconsolidated materials and their properties;
- An illustration of how geological conditions affect engineering activities, and how they can simplify or complicate these activities;
- An appreciation of the richness of underground resources and an understanding of how to manage them, geared toward modern engineers seeking sustainable development in an interdisciplinary environment.

This book is a presentation of the scientific basics for engineering geology. The discussion of geological fundamentals is simplified to meet the specific needs of engineers. Other more specialized texts cover the subject of engineering geology in greater detail.

In this introductory chapter, we discuss the areas of geology of interest to engineers and illustrate how geology contributes to the resolution of civil engineering and environmental problems. We will also examine the ancient origin of Earth science as early man tried to answer questions raised by his observation of the land.

[10, 18, 21, 78, 109, 141, 166, 234, 305]

1.1   Areas of engineering geology

The International Association of Engineering Geology and the Environment (IAEG) gives a detailed definition of engineering geology:

"Engineering Geology is the science devoted to the investigation, study and solution of the engineering and environmental problems which may arise as the result of the interaction between geology and the works and activities of man as well as to the prediction of and the development of measures for prevention or remediation of geological hazards.

Engineering Geology embraces:

- the definition of the geomorphology, structure, stratigraphy, lithology and groundwater conditions of geological formations;
- the characterisation of the mineralogical, physico-geomechanical, chemical and hydraulic properties of all earth materials involved in construction, resource recovery and environmental change;
- the assessment of the mechanical and hydrologic behaviour of soil and rock masses;
- the prediction of changes to the above properties with time;
- the determination of the parameters to be considered in the stability analysis of engineering works and of earth masses; and
- the improvement and maintenance of the environmental condition and of the properties of the terrain."

This book will use concrete examples to illustrate the importance of geology in these areas.

It should be noted that the fundamentals of geology are one of the curriculum basics in the following branches of engineering:

- Civil engineering
- Environmental sciences and engineering
- Soil and rock mechanics
- Hydrology
- Hydrogeology
- Pedology

Beyond the scientific basis, the study of Earth science influences the way an engineer thinks, both in the exercise of his profession and in his culture, as follows:

- Greater sensitivity to the siting of an engineering project (the relationship between the structure and its environment);
- An understanding of different scales of space and time;
- A broader qualitative vision of problems that need resolution, along with quantitative notions;
- Encouragement of an interdisciplinary focus;
- Critical thinking when using local measurements as a base for a regionalization and the construction of simulation models;
- An introduction to modern environmental challenges involving the subsurface (for example, the geological storage of wastes, contaminated sites. See Chapt. 14.).

In this book, the areas where geology and engineering intersect are presented as four large groups, as shown below (§ 1.1.1 to 1.1.4).

1.1.1 Project construction

The geologist collaborates with the geotechnical engineer to determine if a piece of land proposed for construction has the necessary stability conditions. The two professions complement each other in the following ways:

- The *geologist* is a scientist who analyzes the subsurface structure of a site on the basis of his experience of the region and by surveys and borings. He then makes predictions about the nature of the subsurface materials to be encountered in the excavations and describes them in terms of characteristics that are useful to the engineer. He evaluates the worksite's risk to the environment.
- The *geotechnical engineer* is a civil engineer specialized in the measurement of mechanical properties of foundation soils. He orders laboratory tests and interprets their results. He calculates the dimensions of the foundation in collaboration with the design engineer.

[44, 130, 156, 157, 170, 186, 284]

This distribution of tasks between the geologist and geotechnical engineer varies from country to country. In practice, geological conditions will play the major role in the choice of the foundation type for any project (Fig. 1.1).

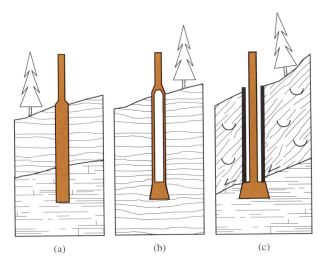

(a)   (b)   (c)

Fig. 1.1 Bridge foundation piers; (a) simple concrete filled piers in stable overburden for a foundation on a substrate with good bearing capacity; (b) light pier, hollow, with an elephant-foot base for stable terrain but with low bearing capacity; (c) cased pier for overburden that is undergoing active sliding.

1.1.2 Natural hazards

Numerous natural hazards threaten man, his structures, and nature. They cause costly catastrophes that destroy property and human life. The financial risk of minor events can be covered by insurance companies; however, this is not possible for major disasters such as large earthquakes.

Fig. 1.2 Example of the Chenaulaz landslide (1983); destruction of the Pully-Belmont road, east of Lausanne, Switzerland. GEOLEP Photo, J.H. Gabus.

Natural hazards (Fig. 1.2) are often the result of geologic phenomena such as slope erosion, earthquakes, and volcanism.

The geologist and engineer work in concert to identify, evaluate, and map these dangers and to develop measures to protect against them.

[48, 127, 151]

1.1.3 Geological resources

The subsurface contains fundamental resources for human life. These resources are necessary for man's very existence (groundwater), or for human socioeconomic development (geomaterials, minerals, energy).

Geomaterials are defined as unconsolidated or rocky geological material. Raw and processed geomaterials are used for industrial applications and for construction. Untreated gravel is widely used for foundations. Argillaceous limestone is processed to make Portland cement.

In collaboration with the engineer, a geologist discovers and evaluates mineral deposits, proposes quarry sites, studies the chemical and mineralogical composition of geomaterials, their resistance to weathering, their continuity within a deposit, the possible impacts of mining on water resources and the stability of the surrounding land. He proposes an appropriate reclamation plan. Problem 1.1 concerns these issues at a cement plant.

[43, 63, 64, 155, 176, 196, 208, 209, 301, 314]

PROBLEM 1.1

There was a cement factory north of the village of Roche (Switzerland) that used to mine two quarries. You have been asked to study the possible reactivation of a Portland cement factory at the site (fictitious case).

The quarries contain limestone and argillaceous limestone (Fig. 1.3).

Required composition for a mixture of calcined rock to produce Portland cement

CaO	35–42 (wt. %)
Al_2O_3	3–8.5 (wt. %)
Fe_2O_3	0–4 (wt. %)
SiO_2	15–25 (wt. %)
CaO/SiO_2	1.25–2.5[–]
$SiO_2/(Al_2O_3+Fe_2O_3)$	2–4[–]
Al_2O_3/Fe_2O_3	≥2[–]
MgO	maximum 5 (wt. %)
SO_3	maximum 3.5 (wt. %)

Geochemical analysis of two types of calcined rock (wt. %)

	Argillaceous limestone Red beds	Massive limestones Malm
SiO_2	23.10	0.85
Al_2O_3	4.90	0.39
Fe_2O_3	1.40	0.13
CaO	35.60	52.96
MgO	0.51	1.80
SO_3	0.01	0.06
K_2O	0.79	0.3
Na_2O	0.24	0.1
P_2O_5	0.09	0.0
Loss by fire*	32.70	43.01
Total	99.34	99.6

*compounds volatilized during calcination.

Questions

(a) What are the different possible mixtures of these two rocks that would meet the chemical criteria for raw material in this process?

(b) Identify other criteria specific to the site that may influence the choice of the mixture.

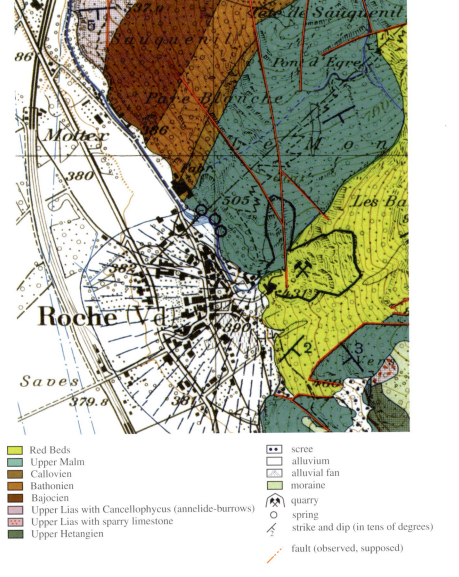

Fig. 1.3 Location of the Roche quarries (Canton of Vaud, Switzerland) showing the boundaries of the geological formations. Extract from the Geologic Atlas of Switzerland 1:25,000, 1264 Montreux Sheet. Reproduced with authorization of the Federal Office of Water and Geology (OFEG), 4/11/2005, and Swisstopo (BA056985).

Look at the solution on the DVD

1.1.4   Environmental pollution

Numerous areas of engineering activity impact the environment. Geological issues are the overriding concern for many projects such as geological waste storage, management of contaminated sites, and engineering project impact studies (Fig. 1.4).

Fig. 1.4 Example of a waste dump showing its direct influence on surface water and groundwater. GEOLEP photo, M. Looser.

As an example, the search for burial sites for nuclear waste, which is a serious problem for society, employs thousands of people. The data acquired and the tools developed during this research have led to impressive advances in geological knowledge in recent decades.

[92, 107, 279, 285, 291, 310]

This discussion will be continued in chapter 14 where we will focus on the significant challenges our society faces in the future and on the role geology may play in their resolution.

1.2   The history of geology

Writings produced by ancient civilizations have been influential in the development of human thought throughout history. These records include observations that awoke the curiosity of early travelers, who, for example, noted rocks composed of shells identical to seashells (Fig. 1.5). They imagined that floods had uplifted these near-shore rocks at some time in the past. But when it became known that the same type of rocks was also present in mountains far from the sea, their existence became a mystery. It is an amazing paradox that the summit of Mt. Everest (Fig. 1.6) is made up of marine sediments.

We will point out some milestones in the history of geology from its origins in Antiquity to show the evolution of thought about the Earth. These ideas sometimes upset dogma and the most tenacious beliefs. The principal stages are summarized in table 1.1.

Let us recall the creation story in Genesis I (a text that in its present form probably dates from the 6th century BC) "In the beginning, God created the heavens and the Earth [...]. God said: Let the water under the heavens be gathered together in one place, and let the dry land appear. And it was so. God called the dry land Earth and the waters that were gathered together he called Seas."

Fig. 1.5 *Lumachelles* are rocks composed predominantly of fossilized marine shells. GEOLEP photo, G. Grosjean.

Fig. 1.6 The roof of the world (8848 m) is made of limestone from marine origin. GEOLEP photo, P. Christe.

Table 1.1 Significant events in the history of geosciences

Century	Author	New Idea
13th BC	Moses	Divine creation of the Earth, seas and living beings
6th BC	Xenophanes of Colophon	Fossils are ancient marine organisms
4th BC	Aristotle	Cyclic movement of Earth and seas
1st	Strabo	Uplift of wedges of Earth's crust (beginning of tectonics)
16th	Leonardo da Vinci	Mountains are ancient shorelines that have been uplifted
18th	Buffon	Geologic time and thickness of sediments
19th	Cuvier	Biologic transformation as a result of catastrophes
19th	Darwin	Slow evolution by natural selection
20th	Wegener	Continental drift

Later comes the story of the creation of man and woman in the form of Adam and Eve. This notion of divine creation, considered by some believers as a universal and indisputable truth, has dominated the centuries that followed. It has established constraints within which scientists of all ages have had to struggle.

The oldest observation that has come down to us (6th century BC) is that of the philosopher Xenophanes of Colophon, which has been reported to us as follows:

"Xenophanes believed there was mixture of Earth and sea and that it was time that caused the separation. As proof he offers that there are shells in the ground and in the mountains, that in the quarries of Syracuse, there are imprints of fish and seals, at Paros there is an imprint of anchovies in the middle of a stone, in Malta, there are seals and all kinds of things from the sea. This happens, he says, because in the past everything was mud, and when the mud dried, the imprints were preserved" (according to St. Hippolytus).

Thus Xenophanes recognized that the fossils of ancient marine organisms show that at one time everything was a sea.

Two centuries later, Aristotle proposed the basics of geodynamics by discussing the upheavals that the Earth must have experienced over time.

"It is not always the same parts of the Earth that find themselves under sea, nor the same parts that are dry. There is a permutation between the continent and the sea. These places do not always remain sea, nor do the others always remain terra firma. Where once there was land, a sea has now formed. There where the sea extends today, the land will reappear. We must think that these transformations occur in a certain order and that they follow a certain cycle." Aristotle, Meteors.

With the hypothesis that the creation of mountain chains is cyclic, Aristotle is a precursor of Wilson's theory of cycles (Chapt. 3).

In the 1st century BC, Strabo of Cappadocia introduced the notion of transgression for marine fossils, an important geodynamic element.

"How can it be that in places that are in the middle of land and separated from the sea by two or three thousand stadia, we see a numerous piles of oyster shells and cheramides and in stagnant lakes where the water is salty? It must be that the bottom rises at times and falls at other times, the sea then rises and falls at the same time as this bottom. When it rises, it inundates the coastal areas, and when it lowers, it goes back inside its bed" Strabo, Geography.

Strabo thus establishes a link between tectonics (raising of a wedge of the Earth's crust) and the sea's invasion of the continents.

The rise of Christianity marked the beginning of a period of scientific repression that continued, as in astronomy, until the 16th century.

The long and gloomy Middle Ages were enlightened all the same by a few "lights" such as Avicenna, in the 11th century, who made interesting statements about erosion and mountain formation:

"Mountains can result from two causes, either upheaval of the ground, as during earthquakes, or by the effects of running water and wind, which carve valleys in soft rocks and makes hard rocks appear, which is the process that has made numerous hills. These changes have taken a very long time and it is possible that the sizes of mountains are now in the process of decreasing." Avicenna, Writings.

Religion's opposition to change was first challenged by Leonardo da Vinci at the beginning of the 16th century. He did not mince words:

"One can only admire the foolishness and simplicity of those who say these shells (= fossils) have been transported by the Deluge [...]. If that was the case, they would have been strewn about randomly, mixed up with other objects, all at the same level. But the shells have been deposited in successive stages: they are found at the base of the mountain and at its summit [...].

The mountains where shells are found were formerly riverbanks beaten by waves, and since then they have been elevated to the height where we see them today." Leonardo da Vinci, 1508 (notes).

Two centuries later Buffon made a first incursion into the problem of time and the age of the Earth, through his observations on tidal areas.

"Let us suppose that each tide deposits an amount of sediment equal to 1/12th of a line of thickness [...] the deposit would increase to a line in 6 days [...] as a result, about 5 inches in a year; which means it takes more than 14,000 years to build a clay hill a thousand toises high (about 2,000 m)." Buffon, Epochs of Nature.

On the basis of these considerations, he estimated the age of the Earth to be at least 75,000 years, which exceeded man's understanding at this time (in the 17th century it was thought that the Earth was between six and eight thousand years old).

The paleontologists of the 19th century made significant contributions to the understanding of geology and the history of life.

Lamarck, precursor of evolution, wrote in 1809 that animals descend one from another according to a plan defined by the Creator and that the evolution of beings is a result of their adaptation to their life environment: "Function creates the organism." A useless organ ends by atrophying then disappearing.

Charles Lyell, one of the "fathers of geology" extended this idea. He introduced the idea of the principal of *uniformitarianism*: "the present is the key to the past." This means that the processes that have modified the appearance of the Earth in the past are the same processes that are operating today. He associated biological stability to this physical stability: the changes that fossils undergo over Earth history take place gradually. Some species disappear, and others are created. The new one does not differ fundamentally from the previous species. By studying the cycles of modifications in living species, Lyell was able to estimate the date of the beginning of the Paleozoic at 240 million years ago (in fact, it is 590 million years, Chapt. 3).

This opinion was harshly challenged by the partisans of *catastrophism*. This group, including the famous paleontologist George Cuvier, thought that catastrophic events have been the major factor in modifying the Earth and causing biological changes at its surface.

Finally, Darwin, a student of Lyell, took the first steps toward the theory of *evolution* of life on Earth through natural selection.

Toward the end of the 19th century, the physicist Kelvin suddenly cast doubt on the long life of the Earth. On the basis of its cooling rate, he calculated the age of the Earth to be somewhere between 20 and 200 million years.

The world would have to wait for Becquerel's discovery of radioactivity in 1896 to understand Kelvin's error, which was the failure to consider the Earth's initial heat only. Since then, numerous scientists have studied radioactivity but it was not until 1955 that Patterson's studies, based on the ratio of lead isotopes in meteorites, set the probable age of the Earth at 4.6 Ma.

In 1912 Wegener proposed his famous theory of continental drift. This fascinating hypothesis found favor among certain geologists, but equally fierce opposition among geophysicists who challenged the idea that the Earth's crust was capable of movement. It wasn't until the 1960s that geophysical surveys of the ocean floors caused geophysicists to consider this theory. This was undoubtedly the last major revolution in the history of geology.

In Switzerland, three scientists of the middle of the 20th century merit citation because of the importance of their discoveries in relation to this book and because they have been professors in Swiss universities:

- Emile Argand, Heim and Lugeon, of Neuchâtel, for the theory of tectonic nappes;
- Albert Heim, of Zurich, for his work on large landslides (Fig. 1.7);

- Maurice Lugeon, of Lausanne, for engineering geology (particularly dams), before it   📖
  was called Engineering Geology.   [59]

We could not end this chapter without honoring these precursors among others and their courage for having proposed theories that were often shocking at the time.

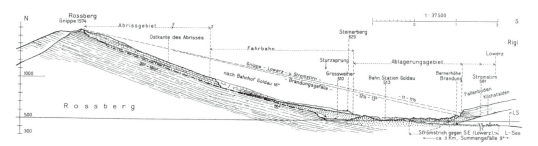

Fig. 1.7 Drawing from Albert Heim (1932) of the rock avalanche of Arth-Goldau (Swiss Alps). When a mass of falling or sliding rock becomes very important, other mechanisms than common fall lead to an exceptional widespread distribution of the debris. The very large kinetic energy due to collisions between particles (velocities up to several hundreds m/s) allow them to be transported very far from their source. The common height/length ratio of the debris trajectory have exceptionally low values. This kind of phenomenon was first observed and defined by Albert Heim in the Alps (called in German "Sturzstrom"). Every year, on September 2nd at 5 pm, the great clocks of Goldau's church ring to recall the dramatic 1806 Sturzstrom which destroyed in a few seconds the valley between Rigi and Rossberg. The Rossberg massive conglomerate summit slid on a very regular strata of marls dipping of about 20 degrees. The mass was about 40 million of m³. The debris covered an area of 6.6 km². The rock avalanche destroyed 111 houses, 4 churches and 220 stables. 700 persons lost their lifes. Heim measured on his profile an longitudinal angle of only 11° to 13°, which is typical for rock avalanche.

Chapter Review

Geology is a diverse and complex natural science involved in many engineering activities. Engineering geology helps the engineer find solutions adapted to geologic conditions. Although they have different mentalities and sensibilities, the engineer and the geologist must understand each other well for their collaboration to be effective. This requires a mutual approach, and it is hoped that this book will contribute to the partnership of these two disciplines.

2 The Earth in Space

Before focusing our attention on the Earth itself, let us examine the Earth's place in the Universe, and particularly in the solar system. This will give us a sense of the scale of distances and will place the Earth in its spatial environment. Is the Earth unique?

2.1 The solar system

In the preceding chapter we reviewed the evolution of scientific thinking about the processes that have shaped the Earth's surface. A similar evolution of thought has produced the current model of the Universe and we can be certain that this model will continue to change in the centuries to come. The model of the solar system will probably not be modified significantly in the future. Let us look at some major stages of thought that have influenced the history of science.

2.1.1 Historical representations

Ancient civilizations originally believed the Earth was flat, as shown by Hecataeus's circular map, for example (Fig. 2.1). What lies outside this map is beyond the imagination. This model was refuted early in history but it has persisted in the minds of men for centuries.

Fig. 2.1 Hecataeus's map from the 5th century BC (from [161]).

In the second century BC, Eratosthenes was the first to demonstrate the spherical nature of the Earth and calculate its circumference (Fig. 2.2). We will discuss this discovery in Problem 2.1.

The first representations of the solar system are geocentric, just as the first representations of the Earth were flat. Ptolemy refined this erroneous model in the second century by correcting aberrations through scholarly calculations (Fig. 2.3).

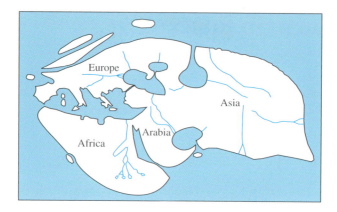

Fig. 2.2 Eratosthenes' sphere (from [221]).

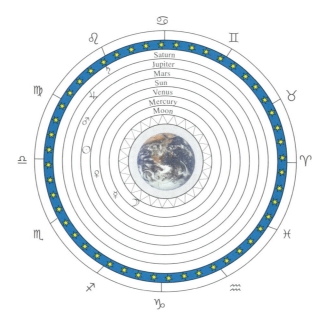

Fig. 2.3 Representation of the *Geocentric model*.

In spite of attempts during the Middle Ages to challenge geocentrism, the world had to wait until the 16th century for the Polish scientist Copernicus to demonstrate the foundations of the current model (Fig. 2.4).

[59]   Later astronomical research has shown that our solar system is but one among a hundred billion systems in our galaxy and that the Universe must contain at least one billion galaxies.

2.1.2   General structure

Let us return to our star and examine the structure of our solar system (Fig. 2.5) to determine the Earth's place in space.

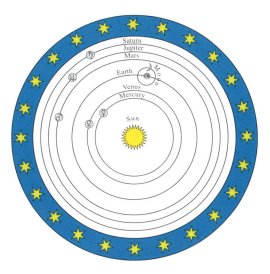

Fig. 2.4 First representation of the **Heliocentric model**, here by Copernicus.

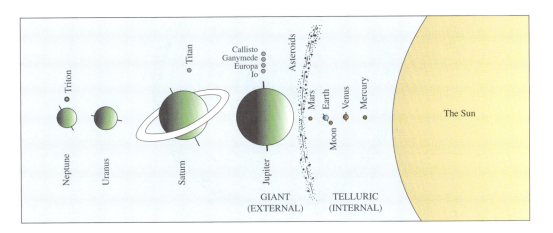

Fig. 2.5 Current schematic representation of the solar system. Only the relative sizes of the planets are shown.

The Sun is surrounded by bodies of various types that follow elliptical orbits as they revolve around the Sun. The Sun occupies one of the foci of the orbits. The bodies are:

- eight planets with their satellites
- a multitude of asteroids
- comets

The four planets closest to the Sun are called telluric (or internal). The outside planets are called giant (or external).

The trajectories of the planets lie very close to the same plane. Pluto, which was considered to be a planet until 2006, deviates from the plane by 20°. All of the revolutions have slight eccentricities. The planet's revolution around the Sun defines the length of its year and its own rotation defines the length of its day.

In the 17th century Kepler (Fig. 2.6) wrote three fundamental equations describing the movement of the planets around the Sun.

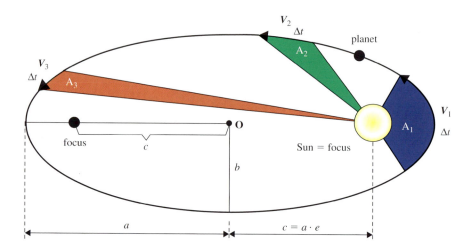

Fig. 2.6 Kepler's three laws. (I) The Sun occupies one of the foci of the ellipse. (II) A line drawn from the Sun to the planet sweeps out equal areas in equal times. This means also that the ratio dA/dt is constant → maximum velocity near the sun. (III) The period T is related to the size of the ellipse by the ratio (T^2/a^3) which is constant → the length of the year increases faster than the size of the ellipse. Eccentricity is given by the ratio: $e = \sqrt{a^2 - b^2}/a$. Note: the eccentricity in the diagram is greatly exaggerated.

2.1.3   The Earth's revolution

Kepler's laws apply to the Earth's revolution. The Earth rotates on its axis, which is inclined at an angle of approximately 23.5° from the perpendicular to its trajectory around the Sun (*ecliptic plane*). This obliqueness to the ecliptic creates Earth's seasons. As shown in figure 2.7, the orientation of the axis in space remains almost constant during the Earth's revolution around the Sun.

In reality, this axis rotates very slowly, completing a turn every 26,000 years. This is called the *precession of the equinoxes*.

The two positions of the solstices explain the extreme seasons for the two hemispheres (Fig. 2.8), the definition of the tropics, and the polar circles.

The Earth takes 365.256 days to revolve around the Sun, which necessitates one leap year every four years. The length of the year has probably not changed significantly during the history of the Earth. This is not true for the Earth's rotation on its axis. Today, the day is 23.934 hours long. The study of coral growth from its appearance in the mid Paleozoic until today shows that the number of daily cycles in the seasonal cycle of older fossils (Fig. 2.9) is greater than that of younger fossils. This means that the Earth's rotation has slowed down over the course of geologic time, as a result of braking due to energy consumption by the tidal deformation of the Earth (especially by the Moon's attraction, Chapt. 9).

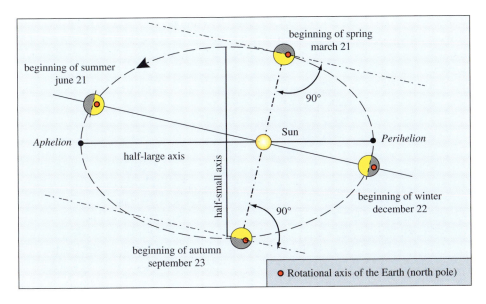

Fig. 2.7 Elliptical trajectory of the Earth around the Sun and location of solstices and equinoxes.

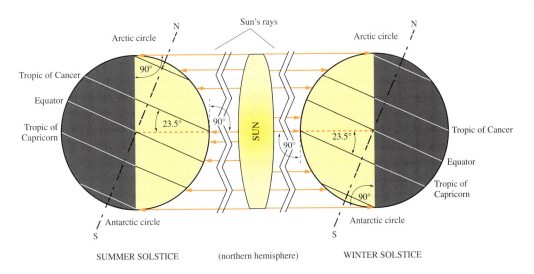

Fig. 2.8 Earth's seasons, definition of the tropics and polar circles (from [146]).

The astronomy of the solar system is now well understood, but the same is not true of the nature of the bodies that circle the Sun. Although still in its infancy, planetary geology has made considerable progress since the construction of space probes in the 1950s. The Moon landing in 1969 was a defining event.

The calculation of planet densities led very early to a classification of the planets into two groups: the heavy ones and the light ones. Table 2.1 summarizes the principal characteristics of the planets and the Moon.

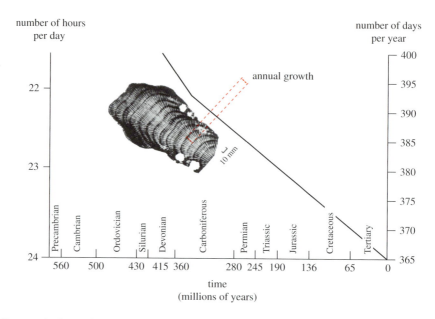

Fig. 2.9 Decrease in the number of days per year as shown by coral fossils. Photo: radiograph of a slice of a recent coral showing annual growth rings.

Table 2.1 General characteristics of the planets of the solar system and the Moon

		Telluric planets					Giant planets				
	Units	Mercury	Venus	Earth	Mars	Moon	Jupiter	Saturn	Uranus	Neptune	Pluto
Average distance from the Sun (except the Moon)	10^6 km	57.895	108.161	149.600	227.841	384404*	778.219	1426.885	2869.478	4496.527	5900.224
Period of sidereal revolution	[d]	87.97	224.70	365.26	686.98	27.3296**	4332.59	10759.22	30685.40	60189.00	90465.00
Period of rotation	[d]	58.65	−243.01	1.00	1.03	27.3296	0.41	0.43	0.72	0.74	6.39
Inclination with respect to Earth's trajectory	[°]	7.00	3.39	0.00	1.85	5.15	1.30	2.49	0.77	1.77	17.20
Diameter at Equator	[km]	4878	12104	12756	6794	3476	142984	120000	51118	49528	env. 2400
Relative Mass (Earth mass = 1)	[−]	0.055	0.815	1.000	0.107	0.0123	317.893	95.147	14.540	17.230	0.002 ?
Average Density	10^3 kg/m^3	5.4	5.3	5.5	3.9	3.3	1.3	0.7	1.3	1.7	2 ?
Gravity at the equator	[m·s^{-2}]	3.78	8.60	9.78	3.72	1.63	24.85	10.54	8.96	11.00	0.44

* Average distance between Earth and Moon [km]

** Period of revolution around the Earth

*** Considered as a planet until 2006

PROBLEM 2.1

Let's go back to the time of Eratosthenes (3rd century BC) in Egypt. His frequent travels between Syene (today Aswan) and Alexandria during different seasons allowed him to make calculations that led to that of the circumference of the Earth.

Eratosthenes had measured the distance between Syene and Alexandria as 5000 stadia. A Greek stadia corresponds to about 1/6 km.

Question

What were his observations, what was his reasoning, and how did he calculate this parameter? Consult figure 2.10.

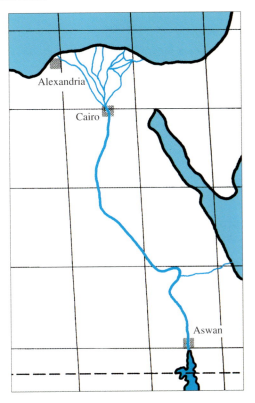

Fig. 2.10 Location map for Problem 2.1.

Look at the solution on the DVD

2.2   The telluric planets

These are rocky planets, like the Earth. They have high densities (ρ between $3.94 \cdot 10^3$ kg/m^3 for Mars and $5.52 \cdot 10^3$ kg/m^3 for Earth) and are small in size. These are the four planets nearest to the Sun: Mercury, Venus, Earth and Mars.

The size relationship between these planets is shown in figure 2.5. To them we can add the asteroids, which are rocky mini-planets of all sizes. Asteroids are usually not spheroidal, and they are mostly located in the *asteroid belt* between Mars and Jupiter.

We will add the Moon to this group of planets because of its strong similarities to the telluric planets. The Moon is somewhat less dense ($\rho = 3.34 \cdot 10^3$ kg/m^3).

Let us examine their similarities and differences by considering their atmospheres, meteorite bombardment, and geologic activity.

2.2.1 Atmospheres and hydrospheres

The only planets that have dense atmospheres today are the Earth and Venus. The Earth's atmosphere is predominantly nitrogen and oxygen. Venus's atmosphere is essentially carbon dioxide with a total pressure 96 times higher than that of Earth.

Mercury no longer has an atmosphere, but Mars has carbon dioxide residues from its original atmosphere. With the intense cold conditions on Mars, these residues form carbon dioxide ice caps that sometimes appear greenish, suggesting the presence of vegetation and even Martians. The belief in life on Mars was so strong in 1802 that the German mathematician Karl Friedrich Gauss wrote messages in giant letters in the Siberian snow to communicate with the inhabitants of that planet.

Images sent back by probes (Mars Global Surveyor and Mars Express, among others) clearly show that a liquid, probably water, has existed on the planet. For example, figure 2.11 shows a river system similar to those on Earth. Figure 2.12 shows a landslide on the side of the

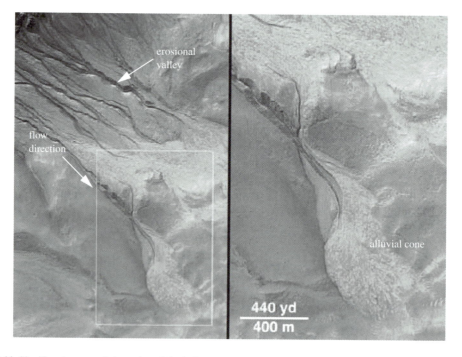

Fig. 2.11 Significant system of channels and detrital cover implying significant liquid flux. North-east of the Gorgonum Chaos impact crater, Mars. Mars Orbiter (NASA/JPL/ Malin Space Science Systems).

Fig. 2.12 Chasma Ophir region. Landslide saturated with pore liquid along the Valles Marineris canyon on Mars. Width approximately 200 km and depth of 6 km. Perspective generated by computer from photos and topographic model. Viking Orbiter (NASA/JPL/Caltech).

Valles Marineris. Today ice sheets can be observed at the two poles and the subsurface of the planet contains solid water in its pores (permafrost, Chapt. 8)

2.2.2 Geology and geologic activity

It is important to be able to compare these planets and to determine the presence or absence of geologic phenomena (Problem 2.2). These considerations will lead to a better understanding of the history of the Earth itself.

2.2.2.1 Internal structure

Like the Earth, the telluric planets are made up of a series of concentric layers with various physical states (solid, liquid), chemical composition, and relative thickness. Analyses of their gravitational fields and the study of meteorites provide an idea of their approximate internal structure (Fig. 2.13). All telluric planets have a solid surface called a lithosphere, composed of light rocks containing primarily silicates. The center of the densest planets must be cores of iron and nickel or metallic sulfides. An intermediary layer of silicate rocks rich in iron and magnesium is generally present.

2.2.2.2 Endogenetic activity

Endogenetic geologic activity is a group of dynamic internal phenomena that affect the surface of the planets. They include widespread tectonic movements and magmatism. Volcanic activity and earthquakes are the most easily observed examples.

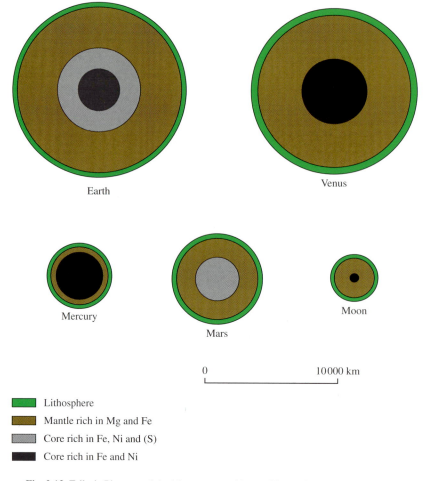

Lithosphere
Mantle rich in Mg and Fe
Core rich in Fe, Ni and (S)
Core rich in Fe and Ni

Fig. 2.13 Telluric Planets and the Moon, composition and internal structure (from [34]).

PROBLEM 2.2

You have a collection of satellite photos of Mars.

Questions

- Study the shot of Mars (Fig. 2.14).
- Draw the principal morphological characteristics.
- What geological phenomena do they suggest?
- What craters are astroblemes, which are not, and how can you tell the difference?

Note: Follow the discussion in the book for suggestions.

20 km

Fig. 2.14 Photo of Mars. Viking Orbiter (NASA/JPL/Caltech).

Look at the solution on the DVD

Among the telluric planets, the Earth is probably the only one that is still geologically active today. Volcanoes on Venus may still be active, but this has not been proven yet. Among the satellites of planets, Io (see Fig. 2.20 and 2.21), a satellite of Jupiter, is very interesting in terms of volcanic activity. One of the satellites of Neptune, Triton, may also have had recent activity. Seismic tremors of magnitude 2 on the Richter scale have been measured on the Moon, probably indicating slight residual magmatic activity.

All the heavy planets and the Moon had an endogenic activity in the past. This can be proven by studying their morphology (Fig. 2.15).

The planets that are no longer volcanically active contained only small quantities of radioactive minerals at the time of their accretion. Their internal "geologic engines" stopped when the "radioactive fuel" was used up. Only external geologic processes now occur on these planets.

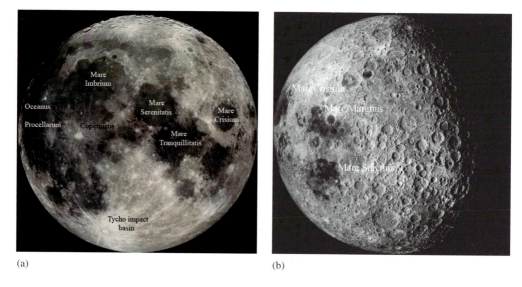

(a)   (b)

Fig. 2.15 Morphology of the Moon. (a) Part of the Moon visible from Earth; the dark zones represent the "maria" (lava flow basins). Galileo Orbiter (NASA/JPL/Caltech). (b) Two thirds of the hidden part of the Moon, characterized by its high impact density and the almost complete absence of "maria". The two dark zones seen on the left third, the Mare Crisium and Mare Smythii, are both visible from Earth. Apollo 16 (NASA/JPL/Caltech). An animation on the DVD shows the first steps of Man on the Moon.

2.2.2.3   Exogenetic activity

Exogenetic geological activity is the transformation of the surface by agents external to the planet. They are classified into three large groups of phenomena: erosion (a process that generates particles of different sizes), the transport of these particles, and their deposition. These exogenetic activities interact in a major way with external manifestations of endogenetic phenomena.

Erosion, transport, and sedimentation

In the absence of water, erosion is primarily the result of meteoritic bombardment and temperature contrasts (for example, on Mercury the daily temperature varies between 427°C and −173°C). Minerals have different dilation coefficients (§ 13.1.1) that cause the mechanical decomposition of rocks as a result of temperature changes. Fine particles are swept away by violent winds and they themselves become agents of erosion. In calmer areas, eolian deposits form furrows and dunes, as in our deserts (§ 8.7). On Mars there is evidence of ancient eolian activity that has diminished over time but still persists today (Fig. 2.16).

Water erosion (by runoff) is illustrated in figures 2.11 and 2.12. This phenomenon was significant on the surface of Mars in ancient times.

Bombardment by meteorites

Planets that lack atmospheres and have been subjected to the most intense bombardment (Mercury and Mars in particular). Images collected by various probes clearly show the morphology of the terrain after these impacts. Figure 2.17 is a photo of Mercury showing recent impacts (clearly formed craters that are also called *astroblemes*), and surrounding lines of *ejecta*. With time, regular bombardment by finer material progressively erases the topography.

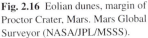

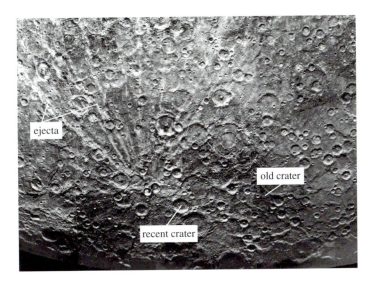

Fig. 2.16 Eolian dunes, margin of Proctor Crater, Mars. Mars Global Surveyor (NASA/JPL/MSSS).

Fig. 2.17 Photograph by the Mariner 10 probe to Mercury showing meteor impact craters of various ages (NASA/ JPL/Caltech).

Similar events have also occurred on Earth although the atmosphere slows down the falling objects and causes them to explode. Exogenetic geological activity on Earth rapidly erases their traces. The Meteor Crater in Arizona is spectacular recent evidence of this phenomenon (Fig. 2.18). Meteorite bombardment continues today. The international press regularly reports on meteorites falling on houses. Let us mention also the explosion of a probable comet piece on Tunguska in Siberia that devastated a forest in 1908.

2.3 Giant planets

The giant planets (Fig. 2.19) are less interesting for those who study the Earth. Although they are also composed of concentric layers, their physical and chemical nature is completely different and their structure is less well known. Their density is much lower, and the pressure and temperature conditions at their surface suggest a liquid surface composed of molecular hydrogen, a middle layer of metallic hydrogen and a core that is thought to resemble the internal part of the heavy planets (Fig. 2.13). The giant planets are surrounded by opaque atmospheres where violent storms occur (Fig. 2.20).

We cannot leave these distant planets without mentioning their numerous and curious satellites (especially for Earthlings). The Galilean satellites of Jupiter (Fig. 2.21) and Io attract our attention in particular. The Voyager 1 probe was able to photograph the plume of a volcanic eruption on Io that extended to a height of 140 km.

Fig. 2.18 Meteor Crater, Arizona. The crater, 1200 m in diameter, was formed approximately 50,000 years ago as the result of the impact of a meteorite that must have had a diameter of about fifty meters and a speed of 100,000 km/hr. This crater was the first to be recognized as an impact crater and is the best preserved meteorite crater on Earth. (HASA/JPL, from the Smithsonian Scientific Series (1938)).

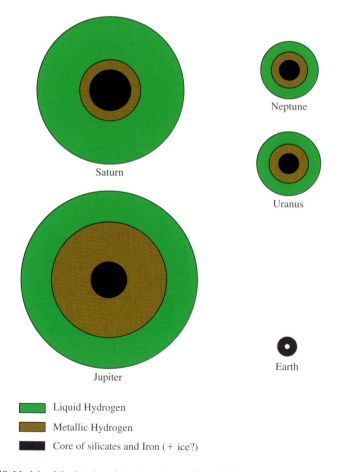

Fig. 2.19 Models of the interior of the giant planets (from [34]). The Earth provides the scale.

Fig. 2.20 Example of a giant planet. Jupiter with its anticyclonic storms and one of its four satellites, Io, whose size is approximately that of the Moon. Cassini Orbiter (NASA/JPL/University of Arizona).

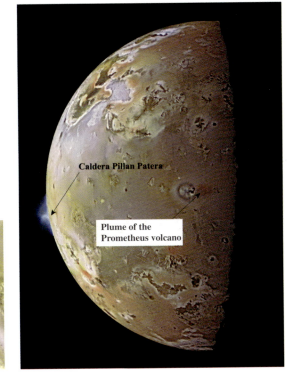

Fig. 2.21 Volcanic eruptions on the surface of Io, satellite of Jupiter. The plume from Pillan Patera is 140 km high. Galileo Orbiter 1 (NASA/JPL/Caltech).

2.4   Comets

Comets are distinguished from other objects orbiting the Sun by their very eccentric ellip-tical trajectories. They are not confined to the plane common to the planets. Their period is variable but always long. For example, Halley's comet, one of the fastest, passes near the Earth only every 76 years.

Comets are easily observable only briefly when they are orbiting near the Sun. As a result of the Sun's heating, a large sphere of light, the *coma*, can be seen followed by a tail of ions. This tail is not distorted by the comet's trajectory, which is directed away from the Sun. The Hale Bopp comet of early 1997 had a similar coma. A tail of dust was also released, but its direction curved along the ellipse (Fig. 2.22).

Comets are interesting geologically because they provide numerous meteorites. Studies on Halley's comet as it passed in 1986 confirmed that its solid body, the *core*, is much smaller (about 10 km in diameter) than the coma (1 million km in diameter). The shape of the core is elongated. Mass spectrometry analysis by the Giotto probe revealed that its core is made up of a mixture of ice and rocky material.

Fig. 2.22 Hale-Bopp Comet with its double tail. Photo E. Girardin.

2.5   Meteorites

Meteorites come from the debris of planets, asteroids or comets. The majority of these pieces are about 4.6 billion years old, the date of the accretion of the solar system. Meteorites are distributed heterogeneously in space. Some meteorite "clouds" are clearly from comets. As comets approach the Earth they produce showers of shooting stars. Meteorites differ in compo-sition and appearance depending on whether they are from the internal or external parts of the parent body. There are two principal categories.

2.5.1 Stony meteorites

Stony meteorites are the most common type (96%). They are very similar to terrestrial mafic rocks (Chapt. 6) and contain principally olivine and pyroxenes. ***Chondrites*** are primarily composed of these minerals and contain chondrules (small granular-shaped silicate inclusions showing evidence of melting, of a chemical and mineralogical composition similar to the rest of the rock), figure 2.23. Carbonaceous chondrites contain magmatic carbon and thus support the idea that they are composed of primitive materials that must have crystallized before the processes of magmatic differentiation occurred.

Fig. 2.23 Chondritic meteorite. GEOLEP, Photo. G. Grosjean.

2.5.2 Metallic meteorites

Metallic meteorites are very heavy (density between $7.3 \cdot 10^3 \, \text{kg/m}^3$ and $8.2 \cdot 10^3 \, \text{kg/m}^3$). Their metallic appearance is very typical (Fig. 2.24). They are essentially composed of iron and nickel, similar to the core of the Earth and they make up 3% of all meteorites.

There is an intermediate between these two categories: the stony-iron meteorites. They are rare (1% of all meteorites).

The impacts of bodies from space are often associated with geochemical and mineralogical anomalies. Metallic anomalies, particularly iridium, have been detected on the Earth. The metallic anomaly is often one of the only ways of detecting old impacts on Earth, because the morphology has been highly altered by erosion and other geological phenomena. A metallic anomaly led to the suspicion of a large astrobleme on the Yucatan peninsula. This impact may have caused the biological catastrophe at the end of the Cretaceous, and the disappearance of the dinosaurs and other organisms. (see § 3.3.3)

[11, 21, 142, 146, 276, 307]

Fig. 2.24 Metallic meteorite. GEOLEP, Photo. G. Grosjean. Collection of the Cantonal Geology Museum of Lausanne.

Chapter review

The Earth is a planet that is similar today to other telluric planets in regards to:

- volcanic phenomena
- tectonic phenomena

The Earth is an unusual planet because:

- it has oxygen and liquid water
- its internal "engine" has not shut down

Cosmic geology tells us:

- less erosion processes result in fresh geologic structures in spite of the age of these structures
- information about the nature of the centers of planets

3 The Earth in Time

Before turning our attention to the Earth, let us look at it in terms of the time scale of the Universe, just as we looked at the Earth's position in space in chapter 2. The Universe is the source of matter that makes up minerals, rocks, and life on Earth.

How do we plot the great events of the Universe and the Earth on a time scale?

Geochronology provides answers to these questions through a variety of methods, related to physics and biology. Let us look at the main principles.

3.1 Measurement of geologic time

Geochronology is used to assign dates to geological events. It is based on four fundamentally different methods:

- Radiometric methods that give an absolute age (primarily for crystalline and metamorphic rocks);
- Stratigraphic methods that give a relative age (sedimentary and metamorphic rocks);
- Paleontological methods that use fossils to give a biologic age that is calibrated to absolute ages (sedimentary rocks);
- Paleomagnetic dating methods, particularly applicable to oceanic basalts, as discussed in chapter 6.

3.1.1 Radiometric methods

These methods, also called isotopic, are based on the nuclear decay of radionuclides. Physics tells us that radioactive parent atoms decay to daughter atoms according to an exponential law that is invariable for a particular pair of atoms (Fig. 3.1):

$$N_t = N_0 \cdot e^{-\lambda t} \tag{3.1}$$

This expression gives the number of parent atoms N_t existing at time t, given that the number of parent atoms at the beginning was N_0. λ is the radioactive decay constant characteristic of a radionuclide (unit: a^{-1}, a = annum = year).

$$\lambda_r = \frac{\ln 2}{T_{1/2}} \tag{3.2}$$

The half-life $T_{1/2}$ is the time necessary for the number of radioactive atoms to decrease by half.

The dating principle is based on the ratio of daughter atoms to parent atoms, which is determined by mass spectrometry. This ratio gives us a point on the disintegration curve by which we can determine the age of formation of the mineral, and thus, of the rock. This is true only if the system is closed, that is, if there is no contribution or loss of the atoms concerned, which may occur during melting accompanied by magmatic segregation, for example.

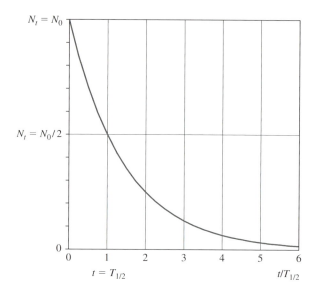

Fig. 3.1 Radioactive decay curve of a radionuclide.

Several disintegration pairs are used in geology. For short periods of time (Quaternary geology or anthropology), carbon-14 is used (Application in Problem 3.1). It has a half-life of about 5730 years, and the age limit of this method is about 45,000 years. For much older ages, for example, the ages of the Earth or the Universe, the following relationships are used:

$$^{40}K \rightarrow {}^{40}Ar + \gamma \qquad T_{1/2} = 1.2\,Ga \tag{3.3}$$

$$^{238}U \rightarrow {}^{206}Pb + {}^{84}He \quad T_{1/2} = 4.5\,Ga \tag{3.4}$$

$$^{87}Rb \rightarrow {}^{87}Sr + \beta \qquad T_{1/2} = 49\,Ga \tag{3.5}$$

[16, 88, 139, 244]

3.1.2   Stratigraphic methods

Age relations can be inferred from geologic relationships. To understand this concept, we must look at three fundamental concepts set forth by Steno in 1669 (Fig. 3.2). These principles are applied to particles deposited in sedimentary basins, as in seas or lakes that are far from coastlines.

Principle of original horizontality

Sediments are deposited in nearly horizontal layers in the same way as particles of sand in a beaker. Oblique strata or those in vertical positions are evidence of deformation after deposition.

Principle of lateral continuity

During a given time period, the same type of sediment is deposited throughout a basin. The strata deposited during one period of time are thus laterally continuous.

PROBLEM 3.1

While digging an excavation for a foundation on a slope, a mass of rock debris was found enclosed in an argillaceous matrix. In the mass, the remains of a shrub were found and dated by the Carbon-14 method. The amount of residual Carbon-14 in the shrub corresponded to 17% of the initial quantity.

The λ_r of ^{14}C is $1.21 \cdot 10^{-4}$ [a^{-1}].

Questions

- Describe the relationship between the half-life and the disintegration constant.
- Calculate the half-life.
- How old is this shrub?
- Explain why this wood debris was present in this material.
- What does this mean for the engineer in charge of this construction?

Look at the solution on the DVD

Principle of superposition

In a sedimentary basin, recent contributions are deposited on top of older sediments. Thus, the deeper the strata, the older it is. Here we have the first age relationship between strata.

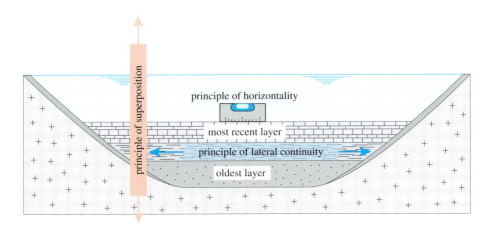

Fig. 3.2 The three basic principles of stratigraphy. See also the animation on the DVD.

We owe to Hutton two supplementary principles that allow us to establish an age relationship between different formations present in a geologic structure (Fig. 3.3). In addition to deposition, these principles introduce the notion of an erosional phase separating two events.

Principle of crosscutting relationships

When a dike cuts across a rock, the rock was present before the dike. If the dike does not extend into a second stratum located on top of the first one, the second stratum is more recent than the dike. An erosional period preceded its deposition.

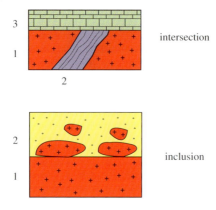

Fig. 3.3 Hutton's two supplementary principles.

Principle of inclusion

If a detrital rock (made up of grains resulting from the breakup of other rocks) contains particles from a recognizable rock, that rock predates the detrital rock.

These simple principles are fundamental to the understanding of sedimentary and tectonic phenomena (Problem 3.2).

3.1.3 Paleontological methods

Paleontological dating is based on the fact that some biological species lived only during a certain period of time. If their fossilized remains are found in a rock, it signifies that the rock formed during this same period. Absolute age limits for these periods can be determined by using paleontological methods in conjunction with radiometric dating.

Obviously this method is applicable only when living species left trace fossils.

3.1.3.1 Fossilization

Fossilization is a group of processes that slowly mineralize the components of living organisms and leave traces in rocks. These processes are part of the general mechanism of diagenesis, that is, the transformation of sediment into rock (Chapt. 10).

Internal or external skeletons undergo slight transformation by recrystallization; this is common in the case of calcareous or siliceous shells. The soft organic parts of animals or plants are mineralized by biochemical processes. Organic chains are broken and transformed into gas or ions that dissolve in water. Aerobic mineralization is rapid; it creates compounds such as CO_2, PO_4^{3-}, NO_3^- which are lost to the environment. All traces of the soft parts of the organism disappear after several months or years. In anaerobic mineralization, the transformation is slower. It leads to the formation of reduced species such as CH_4, liquid hydrocarbons, coal, NH_3, and H_2S; the gaseous and liquid parts migrate into the geologic environment; the solid part remains in place and forms a fossil of graphite and sulfides, particularly pyrite (FeS_2).

It is obvious then that not all dead organisms become fossils. In fact, fossilization is the exception to the rule. For fossilization to happen, the co-occurrence of particular circumstances must take place, such as reducing conditions and rapid burial under fine-grained sediments (clays for example). For the fossil to be useful in stratigraphy, it must be recognizable, which is often not the case because of imperfect fossilization or deformation of the rock.

Examples of fossilization

Figures 3.4 through 3.7 show different types of fossilization, from the most common to the rarest.

Fig. 3.4 Plant fossil, a Carboniferous fern, 295 Ma. The plant material has turned to graphite. Each detail of the plant structure is preserved. GEOLEP Photo, S. Bilgot.

(a)   (b)

Fig. 3.5 Fossilization of plants does not always produce coal. When silica-rich groundwater circulates in the pores of plants, the silica replaces the organic matter and it often faithfully preserves all the structural details of the plant. The Sahara is home to many silicified trunks from the Cretaceous, providing evidence of a climate very different from that of today. (a) Example of the silicified trunk of a conifer of Triassic age discovered in southern Brazil, near to the city of Mata. These trees are ancestors of the Araucaria, the large South American conifers, but their size was clearly much smaller. (b) Their trunks are often cut for paving slabs. GEOLEP Photo, A. Parriaux.

Fig. 3.6 Fossilization of an internal skeleton, the reptile *Eurypterygius* (Ichthyosaurian) from the lower Jurassic (220 Ma). Photo from the Collection of the Cantonal Geologic Museum of Lausanne.

Fig. 3.7 Fossilization of an external skeleton, an ammonite shell. Calcium carbonate, in the form of aragonite (unstable at ordinary temperatures) has been transformed into calcite. GEOLEP Photo, G. Grosjean. Collection of the Cantonal Geologic Museum of Lausanne.

The term "fossilization by freezing" is sometimes used (Fig. 3.8 and Fig. 3.9), but this term is incorrect because the material is not irreversibly transformed.

Trace fossils

Imprints are a special type of fossil. Only the tracks of the organism are preserved. It is generally possible to identify the organism responsible for the tracks. For example, large dinosaur tracks are well known (Fig. 3.10).

Fig. 3.8 Mammoth tibia with flesh "fossilized by freezing", discovered in the permafrost of Igarka, Siberia. GEOLEP Photo, A. Parriaux.

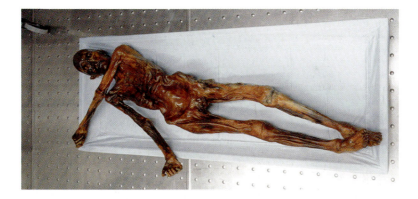

Fig. 3.9 "Fossilization" of an entire man by ice, Ötzi, Tyrolean hunter, 6000 years old. © Fotoarchivio Museo Archeologico dell'Alto Adige – www.iceman.it

Fig. 3.10 Tracks of *Tyrannosaurus Rex* from the El Molino Formation, Sucre, Bolivia. Each imprint is about 50 cm long. GEOLEP Photo, B. Matti.

3.1.3.2 Specificity of fossilized organisms

Fossils provide information on sediment ages and environmental conditions that existed during the period (paleoenvironment) (Tab. 3.1).

A good *index fossil* provides a precise age because it evolved rapidly. For an index fossil to represent a large area, it should not be sensitive to the environment. Pelagic organisms (that float at the whim of ocean currents) meet this criterion: for example, ammonites (Fig. 3.7) or planktonic foraminifera. In reality, scientists use biostratigraphic associations of species that make up *biozones*. These are a type of "biological time unit" distinguished by the arrival or disappearance of one or more species.

A good *facies fossil*, on the contrary, is one that has evolved slowly over time and can serve as long as possible. It must be very sensitive to its environment in order to be a good indicator of paleogeographic and ecological conditions. For example, these corals are characteristic of shallow seas (less than 10 m). Because they have existed for 400 Ma, this corals can be used to demonstrate shallow marine conditions in calcareous rocks of all ages.

Table 3.1 Properties of good index and facies fossils.

Good fossil	
Chronological (dating)	**Environmental (paleoecology)**
Fast evolution	Slow evolution
Not very sensitive to the environment	Narrow biotope
Example: ammonites	Example: corals

3.1.4 Summary of dating methods

By using all the methods described above, it has been possible to establish a chronology of events that have affected the Earth throughout its long history. The very different dating methods complement each other. For example, the combination of stratigraphic criteria, and paleontological and radiometric dating makes it possible to assign an absolute date to the evolution of life on Earth (Application in Problem 3.2). Using observations made around the world and all dating methods, it has been possible to establish a geologic time scale to serve as a chronostratigraphic base (Fig. 3.11). Time is divided into Eras, Periods, and Epochs. Periods and Epochs bear the names of places where the type biostratigraphic profiles have been described. These terms, which denote an age, are often used by extension to describe the rock that formed during that time.

A time axis (absolute age) is correlated with this chronostratigraphic series. Old stratigraphic units cover much longer time periods than modern ones, because we know less about the events that took place during early Earth history. This is because methods for dating older rocks are less precise and there are fewer fossils in older sediments.

[14, 190, 218, 241, 254, 283, 295, 303]

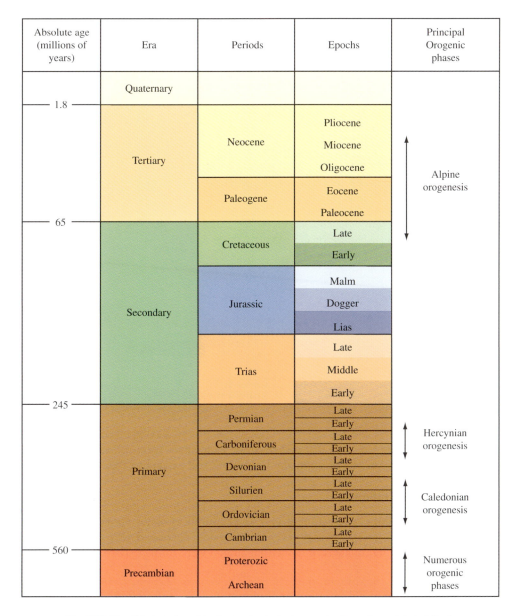

Absolute age (millions of years)	Era	Periods	Epochs	Principal Orogenic phases
	Quaternary			
— 1.8 —	Tertiary	Neocene	Pliocene	Alpine orogenesis
			Miocene	
			Oligocene	
		Paleogene	Eocene	
			Paleocene	
— 65 —	Secondary	Cretaceous	Late	
			Early	
		Jurassic	Malm	
			Dogger	
			Lias	
		Trias	Late	
			Middle	
			Early	
— 245 —	Primary	Permian	Late	Hercynian orogenesis
			Early	
		Carboniferous	Late	
			Early	
		Devonian	Late	
			Early	
		Silurien	Late	Caledonian orogenesis
			Early	
		Ordovician	Late	
			Early	
		Cambrian	Late	
			Early	
— 560 —	Precambrian	Proterozic		Numerous orogenic phases
		Archean		

Fig. 3.11 Simplified stratigraphic chart and principal orogenic phases (mountain building).

3.2 Origin of the Universe and matter

We will briefly review the ***Big Bang theory***, a product of the work of numerous astrophysicists, including Newton, Doppler, Einstein, Friedman and Hubble, to mention only the principal ones. We will examine the major stages that led to the creation of chemical elements that make up the Earth's materials.

Isotopic dating allows us to assign an age of approximately 15 Ga to the birth of the Universe, which is also the time when elements began to be created by nucleosynthesis. Figure 3.12 is a simplified diagram of how these events occurred.

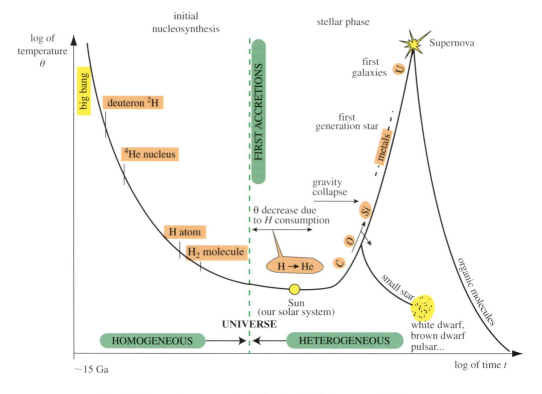

Fig. 3.12 Stages of matter creation following the Big Bang (unscaled diagram).

Extraordinary density, energy, and temperature conditions accompanied the Big Bang, a gigantic explosion in the Universe that dispersed the elementary forms of matter into space at high energy. This explosion explains why the Universe is still expanding today.

3.2.1   Initial nucleosynthesis

This first phase includes the creation of very simple forms of matter (protons and neutrons). After the Big Bang, the internal energy level of the system decreased very rapidly. After several seconds, deuterons ^{2}H (nucleus of deuterium atoms) were formed. Then in succession, the following appeared:

- Helium nuclei ^{4}He,
- Hydrogen atom H (10^6 years after the Big Bang),
- Hydrogen molecule H_2.

The Universe was homogeneous, not diversified.

3.2.2   Stellar phase

After several million years, when the general energy level had decreased sufficiently, masses of matter began to congregate locally, by accretion, according to the law of universal attraction (Chapt. 4).

This phase was followed by collisions that gradually increased the internal energy of the mass, finally forming a first generation star and the first galaxies.

Accretion caused the formation of heavier elements than hydrogen through nuclear fusion. The fusion of hydrogen atoms, two by two, led to the formation of helium atoms; this reaction increased the thermal energy within the star, which is opposed to the gravitational energy. The star was thus in equilibrium. Little by little the hydrogen was used up, and as a result, helium formation slowed down, causing a temperature drop at the center of the star. Gravitational energy became predominant. The star collapsed and so on itself, contracts, and its central part become increasingly dense. This contraction led to another temperature increase; if the temperature rises enough, the previously formed helium will be used to synthesize carbon and oxygen, which are necessary for life, and then the first geologic element, silicon (Fig. 3.13).

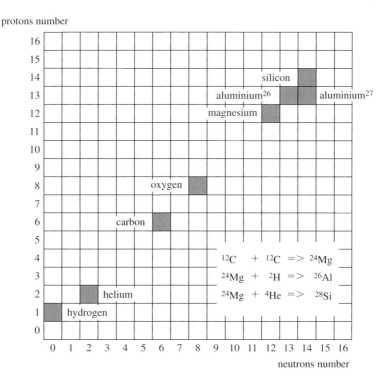

Fig. 3.13 Generation of new atoms by fusion (from [229]).

The majority of stars will stop at the carbon creation stage. Gravity causes them to collapse abruptly with a prodigious density increase, creating a white dwarf (a dead star, no nuclear reaction, but brilliant due to its heat), a brown dwarf (a body much smaller in size and luminosity than a white dwarf), a pulsar (a rapidly turning neutron star that produces electromagnetic rays at irregular intervals), or other types of celestial objects.

If the converging mass of matter is very large (massive stars), there is then sufficient energy to synthesize all elements in the periodic table, including the heaviest, such as uranium (U). The mass involved and a temperature of millions of degrees will lead to a more massive collapse much larger than that of a white dwarf, finally resulting in a gigantic explosion (supernova) that will disperse these atoms throughout the Universe, thus sowing chemicals throughout space. This matter may again participate in the accretion process, and form a second generation star, and so on. Smaller stars like our Sun will have a limited thermal increase and will end their lives as white dwarfs.

The Universe is also the birthplace of low energy organic substances such as the simple molecules like CH_4 and NH_3. But the Universe also produces complex molecules (alcohols and organic acids). These molecules exist today in interstellar space. They have certainly played a fundamental role in the Earth's first atmosphere and in the appearance of life.

3.2.3   Accretion of the planets

The formation of our solar system is probably the result of the accretion of matter. About 4.6 Ga ago, a part of our galaxy underwent gravitational accretion that gave rise to our solar system. A cloud of particles condensed as it revolved around an axis passing through the point of maximum concentration, which became our Sun.

At some distance from the rotation axis, particles of dust and gas were subjected to two types of forces: centrifugal force and gravitational force according to the law of universal attraction (Fig. 3.14).

As a result of these forces, matter reached a state of equilibrium along a plane perpendicular to the axis passing through the proto-Sun. Accretions occurring on this plane formed the proto-planets, the raw material of future planets. The proto-planets were attracted to each other, generating larger bodies that became planets.

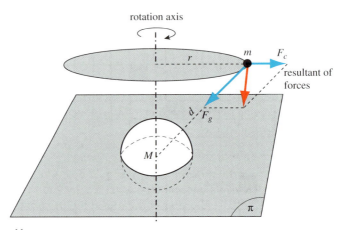

M, m : mass
d : distance between m and M
v : tangential velocity
ω : angular velocity
F_g : gravitational force
F_c : centrifugal force
π : future plane of planetary revolution
G : universal constant $= 6.67428 \cdot 10^{-11}$ [N $\cdot$ m^2 $\cdot$ kg^{-2}]

$$F_g = G \cdot \frac{M \cdot m}{d^2} \qquad F_c = m \cdot \omega^2 \cdot r = m \cdot \frac{v^2}{r}$$

[58, 230, 231] **Fig. 3.14** Forces acting on a particle of mass m revolving around an axis passing through the proto-Sun of mass M (the frame of reference is the particle).

3.3   Voyage through time: From the Precambrian to the Quaternary

In this section we will discuss the great changes the Earth has seen from the time of its accretion until today, in all its components: geosphere, atmosphere, hydrosphere and biosphere. In terms of the geosphere, we will talk about geologic mechanisms that have not yet been presented at this stage of the book, but will be discussed in detail later. This chapter will serve as a space-time framework for the more specialized chapters to follow.

The time scale will be divided into large time spans in order to preserve the general nature of this chapter. We will use era and periods (Fig. 3.11) as established by the International Commission on Stratigraphy (ICS).

3.3.1   Precambrian Era

The Precambrian Era extends from the formation of the Earth to the beginning of the Paleozoic (4.55 Ga to 560 Ma); it covers eight ninths (8/9) of Earth history. The knowledge of this time span is limited because of the following:

- Scarcity of fossils,
- Erosion (the disappearance of many rocks),
- Transformation of rocks by metamorphism and magmatism.

Nevertheless, it was during the Precambrian that major geologic events took place, that have determined the Earth's history.
The Precambrian Era is divided into two periods:

- The Archean, from the origin of the Earth until 2.5 Ga, is a poorly understood period because of the lack of evidence.
- The Proterozoic (from 2.5 Ga until 560 Ma) is better documented and more easily understood because of its plate tectonics and orogenies (creation of mountain chains) and to the presence of sedimentary rocks.

During the Archean, the Earth that had originally aggregated by gravitational accretion still had a high level of energy due to collision energy (Fig. 3.15). The Earth differentiated as a result of gravity; the heavy elements concentrated at the center and the lighter elements were left near the surface, and the volatile elements concentrated into a gaseous envelope around the Earth. At that time, the Earth was a molten mass surrounded by a primitive gaseous reducing atmosphere (Point 1, Fig. 3.16) whose composition resembled that of interstellar gas. In time, the collision energy dissipated and the energy curve took on the shape of a radioactive disintegration curve (Fig. 3.15) as we have seen. This explains the production of internal energy by nuclear reactions inside the Earth.

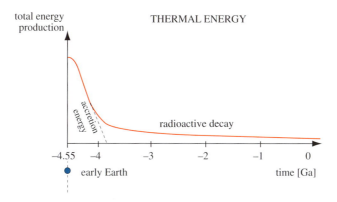

Fig. 3.15 Variation of the Earth's thermal energy through time.

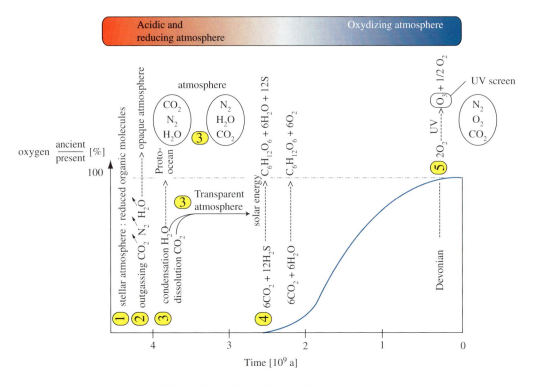

Fig. 3.16 Evolutionary history of the atmosphere.

Around 4 Ga as the slow cooling of the Earth was taking place, primitive magmatic rocks crystallized to form a primitive solid crust and the outline of the cratons (future continental plates). The *liquid hydrosphere* was created gradually by the condensation of water vapor and gases (released by convection currents and volcanoes). An extraterrestrial origin has also been proposed: water would take its origin from the core of a comet. The *proto-ocean* covered a substantial portion of the globe (Point 3, Fig. 3.16). Density and chemical reactions caused magmatic segregation which organized the internal structure of the Earth and its envelopes (Fig. 6.1).

The relative chronology of this period was established by the timing of a series of orogenies. The last and most notable phase of deformation and metamorphism was the Saamian orogeny (3.75–3.5 Ga).

At the beginning of the Proterozoic (2500–1950 Ma), the process of magmatic segregation (Chapt. 6) continued in the Earth's crust. The continents were composed of light rocks enriched in silicon (granitic crust). The later Proterozoic saw the beginning of plate tectonics as we know it: long distance movements of a small number of thick, almost undeformed lithospheric plates. The ocean began to open (*rifting*); the first continental plates began to move due to deep convection currents. Their collisions led to the formation of the first mountains and the formation of the vast supercontinent *Rodinia*, which will break up at the end of the Precambrian.

Afterwards, the mountains disappeared as a result of continental erosion and several phases of glaciation. Sediments were created, followed by sedimentary rocks. Internal and external geodynamic cycles were underway. During the three billion years at the end of the Precambrian, orogenies created *continental plates* that will form the jigsaw puzzle of the beginning of the

Paleozoic. Today's world geologic map shows many old Precambrian basements constituting the backbones of the continents (Fig. 3.17). These ***cratons*** are characterized by their near horizontality because they have not been modified by significant deformations after their emplacement. They are often overlain by unaltered Precambrian sedimentary series that have remained in a quasi-horizontal position for more than a billion years (Fig. 3.18).

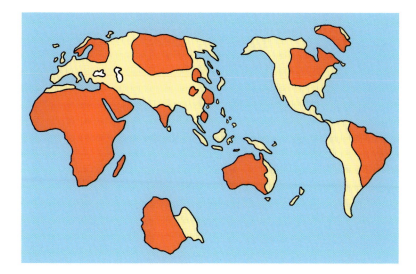

Fig. 3.17 Old Precambrian basements (in red) under today's continents (from [180]).

Fig. 3.18 In some places on Earth, ancient rocks have escaped transformations. One example is the Proterozoic dolomites of the Yanking-Huailai basin (Mt. Jiming, 150 km to the NW of Beijing). Their fresh appearance belies their great age. GEOLEP Photo, A. Parriaux.

The climatic conditions of the Precambrian were relatively hostile to all forms of life. The original atmosphere was essentially made up of chemically-reduced organic molecules (Point 1, Fig. 3.16). Surface temperatures were high. The first endogenous geologic activities progressively modified this atmosphere by introducing large quantities of water vapor, nitrogen, and carbon dioxide from volcanoes. This created a cloudy cover that was opaque to the Sun's rays, thus producing a greenhouse effect (Point 2, Fig. 3.16). High concentrations of CO_2 of magmatic origin made the atmosphere highly acidic. The proto-ocean lowered this concentration by dissolving the acid in water. Gradually the atmosphere became transparent to the Sun's rays (Point 3, Fig. 3.16).

Two major events in Earth history occurred at this point: the appearance of life around 3.85 Ga from the first organic molecules (most likely stimulated by electrical discharges in the atmosphere) and photosynthesis (2.7 Ga) (Point 4, Fig. 3.16). In the beginning, photosynthesis was anaerobic, that is, it did not produce free oxygen (3.6):

$$6\,CO_2 + 12\,H_2S = C_6H_{12}O_6 + 6\,H_2O + 12\,S \tag{3.6}$$

The carbon in the CO_2 was reduced into glucose and the sulfur present in H_2S was transformed into S by oxidation.

Slightly later, aerobic photosynthesis appeared, producing free oxygen.

$$6\,CO_2 + 6\,H_2O = C_6H_{12}O_6 + 6\,O_2 \tag{3.7}$$

This phenomenon is explained by the fact that the minor amount of free oxygen that was produced was immediately and preferentially used for the oxidation of certain elements such as iron. We have sedimentological evidence of this in the form of uraninite UO_2 deposits (up until 2.3 Ga) and the Banded Iron Formations (BIF) rich in iron oxides (from 2.8 until 1.8 Ga).

Once the oxidizable material was used up, free oxygen could be released to progressively enrich the atmosphere. The reaction accelerated as a result of an increase of marine plants. However, it was not until the middle of the Paleozoic that terrestrial photosynthesis became established and reached the concentrations we see today (Point 5, Fig. 3.16).

The appearance of oxygen had a catastrophic consequence on the development of life, which had been anaerobic up to that point. Free oxygen was highly toxic to existing organisms. Only rare organisms were resistant enough to adapt to this drastic change. This event was the first massive extinction in the biosphere.

Around 1.5 Ga the first molecules of ozone formed, but it wasn't until the Carboniferous that the concentration of free oxygen was sufficient to create a real ozone layer that would permit significant development of life on Earth.

There are few fossils from this distant time because of the scarcity of life, the microscopic nature of the embryonic organisms, and the absence of skeletons. In addition, rocks were often transformed and all traces of life were lost. The origin of life is set at about 3.85 Ga, the date of the first rocks and the first traces of life found in southern Greenland. The two types of fossils at this time were microfossils and stromatolites that date from about 3.5–3.4 Ga. Globally, the Precambrian was a time of significant evolution of organisms. The first two billion years of life on Earth are represented by a single-cell microbial life form called a ***prokaryotic*** cell (organism that has no nucleus or organelles). The ***blue algae*** (Cyanophytes) appear and form the first calcretizations, the ***stromatolites***. This algae still exists today (Fig. 3.19). Later, around 1.5 Ga, the unicellular ***eukaryotic*** cell appeared; this organism has cells with a real nucleus covered by an envelope that contains genetic material. From 1 Ga on, the ***metazoans*** (multi-cellular mobile

Fig. 3.19 Present day Stromatolites, Shark Bay, Western Australia. Photosynthetic activity of cyanobacteria causes the absorption of carbon dioxide and as a result, the precipitation of calcium carbonate in superposed layers. Photo E. Davaud, University of Geneva.

and heterotrophic eukaryotic organisms) contribute to Precambrian evolution through the development of the Ediacara (soft-body fauna) that appeared suddenly around 670 Ma as a prelude to the significant diversification of Paleozoic fauna.

3.3.2   Primary Era (Paleozoic)

This period extends from 560 to 246 Ma. Paleogeography at the beginning of the Paleozoic is marked by the fragmentation of the supercontinent ***Rodinia***. Continental plate movements that followed are difficult to reconstruct because of missing elements. We know that the continents moved because accretion and collision events led to several orogenies. In particular, the widespread ***Hercynian Orogeny*** built mountain chains in Europe, America and Africa (Fig. 3.11). We find evidence of this orogeny in ancient rocks in Brittany, the Vosges, and the Black Forest, for example. These old massifs were tall mountain chains that have been deeply eroded to the much lower elevations we see today. The end of the Paleozoic corresponds with the end of a ***Wilson cycle*** (cycle of oceanic opening and closing events) and the reuniting of all the emergent lands to form a single continent, ***Pangea***, and one large ocean, ***Panthalassa*** (Fig. 3.20).

Several major factors affected the Paleozoic climate. Glaciations made their appearance in the late Ordovician and continued until the end of the era (Fig. 3.21). The alternation of warm and cold periods influenced paleogeography as well as paleontology. As the volume of solid water in glaciers fluctuated throughout the Paleozoic, broad ***transgressions*** (sea level rise) occurred. Finally, an essential event occurred during the Carboniferous: a significant increase in oxygen concentration leading to the formation of a real ***ozone layer*** that kept out ultraviolet (UV) radiation (Point 5, Fig. 3.16). Up until then there had been an increase in the free oxygen content in the atmosphere, but it was not enough to allow terrestrial life to develop pulmonary respiration or to form an ozone layer to prevent bombardment by short wave length UV (<200 nm), which is lethal to living creatures. These rays are absorbed by water, so they did not prevent life in the oceans before the Carboniferous.

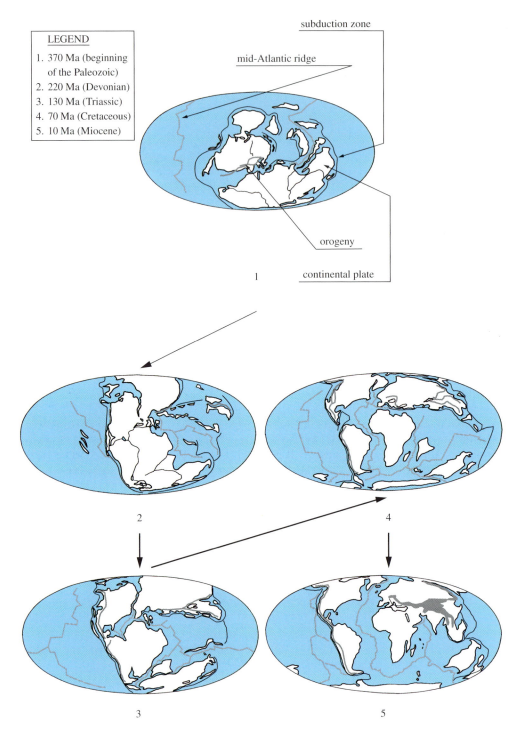

LEGEND
1. 370 Ma (beginning of the Paleozoic)
2. 220 Ma (Devonian)
3. 130 Ma (Triassic)
4. 70 Ma (Cretaceous)
5. 10 Ma (Miocene)

Fig. 3.20 Plate tectonics from the Paleozoic until the end of the Tertiary (Northern Arizona University)

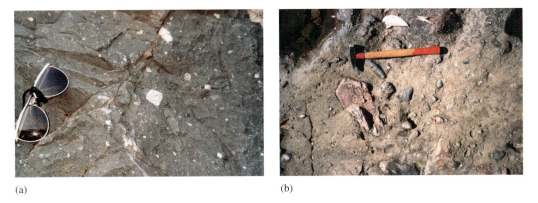

(a) (b)

Fig. 3.21 There is evidence of ancient glaciation during the Paleozoic on the continent of Gondwana, which has now broken up. Ancient moraines called *tillites* show structures very similar to lodgment tills of the Quaternary (Chapt. 8 for comparison). (a) Durban tillite (South Africa), Dwyka Group; very hard and uniform rock that outcrops on a vertical cliff almost 100 m high. (b) Southern Brazil tillite, Sao Gabriel region, Itararé Group, the more weathered matrix makes it possible to see cobbles that have a typically glacial morphology. These two rocks date from lower Permian and are part of the same glaciation that covered southern Gondwana about 300 Ma. GEOLEP Photos, A. Parriaux.

During the lower Cambrian, when the Ediacara fauna became extinct, the first fossils of shells appeared (Small Shelly Fossils S.S.F). These are two major pulses of the emergence of macroscopic animals, but one of the most important paleontological events in history is the ***Cambrian explosion***. It marks the start of fossiliferous geologic time with the first appearance of mineralization of certain tissues and an increase in diversity through biological innovation (gills, jaws, intestines, etc.). These properties may have been instigated by intrinsic factors such as genetic flexibility or by extrinsic factors such as an increase in oxygen content or modifications of the chemistry of the oceans etc. It was a time of experimentation and diversification of species. Marine biodiversity literally exploded (corals, jellyfish, sponges, etc.). All the major groups of Invertebrates existed by the end of the Cambrian. ***Trilobites***, marine Arthropods (Fig. 3.22),

Fig. 3.22 Specimen of *Phacops*, species of Paleozoic trilobite (Early Cambrian–Late Devonian). Length 8 cm. Photo Collection of the Cantonal Geologic Museum of Lausanne.

dominated the Cambrian. In contrast, the first Vertebrates did not appear for another 50 Ma (for example, Fish during the late Ordovician).

During the Devonian, and even more so than during the Carboniferous, there was spectacular colonization of continents by living organisms, indicating significant morphological and physiological modifications in different species. Plants first tamed the continent (Ferns, for example); particularly abundant in the Carboniferous, the first coal deposits developed as shown today by coal mines distributed on all continents. Then Vertebrates arrived, marked by significant physiological changes such as gills in Fish which later evolved into lungs to allow the appearance of Amphibians.

This development of the living world was interrupted by several crises whose causes are still under discussion. Tectonic upheaval (formation of Pangea) and climatic upheaval had a major impact on paleogeography, which may explain a large number of adaptive radiations but also massive extinctions. For instance, the loss of bentic species (fastened to the seafloor) follows the disappearance or drastic reduction of their living environment. The end of the era was marked by a true ecological catastrophe, followed by a significant renewal of fauna.

3.3.3   Secondary Era (Mesozoic)

During the Mesozoic (245 to 65 Ma), large tectonic events drastically changed the geography and dynamics of the Earth (Fig. 3.20).

The breakup of Pangea generated a general phase of transgression. A sea opened between Africa (Gondwana) and Europe (Laurasia): the ***Tethys*** (the sedimentary cradle of the future Alps). The change from a continental environment to a marine environment took place gradually with the creation of shallow basins (Fig. 3.23). The different pieces of Gondwana that had

Fig. 3.23 Ripple marks from ancient beaches have been often fossilized as in these Triassic (225 Ma) sandstones in the sedimentary cover of the crystalline massif of Fully (Swiss Alps). This is the first evidence of marine transgression following the Permo-Carboniferous continental phase. GEOLEP Photo, A. Parriaux.

already been in the process of breaking up since the end of the Permian as a result of generalized distension, moved separately over large distances. These pieces were the Brazilian, Guyanan, African, Indian, Australian and Antarctic shields and platforms. On the global scene, oceans were expanding, and the drift of the Americas created the Atlantic during the Cretaceous (Fig. 3.20). Oceanic expansions led to plate movements that caused compression, followed by subduction and collision, and then orogenies (Andes, Sierra Nevada in America, beginning of Alps in Europe and Himalayas in Asia).

In the Mesozoic, there are fewer climatic fluctuations and the climate is warmer than at the end of the Paleozoic. Oceanic life is characterized by the flourishing of the ***ammonites***, followed by their disappearance at the end of the Mesozoic (65 Ma); this evolution makes them excellent index fossils for the Mesozoic. On the continents, major diversification in the living world is accompanied by significant new additions. Mammals appeared 200 Ma (Late Triassic), Birds in the Late Jurassic, and Angiosperms around 150 Ma. The most spectacular event is the diversification of reptiles leading to the gigantic size of the famous ***dinosaurs*** which dominated from the Jurassic until the Cretaceous and colonized all environments (Fig. 3.24). However, the ammonites and dinosaurs would not survive a new biological catastrophe at the end of the era.

The Mesozoic ended with a serious biological crisis that would led to a widespread renaissance in the living world after the extinction or the reduction of archaic groups. Terrestrial flora and fauna changed: more than half of the continental and marine plant and animal species became extinct. However, the Mammals survived. This catastrophe has excited the curiosity of scientists for many years. Numerous hypotheses have been proposed to explain it. Many of

Apatosaurus, previously known as *Brontosaurus*
(Late Jurassic)

Tyrannosaurus
(Late Cretaceous)

Ichthyosaurus
(Early Jurassic)

Fig. 3.24 Several dinosaurs, major actors of the living world during the Mesozoic.

these theories seem valid at first glance, and the lack of proof means that all kinds of scenarios are possible. Here we present two currently competing hypotheses, without bias; both provide their own interpretation of the presence of high concentrations of iridium (Ir) in rocks formed at the time of the extinctions.

• Large meteorite impact: According to this hypothesis, very rapid extinction (50 years to several thousands of years) resulted from the impact of an enormous meteorite in the Yucatan peninsula. The impact created enough dust to make the atmosphere opaque, causing difficult conditions for plant and animal life on Earth. This "winter" caused by the impact was followed by greenhouse warming. During the warming, nitrogen oxides formed and decimated aquatic life while continental-scale fires threatened terrestrial life. This scenario is based principally on the presence of high concentrations of iridium in meteorites and the surrounding circular structure of the Yucatan.
• Intense volcanic activity: Iridium is also concentrated in volcanic materials; for example, it is found today in the Kilanga (Tanzania) and Piton-de-la-Fournaise (Reunion Island) lavas. According to this hypothesis, the activity lasted 500,000 years, compatible with the volcanic phenomena of this period. In fact, significant quantities of volcanic material were extruded onto the Earth's surface at the end of the Cretaceous, notably the Deccan Traps basalts in India. This theory proposes that iridium was brought to the surface by the continental eruptions as a result of magma plumes rising convectively under continents (Chapt. 6).

These two hypotheses lead to one and the same conclusion: a clear increase in fine particles in the atmosphere creating a shield against solar radiation. This sudden change caused the death of many animal and plant species that were not able to adapt to the lack of light and food. It should be noted that some species had begun to disappear well before the appearance of the high level of iridium. The extinction process is complex and continuous according to some scientists for more than 15 Ma. It should be noted that one hypothesis does not necessarily exclude the other. Nothing would prevent a meteorite crashing into the Earth during a period of intense volcanic activity, with climatic conditions being influenced by the two phenomena.

3.3.4 Tertiary Era (Cenozoic)

The Tertiary Era extends from 65 to 1.8 Ma. The intense tectonic activity and continental movement that began in the Mesozoic terminated the Alpine cycle in the Tertiary with a double continental collision and the genesis of the Alpine mountain chains along the Gibraltar-Burma axis (Fig. 3.20).

• India (Gondwana fragment) collided with the Eurasian continent (Laurasia) and was subducted under it, creating the Himalayan uplift.
• The African plate (Gondwana fragment) collided with the Eurasiatic plate, which was subducted under Africa, causing the Tethys to disappear and creating the Alps (in the broad sense, that is, the Atlas-Pyrenees-Alps-Carpathian- Balkans-Caucasian arc).

The genesis of these mountain chains is explained by the disappearance of an oceanic territory by subduction, then by the collision of continental margins.

It is important to note that the duration and effects of the large scale orogenies vary according to regions or continents. They undergo a complex evolution of collision, folding, magmatism, and uplift. Orogenesis includes a first phase, the collision of lithospheric plates. This collision creates a significant thickness of light rocks that leads to the second phase, uplift and high elevations due to isostasy (Chapt. 4). An orogeny occurs in regions of uplift, but within these regions, there are also basins that show the opposite tendency. Subsidence leads to the deposition of thick detrital series contributed by erosion of the mountain chain. A present day example is the Pô plain (Italy). The subsidence rate for this type of basin can be calculated through stratigraphic records to be on the order of one millimeter per year (present-day settlement is much higher due to anthropogenic causes, essentially groundwater pumping and agricultural drainage).

Starting in the Tertiary, we find evidence of very hot climates. For example, in the northern Alps, pedogenesis (soil formation) created ferruginous clays that bear witness to a tropical humid climate in the Eocene. This climate continued until the Oligocene, as shown by reddish marls and evaporites. Evidence of this hot climate is further supported by fauna (marine turtles) and flora (palms, laurels) that are found in the alpine molasse. Beginning in the Miocene, the climate began to cool, perhaps caused by drastic changes of oceanic currents. The great equatorial current that had flowed though the Tethys and between the two Americas was blocked by the closure of these straits (Fig. 3.20). This circulation was replaced by local currents that had much lesser consequences on the global climate of the Earth. In addition, as the strait separating South America and Antarctica widened, a great circumpolar current isolated this cold region of the planet and contributed to increased glaciation. These changes increased the latitudinal climate gradient, a tendency that was reinforced during the Pliocene by the appearance of arctic ice floes.

The principal upheaval during the second half of the Tertiary took place in the Mediterranean. The Tethys was held as if in pincers between Africa in the south, Europe and Arabia in the north. These tectonic constraints had major impacts on the position of shores of the Tethys but also on the rise and fall of its floor. At the beginning of the Miocene, the sea was largely open to the Indian Ocean to the east and the Atlantic to the west. Then it became progressively isolated, first by a closing in the location of the present day Red Sea, then by the raising of the thresholds that connected it to the Atlantic through Morocco and Spain. The Tethys became a closed saline basin where evaporation was no longer offset by decreasing contributions from the Atlantic. This critical situation culminated in the Messinian (9 Ma) when evaporation had practically dried up the sea at an elevation of 1500 m below sea level. One of the largest salt masses ever seen on Earth (more than 1000 m in thickness) formed during this short period. To create this thickness of evaporites, there must have been repeated phases of water flow into the depression followed by evaporation phases. This was possible because sea levels fluctuated with variations in the size of the polar ice sheets. At the end of the Messinian, the Strait of Gibraltar opened, and the increased contributions from the Atlantic ended the salinity crisis. However, the water dilution deficit of the Mediterranean continues today, as shown by the fact that it has higher salinity than the oceans.

On the continents, all the large present-day biological groups are well represented beginning in the Paleogene, but it is the Mammals whose remarkable development characterizes this era (Fig. 3.25).

In particular, rodents are an excellent clock for the rocks of the molasse because they evolved rapidly and their teeth are well preserved. After two large radiations during the

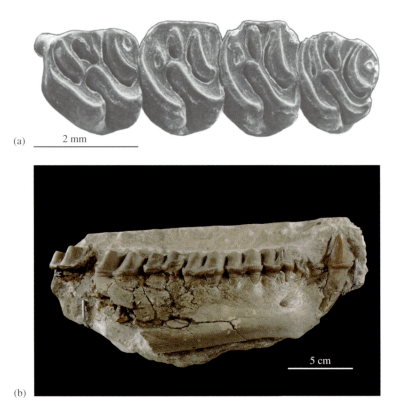

(a)   2 mm

(b)

Fig. 3.25 Some Tertiary fossils (a) Rodent teeth; (b) Rhinoceros jaw. GEOLEP Photo, G. Grosjean. Collection of the Cantonal Geologic Museum of Lausanne, M. Weidmann.

Paleocene and again in the Eocene, the Mammals had established themselves in all environments and latitudes due to their capacity to adapt (swimming, flight, etc.). At the end of the Eocene, most of the modern groups were present. Significant adaptive radiations took place among the Birds. Primates began to evolve in the Paleocene and rapid evolution of monkeys occurred during the Oligocene. The ancestors of man were already living in the great African rift by 7 Ma.

In the seas, modern fauna was in place following the disappearance of the ammonites at the end of the Mesozoic. It should be noted that with the narrowing of the Tethys, the partitioning of sedimentary basins makes biostratigraphic correlations more difficult because of increasing *endemism* (development limited to a confined area) in the various basins.

3.3.5 Quaternary Era (Anthropozoic)

Anthropozoic extends from 1.8 Ma until the present. Geologically, the Quaternary is characterized by the shaping of the northern hemisphere continents and its effects on the life of our ancestors: the ***big glaciations*** caused by intense, rapid and repeated climatic variations. These glaciations caused repeated marine regressive–transgressive cycles and erosion. Glaciation is of course not exclusive to the Quaternary, since it also occurred on the supercontinent Gondwana during the Paleozoic. Early naturalists attributed the presence of numerous erratic blocks in areas far from their original areas to the ***Diluvium***, a deluge similar to that described in the

Bible. It was not until the 19th century that scientists presented the hypothesis of ice caps whose extent far surpassed those of today.

In the Quaternary, in response to climate pulsations, polar climatic zones generally moved toward the Equator during periods of glaciation and away from it during interglacial periods. The significant climate changes that possibly led to glaciation include temperature, precipitation, and as a result, atmospheric circulation. For example in Europe, two regions are particularly affected by this phenomenon; (1) northern European glaciers (latitudinal effect) that made spectacular advances to the south at this time: the Scandinavian ice sheet reached northern Germany and Poland, and left many moraines, and (2) the alpine glaciers (altitudinal effect) with their tongues reaching down into large valleys and invading flat areas at the foot of the mountains. As an example, the Rhone glacier reached Lyons and covered almost all of the Molassic Plateau during the course of its major extent (Riss glaciation); a small ice cap covered the Haut-Jura.

These cold periods were interrupted by warmer episodes (*interglacial* periods) during which the large glaciers rapidly disappeared. The last glacial period (the Würm) ended only about 14,000 years ago in the Alps. When the present site of the Federal Institute of Technology Lausanne was still covered by 700 m of ice 18,000 years ago (Fig. 3.26), the Rhone glacier retreated from Solothurn to its present location in three thousand years.

The drastic change in the mass of water taken up by the ice had important consequences on coastal zones where regression (glaciation)–transgression (interglacial) phenomena took place.

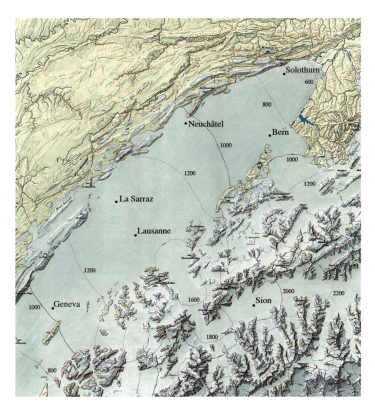

Fig. 3.26 Western Switzerland at the time of the Würm glacial maximum. Extract from the Atlas of Switzerland, plate 6, 1970 edition, prepared by Heinrich Jäckli. Reproduced with authorization of Swisstopo (BA056911).

During glacial maxima, sea level was about one hundred meters lower than today. Ancient valleys and marine terraces on the continental plateau between France and England (Fig. 3.27) and numerous karst springs in the Mediterranean that flow below sea level give evidence of the former low sea level. Variations in marine fauna are also good indicators (migration was a result of temperature, climate, etc.).

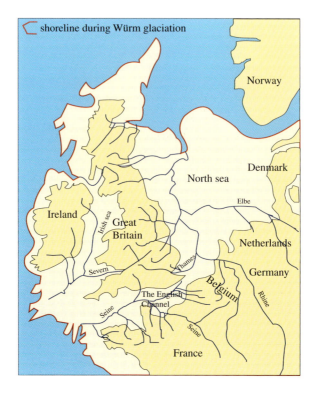

Fig. 3.27 During glacial periods, England and Ireland were attached to the continent (The English Channel opening dates from about 7000 BC). There are traces of the old river valleys under the sea today.

The cause of the glaciation is not known for certain. It appears that climatic variations leading to glaciation were due to a combination of geologic and astronomical factors. Continental drift and the breakup of Gondwana caused Eurasia and North America to move to a more northern latitude. This slow movement, coupled with changes in oceanic and atmospheric circulation, put the continents in a position of potential glaciation; the orogenies may also have modified this atmospheric circulation. Glacial cycles are generally attributed to astronomic causes. According to Milankovitch's theory, they could have resulted from variations in solar activity and the effects of three cycles of different periods all acting on the solar flux at the surface of the continents (see problem 3.3). An animation on the DVD shows these variations.

Paleogeography was strongly influenced by various closely related phenomena such as climatic variation, eustatics (sea level fluctuation), and tectonics.

The name Anthropozoic highlights that the present era is the one of **Man** and his simian ancestors (Fig. 3.28). We have seen that in reality, Man's ancestors date from about 7 Ma. The Quaternary is also characterized by the *Bos-Camelus-Elephas-Equus* association which provides good biostratigraphic and paleontological dates and by the development of large mammals such as the mammoth (Fig. 3.29), typical of cold climates.

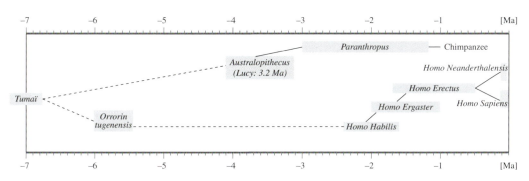

Fig. 3.28 Discoveries of fossil remains have made it possible to reconstruct the origin and evolution of Man. The advance in knowledge remains relative because all the Hominid fossils have been found in eastern Africa, which does not exclude a prior existence elsewhere. In 1974, a team of paleoanthropologists in Ethiopia discovered the oldest hominid skeleton ever identified: Lucy, a feminine *Australopithecus* 3.2 Ma old. However, the discovery of *Orrorin Tugenensis* in 2000 pushed back the age to 6 Ma. This discovery has a significant implication that has caused rethinking of the phylogenetic tree developed up until then. Several characteristics of Orrorin place it closer to humans than to Australopithecus; it is probable that this genus is the closest to the origin of the ancient human lineage and it is very distinct from *Australopithecus* and chimpanzees. In 2001, the discovery of *Tumaï Sahelanthropus tchadensis*, 7 Ma years old, contributed new information: it is a hominid, but its characteristics which are also simian, may allow the hypothesis of a common ancestor.

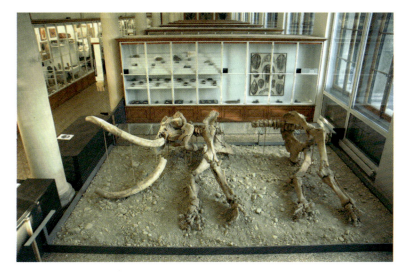

Fig. 3.29 The famous and ephemeral mammal. This almost complete skeleton of the Praz-Rodet mammoth was discovered in 1969 during gravel mining in Vallée-de-Joux (Jura range). Photo Collection of the Cantonal Geologic Museum of Lausanne, R. Marchand. An animation on the DVD shows the discovery and the treatment of this mammoth.

This era is very important for the engineer because during this time the present-day topography was formed by glacial erosion and the deposition of moraines. Many foundation projects are built on these recent deposits. An understanding of the phenomena that gave rise to them will help the engineer understand their nature and structure.

[104, 116, 147, 282, 313]

PROBLEM 3.2

The structure of a rock massif is a series of formations as shown in figure 3.30.
The eight formations are the following:

1. Schists containing trilobites of the species *Phacops*,
2. Anthracitic coal,
3. Crushed rock (cataclastic rock),
4. Conglomerate with granitic elements followed by sandstones and marls with rodent teeth,
5. Micaschists,
6. Granite,
7. Aplitic dike
8. Gold-bearing dike.

The feldspars in the aplitic dikes have been dated by the rubidium-strontium method at 590 million years.

Questions

- Give a relative age for the different formations on the basis of the geometric structure, chronostratigraphic criteria, and the fossil content of the rocks.
- Describe the geological events that can explain the present juxtaposition of the formations.

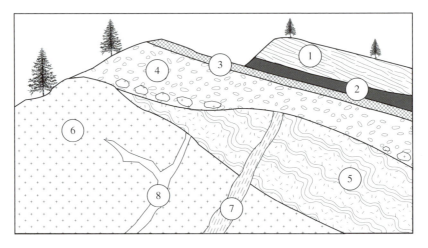

Fig. 3.30 Geologic profile for Problem 3.2.

Look at the solution on the DVD

PROBLEM 3.3

Milankovitch's theory states that the glaciations resulted from a variation in solar insolation of the Earth's surface.

This theory attributes this variation to the insolation at the Earth's position relative to the Sun, which is controlled by three astronomic phenomena:

1. Variation in eccentricity: figure 3.31a shows the variation of eccentricity of the Earth's orbit (only the semiminor axis varies)
2. Obliqueness of the ecliptic: figure 3.31b shows the variation of the angle formed by the polar axis with the perpendicular to the plane of the Earth's trajectory around the Sun.
3. The precession of the equinoxes: figure 3.31c shows the precession expressed as the angle of the Earth-Sun line and angle of the Sun-perihelion line at the time of the spring equinox, measured from the perihelion.

An animation on the DVD shows these variations.

Over the same period of 750,000 years, figure 3.31d shows the variation in oxygen ^{18}O measured in carbonate-forming shells of planktonic marine organisms (foraminifera). The oxygen present in a molecule of water is composed of two isotopes: ^{16}O and ^{18}O. When water evaporates, the lighter isotope (^{16}O) enters easily into the vapor phase, and the ^{18}O tends to remain in the water. There is thus a fractionation of oxygen, with the result that $^{16}O/^{18}O$ ratios in clouds and ocean water are different.

To quantify this fractionation, the isotopic ratios are compared to a standard to place the isotopic variations to one side or the other of the standard. In the case of water, this standard is the SMOW (Standard Mean Ocean Water):

$$\delta^{18}O = \left(\frac{Re - Rs}{Rs} \cdot 1000 \right)$$

Re = Abundance of ^{18}O in the sample/Abundance of ^{16}O in the sample.
Rs = Abundance of ^{18}O in the standard/Abundance of ^{16}O in the standard.

On figure 3.31d the portions of the curves marked G indicate the times when polar glaciers advanced on the American Midwest.

Questions

- Why does the isotopic ratio of oxygen in the carbonates vary?
- Consider the three curves (Fig. 3.31a, b, and c and by simply examining their shapes, determine the contributions of each of them to the explanation of the climatic variations observed in figure 3.31d.
- Draw a diagram for each of the three phenomena. From these drawings, discuss the influence of each of these parameters on the climate at moderate latitudes.
- Using the results of the previous step, is it possible to forecast whether the temperature is going to increase or decrease in the coming centuries?

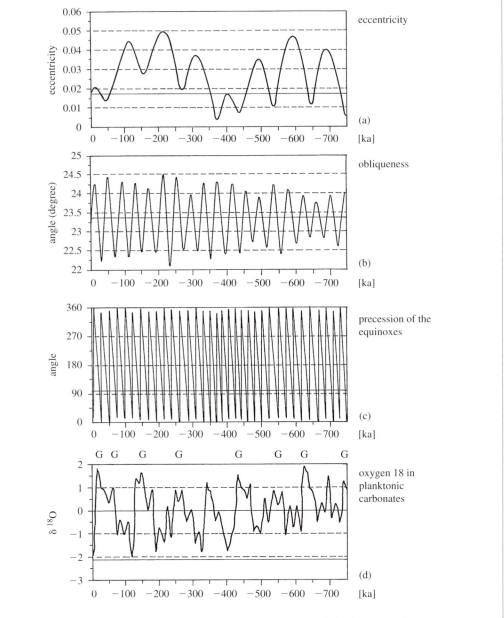

Fig. 3.31 (a), (b) and (c): Time functions of Milankovitch's theory. (d): Variation in oxygen 18.

Look at the solution on the DVD.

Chapter review

The modern Earth is the result of many processes that have been active over a long history. Each has left its imprint and each helps explain the properties of the geological material the engineer has to work with. Knowing the spatial and temporal framework of the Earth helps man better evaluate the consequences of his activity.

4 Physics of the Globe

Now we will examine the physical properties of the materials that make up the Earth. We will first focus our attention on the parts of the Earth's surface where engineering work is conducted. In chapter 5 we will look at the components of rocks on a microscopic scale. Prior to that we must discuss the broad outlines of the Earth's structure to establish the general distribution of rocks and magma. First we will look at geophysical properties in a summary fashion to understand their near-surface effects and their influence on human activities.

This chapter is presented in a targeted way: properties that have only a minor practical importance to engineers will be discussed in a rudimentary way; important topics will be discussed in more detail.

Four types of geophysical parameters are of prime importance: the field of mechanical stresses, density, magnetic properties, and temperature.

4.1 Seismology

Seismology is the science that analyzes the causes of earthquakes and the propagation of waves within the Earth and on its surface.

We have already made a first incursion into this topic with our discussion of the Earth's tectonic movements throughout its history (Chapt. 3). Obviously, the movements of lithospheric plates and the collisions that built mountain chains did not occur quietly. In fact, the entire surface of the globe experiences mechanical stresses that cause tension in places, compression in other places, and shearing elsewhere.

Throughout history, man has suffered major catastrophes as a result of earthquakes (Tab. 4.1).

Table 4.1 Significant earthquakes of the recent history (since 1990) and their consequences

Date	Country	Location	Magnitude	Victims
4.6.2009	Italy	Abruzzo	6.3	290
10.28.2008	Pakistan	Quetta	6.4	400
10.5.2008	Kyrgyzstan	Nura	6.6	75
5.12.2008	China	Sichuan	8.0	69 000
2.3.2008	Congo	Lake Kivu	6.0	40
9.12.2007	Indonesia	Sumatra	8.5	23
8.15.2007	Peru	Chincha Alta	8.0	500
7.17.2006	Indonesia	Java	7.7	400
5.27.2006	Indonesia	Java	6.3	6200
10.8.2005	Pakistan	Kashmir	7.6	100 000
3.28.2005	Southeast Asia	Sumatra	8.7	>1000
2.23.2005	Iran	Zarand	6.4	612
12.26.2004	Southeast Asia	Sumatra	9	290 000
10.24.2004	Japan	Niigata	6.8	25

(Continued)

Table 4.1 (*Continued*)

Date	Country	Location	Magnitude	Victims
5.28.2004	Iran	Northern Region	6.9	35
12.26.2003	Iran	Bam	6.6	30000
5.21.2003	Algeria	Boumerdes	6.8	3000
1.21.2003	Mexico	Colima	8	9500
3.3.2002	Afghanistan	Hindu Kush Region	7.4	100
1.26.2001	India	Gujarat	7.6	20000
8.17.1999	Turkey	Izmit	7.6	30000

Source: USGS

4.1.1   Rupture mechanisms

In the chapter on tectonics (Chapt. 12), we will take a detailed look at rock behavior in response to stress. On the scale of the Earth, we can make the simplifying assumption that the Earth is an elastic solid, except for the outer core, which behaves like a fluid (§ 4.1.3.3). The elastic solid is characterized by a linear stress-strain relationship whose slope defines the modulus of elasticity (§ 12.2.1). In practice, elasticity implies that stress release cancels deformation. When the stress is too great, the rock ruptures and causes an earthquake. The stresses are thus released, at least for a while. We see an example of this mechanism along a fault that is famous for its earthquakes: the San Andreas Fault in California (Fig. 4.1)

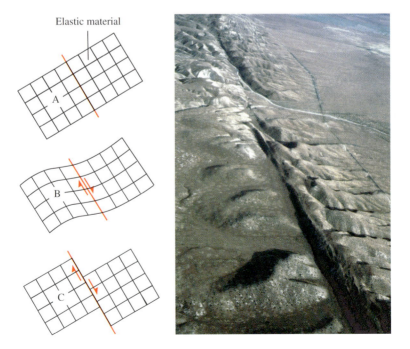

Fig. 4.1 Progressive deformation leading to rupture (A–C). Rupture along the San Andreas Fault in California. Photo Earth Sciences Photographic Archive, U.S. Geological Survey.

Let us assume that the sides of the fault move by shearing at a constant velocity. The fault surface is rough, and at the beginning, movement causes elastic deformation of the rock near the fault. Once the shearing strength of the fault surface is reached, rupture occurs abruptly and the stresses are released: this is an ***earthquake***. Then the process begins again. Generally, the longer the time between two earthquakes, the more violent the next earthquake will be, because of the energy accumulated by elastic deformation prior to rupture. The San Andreas Fault has an average shear velocity of 35 m in 1000 years. In 1857 it moved 6 m along more than 300 km in one movement, causing a magnitude 8.3 earthquake (see § 4.1.5.3 for magnitude scale). At present, the movement is blocked in some places, and is about 5 m behind the average velocity. This is why some inhabitants are expecting to see a very large amplitude earthquake soon: the Big One. Earthquakes are often followed by secondary, less intense shocks due to readjustments of portions of the crust under the new stress regime: these are called ***aftershocks***.

In seismology, the source of the earthquake is called the ***focus*** or ***hypocenter*** (see Fig. 4.5). The intersection of the wave surfaces (which are spherical if the media is homogeneous and isotropic) with a supposedly planar topography gives wave circles. The center of these circles, placed vertically above the focus, is called the ***epicenter*** (Fig. 4.5).

Later we will study the numerous reasons for earthquakes. For the moment, let us see how seismic waves propagate in a homogeneous and isotropic elastic material, similar to a metal beam for example. This will enable us to identify several types of waves.

4.1.2   Types of seismic waves

Let's create a shock at the end of a beam and examine the propagation of deformation along the beam (Fig. 4.2). There are two ways to create initial deformation: by compression or by shear. Some waves travel through the body of a solid, and others remain at the surface.

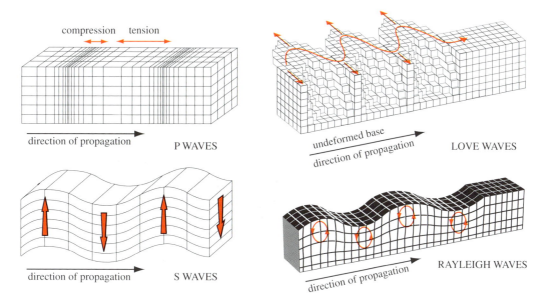

Fig. 4.2 Different types of waves in an elastic beam.

4.1.2.1   Body waves

Body waves can be generated by two types of stresses.

Compressional stress

If we strike the end of a beam, deformation propagates along the entire volume of the beam as a longitudinal wave: each point on the beam vibrates in a sinusoidal movement in the direction of wave propagation. The points parallel to the stressed surface vibrate in phase: they are the wave surface. If we photograph the beam at a particular moment, we may see that the part farthest from the blow has not been deformed: the wave front has not yet arrived. The stressed area is composed of a succession of compressional areas (nodes) and extensional areas (antinodes). When these waves move through the subsurface they are the first waves perceived after an earthquake, and are called *primary waves* or *P waves* (Fig. 4.2).

Shear stress

This time we apply a stress to the upper surface of the beam. The points oscillate perpendicularly to the direction of wave propagation: this is a transverse wave. Two neighboring elementary volumes move independently of each other and perpendicularly to the propagation direction. There is a shear stress between these two volumes. For this reason, these waves are absorbed in liquids. Transverse waves travel more slowly than primary waves; they are called *secondary waves* or *S waves* (Fig. 4.2).

4.1.2.2   Surface waves

Blows near the side of a beam also cause deformation, limited this time to the surface of the solid. There are several types of movements for different types of waves. For example, Rayleigh waves generate a sort of swell on the solid surface. Love waves are transverse shear waves on a horizontal surface. These waves are called surface waves because they are felt only near the surface. However, they can also travel along the boundary between two media. They play a significant role in the destruction of buildings during earthquakes. They have a slower velocity than S waves.

4.1.2.3   Wave propagation velocity

Timoshenko and Gooder [287] established the relationship between the velocity of P waves (v_p) and two parameters that can be easily determined in the laboratory by subjecting a rock sample to uniaxial stress (Chapt. 12): elasticity modulus (or Young's modulus) and Poisson's coefficient:

$$v_p = \sqrt{\frac{E(1-\nu)}{\rho(1+\nu)(1-2\nu)}} \quad [\text{m/s}] \tag{4.1}$$

$$E = \text{Young's modulus} = \frac{\Delta \sigma_z}{\Delta \varepsilon_z} [\text{N/m}^2]$$

$$\nu = \text{Poisson's coefficient} = \frac{\Delta x/x}{\Delta z/z} = \frac{\Delta \varepsilon_x}{\Delta \varepsilon_z}$$

$\Delta \varepsilon$ = Relative strain [–]
ρ = Density [kg/m^3]
σ = Stress [N/m^2]
z = Length of sample [m]
x = Width of sample [m]

We obtain the functions graphed in figure 4.3.

The waves' velocities in the different regions of the Earth are shown in Fig. 4.4. It should be noted that these velocities increase everywhere with depth in the crust and mantle, but decrease when entering the core.

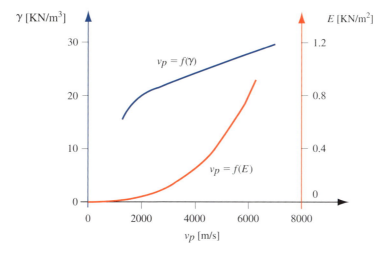

Fig. 4.3 Graphical relationship of $v_p = f(E, \gamma)$ (According to [87]). The specific weight $\gamma = \rho \cdot g$, where ρ is the density and g the gravitational acceleration.

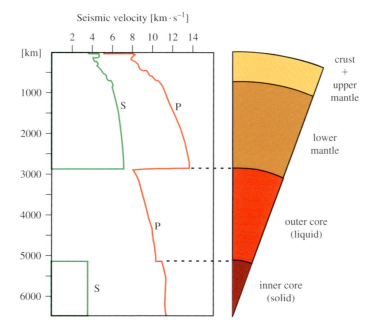

Fig. 4.4 Velocity of body waves within the Earth, as function of depth.

Rocks near the Earth's surface have lower velocities than deeper rocks (Tab. 4.2). Rocks and unconsolidated material cited in this table are defined in later chapters where their formation is discussed.

Table 4.2 Velocity ranges for P waves at shallow depths in unconsolidated materials and rocks.

Unconsolidated material	Unsaturated [m/s]	Water–Saturated [m/s]
Lodgement till	1800–2500	1800–2500
Sandy gravel	700–1500	1800–2500
Middle sand	500–1200	1500–2000
Clayey silt	500–1200	700–1600
Peat	300–500	500–1000

Rocks (the degree of saturation do not change noticeably the velocity)	[m/s]
crystalline rocks	4000–6000
limestones	2000–4000
Mesozoic marls	2000–3000
molassic conglomerate	2500–3500
molassic sandstone	2000–3000
molassic marls	1800–2500

4.1.3   Laws of wave propagation

Although seismic waves are very different from electromagnetic waves in general and light waves in particular, they share some properties.

4.1.3.1   Propagation in a single homogeneous and isotropic medium

Similar to light in optics, a point source of mechanical shock establishes spherical wave surfaces around itself (Fig. 4.5). A wave surface is the geometric place of points that have been simultaneously affected by the deformation and are vibrating in phase. By analogy with optics, the rays of a wave are the propagation directions. They radiate outward from the source. In a homogeneous and isotropic medium they are linear and perpendicular to wave surfaces.

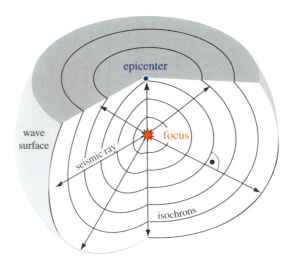

Fig. 4.5 Propagation of seismic waves in an isotropic homogeneous medium with location of the focus (or hypocenter) and the epicenter of an earthquake.

4.1.3.2 Propagation in two media with different velocities.

Three fundamental principles similar to those of optics are applicable.

Huygens principle

Each point where two media with different velocities are in contact behaves as a new source when it receives the vibrational signal.

The Law of reflection

A seismic ray arriving at the interface of two media with different velocities is reflected. The angle of reflection is equal to the angle of incidence (Fig. 4.6).

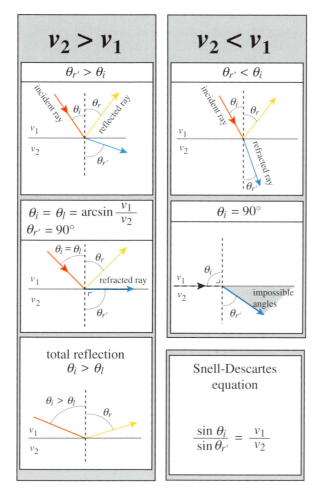

Fig. 4.6 Laws of reflection and refraction applied to seismic rays. v_1 and v_2 are wave velocities in the two media. θ_i = Incident angle, θ_r = Reflected angle, $\theta_{r'}$ = Refracted angle, θ_l = Maximum angle (from [280]).

The Law of refraction (Snell-Descartes Law)

The seismic ray arriving obliquely at the interface between two media with different velocities is refracted (Fig. 4.6). The angle of the refracted ray is a function of the incident angle and of the contrast in the seismic velocities of the media according to the ratio

$$\frac{\sin\theta_i}{\sin\theta_{r'}} = \frac{v_1}{v_2} \tag{4.2}$$

As in optics, it is possible to define the maximum angles of refraction. Two cases must be considered.

Increasing velocity

A limiting situation is reached when the angle of the refracted ray is equal to 90°. Any incident ray that exceeds this critical position is entirely reflected (total reflection).

Decreasing velocity

A limiting situation is reached when the angle of the incident ray approaches 90°. The refracted ray cannot exceed the limiting value of its angle. This determines a shadow zone for elevated values of $\theta_{r'}$.

4.1.3.3 Propagation of seismic waves inside the Earth

The study of seismic responses at different points on the Earth has revealed the geometry and the composition of layers within the Earth. The reconstruction of seismic trajectories is a highly complex discipline due to numerous interferences between the different types of waves. For example, a P wave arriving at a contact will be reflected and refracted as a P wave, but it will also create two new S waves that are polarized on the vertical plane. Each new S wave in turn generates two new P waves at the next contact (not discussed here).

Let us consider an earthquake at the crust-mantle contact and let us first look at what becomes of the P waves that penetrate the interior of the Earth (Fig. 4.7 left side). The rays bend slightly as a result of the velocity increase with depth in the mantle. Of the bundle of rays that diverges from the focus, a part remains in the mantle. Those that exit toward the center of the Earth enter into the outer core on contact. There, a refraction of the "decreasing velocity" type occurs, because the velocity of the core is less than that of the mantle. These rays are thus deviated toward the center of the Earth. As a result, a shadow zone occurs at the surface of the Earth that corresponds to the interval between the last ray that travels through the mantle only, and the first ray to be refracted by the outer core. This **shadow zone** makes a spherical ring of an aperture of about 40° located in the hemisphere opposite the hypocenter.

The paths of S waves (Fig. 4.7, right side) also create a shadow zone, but for a different reason. As we have seen, shear waves are absorbed in liquids, and the outer core is a liquid. This creates a shadow zone even wider than that created by the P waves. It is a spherical cap of 154°, located in the hemisphere opposite the hypocenter.

It was the discovery of shadow zones by geophysicists as they deciphered the signals emitted by earthquakes that led to the understanding of the velocity decrease of P waves in the core and the liquid properties of the outer core.

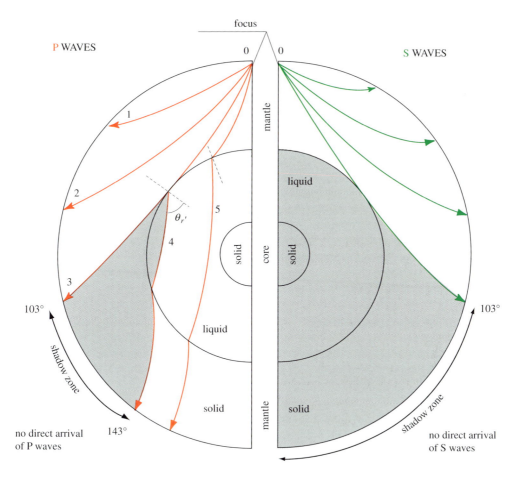

Fig. 4.7 Paths of waves generated by an earthquake through the Earth: P waves (on left) and S waves (on right) (from [262]).

4.1.4 Causes of earthquakes

We must distinguish between earthquakes that affect large regions of the planet and shocks that are more local. Earthquakes of tectonic and volcanic origin affect large areas. Other less energetic earthquakes are the result of human activities (dams, mining, deep pumping, etc.); they are important to the engineer because they can also cause damage to structures.

4.1.4.1 Tectonic origin

Shocks result from abrupt movements along active faults. The epicenters of these earthquakes are located beneath subduction zones, in active mountain chains, or even along large strike-slip faults. The foci are generally between 10 and 60 km deep, but they can be as deep as 700 km in subduction zones. This is the case of the famous Pacific ring of fire (Fig. 4.8).

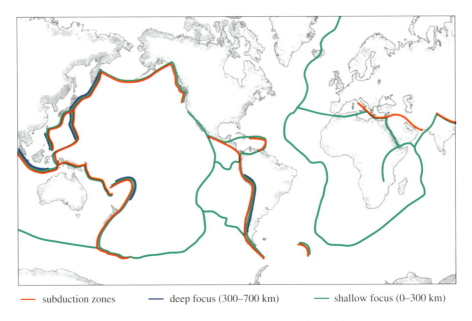

——— subduction zones   ——— deep focus (300–700 km)   ——— shallow focus (0–300 km)

Fig. 4.8 Principal seismic zones around the world

4.1.4.2   Volcanic origin

Rupture zones in the crust are regions where lava is able to reach the surface. Shaking comes principally from mechanical reactions to rising magma, explosion of plugs that block the magmatic conduit, and abrupt outbursts of gas, etc. They can also occur later as a caldera forms due to subsidence of a part of the cone, or to the collapse of the roof of the magma chamber. The volcanic region of Campi Flegrei, west of Naples, is of particular interest to show the link between volcanic activity and earthquakes. The region is known for its Solfatara crater, which gave *solfataras* (emissions of sulfurous gas) their name. It is made up of a series of adjacent volcanoes of different ages, of which the most recent, Monte Nuovo, formed in 1538. These volcanoes tend to collapse to form calderas, causing continuous subsidence and sometimes an uplift of the ground. In the port of Pozzuoli, this movement is apparent from the sites of the roman market and the Temple of Serapis, which laid below sea-level in the 1980s. From 1982 onwards, the sites were raised upwards showing that the tendency was reversed. Uplift suddenly accelerated in October 1983. In a few days, the sites at Pozzuoli rose by 1.8 m, as a violent earthquake shook the historic center (Fig. 4.9). The Temple of Serapis also rose above the water during this event; the presence of lithophage marks shows the partial immersion of the temple's columns. This subsidence and upheaval phenomenon linked to gas and magma pressure in the volcanic chamber is called *bradyseism*.

4.1.4.3   Underground nuclear explosions

Although underground nuclear explosions are rarely conducted nowadays, they have left a seismic record in numerous regions of the USA, China, Russia, Japan and the Pacific. French tests on Mururoa (Fig. 4.10) caused local subsidence of the coral reef as a result of the explosions. The strongest blasts were recorded as far away as New Zealand.

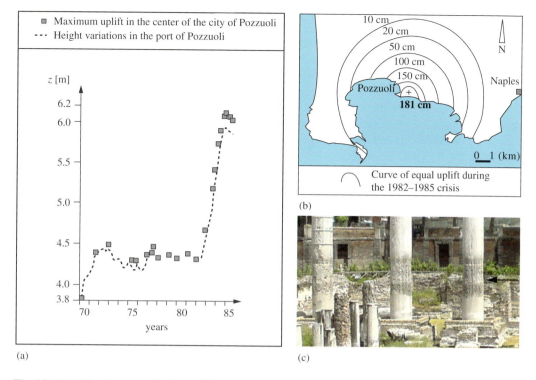

Fig. 4.9 (a) Uplift of the ground at Pozzuoli between 1970 and the rapid change of 1983–1985; the surface is again subsiding since this date. (b) Distribution of this uplift in the Campi Flegrei region. (c) Columns of the Temple of Serapis with the marks of previous sea-levels (GEOLEP photo, A. Parriaux). Osservatorio Vesuviano data.

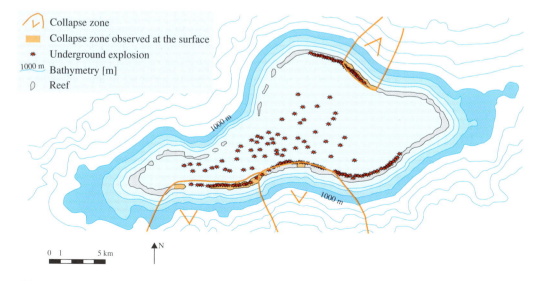

Fig. 4.10 Subsidence of the Mururoa coral reef, probably as a result of underground nuclear explosions (from [86]).

4.1.4.4 Dams

The creation of deep reservoirs significantly modifies the static equilibrium of a region. In addition, they significantly increase the hydraulic head in fissures, causing them to open and diminishing their shear strength. It appears that the Salanfe dam, on the left bank of the Rhone, caused a strong earthquake in the Val-d'Illiez, then the appearance of a thermal spring (Fig. 4.11) through a fault at the bottom of the valley. This cause-and-effect relationship is supported by the occurrence of additional seismic activity and increased discharge of the thermal spring following a grouting campaign around the dam to reduce the water loss and allow the maximum level of the lake to rise by about ten meters.

Another example is an event that occurred at the Monteynard (France) dam on the Drac in April 1963. The pressure exerted by 135 m of water behind the dam caused two magnitude-5 earthquakes soon after it was filled. The earthquakes caused roof tiles to fall and chimneys to collapse in nearby villages and wells to dry up in the neighboring town of Corrençon-en-Vercors.

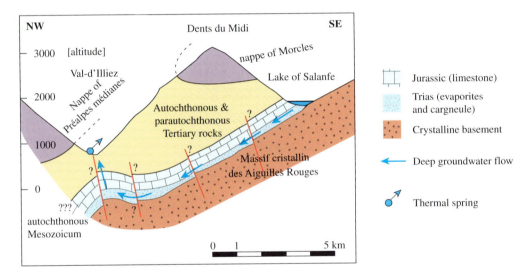

Fig. 4.11 Probable relationship between the Salanfe Dam (Valais) and the appearance of a thermal spring in Val-d'Illiez (according to [26])

4.1.4.5 Drainage by underground projects or deep wells

Groundwater normally drains into tunnels during their construction. Decreased hydrostatic pressure in fissures modifies their hydro-mechanical equilibrium, and may lead to local ruptures and weak tremors. As an example, the Rawil reconnaissance tunnel on the right bank of the Rhone in Valais may have caused subsidence of the foundations of the Tseuzier Dam (Fig. 4.12). Studies have demonstrated that the principal phase of rock deformation was not accompanied by large seismic events. However, a considerable increase in micro-seismic activity was recorded when massive amounts of water flowed into the reconnaissance tunnel. It is still not certain if the earthquakes recorded in this region resulted from massive drainage of groundwater through the tunnel or if they were a natural consequence of the dam being built in a fault zone. Other examples of the relationship between civil engineering projects and groundwater will be given in chapter 7.

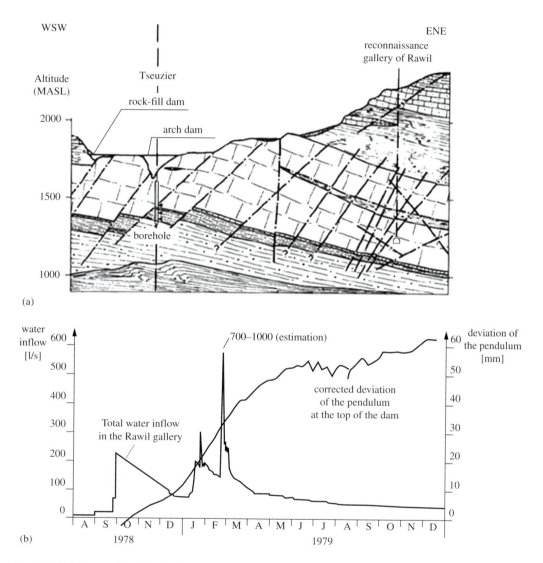

(a)

(b)

Fig. 4.12 Subsidence of the Tseuzier Dam, probably as a result of drainage of groundwater through the Rawil gallery. (a) Location of the dam and the tunnel. (b) Graph of water entry into the tunnel and dam movements (from [250]).

4.1.4.6   Roof collapse into cavities

Natural cavities hollowed out by the dissolution of rocks (limestone and gypsum) are usually found in karst areas. However this phenomenon is not common and usually limited in scope.

Subsidence in mining areas is more serious. For centuries, humans have extracted ores from the subsurface and have left only fragile pillars to hold up the mine roofs. Over the years, the pillars weather, crack, and finally break, causing the collapse of neighboring pillars and the subsidence of large surface areas. This phenomenon may lead to major topographic modifications such as changes in the flow direction of rivers and the creation of lakes (see also Chapt. 14).

4.1.5   Monitoring and treatment of seismic signals

4.1.5.1   Measurement

Seismographs collect and record seismic data from the Earth. These devices contain a mass that tends to remain stable while the housing of the device vibrates with the ground (Fig. 4.13). The mass is connected to a magnet. The housing contains a coil that encircles the magnet. Movement of the magnet and coil induces an electrical current proportional to the velocity of the relative movement of the two objects, thus the velocity of the ground motion. This current is recorded, and when converted it provides a time series of the velocity.

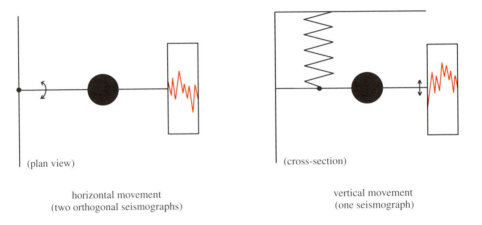

(plan view)   (cross-section)

horizontal movement   vertical movement
(two orthogonal seismographs)   (one seismograph)

Fig. 4.13 Schematic diagram of a seismograph.

To measure the velocity vector of the ground in space, three components must be recorded: two horizontal and one vertical. Some devices record ground acceleration directly for large earthquakes. They are automatically set in motion when the ground acceleration threshold is reached.

Seismographic stations worldwide are linked together to form a network.

4.1.5.2   Signal treatment

We must distinguish here between the simple translation of time series of ground movement (acceleration, velocity, movement) and more elaborate treatments that analyze the spectral content of vibrations.

Treatment of time series

The first signal is either velocity or ground acceleration. It is treated to provide a more elaborate response from the phenomenon. A time series of the missing parameter is calculated (acceleration or velocity) then the movement is calculated by integration (Fig. 4.14). The acceleration curve as a function of time is called an ***accelerogram***.

Here are the ranges of these parameters for an average earthquake (magnitude of about 6.5 or intensity of about VIII) (§ 4.1.5.3):

- Maximum acceleration: 1.5 to $3\,\text{m/s}^2$ or 0.15 to 0.3 g,
- Maximum velocity of ground movement: 0.1 to 0.3 m/s,
- Maximum ground movement: 10 to 30 cm.

Experience shows that the maximum vertical acceleration is equal to about two thirds of the maximum horizontal acceleration.

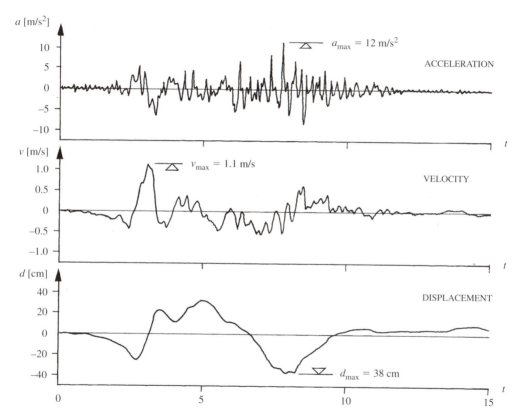

Fig. 4.14 Signal treatment. A seismograph recording the acceleration of the San Fernando earthquake, near Los Angeles in 1971 (the Pacoma dam site is 200 km from the epicenter). This 6.5-magnitude earthquake shows a very strong acceleration peak and a maximum ground displacement of about 40 cm (horizontal component). The time series of the velocity is inferred by the first integration of the accelerogram. Ground displacement is obtained by a second integration (from [15]).

Spectral analysis of vibrations

In engineering it is useful to know the spectral composition of an earthquake in order to determine which vibrations account for the majority of the recorded movements. The spectral composition also determines possible resonances with unconsolidated sediments or structures, which an engineer must take into account when designing a project. The spectral content of the seismic signal (Fourier vibration spectrum) is of interest in itself. Engineers also study its effect on vibrating solids, which are equivalent to unconsolidated surface sediments (site effect) or structures that may resonate with the signal (response spectrum).

To determine the spectral content, the seismogram is subjected to a Fourier decomposition. This complex step, which we will discuss here in principle only, is done by computer. A sinusoid with a previously chosen wavelength, amplitude, and phase difference is considered and this theoretical function is correlated with the experimental function. If this correlation is significant, the sinusoid is said to explain a part of the signal, and it is retained. If not, it is rejected and another slightly different one is tried until a suitable one is found. Then another sinusoidal function is sought that best explains the residual between the first sinusoid and the experimental curve. When the sum of retained sinusoids is sufficiently well correlated with the actual curve, the calculation is stopped (Fig. 4.15). From this a spectral diagram is drawn, as a function of the frequency or period (Fig. 4.16) that shows if the vibration is distributed homogeneously or if there are dominant frequencies. We will see the utility of these spectra in the analysis of seismic risk (§ 4.1.6.4).

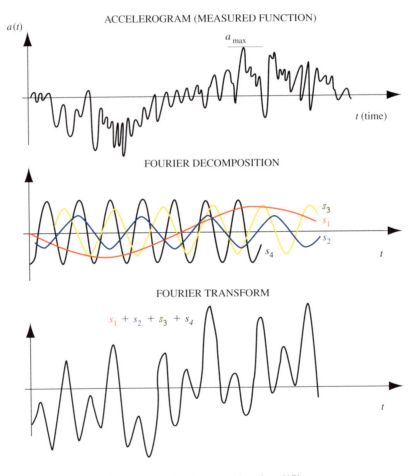

Fig. 4.15 Spectral decomposition (from [17]).

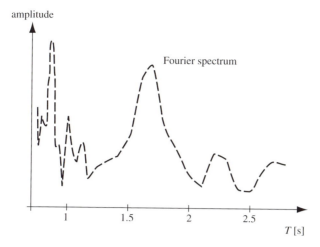

Fig. 4.16 Spectral distribution of an earthquake showing the contribution of different T periods to vibrations. Two dominant periods appear near 0.9 s and 1.7 s (from [17]).

Applications to construction

The vibrational response of a structure depends on three factors:

- The seismic signal that is transmitted by the rocky substrate at depth,
- Its transformation during travel through loose surficial deposits where the first resonances are produced,
- Its transformation within the construction itself where new resonances are produced.

An understanding of the response spectrum enables engineers to design civil engineering projects that take into account the seismic hazard which is specific to the site. This spectrum can be determined experimentally using a series of flexible bars (oscillator) that are attached perpendicularly to a frame (Fig. 4.17). The bars are of variable length and have an identical mass (m) at their end. The frame is then subjected to vibration perpendicular to the bars that is equivalent to the accelerogram of the earthquake under study. Each bar begins to vibrate with a amplitude that varies through time, but always at a set frequency, which is equivalent to its proper frequency f.

This is independent of the excitation accelerogram. It equals:

$$f = \frac{1}{2\pi} \sqrt{\frac{k}{m}}$$

$$[T^{-1}] = [M \cdot T^{-2}]^{1/2} \cdot [M^{-1}]^{1/2} \tag{4.3}$$

where k, the "elastic constant" of the bar, represents the force that must be applied to the bar at rest, at the application point of mass m, to obtain a unit movement perpendicular to the bar.

The time series of movement, velocity, and acceleration of mass is recorded on each bar. This is the response of the frame to the ground vibration. The maximum resonance amplitude is taken from the response of each oscillator. It is reported graphically as a function of the particular frequency of the corresponding oscillator and the response spectrum to a given vibration is obtained.

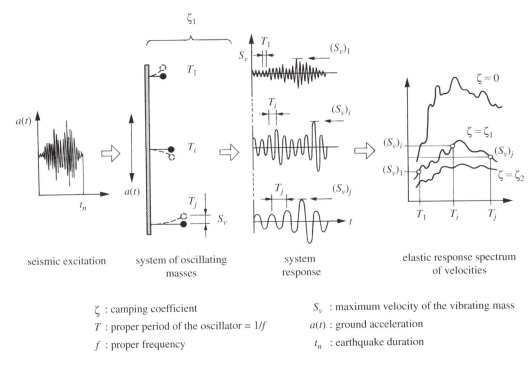

seismic excitation   system of oscillating   system   elastic response spectrum
                              masses   response   of velocities

ζ : camping coefficient S_v : maximum velocity of the vibrating mass

T : proper period of the oscillator = $1/f$ $a(t)$: ground acceleration

f : proper frequency t_n : earthquake duration

Fig. 4.17 Experimental determination of the response spectrum of an earthquake. See the explanation in the text (from [134]).

This experiment is repeated for different dampening coefficients of the bars. The weaker the dampening, the higher the resonance frequency peaks.

We will come back to the notion of dominating frequencies in paragraph 4.1.6.3 with regard to site effects and damage to structures.

4.1.5.3   Quantification scales of earthquakes

Throughout time, man has sought to compare earthquakes in terms of their importance. In the past, before measurement equipment was available, man used descriptive scales based on effects observed at the surface. With new technologies and the development of seismological networks, we have moved forward to more quantitative scales.

Intensity scale

Mercalli first defined an intensity scale in 1902, and Medvedev, Sponheuer, and Karnik modernized it in 1964 (MSK intensity scale). It is useful because it is an expression of the observations of the earthquake effects on structures. Scientifically, however, it has several disadvantages:

- It is qualitative,
- It is not applicable to regions where there are no buildings,
- It does not take into account differences in construction quality (a wall in a city in a developed country does not have the same resistance as a wall in a small village in the Himalaya).

To correct these shortcomings, a scale that contains 12 degrees was developed in Europe (European Macroseismic Scale EMS 1998, from [110]). The intensity is defined on the base of the effects on humans, on objects, on nature and on damage to buildings for different types of constructions (Tab. 4.3, Fig. 4.18, Fig. 4.19, Fig. 4.20).

Table 4.3 Classification of structures on the basis of vulnerability according to EMS 1998.

Type of structure		Vulnerability class					
		A	B	C	D	E	F
masonry	dry stone wall	■					
	adobe (earth brick)	■					
	simple stone		■				
	massive stone			■			
	bricks with concrete panels		■				
	bricks with reinforced concrete slabs			■			
	reinforced brick				■		
reinforced concrete	not designed to earthquake-resistant standards			■			
	with minimum earthquake-resistant design				■		
	with average earthquake-resistant design					■	
	with advanced earthquake-resistant design						■
wood	timber structures				■		

■ Most probable vulnerability class

Magnitude scale

In 1935 Richter introduced a concept that made it possible to define earthquake violence more objectively and to quantify the energy released. Richter was trying to characterize local earthquakes in California. Since the emplacement of a large network of seismographs around the Earth, this notion has been generalized by the introduction of a series of magnitudes, most of which refer to Richter but are far from the historical definition.

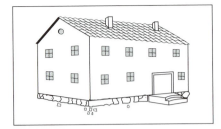

Grade 1:

Negligible to slight damage (no structural damage, slight non-structural damage)

Hair-line cracks in very few walls. Fall of small pieces of plaster only.

Fall of loose stones from upper parts of buildings in very few cases.

Grade 2:

Moderate damage (slight structural damage, moderate non-structural damage)

Cracks in many walls. Fall of fairly large pieces of plaster. Partial collapse of chimneys.

Grade 3:

Substantial to heavy damage (moderate structural damage, heavy non-structural damage)

Large and extensive cracks in most walls. Roof tiles detach. Chimneys fracture at the roof line. Failure of individual non-structural elements (partitions, gable walls).

Grade 4:

Very heavy damage (heavy structural damage, very heavy non-structural damage)

Serious failure of walls; partial structural failure of roofs and floors.

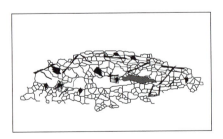

Grade 5:

Destruction (very heavy structural damage)

Total or near total collapse.

Fig. 4.18 Classification of damage to masonry buildings (Tab. 4.3), according to EMS 1998.

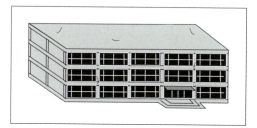

Grade 1:

Negligible to slight damage (no structural damage, slight non-structural damage)

Fine cracks in plaster over frame members or in walls at the base. Fine cracks in partitions and infills.

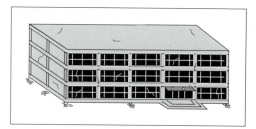

Grade 2:

Moderate damage (slight structural damage, moderate non-structural damage)

Cracks in columns and beams of frames and in structural walls. Cracks in partition and infill walls. Fall of brittle cladding and plaster. Falling mortar from the joints of wall panels.

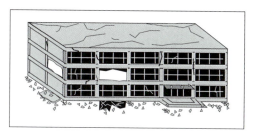

Grade 3:

Substantial to heavy damage (moderate structural damage, heavy non-structural damage)

Cracks in columns and beam column joints of frames at the base and at joints of coupled walls. Spalling of concrete cover, buckling of reinforced rods. Large cracks in partition and infill walls. Failure of individual infill panels.

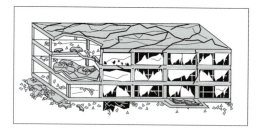

Grade 4:

Very heavy damage (heavy structural damage, very heavy non-structural damage)

Large cracks in structural elements with compression failure of concrete and fracture of rebars, bond failure of beam reinforced bars. Tilting of columns. Collapse of a few columns or of a single upper floor.

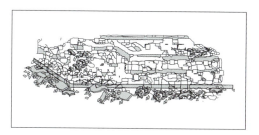

Grade 5:

Destruction (very heavy structural damage)

Collapse of ground floor or parts (e. g. wings) of buildings.

Fig. 4.19 Classification of damage to reinforced concrete buildings (Tab. 4.3), according to EMS 1998.

The intensity scale is based on:
a)   Effects on humans
b)   Effects on objects and on nature
c)   Damage to buildings

Introductory remark:
The various intensity degrees can include the effects of shaking at lower intensity degree(s) even though these effects are not mentioned explicitly.

I. Not felt
a)   Not felt, even under the most favorable circumstances.
b)   No effect.
c)   No damage.

II. Scarcely felt
a)   The tremor is felt only in isolated cases (<1%) of individuals at rest and in particularly receptive positions indoors.
b)   No effect.
c)   No damage.

III. Weak
a)   The earthquake is felt indoors by a few people. People at rest feel a swaying or light trembling.
b)   Hanging objects swing slightly.
c)   No damage.

IV. Widely observed
a)   The earthquake is felt indoors by many and felt outdoors only by very few. A few people are awakened from sleep. The level of vibration is not frightening. The vibration is moderate. Observers feel a slight trembling or swaying of the building, room or bed, chair etc.
b)   China, glasses, windows and doors rattle. Hanging objects swing. Light furniture shakes visibly in a few cases. Woodwork creaks in a few cases.
c)   No damage.

V. Strong
a)   The earthquake is felt indoors by most, outdoors by some. A few people are frightened and run outdoors. Many sleeping people are awakened. Observers feel a strong shaking or rocking of the whole building, room or furniture.
b)   Hanging objects swing considerably. China and glasses clatter together. Small, top-heavy and/or precariously supported objects may be shifted or fall down. Doors and windows swing open or shut. In a few cases window panes break. Liquids oscillate and may spill from nearly full containers. Animals indoors may become uneasy.
c)   Damage of grade 1 to a few buildings of vulnerability class A and B.

VI. Slightly damaging
a)   Felt by most people indoors and by many outdoors. A few persons lose their balance. Many people are frightened and run outdoors.
b)   Small objects of ordinary stability may fall and furniture may move slightly. In few instances dishes and glassware may break. Farm animals (even outdoors) may be frightened.
c)   Damage of grade 1 is sustained by many buildings of vulnerability class A and B; a few of class A and B suffer damage of grade 2; a few of class C suffer damage of grade 1.

Fig. 4.20 Definition of degrees of intensity. Note: The terms used in the quantitative estimates have the following meanings: "some or few" is equivalent to 0–20% of cases, "several" is equal to 10–60%, "the majority" is equal to 50–100%.

VII. Damaging

a)   Most people are frightened and try to run outdoors. Many find it difficult to stand, especially on upper floors.
b)   Furniture is shifted and top-heavy furniture may be overturned. Objects fall from shelves in large numbers. Water splashes from containers, tanks and pools.
c)   Many buildings of vulnerability class A suffer damage of grade 3; a few of grade 4.

   Many buildings of vulnerability class B suffer damage of grade 2; a few of grade 3. A few buildings of vulnerability class C sustain damage of grade 2. A few buildings of vulnerability class D sustain damage of grade 1.

VIII. Heavily damaging

a)   Many people find it difficult to stand, even outdoors.
b)   Furniture may be overturned. Objects like TV sets, typewriters fall to the ground. Tombstones may occasionally be displaced, twisted or overturned. Waves may be seen on very soft ground.
c)   Many buildings of vulnerability class A suffer damage of grade 4; a few of grade 5.

   Many buildings of vulnerability class B suffer damage of grade 3; a few of grade 4. Many buildings of vulnerability class C suffer damage of grade 2; a few of grade 3. A few buildings of vulnerability class D sustain damage of grade 2.

IX. Destructive

a)   General panic. People may be forcibly thrown to the ground.
b)   Many monuments and columns fall or are twisted. Waves are seen on soft ground.
c)   Many buildings of vulnerability class A sustain damage of grade 5.

   Many buildings of vulnerability class B suffer damage of grade 4; a few of grade 5. Many buildings of vulnerability class C suffer damage of grade 3; a few of grade 4. Many buildings of vulnerability class D suffer damage of grade 2; a few of grade 3. A few buildings of vulnerability class E sustain damage of grade 2.

X. Highly destructive

c)   Most buildings of vulnerability class A sustain damage of grade 5.

   Many buildings of vulnerability class B sustain damage of grade 5. Many buildings of vulnerability class C suffer damage of grade 4; a few of grade 5. Many buildings of vulnerability class D suffer damage of grade 3; a few of grade 4. Many buildings of vulnerability class E suffer damage of grade 2; a few of grade 3. A few buildings of vulnerability class F sustain damage of grade 2.

XI. Catastrophic

c)   Most buildings of vulnerability class B sustain damage of grade 5.

   Most buildings of vulnerability class C suffer damage of grade 4; many of grade 5. Many buildings of vulnerability class D suffer damage of grade 4; a few of grade 5. Many buildings of vulnerability class E suffer damage of grade 3; a few of grade 4. Many buildings of vulnerability class F suffer damage of grade 2; a few of grade 3.

XII. Completely catastrophic

c)   All buildings of vulnerability class A, B and practically all of vulnerability class C are destroyed. Most buildings of vulnerability class D, E and F are destroyed. The earthquake effects have reached the maximum conceivable effects.

Fig. 4.20 (*Continued*) Definition of degrees of intensity. Note: The terms used in the quantitative estimates have the following meanings: "some or few" is equivalent to 0–20% of cases, "several" is equal to 10–60%, "the majority" is equal to 50–100%.

The Magnitude m_b (or magnitude of body waves) determined on the basis of the recording at a given station is related to the logarithm of the maximum amplitude in microns of the P waves at this station:

$$m_b = \log \frac{A_1}{T_1} + f_1(\Delta, P) + C_1 \qquad (4.4)$$

where T_1 is the period of the seismic wave having the greatest amplitude after data treatment, A_1 is the maximum amplitude of the wave of period T_1, f_1 is a function of the epicentral distance Δ (§ 4.1.5.4) and the focal depth P and C_1 is a constant unique related to the seismological station and the site effect. It is used only for distant earthquakes.

It is also possible to calculate a magnitude M_S from the amplitude of Rayleigh type surface waves (the identification of the variables is similar to that of m_b):

$$M_S = \log \frac{A_2}{T_2} + 1.66 \log \Delta + 3.30 + C_2 \qquad (4.5)$$

m_b and M_S are related by the empirical formula: $m_b = 0.56\, M_S + 2.9$.

Finally, for the largest earthquakes, the seismic moment magnitude M_w is used:

$$M_w = (\log M_O/1.5) - 10.7 \qquad (4.6)$$

where the seismic moment $M_O = \mu \cdot L \cdot P \cdot D$; μ being the rigidity coefficient of the medium in $[N \cdot m^{-2}]$, L is the length of the fault that generated the earthquake [km], P is the depth [km], and D is the average slip [m] (Fig. 4.21).

Because of its logarithmic relationship with amplitude, magnitude is sensitive only to relative variations: magnitude 6 has amplitudes ten times higher than magnitude 5, magnitude 7, one hundred times higher.

In terms of magnitude and energy ratio, two thirds of the energy released from an earthquake is thermal, and one third is mechanical (the total seismic energy is about 10^{18} [J/a] for the entire planet).

In addition, the relationship between vibrational energy at the focus (ε) and earthquake magnitude can be calculated by an empirical formula of the following type:

$$\log \varepsilon = a + bM \qquad (4.7)$$

Coefficients a and b are defined for each type of magnitude. For example, for M_s, we have:

$$\log \varepsilon = 4.8 + 1.5\, M_S \qquad (4.8)$$

In this case, a magnitude difference of 1 is approximately equivalent to an energy difference factor of 32 at the focus. Thus, an earthquake of magnitude 7 would be $32 \cdot 32 = 1024$ times more energetic than an earthquake of magnitude 5.

We can establish a very approximate relationship between the intensity and magnitude scales (Tab. 4.4).

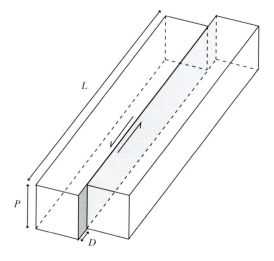

Active fault				
Length L (km)	Depth P (km)	Displacement D (m)	Seismic moment M_0 (N · m)	Magnitude M_w (-)
800	170	8	$3.2 \cdot 10^{22}$	9
250	30	5	10^{21}	8

Fig. 4.21 Parameters used for calculating seismic moment magnitude, according to equation (4.6), with example calculation.

Table 4.4 Approximate relationship between intensity and magnitude scales. Intensity is defined at the epicenter.

MSK	Magnitude
I	2
II	3
III	
IV	4
V	
VI	5
VII	6
VIII	
IX	7
X	
XI	8
XII	

4.1.5.4   Locating the epicenter

This is one of the first tasks of seismological research. Determining the location of an earth-quake should be easy, in principle, if the velocities of the ground between the focus and the station are known. The problem rests in the fact that the exact time of the initial rupture is unknown. It must be determined indirectly by measuring the delay of S waves compared to P waves on the seismographic record. This delay increases linearly with distance. Because the distance from the epicenter is measured on a sphere, non-linear functions are obtained for these time-distance equations (Fig. 4.22). The Jeffreys-Bullen tables give the time-distance ratio for normal velocities (Tab. 4.5). Once time is translated into distance, a minimum of three stations is required to determine the epicenter in an unequivocal manner.

In reality, the position of an earthquake is determined by a series of numerical simulations that take into account the actual structures and velocities of the media through which the waves travel. The calculations are done in three dimensions.

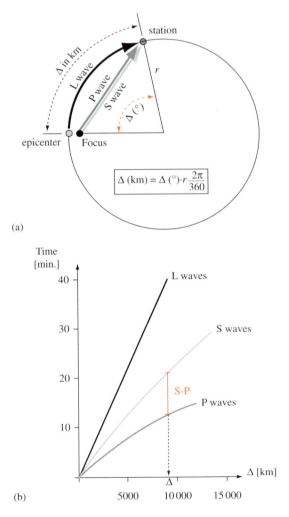

Fig. 4.22   (a) Determination of the distance to the epicenter (given in km or degrees). (b) Functions of P, S and L (Love) waves arriving at a station.

Table 4.5 Jeffreys-Bullen table for the calculation of the distance to the epicenter of an earthquake.

Δ[°] (1° = 111 km)	arrival time of P-waves		delay between P- and S-wave	
	min.	sec.	min.	sec.
0.0		(5.4)		3.8
0.5		10.5		7.6
1.0		17.7		13.1
1.5		24.8		18.7
2.0		32.0		24.1
2.5		39.1		29.7
3.0		46.3		35.2
3.5		53.4		39.9
4.0	1	00.5		46.4
4.5	1	07.6		51.9
5.0	1	14.7		57.4
5.5	1	21.7	1	03.0
6.0	1	28.7	1	08.5
6.5	1	35.8	1	13.9
7.0	1	42.8	1	19.3
7.5	1	49.8	1	24.8
8.0	1	56.7	1	30.3
8.5	2	03.7	1	35.8
9.0	2	10.6	1	41.3
9.5	2	17.5	1	46.8
10.0	2	24.4	1	52.2

PROBLEM 4.1

On May 12, 2005, an earthquake with a local magnitude of 4.1 was recorded. You have seismograms from the Swiss Seismographic Network Stations (Balsthal, Bourrignon, Sultz, Wimmis, Hasliberg, Torny-le-Grand and Muotathal).

Data

Seismograms (Fig. 4.23), location map of seismographic stations (Fig. 4.24).

Question

Determine the position of the epicenter of the earthquake using the Jeffreys-Bullen tables.

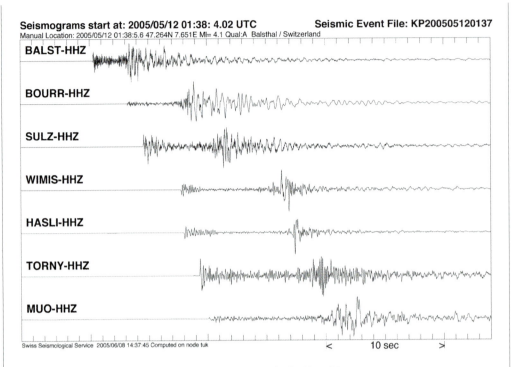

Fig. 4.23 Seismograms for Problem 4.1.

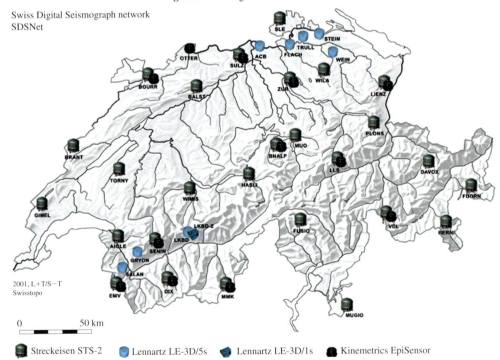

Fig. 4.24 Location map showing the 7 seismographic stations in Problem 4.1.

Look at the solution on the DVD

4.1.6   Seismic risk

Seismic risk is expressed by three fundamental concepts, as is the case for all geological risks.

In seismology, ***hazard*** or ***danger*** is the physical data that characterizes a quake at a point (magnitude, but also amplitude, duration, vibration spectrum, intensity). It is seismic loading and is independent of whether the territory is inhabited. We assign an average frequency of occurrence to a seismic load of a given intensity, similar to the assignment of frequency of occurrence to extreme floods (Section 7.3). This frequency is generally expressed as a return period T expressed in years. As an example, figure 4.25 shows the maximum intensity that should occur on average once every 10,000 years. The frequency of an event can also be expressed by the probability of its occurrence during a reference period of n years, the duration of the lifespan of a structure, for example. The probability P is calculated on the basis of the return period T and the reference period n by the following equation:

$$P = 1 - (1 - 1/T)^n \tag{4.9}$$

Thus, an earthquake with a return period $T = 475$ years would have one chance in ten ($P = 0.1$) of occurring during the life of a structure $n = 50$ years.

Vulnerability is a factor that estimates the extent of damage to endangered objects during a given event. It ranges from zero (no destruction) to one (total destruction). Vulnerability is a factor that is relevant only to the human occupation of the territory (population density, quality of construction, etc.)

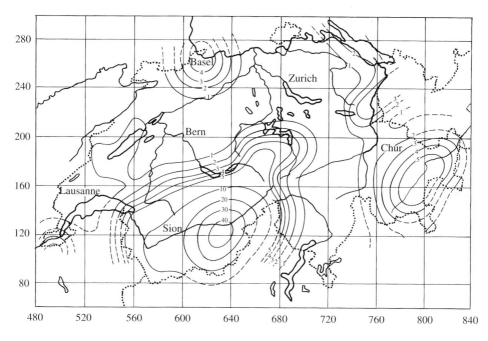

Fig. 4.25 Map of seismic hazard in Switzerland (Swiss coordinate system, in kilometers) edited in 1978 (under revision). Average frequency of occurrence of an earthquake of intensity equal to or higher than VIII, expressed in 10^{-4} events/year. Active regions can be noted: The Valais is most active (probability 0.004 or once every 250 years), followed by Graubünden and the region of the Rhine graben in Basel. The Swiss plateau is not very active (probability in Lausanne 0.0001 or once every 10,000 years) (from [15]).

Property value is the overall financial loss of material possessions and persons subject to risk in the event of total destruction.

Mathematically speaking, seismic risk is the product of the probability of occurrence of an event of intensity x, and of vulnerability and property value. It is expressed as a sum of money per year. It is the most commonly used definition in Anglo-Saxon countries.

Even though the notion of risk is simple in terms of definition, its calculation is very complex in practice. For instance, the calculation of the cost of foreseeable damage and the estimate of the number of victims are highly uncertain.

4.1.6.1 Average frequency of occurrence

To characterize the seismic risk of a given region, the average frequency of an event must be known for that location. This is called a *forecast*. But this number does not tell us when an event will take place. This is a *prediction*. This is why it is also necessary to develop methods to predict events in real time (§ 4.1.6.2).

For hazard forecasts, the frequency of occurrence calculation is first based on the seismic past of a region. Research into historical records makes it possible to date significant events and attribute to them an approximate intensity based on reported damage (on the basis of the MSK scale). From this data, a frequency of occurrence is produced. This research then consists of a systematic analysis of seismograms produced since the installation of seismologic networks, and consideration of the statistical distribution of seismic events, including events of lesser amplitude. Finally, this data is used to construct seismic hazard maps showing the regions of a country most sensitive to these phenomena. Figure 4.25 shows seismic risk for Switzerland; by using a map of this type, it can be estimated for example that an earthquake of intensity VIII should occur in Lausanne only once every 10,000 years.

4.1.6.2 Event prediction

In spite of the large sums invested in earthquake prediction research, especially in Japan and the United States, operational success is still a distant goal. Several research directions are being studied, all based on identifying signs that occur before earthquakes, called *precursors*. The major phenomena being studied are:

- Escalation in seismic activity,
- Increase in volcanic activity,
- Modifications of the topography,
- Unusual variations of water levels in wells,
- Changes in the status of mechanical stresses in underground constructions,
- Variations in natural electric potentials,
- Releases of radon,
- Observation of animal behavior.

It is easy to understand the logic of the first methods, but it is difficult to explain animal behavior. However, the Chinese claim to have observed unusual animal behavior prior to earthquakes and volcanic eruptions. It is well known that animals are often more sensitive to the environment than we are, even with our measurement apparatus.

Overall, these detection methods are often effective, but far from systematic. They all suffer from lack of prediction and false alerts. Let us look at some successes and failures of these methods, considering also the human factors that play a significant role:

- Haicheng (China) 1975: Unusual micro-seismic activity was felt by seismic surveillance personnel who set up protection for the population. As a result, the earthquake that happened just afterward claimed few victims.
- Tangshan (China) 1976: In the same region as the 1975 earthquake, in spite of several precursors (well levels, electrical resistivity, animal behavior), no one believed an earthquake was imminent because no strange vibrations were observed. The subsequent earthquake buried 500,000 people. This case highlights the difficulty of human interpretation of indicators.
- Kobe (Japan) 1995: No precursors were detected. The earthquake ($M_W = 7.2$) caused about 6000 deaths and cost 20 billion dollars.

The first conclusion that can be drawn is the necessity of using all possible detection methods. Research into the geodynamics of earthquakes is a priority for civil protection.

4.1.6.3   Local response to seismicity

Just like an engineering structure, the subsurface reacts in different ways to vibrations that come to it from depth; the subsurface can also exhibit surprising behavior. These local particularities are called **site effects**.

The subsurface as a modulator of vibrations

For the same regional seismic impetus, two terrains of different composition will produce different vibrational responses. In addition, two terrains of identical composition receiving the same signal can produce dissimilar responses as a function of the structure of the subsurface and topography. The media acts on the vibrational signal emitted by the seismic focus and can significantly modify three important vibration factors: amplification, duration and frequency distribution.

Aside from the topographic factor (numerous earthquakes have shown that vibrations are amplified at the top of hills), the dominant factors are geological, hydrogeological, and geotechnical characteristics. These factors can generate concentrations of seismic echoes that amplify vibrations or make them last longer. In Quaternary cover, the following factors can be identified: the type of terrain and its composition, compaction, and thickness, lateral variations in thickness, and the depth of the saturated zone. In the case of a rocky substrate, the important factors are hardness (including the degree of alteration and tectonic deformation), the clay content, distance to a tectonic feature and sub-vertical geologic contacts (surface waves concentrate along the discontinuity). These phenomena are highly complex because of the numerous interacting parameters. Several empirical rules can be drawn from experience, however. Figure 4.26 gives an example of an imaginary region with three typical risk configurations. These factors should be weighted according to their contribution to the site effect prior to *microzoning*, a local mapping effort that represents the various site effects in a given region.

Secondary effects

Secondary effects are non-seismic processes that are initiated or amplified by ground vibrations. We can cite landslides and rockfalls on hillsides in addition to liquefaction of the subsurface. The latter dynamic mechanism merits some explanation. Some sediments are likely to liquefy when they are subjected to earthquakes. There are two different phenomena:

Liquefaction of fine sands

Unconsolidated alluvium with grain size between 0.05 and 1 mm undergoes liquefaction during certain earthquakes if it has a loose texture and is water-saturated. This is caused by an

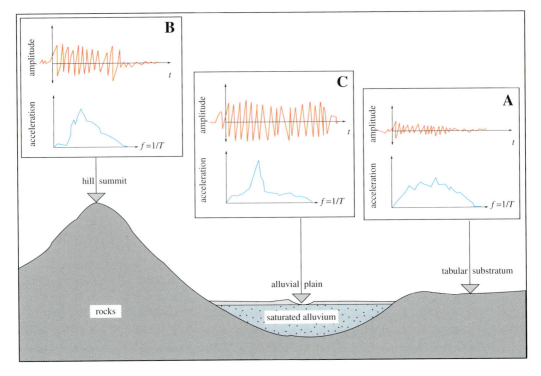

Fig. 4.26 Examples of site effects on the seismic response of the ground. A point located on a flat and homogeneous rocky substrate shows the base case (Point A). The signal has low amplitude, short duration, and the frequency spectrum is widely dispersed. This case is ideal for construction. A point located on the same rocky substrate but at the top of an isolated hill (Point B) has a short signal, but amplified vibration. The Fourrier spectrum shows the appearance of a preferred wave length. This results from the focusing of the waves due to topography. This site is thus clearly less favorable, although it has the same substrate. A point located on an alluvial plain (Point C) made up of a thick fill of loose detrital soils, with a high water content, is unfavorable on all counts: high amplification, prolongation of the signal and presence of a band of dominating frequencies within the range of low frequencies (0.3 to 2 Hz). This phenomenon of a concentration of low frequencies of very soft soils was clearly demonstrated during the 1985 Mexico City earthquake (from [17]).

increase of interstitial pressure between the grains due to the vibration. When the pore pressure exceeds the steady state value, the effective stress corresponding to the grain-on-grain contact forces diminishes. Because of the low permeability of these terrains, the over-pressure caused by the seismic wave is not able to release rapidly. This leads to a decrease in shear strength that is directly correlated to inter-grain friction. Static equilibrium is destroyed and liquefaction causes a sudden rupture. This leads to sinking buildings, rapid landslides, and mudflows.

Thixotropic clays

   Thixotropy is the property of certain materials to behave as solids in a static state and as liquids in a dynamic state. The phenomenon is generally reversible. Although this general definition could also be applied to fine sands, it is in fact reserved for clay-rich deposits in which the phyllosilicate particles are connected by a very specific type of aggregation. The vibrations destroy this structure of the clay, and as a result, the ground behaves as a fluid. Hydrostatic over-pressuring occurs and it lasts even longer than in fine sands because of the extremely low permeability. The fluid state can thus last a long time, which has serious consequences on mudflows caused by this type of instability.

4.1.6.4   Damage caused by earthquakes

Damage on land

Civil engineering structures are the terrestrial objects most dramatically affected by earthquakes. Constructions are above all affected by the vibrations' strong horizontal components. However, usual structures are designed to withstand mainly to gravity, and thus vertical forces. A horizontal acceleration equal to half of the terrestrial acceleration is the same as placing the structure at a 30° angle. This example shows that a building not dimensioned to withstand such forces has no chance of resisting.

Seismic vibrations have a cyclic character that tends to weaken structures. Experience shows that low buildings (1–2 storeys) suffer little damage. Most vulnerable are 3–8 storey buildings, because these are not elastic enough to resist flexural forces. Plastification zones (non-reversible deformation) occur in the structure as a result, and can lead to its destruction. The tallest buildings are exposed to a high deformation, but these keep an elastic character, limiting the damage.

The structures can also behave as amplifiers for vibrations that correspond to their proper frequency. There are methods for calculating the proper frequency f of structures. The following empirical formula gives a range for buildings:

$$f = 10/n \; [\text{Hz}] \qquad (4.10)$$

where n is the number of storeys.

The most devastating frequencies for buildings are between 0.1 and 30 Hz. Low frequencies, which are equivalent to the longest waves, endanger elongated structures such as suspension bridges and towers. Other frequencies affect smaller buildings and more rigid structures. During the 1985 earthquake in Mexico City, the low frequency of ground vibrations was equivalent to the characteristic frequency of 12- to 18-storey buildings during the 80-second duration of the earthquake. The exceptionally long duration of the earthquake was due to loose sediments in the ancient lake beneath Mexico City.

The most dangerous buildings are those made of a succession of slabs minimally attached to vertical beams. Ground vibrations are transmitted to beams, which begin to bulge apart. The slabs come undone and collapse on each other, crushing the inhabitants (Fig. 4.27).

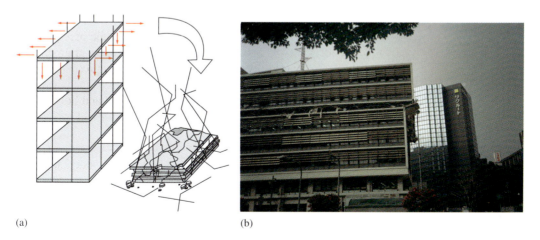

(a)   (b)

Fig. 4.27 (a) Scheme showing the destruction of a multiple story building with slabs fixed to metal beams. (b) Example of a building in Kobe (Japan) where one storey was completely crushed under the weight of the upper storeys. Courtesy National Information Service for Earthquake Engineering, University of California, Berkeley.

Earthquake-resistant design of structures must first consider the seismic hazard of a region. For example, in Switzerland, the SIA 261 Standard classifies the country into four zones and establishes the maximum ground acceleration to be considered (Fig. 4.28). It is necessary to consider also possible site effects and integrate micro-zoning data. From the study of numerous earthquakes, an elastic response spectrum for design has been determined. This is the smoothed envelope of acceleration amplifications resulting mainly from Quaternary deposits, in comparison with raw vibrations on bedrock. Figure 4.29 shows this acceleration amplified as a function of different characteristic frequencies of the Quaternary cover for the four zones of seismic hazard in Switzerland.

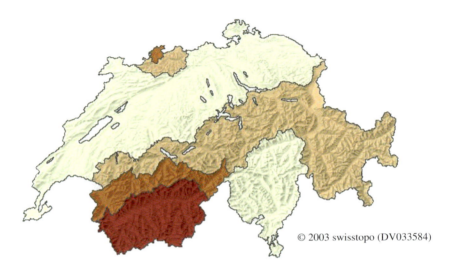

© 2003 swisstopo (DV033584)

Legend	Seismic risk	Ground acceleration a_{gd} [g]
	Zone 1	0.06
	Zone 2	0.10
	Zone 3a	0.13
	Zone 3b	0.16

Fig. 4.28 Simplified seismic regional hazard map for design of structures, according to the Swiss standard SIA 261. Ground acceleration of an earthquake of a frequency of occurrence: $2.5 \cdot 10^{-3} \, a^{-1}$ or $1/400 \, a^{-1}$. See also Fig. 4.29.

When the ground liquefies (non-elastic behavior), the foundation exceeds the bearing capacity and the building sinks and may tip over. The famous Niigata earthquake in 1964 in Japan shows how buildings that were constructed according to earthquake resistance standards for their structure remained intact although they tipped over (Fig. 4.30).

In the case of an earthquake caused by faults that reach the surface, spectacular topographic modifications can occur. For example, the 1819 Kutch earthquake in northwestern India created an escarpment more than 100 km long and a ledge several meters high ("God's Wall"). The southern block sank, and was swallowed up in the Fort Sindree waters, as drawn by Charles Lyell (Fig. 4.31).

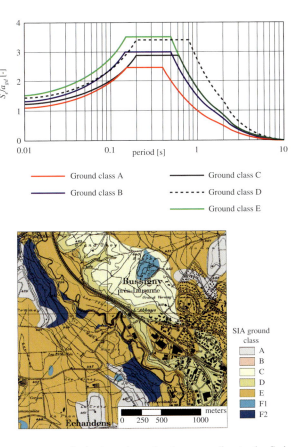

Fig. 4.29 (a) Elastic response spectrum for horizontal acceleration, according to the Swiss SIA-261 standard. The design spectrum is then calculated from the elastic response spectrum with the aid of several factors that depend on the construction and the foundation ground class; Class A: rocks, Class B: over-consolidated ground, Class C: coarse ground, Class D: fine, unconsolidated ground, Class E: Classes C and D less than 30 m thick. S_e = elastic response spectrum value; a_{gd} = horizontal component of ground acceleration, according to regional seismic hazard (Fig. 4.28). Class F1 (palustrine sediments) and F2 (landslide) need special design. (b) Example of a map of foundation grounds (west of the city of Lausanne).

Fig. 4.30 Sinking of buildings due to liquefaction. Niigata Japan earthquake in 1964. Courtesy of National Information Service for Earthquake Engineering, University of California, Berkeley.

Fig. 4.31 "God's Wall" and Fort Sindree, before and after the earthquake. Engraving by Lyell, India (1819). On the map, the hachured zone indicates swamps.

Earthquakes sometimes set off landslides or rockfalls. In Peru, for example, an earthquake whose epicenter was located more than 100 km deep set off the Huascaran debris flow on May 31, 1970 (Fig. 4.32).

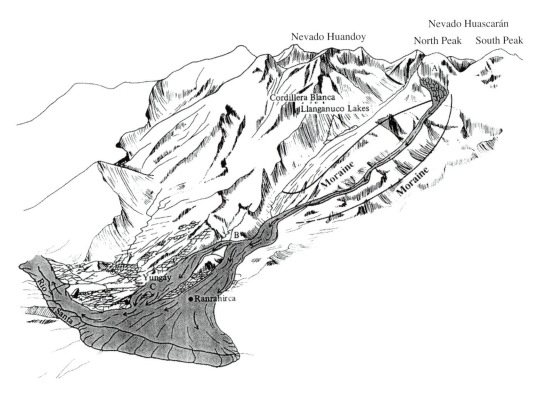

Fig. 4.32 An earthquake triggered a debris flow in Huascarán Peru. The rock and ice debris flowed 11 km from the hill in 3 minutes, at speeds that reached 300 km/hr, burying 30,000 people. Several lucky people found refuge on the little hill of the Yangay cemetery (Point C), which was spared because of its location between the two lobes of the flow (from [47]).

Damage in water

An abrupt rise of the ocean bottom can cause a very rapid wave called a *tsunami* ("wave in the port" in Japanese) (Fig. 4.33a). The Pacific Ocean is the more often affected by tsunamis because of the circum-Pacific subduction zone that causes numerous high magnitude earthquakes. The Kamchatka peninsula, Japan and the Hawaiian Islands are particularly affected by this phenomenon. On December 26, 2004, the Indian Ocean was affected by a tsunami as a result of a 9.0 magnitude earthquake (Fig. 4.33b) The wave reached a velocity of 800 km/hr in the open sea and spread devastation along the neighboring coasts of Indonesia, Thailand, India and Sri Lanka, and even as far as Africa, more than 5000 km from the epicenter of the earthquake. Because of its low amplitude and its long wavelength, the wave was almost undetectable

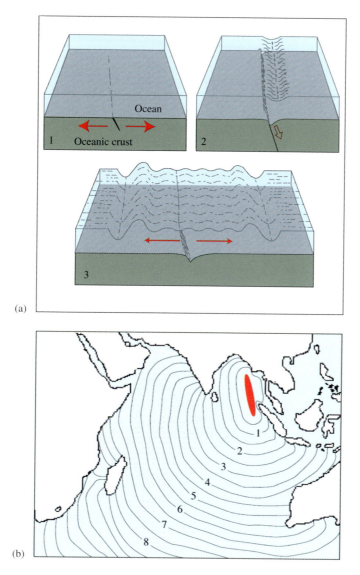

(a)

(b)

Fig. 4.33 (a) Example of the tsunami formation process. 1. The oceanic crust is subject to tensile stress. 2. Subsidence of the right-hand block and a sudden lowering of sea level. 3. Propagation of the tsunami wave. (b) Example of a tsunami caused by the December 26, 2004 earthquake in the Indian Ocean. Isochrons of the wave are in hours.

by boats in the open sea. As the wave approached the coasts, the shallows exerted a significant frictional force on the wave. As the wave was slowed down, its height increased spectacularly as it reached the shore. Because of its high kinetic energy, it was able to penetrate several kilometers into the interior of the land in areas of low relief. It killed at least 285,000 people. The death toll could have been much lower if there had been an observation and warning network.

Because of their large geographical extent, tsunamis are often more destructive than continental earthquakes. The history of these catastrophes provides a sad record of this type of phenomenon:

- November 1, 1755: an earthquake in the eastern Atlantic caused a wave that reached a height of 5 to 10 meters as it struck Lisbon, resulting in approximately 60,000 deaths.
- August 27, 1883: An earthquake at the Krakatoa volcano caused a tsunami along the coasts of Java and Sumatra that killed 36,000 people.
- March 2, 1933: In Honshu, Japan, a 20-meter wave along the coast killed 3,022.
- May 22, 1960: A magnitude 8.5 earthquake at the southern end of the Andean Cordillera created a wave that destroyed all the coastal cities of Chile over a distance of 900 km. It traveled as far as Hawaii, 10,000 km from the source, and made its final strike in Japan and the Philippines, twenty-two hours later, 15,000 km from the source. It killed a total of 2,000. As a result of this catastrophe, the first tsunami surveillance network was created in the Pacific in 1964, considerably reducing the number of victims of subsequent tsunamis.

4.1.6.5 Prevention

Seismic hazard prevention is a long-term mission. It rests on the following planning measures:

- Installation of a dense seismologic network in areas of high risk,
- Good understanding of the geology present in the territory (systematic geological mapping),
- Preparation and updating of regional scale probability maps that take site effects into account (seismic micro-zone maps),
- Detailed observations of active faults,
- Installation of a precursor screening system that is connected to an alert system.

Regarding civil engineering, prevention includes seismic resistant construction in high probability regions (earthquake-resistant design). This entails relatively simple measures: buildings with compact shapes, structural reinforcement elements, foundations on piles, etc. For tall buildings, sophisticated foundations on vibration-dampening springs or jacks are provided (Fig. 4.34). Books on structural dynamics provide the details of seismic resistant design, where resonance calculations play an important role.

It may also be possible to modify the shear strength of a fault that is the source of earthquakes. The case of active faults on the eastern edge of the Rocky Mountains is very instructive, although the goal was not earthquake prevention. In the 1960s the city of Denver (Colorado) began injecting wastewater into 4 km-deep disposal wells. The result has been a series of small earthquakes around the wells (Fig. 4.35). In the same region, oil companies have injected pressurized water into the rock in order to enhance petroleum recovery from the pores, with the same result. The earthquakes are the net result of pressurized water spreading apart of the edges of the active shear planes.

Would we dare use the same process in the blocked parts of the San Andreas Fault to relax the strain before it becomes too high?

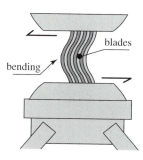

(a)

(b)

(c)

Fig. 4.34 Earthquake resistant techniques developed in Mexico City after the September 19, 1985 earthquake. (a) For existing buildings, a mechanism to absorb horizontal vibration. Between the metal weight-bearing structure and the slab, a series of metal blades has been added which are capable of deforming by bending when the slab vibrates in a different manner from the bearing structure. The x shape of the blade ensures that the deformation takes place at the midpoint between the slab and the bearing structure. The social security building, secured after the earthquake. (b) Metallic frames on the interior of the structure (Social Security building). (c) Buttresses placed outside the building to support the walls (hospital). GEOLEP photos, A. Parriaux.

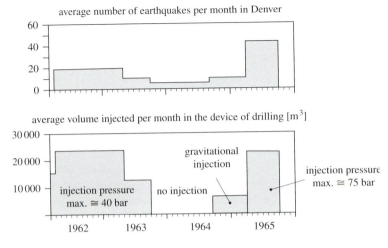

[15, 152, 237, 300, 311]

Fig. 4.35 The effect of water injection along a fault on earthquake frequency, Colorado (from [84]).

PROBLEM 4.2

Californians fear that a sudden liberation of the shear stress on the San Andreas Fault and its related faults may cause a very large earthquake (the "Big One"). In fact, significant stress has probably accumulated through the years because it has not been released by small earthquakes.

A new mayor has been elected in San Francisco. He wants to address this problem in all its components. He has established a task force that includes:

- The head of public safety,
- An insurance expert on construction in zones of natural hazards,
- A geologist who is an advocate for a project to lubricate the San Andreas Fault by deep-well injection of wastewater or oil.

Question

Put yourself in the role of each of the participants of this meeting and study the project from the point of view of each specific discipline. In the role of mayor, prepare a final feasibility assessment of the project.

Look at the solution on the DVD

4.1.7   Seismic prospecting

The study of the propagation of mechanical waves in the Earth has created induced seismology, which is used to identify geologic structures in the subsurface and indirectly, the nature of the ground. The principle is identical to natural seismology, but it uses a man-made seismic source such as an explosion in a shallow borehole or a vibrational device. The signal is then captured by geophones distributed on the surface. Two principle methods can be distinguished (see also the animations on the DVD):

- *Seismic Refraction* (Fig. 4.36): This method is the most commonly used because it is inexpensive and easy to use. It is based on the fact that "rapid" ground at depth accelerates

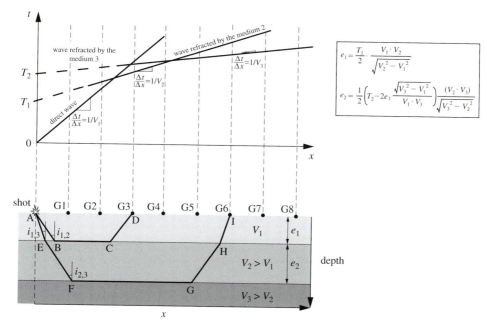

$$e_1 = \frac{T_1}{2} \cdot \frac{V_1 \cdot V_2}{\sqrt{V_2^2 - V_1^2}}$$

$$e_2 = \frac{1}{2}\left(T_2 - 2e_1 \frac{\sqrt{V_3^2 - V_1^2}}{V_1 \cdot V_3}\right) \frac{(V_2 \cdot V_3)}{\sqrt{V_3^2 - V_2^2}}$$

Fig. 4.36 Seismic refraction. In the case of an energy source at the surface and assuming three horizontally stratified layers and increasing velocity with depth, the function on the time-distance curve begins with a first straight-line segment showing the direct wave in the first strata. The direct wave follows the surface (straight-line ADI). It arrives first in the case of geophones near the source because its trajectory is much shorter than waves refracted by deep strata. At some distance, the direct waves are passed by waves refracted at the second strata (trajectory ABCD). This phenomenon is shown on the dromochronic plot by a second straight-line segment that does not pass through the origin and has a flatter slope. In their turn, the waves refracted by the second stratum are passed by waves refracted by the third stratum (trajectory AEFGHI), creating a third straight-line segment. The velocities of the strata can be determined from the slopes of the segments on the time-distance plot. The depth of the contacts can be calculated from the y-axis at the origin of the refracted segments ("intercept time") or the position of the intersection points of the various straight-line segments. The case of non-horizontal planar contacts can be calculated without difficulty. Configurations of decreasing velocity with depth or non-linear or discontinuous contacts are much more difficult to interpret and can be a source of errors. Unfortunately, these cases occur frequently in nature. [8, 256]

the wave propagation and sends it back to the surface. The wave arrival times are noted on each geophone. A time-distance curve (called a **dromochronic** plot) can be made showing the arrival time as a function of the distance between the geophone and the source. It is possible to deduce the speed of the waves in different media and the structure of the various media. This method is used to find shallow structures (bedrock under landslides, thickness of a filtering bed on top of aquifers etc.).

• **Seismic Reflection** (Fig. 4.37): In this method, the surface devices record the arrival of waves that have been reflected at the interfaces of deep strata. As in the preceding method, the arrival times are recorded at each receptor, and then a time-distance curve is drawn to obtain the wave velocity in different media. This method is more complex to process but it gives excellent results for deep reflectors. It is used particularly in petroleum exploration to research tectonic structures likely to trap hydrocarbons. On a smaller scale, it makes it possible to prospect large alluvial valleys (Fig. 4.38).

In both cases, seismic profiles are obtained that make it possible to represent surfaces as a result of obvious seismic velocity contrasts between different media.

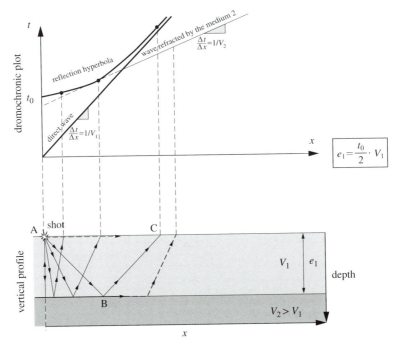

Fig. 4.37 Seismic Reflection. A contact between two strata reflects a portion of the seismic energy ("seismic mirror"). On the time-distance plot, the arrival of reflected waves is shown as a hyperbola in the case of a horizontal plane. The shape of the hyperbola and its intersection with the time axis shows the velocity of the stratum in which the reflection took place and the position of the mirror. Because reflected waves have a long trajectory with slow velocities, they arrive very late, well after the first arrivals studied in seismic refraction. Their study requires longer recording. But the geophones also record all sorts of "parasite" waves that are slower than the first arrival: direct waves, refracted waves, S waves, surface waves, etc. Seismic reflection thus requires extremely complex treatment of signals to identify the reflections among the noise.

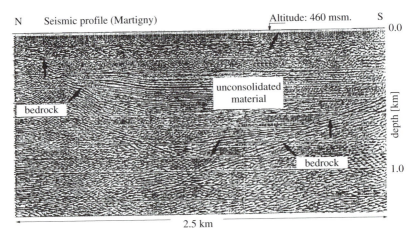

Fig. 4.38 Example of seismic reflection measurements showing the base of Quaternary deposits and the principal structures of the bedrock. Example of the Rhone Valley at Martigny (from [124]).

4.2   Gravimetry

Gravimetry is both a theoretical discipline in academic geology and a geophysical technique for resource exploration (groundwater, minerals, etc.). We will first examine the method in general, then the applications that involve engineering.

4.2.1   Connection with mechanics

In 1687 Newton established that there is a force F that attracts two masses M and m at a distance d (Fig. 4.39). It has the following value:

$$F = \text{G} \cdot \frac{m \cdot M}{d^2} \qquad (4.11)$$

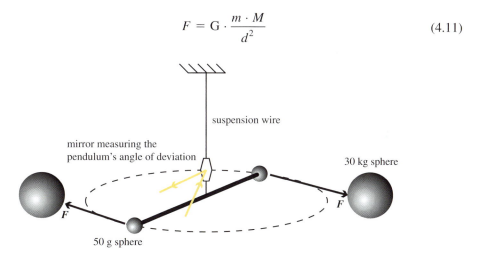

Fig. 4.39 The force of attraction can be shown by Cavendish's experiment. The force was calculated from the torsion angle and the torsion constant of the wire.

He estimated the constant G by studying the Earth–Moon system. In 1798, Cavendish conducted a simple experiment (Fig. 4.39) that allowed him to measure this universal constant precisely:

$$\text{G} = 6.67428 \cdot 10^{-11} \ \text{Nm}^2\text{kg}^{-2} \qquad (4.12)$$

This law is thus applied to the case of a mass m that is attracted by the mass of the Earth ($M = 5.9742 \cdot 10^{24} \, \text{kg}$).

The force of gravity is thus:

$$\boldsymbol{F} = m \cdot \boldsymbol{a} \qquad (4.13)$$

where $\boldsymbol{a}$ = the Earth's gravitational acceleration.

Assuming the Earth is a sphere of radius $r = 6371 \, \text{km}$, we obtain:

$$\boldsymbol{a} = \frac{\text{G} \cdot M}{r^2} = 9.81 \, \text{m/s}^2 \qquad (4.14)$$

A unit mass $m = 1 \, \text{kg}$ on the Earth subjected to an average acceleration of 9.81 m/s² experiences an attractive force of 9.81 N.

In reality, the weight W of an object located on the Earth far from the poles is the result of Newton's gravitation force (F_g), also called **gravity**, and the centrifugal force (F_c) due to the Earth's rotation (centrifugal force is the inertial force that must be taken into consideration when the reference is set at the Earth's surface):

$$W = F_g + F_c \qquad (4.15)$$

At the Equator, these forces are parallel and opposite.

A spring measures the sum W of the forces acting on the suspended mass.

The relation becomes $W = m \cdot g$, where g is the non-corrected gravitational acceleration.

This non-corrected gravitational acceleration results from Newton's gravitational force minus centrifugal force. In gravimetry only the true gravitational acceleration is of interest, so corrections must be made to account for the latitude of the point (§ 4.2.2).

The attraction of other bodies in the solar system must also be taken into account, especially the Moon and Sun. The effects of these two bodies vary depending on their relative positions. This is the origin of the lunar-solar tides (Chapt. 9).

Geophysicists traditionally use the CGS (centimeter-gram-second) unit, the milligal, which equals:

$$1\,\text{mGal} = 10^{-5}\,\text{m/s}^2$$
$$(1\,\text{cm/s}^2 = 1\,\text{Gal}) \qquad (4.16)$$

There is also a more modern SI unit, the g.u. ("gravity unit"):

$$1\,\text{g.u.} = 10^{-6}\,\text{m/s}^2 = 0.1\,\text{mGal} \qquad (4.17)$$

4.2.2   Notion of a geoid

If the Earth was perfectly spherical, if it did not rotate, and if its internal structure had complete rotational symmetry, gravitational acceleration would be identical everywhere. The surface of this ideal Earth would be a gravitational equipotential. In fact, for many reasons, this is not the case.

First, the idea that the Earth is spherical should be dismissed immediately because it doesn't correspond to reality. However, it is necessary to define a mathematically simple surface that closely approximates the Earth. Because of its rotation on its axis, the Earth undergoes centrifugal bulging, most pronounced at the Equator. If the Earth was homogeneous, it would be a slightly flattened rotational ellipsoid. This is called a **universal reference ellipsoid**. It establishes an Earth-scale reference for the determination of latitudes and longitudes.

The reference ellipsoid adopted by the International Union of Geodesy and Geophysics in 1980 has the following parameters:

- $a = 6{,}378.137$ km (large axis)
- $b = 6{,}356.7523141$ km (small axis)
- $f = (a - b)/a = 0.00335281068118$ (flattening)

This type of reference makes it possible to assign a latitude L (in degrees), a theoretical or normal gravity (g_0) that includes the effect of centrifugal force (international gravity formula, IUGG, 1980):

$$g_0(L) = 9.780327\,(1 + 0.0053024\,\sin^2 L - 0.0000058\,\sin^2 2L)\,\text{m/s}^2 \qquad (4.18)$$

The value of $9{,}780{,}327\,\text{m/s}^2$ is the average gravitational acceleration at the Equator.

It should be noted that gravitational acceleration is minimal at the Equator (978 Gal) because of the Earth's ellipsoidal shape (it is farther from the center). At the poles, it is 983 Gal.

Let us consider a surface that should represent a gravitational equipotential at every point: the ocean surface. The low viscosity of water effectively prevents disequilibrium from persisting over time (which is not the case for rocks and magma, as we will see later). Satellite measurements now allow us to integrate all the temporal variations (waves, tides) to achieve very precise knowledge of the average water level at a point. The geometric place of these points is called the **geoid**.

In the past this skewed ocean surface was used by cartographers as a reference, as the zero level for so-called normal altitudes measured on land surfaces by topographic surveys that started at the coasts. If we subtract these altitudes from those of the ellipsoid, we obtain the difference between the two reference surfaces, or in other words, the level of the geoid on the ellipsoid. This difference, called **geoid height** or idem for **undulation** of the geoid, is also observed under continents. A distribution map of these differences shows high and low areas with values that are as high as several tens of meters (Fig. 4.40). South of India, this difference exceeds 100 meters. These variations are explained by the fact that the subsurface does not have the same average density everywhere.

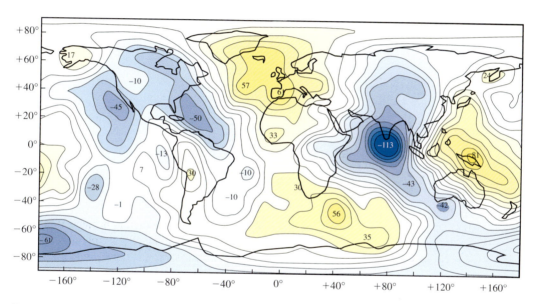

Fig. 4.40 Map showing differences (in meters) between the geoid and the reference ellipsoid. Positive difference: the geoid is higher than the reference ellipsoid.

4.2.3 Measurement of the gravitational field and its treatment

Because the Earth is heterogeneous, we can use variations in gravitational acceleration to learn more about the structure and the nature of the subsurface. Equipment was developed very early to measure gravitational acceleration everywhere on the globe.

4.2.3.1 Gravimeter

The gravimeter is an instrument for measuring gravitational differences between a reference station and points measured in the field. The principle of the gravimeter is very simple (Fig. 4.41).

A mass is suspended from the housing of the gravimeter by a spring that has a small drift over time. A mirror is placed on both the fixed solid arm of the housing and the suspended mass. At the reference station, the length of the spring is set using a micrometric screw so that the luminous bundle reflected by the mirror arrives right in the middle of the ocular. When the instrument is moved to a point in the field, a gravitational difference requires a modification of the spring length. The luminous bundle no longer reaches the ocular. The initial position of the bundle is re-established by moving the micrometric screw. The adjustment of the micrometric screw is converted to the gravity difference.

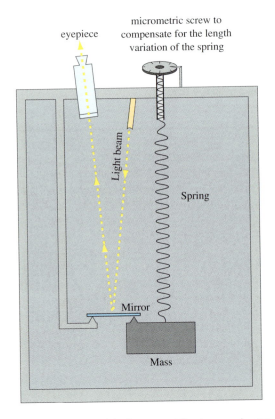

Fig. 4.41 Principle of the Lacoste and Romberg gravimeter.

In reality, things are more complicated because the measurement of the very slight differences requires a very high sensitivity. However, a device with the necessary sensitivity is also sensitive to perturbations (drift of the spring, temperature effects and shocks, imperfect support of the housing at the measuring point, etc). This is why silica springs, a vacuum box, and thermal regulation are used.

In practice, measurements are made at places of known altitude (necessary for corrections). Frequent trips to reference stations make it possible to determine the drift of the instrumental due to spring fatigue, which is subtracted from the measurements. The lunar-solar tidal effect, determined in real time by astronomical tables, is also deducted.

4.2.3.2 Corrections and the Bouguer anomaly

Bouguer, the famous French physicist of the 18th century and a pioneer in gravimetry, devised the basic methods of gravimetric measurement. The goal is to use field measurements to identify regions where the substratum is abnormally dense or abnormally light. The "normal" case is defined by the value of theoretical gravity g_0 on the reference ellipsoid. To compare measured values to theoretical values, characteristics unique to the measurement point must be taken into account because they affect the measured values g_m. The principle consists of transposing the value of g_0 (on the ellipsoid) to the conditions that exist at the measurement point (altitude h, surrounding topography, assumed geological substrate). In this way, we obtain a corrected theoretical value g_0 (h) which is comparable to the measured value g_m. The difference between the two values is the Bouguer anomaly:

$$\Delta g = g_m - g_0\,(h) \qquad\qquad (4.19)$$

The Bouguer anomaly thus depends only on the actual density of the geologic substrate. If the anomaly is positive, it means that the rocks under the point are denser than those chosen for the correction. If it is negative, the opposite is true.

To reduce the theoretical value to the conditions at measuring point P, a series of corrections are performed. They are done in three stages (Fig. 4.42).

Altitude correction

The value of g_0 on the ellipsoid is calculated by equation (4.19) from the latitude L at point P (Fig. 4.42). To migrate this value at the altitude h of the point, the effect of distance from the center of the Earth ($r + h$ instead of r) must be subtracted. This correction is also called the *"free-air correction"* (Δg_A) because it is as if the point was moved vertically in the air. This correction is the most important. It is equal to 0.3086 mGal/m.

Bouguer correction

To compare g_m at point P to g_0 corrected to free-air, it is necessary to add to g_0 the effect of a horizontal rocky plateau passing through P.

We ignore the morphology around the point. The rock mass that constitutes the plateau of $z = 0$ to $z = h$ exerts an attractive force (Δg_p) on the point. This force must be added to the theoretical value that resulted from the first correction (Fig. 4.42). This operation is not as well defined as the first one since assumptions must be made about the density of the rocks that make up the plateau. In principle, the plateau is considered to be a single stratum of the most likely density, given its geologic context. For a standard density of $2.67 \cdot 10^3 \, kg/m^3$ (density of granite and the upper part of the continental crust), the correction is about 0.1 mGal/m. Often several densities will be used to test their consistency with the result. Density is determined by laboratory measurement on rock samples or by geophysical measurements in boreholes (called "gamma=gamma logs) (Tab. 4.6).

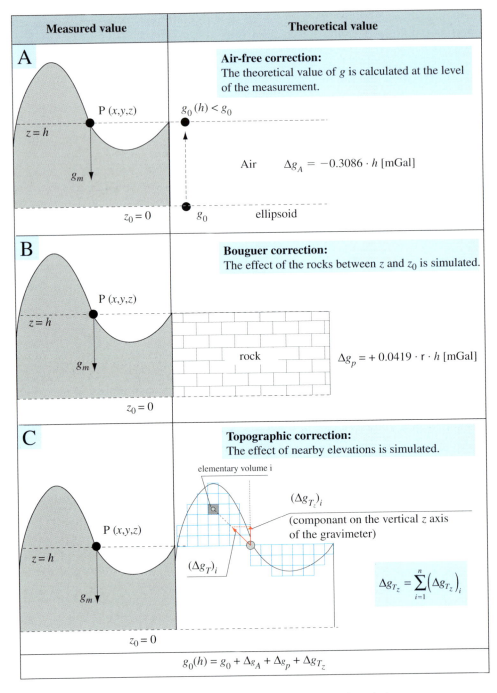

Fig. 4.42 The three stages of corrections for obtaining theoretical g.

Table 4.6 Densities of some rocks and unconsolidated materials

Rocks	density ($\rho \times 10^3$ kg/m³)
Granite	25 to 2.7
Gneiss	2.6 to 2.7
Limestone, marls	2.6
Basalt	2.9 to 3
Gabbro	2.6 to 2.9
Unconsolidated material (saturated)	**density ($\rho \times 10^3$ kg/m³)**
Fluvial alluvium	2
Peat	1.3
Superficial till	1.9
Lodgement till	2.2
Clay	1.5 to 1.9
Sand	1.7 to 2.1

Topographic correction

While the calculations of the first two corrections are simple, the same is not true for topographic correction. Topography significantly influences the measurements in both intensity and direction, up to distances 20 km from the point of measurement. The topographic correction (Δg_{T_z}) consists of discretizing the relief and calculating the gravimetric effect of all the elementary solids thus defined at the measuring point P, taking into account their distance from the station and the density of the rocks that make up these blocks (Fig. 4.42). This effect is added to the theoretical value obtained from the Bouguer correction. Luckily, numerical models of the ground (digital topography) simplify the calculation.

PROBLEM 4.3

Criticize the following statement: "Topographic correction at point M is null because the effects of high and low elevation cancel each other." (Fig. 4.43).

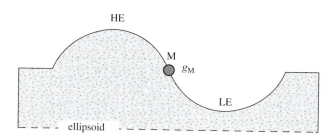

Fig. 4.43. Position of the measuring points for problem 4.3. High elevation (HE) and low elevation (LE).

Look at the solution on the DVD

4.2.4 Interpretation of anomalies

At each point measured during exploration, the difference between the measured value g_m and the corrected theoretical value $g_0(h)$ is calculated. Deviations are reported at the relevant points on maps. In this way, a plan view image of the Bouguer anomaly is created. The null anomaly curve is traced on this map. It shows regions where the acceleration is higher than the theoretical value (positive Bouguer anomalies) and areas of lower values (negative Bouguer anomalies). Positive Bouguer anomalies are regions of dense rocks and negative anomalies indicate lighter terrains.

Anomaly interpretation is an inverse method: it was traditionally based on graphs showing the form and intensity of an anomaly of a simple geometric shape (horizontal cylinder, sphere, fault plane, etc.) at different depths. Nowadays, finite element methods make it possible to calculate the effect of shapes that are more geologically realistic. Figure 4.44 shows several qualitative examples of regional and local anomalies, positive and negative. This geophysical method is very useful in exploration for natural resources. For example, groundwater reservoirs, which are particularly interesting for engineers (Fig. 4.44, case 3) are indicated by a significant deficit of mass due to their porosity.

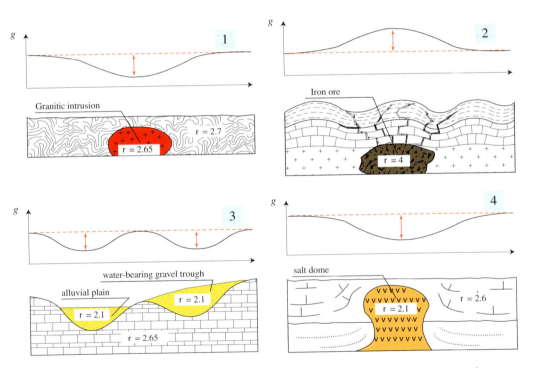

Fig. 4.44 Examples of geologic structures that may result in gravity anomalies. Densities are in t/m³.

Microgravimetric methods are also very useful to the engineer. Very precise measurements with particularly sensitive gravimeters over several hectares make it possible to identify natural and artificial underground caverns. It is also possible to identify heterogeneities in the subsurface that may cause difficulties in tunnel excavation (Fig. 4.45).

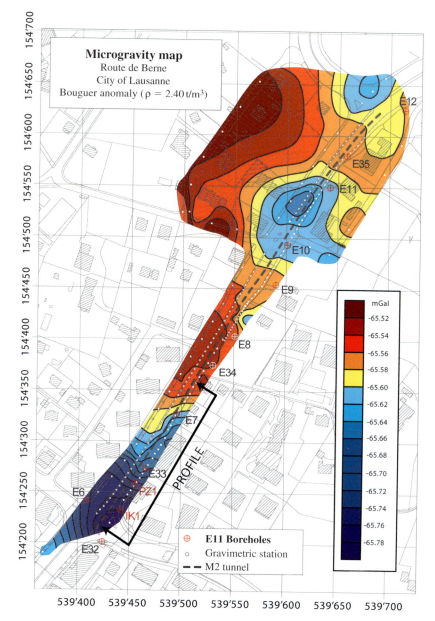

Fig. 4.45 Examples of an application of microgravimetry to determine the top of bedrock under unconsolidated deposits. The city of Lausanne has recently constructed a 6-km long stretch subway M2 from the shore of Lake Geneva to connect to the north suburb. The tunnels are excavated primarily in sandstone and molassic marls that pose few stability problems. However, the overlying deposits are often unstable. Knowledge of the top of the bedrock is thus a determining element in the feasibility of the project. In the first boring campaign, the elevation of the top of the rock was determined precisely at these points. The segment of the Route de Berne showed that the unconsolidated cover was important and that the future tunnel would be only slightly below this contact. At this point a microgravimetric campaign was conducted along this section. Gravity was measured with extreme precision (Scintrex CG-5 gravimeter). The map shows the Bouguer anomaly resulting from a comparison of these measurements to those calculated from the unique density subsurface model. Variations in the anomaly are associated to heterogeneities in the subsurface shown by the density contrast. On the map, the blue zone shows greater thicknesses of unconsolidated material.

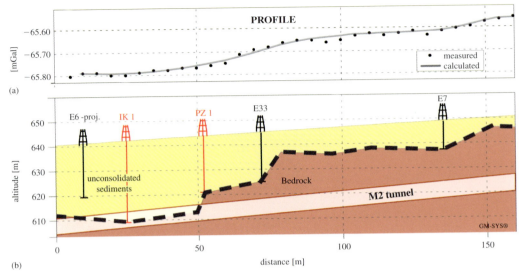

Fig. 4.45 (*Continued*) The profile shows the interpretation of an anomaly in terms of rock depth for the southern part of the map. The upper graph shows the anomaly along the axis of the tunnel (the divergent points along the curve of the gravity anomaly are caused by local anthropogenic effects due to small structures such as conduits, walls, talus, and underground buildings that were not entered into the model. For three-dimensional gravity modeling, the thickness of the cover is calculated by making a density contrast of $0.4 \cdot 10^3 \, kg/m^3$ between the two formations. These thicknesses are calibrated to the boreholes from the first campaign (in black on the profile). The resulting geologic profile clearly shows that the tunnel may emerge from the bedrock between boreholes E6 and E33. A second drilling program (in red on the profile) confirms that this risk is real. Finally, the engineer was able to take this new element into account and modify the construction technique to adapt to this geologic difficulty. (Study, P.V. Radogna, R.J. Oliver, Geophysics Institute of the University of Lausanne).

PROBLEM 4.4

Gravimetric measurements were done for the construction of the Swiss highway A1 several kilometers south of Yverdon-les-Bains, on the Pomy-Sermuz plateau. After the usual corrections, the Bouguer anomaly on the A-B profile shows a hollow that must be interpreted. Knowing that at the time of the measurements only the southern tube of the Pomy-Sermuz highway tunnel had been dug, could it be the source of the measured anomaly?

Data:

- Location map showing the A-B Profile (Fig. 4.46),
- Bouguer anomaly along the A-B profile,
- Density of the subsurface (molassic rocks) : $2.43 \cdot 10^3 \, kg \cdot m^{-3}$
- Radius of the southern tunnel: 5.13 m,
- Depth of the axis of the southern tunnel: 22.5 m.

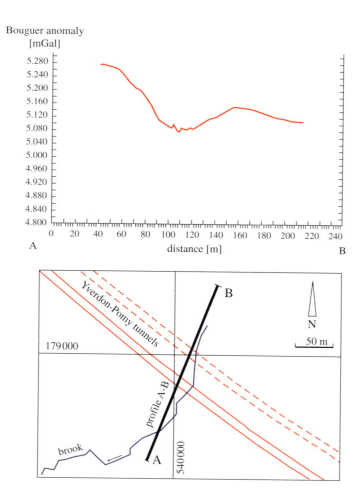

Fig. 4.46 Bouguer anomaly along the A-B profile and location map of the A-B profile for problem 4.4. The data were taken from the postgraduate work of A. Rosselli in GEOLEP, in collaboration with the Institute for Geophysics of the University of Lausanne.

Look at the solution on the DVD

4.2.5 Principles of isostasy

In 1745 Bouguer had already used gravimetric measurements to show that the Andes did not have the gravity effect that had been predicted by their elevation. The following year, he made the same observation on the Himalayas. He concluded that mountain chains always have a strong negative anomaly in the center that decreases toward the margins. It was inferred that rocks were lighter at the core of mountains. But what kind of geometry would lead to that?

In 1855 Airy proposed a model that turns out to correspond well with more recent observations (Fig. 4.47). This anomaly comes from the light continental crust in the roots of mountains.

Calculations of gravitational equilibrium of lithospheric plates rest on two physical principles, one for permanent conditions, another for dynamic cases where geologic events modify the boundary conditions.

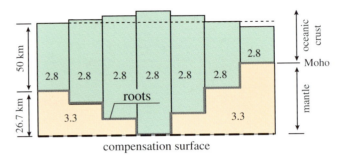

Fig. 4.47 Airy's density model. In an orogenic belt, the crust may be compared to a series of columns. Each column has an equal density, but a different height. The relief is thus compensated by a root of light rocks in the upper mantle (from [51]). The numbers indicate the density in t/m^3.

4.2.5.1 First principle

"Any spherical surface passing below the deepest roots of the crust (in the upper mantle) is a compensation surface on which light columns of rock (density crust $\rho_c = 2.8 \cdot 10^3\,kg/m^3$) resting on heavier ones (density upper mantle $\rho_{um} = 3.3 \cdot 10^3\,kg/m^3$) are in equilibrium". A comparison can be made with light rafts floating on a heavier liquid mass (Fig. 4.47).

4.2.5.2 Second principle

"If the elevation changes, the resulting weight variation will be compensated by a change in the elevation of the light rock–heavy rock contact at depth, such that at the compensation surface, the load will remain unchanged."

This principle can be understood easily by the analogy of a ship being loaded and unloaded (Fig. 4.48). In this case equilibrium occurs almost instantaneously because the low-viscosity water responds rapidly to changes in the load. It is obvious that if the boat is floating in mud, it will take longer to achieve balance. How do we relate this scenario to materials that underlie the Earth's crust? The crust, the upper part of the lithosphere, is made up of rock that is not ductile, and thus is not capable of flowing or deforming over time (Chapt. 12). The upper mantle, which forms the base of the lithospheric plates, is also rigid. The ship can thus be considered equivalent to these two layers together. As we will see in chapter 6, there is a layer in the mantle called the asthenosphere. It is the location of partial fusion of the mantle, which gives it greater plasticity. S waves can pass through it, so it is not completely liquid. However, it flows enough to be considered as the medium where equilibration with the material under the "lithospheric ship" takes place. When a plate is loaded, by an ice cap for example, the asthenosphere pushes aside rock to let the lithosphere sink. When the ice melts, the rock flows back under the continent again, exactly like water under the ship. The velocity of the equilibration is several seconds in the case of a ship in water, several days in the case of a boat in mud, and several thousand years in the case of the asthenosphere.

Let's take the case of Scandinavia (Fig. 4.49). Scandinavia was freed from its thick Würm ice sheet about 15,000 years ago. Elevation measurements on this old basement show that it is still rising today, at a velocity of about 9 mm/year at its center. The equilibration will likely be completed only after several thousands or tens of thousands of years. But between now and then another ice sheet may appear. This example shows that isostatic equilibration on Earth is so slow that it may outlast the cause that gave rise to it. This phase difference complicates the interpretation of movements actually measured on the ground. Calculation models attempt to reconstruct

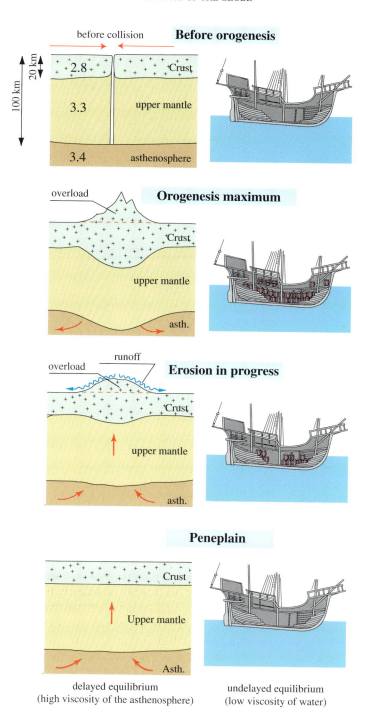

Fig. 4.48 Second principle of isostasy applied by analogy to a ship. The overburden created by a mountain chain causes the progressive sinking into the asthenosphere, just as a loaded ship sinks in the water. Erosion eliminates the excess weight and the continental plate rises again.

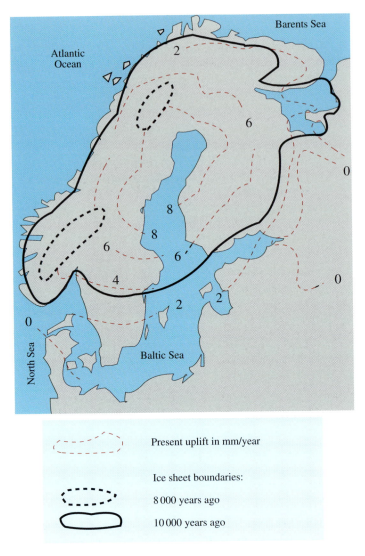

Atlantic
Ocean

Barents Sea

North Sea

Baltic Sea

- - - - - Present uplift in mm/year

Ice sheet boundaries:

- - - - - 8 000 years ago

—————— 10 000 years ago

Fig. 4.49 Isostatic rebound in Scandinavia after the Würm glaciers.

these phenomena. The credibility of these results depends on the parameters chosen to characterize the material. One factor that is poorly known is the viscosity of the asthenosphere.

These vertical movements, although minor on the human scale, have important implications for the engineer. For example, the isostatic rebound of Scotland, which was glaciated during the Würm, is thought to be the cause of a compensatory effect that causes the southern part of Great Britain to sink, similar to the bending of a beam. The London basin is sinking at a velocity of 3 mm/year. This slow movement, coupled with very strong tides in the English Channel, have made the city of London vulnerable to floods on the Thames. In 1663 an exceptional tide devastated the capital. In 1981 engineers constructed the largest pivoting dam in the world made of 10 semi-cylindrical gates to hold back the estuary water during critical periods (Fig. 4.50). It is 570 m long and the gates are 15 m high. The gates can be put in place in 30 minutes. Isostasy is the cause of this spectacular construction. If we consider the problem of rising sea level due to climate change, it is obvious that this situation will become increasingly problematic.

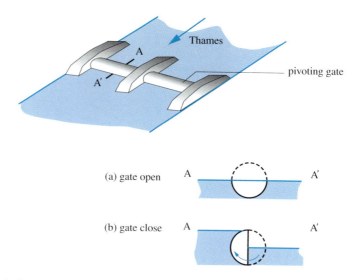

Fig. 4.50 Schematic diagram of the pivoting dam on the Thames. (a) Normal conditions, the gate rests in its housing at the bottom of the river. (b) When the tide threatens to rise, the gate pivots upward to form the dam.

PROBLEM 4.5

During the extensive Quaternary glaciations, the central part of the Alps was covered by an ice sheet. You are asked to estimate the approximate average thickness of ice in this region by comparing the map showing the extent of the last period of glaciation (see Fig. 3.26) with the present day topographic map (Fig. 4.51).

Question

If this glacial period had lasted long enough to reach a new isostatic equilibrium, what movement of the crust would have been observed?

You can calculate this movement by knowing that:

- Seismic measurements done by the National Research Program PNR 20 have determined the geologic structure shown in figure 4.52.
- The density of ice is $0.91 \cdot 10^3 \, \text{kg/m}^3$.

In the book, find the parameters necessary to do the calculation.

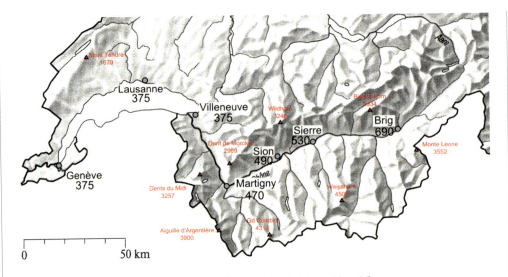

Fig. 4.51 Topographic base simplified for problem 4.5.

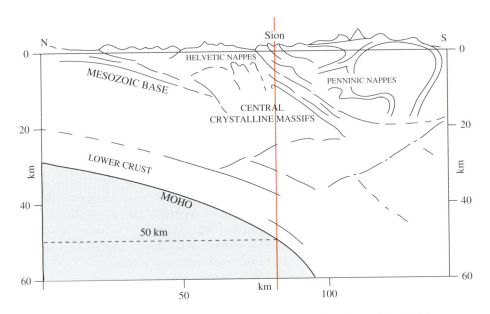

Fig. 4.52 Profile across the Alps showing the base of the continental crust (from [100]).

Look at the solution on the DVD

4.2.6 Density of Earth's rocks

Table 4.6 shows a range of the density of unconsolidated deposits and rocks near the Earth's surface. Gravimetric measurements and other geophysical methods have made it possible, in addition, to determine the density of rocks and magma in the interior of the Earth. Density increases toward the Earth's center (Fig. 4.53). The reason for this increase is the geochemical composition of these deep layers, which are increasingly rich in metals: ferromagnesian silicates in the mantle, metallic iron and nickel in the core.

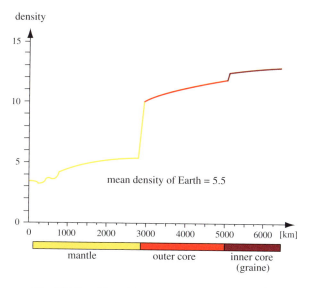

[234]

Fig. 4.53 Densities within the Earth (expressed in t/m^3).

4.3 Magnetism

We saw in chapter 2 that the Earth is a decidedly hospitable planet compared to the other telluric planets. This characteristic is reinforced by the presence of a strong magnetic field, about one hundred times stronger than that of Mercury and Mars (Venus does not have one at all) (Fig. 4.54). This difference is not surprising because the Earth's magnetic field is linked to current geologic activity.

The Earth can be compared to a magnetic dipole whose axis is near its axis of rotation (the axis of rotation makes an angle of 11 degrees with the Earth's axis). The negative pole is near the geographic north pole, and vice versa.

The presence of a strong magnetic field is another condition favorable to life on Earth. The *magnetosphere* is a type of umbrella that protects the Earth from high-energy solar radiation. The magnetopause is the border of the magnetosphere. The forces of the magnetic field deflect this radiation.

Magnetism is also the oldest method of geophysical prospecting. Prospectors for iron deposits in the Middle Ages in Sweden detected local deviations using a magnetized needle, the precursor to the compass.

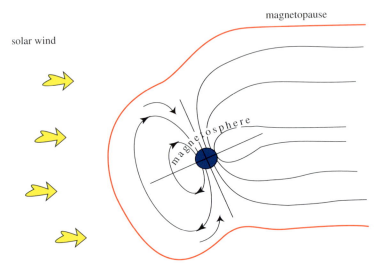

Fig. 4.54 Protective effect of the magnetosphere.

4.3.1   Causes of the Earth's magnetic field

Let us look in detail at how the Earth generates a magnetic field as a result of its internal activity. There is no permanent cylindrical magnet at the center of the Earth. Of course the core contains a lot of iron, but its temperature is clearly above the Curie temperature. These iron masses thus cannot give rise to permanent magnetism, the pre-condition for a large magnet. To understand this, let us remember several principles of magnetism.

Magnetism is a manifestation of moving electrical charges. It is detectable by its action on other moving electrical charges. This is why the magnetic and electrical interactions should be considered as a unit, under the more general name of electromagnetic interactions. Magnetic poles of the same sign repel each other and poles of opposite magnetic signs attract.

A magnetic field can magnetize another body. In other words, it generates a force that orients the particles of matter in a single direction. The electronic orbits of this matter are distorted by the presence of the field, which gives rise to magnetic polarization.

Physics shows that a magnetic field can be obtained without a permanent magnet, by making an electrical current circulate within a wire. In a wire shaped as a circular loop, the magnetic field is perpendicular to the plane of the loop. This field can be reinforced by rolling several loops or turns around a cylindrical body. This current produces a magnetic field B_0 perpendicular to the turns, proportional to the current and the density of the wire coil (Fig. 4.55). If the cylinder is in a vacuum, its intensity is equal to the following, per unit of length:

$$B_0 = \mu_0 \cdot n \cdot i \tag{4.20}$$

n = number of turns per unit length of the cylinder [1/m]
i = intensity of electrical current in the turns [A]

$$\mu_0 = 4 \cdot \pi \cdot 10^{-7} \, [\text{Wb/A} \cdot \text{m}] \tag{4.21}$$

μ_0 = magnetic permeability of the vacuum

Placing a magnetic body inside the coil gives rise to a second field $\boldsymbol{B}_m$ such that:

$$\boldsymbol{B}_m = \mu_0 \cdot \boldsymbol{M} \tag{4.22}$$

$\boldsymbol{M}$ is the magnetizing vector. Its intensity is equal to the magnetic moment per unit volume of the magnetic substance.

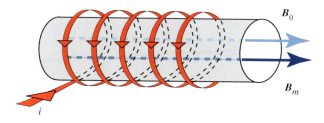

Fig. 4.55 The two components of a magnetic field. One $\boldsymbol{B}_0$, created by electrical current in the loops of the coil (free electrons), and the other, $\boldsymbol{B}_m$, due to magnetization of the material in the cylinder (electrons linked to atoms).

The total field is thus

$$\boldsymbol{B} = \boldsymbol{B}_0 + \boldsymbol{B}_m = \mu_0 \, (\boldsymbol{H} + \boldsymbol{M}) \tag{4.23}$$

from which we obtain $\boldsymbol{H}$, the *force of the magnetic field*:

$$\boldsymbol{H} = (\boldsymbol{B}/\mu_0) - \boldsymbol{M} \tag{4.24}$$

It is independent of the substance and proportional to the electrical current i. Its intensity is equal to

$$H = n \cdot i \tag{4.25}$$

$\boldsymbol{H}$ and $\boldsymbol{M}$ have the same dimension and are generally expressed as [A/m].

The unit of magnetic field is the *tesla* [T]. A tesla is the value of magnetic field that produces the force of 1 Newton on an electrical charge of 1 Coulomb moving perpendicularly to the field at a velocity of 1 m/s. It is expressed as

$$[\text{N/(C} \cdot \text{m/s)}] \text{ or } [\text{kg} \cdot \text{s}^{-1}\text{C}^{-1}] \tag{4.26}$$

Geophysics uses the nannotesla [nT], also called the "gamma" [γ]. The *weber* [Wb] is the unit of magnetic flux. It is equivalent to a field of 1 T across a surface of 1 m^2.

The majority of magnetic substances only become magnetized when they are exposed to a primary magnetic field (magnetizing field). Their own magnetic field is thus weak compared to the magnetizing field (about 1/1000). This weak magnetic field is linked to the primary field by the following ratio:

$$\boldsymbol{M} = \chi \cdot \boldsymbol{H} \tag{4.27}$$

where $[\chi]$ is the ***magnetic susceptibility***.

Magnetic susceptibility, a dimensionless number, is a property of a substance. It is positive for ***paramagnetic*** substances (iron salts, metallic aluminum). The field itself has the same direction as the magnetizing field. It is negative (secondary magnetic field opposite to the primary field) for ***diamagnetic*** substances (water, salt, anhydrite, calcite and quartz) (Chapt. 5).

Few natural substances have magnetic properties. Some rare minerals have strong magnetic susceptibility, which varies nonlinearly according to the intensity of the primary magnetic field. When the field is cancelled, these minerals keep their remanent magnetism. They are called ferromagnetic. These are basically magnetite (Fe_3O_4) and secondarily ilmenite ($FeTiO_3$) (Chapt. 5). This fossilized magnetism is destroyed when the material exceeds the Curie temperature, which is 585°C for magnetite.

The magnetic susceptibility of rocks depends principally on their magnetite content. Table 4.7 gives an idea of the magnetic susceptibility of common rocks. It should be noted that the majority of them do not have remanent magnetism. However, certain mafic rocks, basaltic lavas in particular, preserve a record of the magnetic field that existed at the time of their solidification in the orientation of their magnetite crystals (Chapt. 6). In the same way, some lacustrine or marine sediments have small magnetic particles oriented along the direction of the ambient field at the time of their deposition, and these can cause detrital remanent magnetism.

Table 4.7 Magnetic susceptibility of some rocks. The susceptibilities of magnetite and ilmenite are given as comparison.

Rocks	χ [-]
Dolomite	$0{-}20 \cdot 10^{-6}$
Limestones	$0{-}50 \cdot 10^{-6}$
Sandstone	$0{-}100 \cdot 10^{-6}$
Clay	$0{-}200 \cdot 10^{-6}$
Metamorphic schists	$0{-}500 \cdot 10^{-6}$
Gneiss	$0{-}1000 \cdot 10^{-6}$
Granite	$100{-}1500 \cdot 10^{-6}$
Lavas	$500{-}10\,000 \cdot 10^{-6}$
Ilmenite	$20\,000{-}50\,000 \cdot 10^{-6}$
Magnetite	min. $100\,000 \cdot 10^{-6}$

This incursion into general physics shows the Earth's magnetic field does not require a magnet inside the Earth. We have seen that the outer core is fluid and rich in iron. Geophysicists think that the Earth's rotation causes currents in the outer core. Because of the low viscosity of this material, these currents would be about one million times more rapid than in the mantle, or several kilometers per year. These fluxes of ferrous material could give rise to an electrical current more or less perpendicular to the Earth's rotational axis and thus create the Earth's magnetic field.

4.3.2   Components of the magnetic field at a point on the Earth's surface

The magnetic field measured at the Earth's surface is the sum of three different magnetic fields.

4.3.2.1   Internal magnetic field

The Earth's internal magnetic field is represented by lines of force. It is tangent to the lines of force at all points. These lines are very densely grouped around the poles where they are close to vertical. They are more widely spaced and tangent to the Earth's surface near the equator (Fig. 4.56).

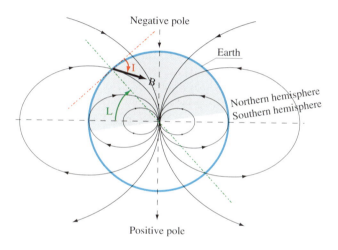

Fig. 4.56 Lines of magnetic force around the Earth.

The two points where the field is perpendicular to the Earth's surface are the magnetic poles. The negative magnetic pole, that is, the place where the field lines "enter" is in northern Canada, at a latitude of ~80°N. The positive magnetic pole is in Adélie Land (Antarctica). The plane orthogonal to the line of the poles is the magnetic equator. The design of the field lines has rotational symmetry around the axis of the magnetic poles.

At all points on the Earth's surface, the magnetic field is defined by its intensity and two angles (Fig. 4.57):

- **Declination** (*D*): the angle between the projection of the magnetic field vector on the local horizon at the given point and the geographical meridian passing though this point. This angle is low for points that are far from polar regions. As one approaches the poles, the declination values increase, and it is not as easy to locate oneself with a compass (Fig. 4.58). Over time, the magnetic poles move slightly, although they remain in more or less the same region. The declination values thus vary slightly. For example, in Switzerland, the declination was zero in 1994. Since that time, it has changed about 10 minutes of an angle per year. On topographic maps the declination is generally indicated for a given date. This declination is used for adjusting the compass to the work area.

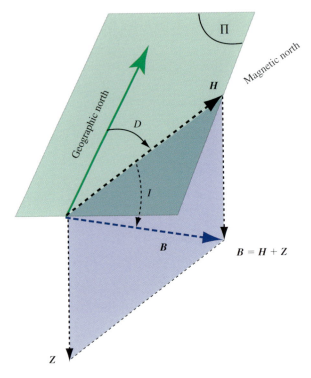

D = declination (> 0 toward east; < 0 toward west)
I = inclination (> 0 in the northern hemisphere
   < 0 in the southern hemisphere)
B = vector of the terrestrial magnetic field
H and Z = horizontal and vertical components of B respectively
Π = horizontal plane of the place

Fig. 4.57 Local decomposition of the magnetic field vector.

- **Inclination** (*I*): this is the angle between the magnetic field vector and the horizon of the place of interest. It can be seen on figure 4.56 that it is strongly linked to the magnetic latitude *L* of the point. An approximate relationship between these two factors has been established:

$$\mathrm{tg}I = 2\,\mathrm{tg}L \qquad\qquad (4.28)$$

The intensity is at a maximum at the poles (70,000 nT at the South Pole). It is weakest at the equator (33,000 nT). At central European latitudes, it is about 45,000 nT. It changes slightly over time (Chapt. 6).

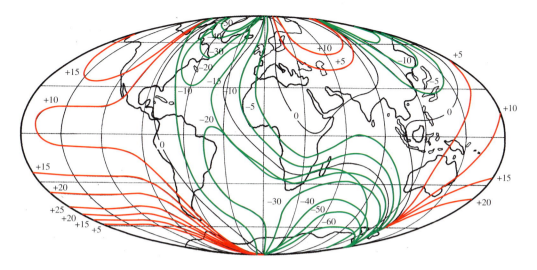

Fig. 4.58 World map of magnetic declination in degrees (situation in 2005). Red contours are positive (east). Green contours are negative (west). From NOAA.

4.3.2.2 Magnetic field of rocks in the lithosphere

This is the field produced by the magnetization of rocks in the lithosphere. The field varies according to local geologic conditions. There are two types of magnetization:

- Magnetization induced by the actual field: it is parallel to the ambient field and obeys the law (4.18);
- Remanent magnetism: it is not linked to the present-day field in either direction or intensity; it is a result of geologic history.

This is the component that the geophysicist looks for, both in paleomagnetic studies (§ 4.3.4) and in mineral exploration. At 30 to 40 km of depth, temperature and pressure conditions are such that the remanent magnetism of rocks is destroyed.

4.3.2.3 External magnetic field

This is a generally weak component that comes from currents in the ionosphere. It has diurnal variations that can become intense during periods of "magnetic storms" linked to solar activity.

4.3.3 Measurement of the magnetic field

Geomagnetic reconnaissance is a common practice in scientific research and mineral exploration. Nuclear magnetometers measure the intensity of the magnetic field by the effect of the magnetic field on the rotation of a hydrogen core. If there is no magnetic field, the orientation of the rotational axis does not move. In the presence of a magnetic field, the axis is activated by a precession movement whose velocity depends on the intensity of the field. The precision of this measurement method is about 1 nT.

The magnetic field is usually measured by aircraft (rarely on the ground). A magnetometer is placed far enough from the metallic mass of the aircraft (in a gondola for example) so that it won't disturb the measurement. The magnetometer determines the three components of the magnetic field vector.

4.3.4   Paleomagnetism

We have seen that the intensity of the magnetic field varies slowly over time. The study of remanent magnetism in samples of relatively recent basaltic lava (less than 5 million years) reveals that magnetite crystals in the lava line up according to existing lines of force. By studying the magnetism of older lavas, it has been observed that the orientation of magnetites is sometimes opposite that of the present-day field. This has been observed at numerous places on the Earth, making it possible to determine that the magnetic field has reversed its polarity many times over the course of the Earth's history. This fact makes it possible to use remanent magnetism as a new dating method: the paleomagnetic scale (Chapt. 3). We will see applications in plate tectonics and the rate of creation of oceanic crust in § 6.1.2.

These reversals do not occur over regular cycles. The major periods last a long time (several hundred thousand years to several million years, on average 500,000 years). Aside from these periods, there are shorter reversals (several thousand years). Scientists think that at the end of stable periods, the intensity of the field diminishes progressively, reaching zero in about 5000 years. Then, a period without a well-defined pole follows, during which the biosphere would be exposed more strongly to the solar wind. Next, a strong reversed field becomes rapidly established. At present, the magnetic intensity is decreasing, perhaps signifying a prelude to a reversal, which could happen within 2000 years.

The reason for these reversals is not yet known; it is probably to be found in variations in the movement of material in the outer core.

4.4   Geothermics

The thermal properties of the Earth have direct consequences for the engineer. They require him to design aeration for deep tunnel worksites and allow him to exploit this ecological energy to heat buildings.

4.4.1   Thermal transfer within the Earth

We saw in chapter 3 that the energy emitted by the Earth comes from nuclear reactions that take place within it. Figure 4.59 gives an idea of the ranges of magnitude of energy fluxes coming from the core and the mantle. Their effects on magmatism are treated in chapter 6.

4.4.1.1   Mechanisms of heat transfer

Heat is transported in rocks and magma by two different processes.

Conduction is the transport of heat by simple contact of a hot body and a cold body, without transfer of material. $d\theta$ is the difference of temperature and dx the distance between the hot and the cold body. The characteristic that describes the capacity of rocks to conduct heat is thermal conductivity λ. It is defined by Fourier's Law:

$$q_T = -\lambda \cdot d\theta/dx \tag{4.29}$$

q_T being the flux in $[J/(m^2 \cdot s)]$; λ is expressed as $[J/s \cdot m \cdot K]$ in SI units.

Figure 4.60 illustrates this calculation and gives the range of thermal conductivity in common rocks and unconsolidated materials.

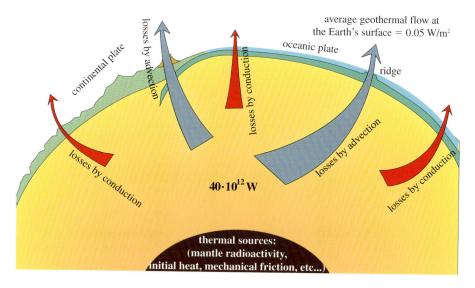

Fig. 4.59 Heat flux in the Earth.

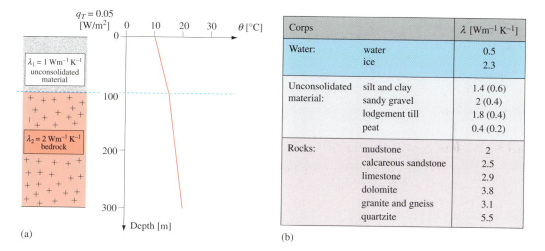

Corps		λ [Wm^{-1} K^{-1}]
Water:	water	0.5
	ice	2.3
Unconsolidated material:	silt and clay	1.4 (0.6)
	sandy gravel	2 (0.4)
	lodgement till	1.8 (0.4)
	peat	0.4 (0.2)
Rocks:	mudstone	2
	calcareous sandstone	2.5
	limestone	2.9
	dolomite	3.8
	granite and gneiss	3.1
	quartzite	5.5

Fig. 4.60 Heat transfer by conduction. (a) Effect of thermal conductivity on terrains under geothermal gradient at shallow depth. (b) Mean thermal conductivity for the principal terrains at saturation. In brackets, values for unsaturated media. Compilation of GEOLEP for mapping of the geothermal potential.

Advection, a fluid displacement process, is capable of transporting thermal energy if the fluid is hotter than its destination. This mechanism follows the laws of fluid flow based on density gradient (natural thermal convection) or gravity potential. The movement of hot material is important for rocks that become ductile with temperature, and for magmas and gasses. Near the surface, it affects groundwater. The advection of thermal water also contributes to the heat losses from the Earth in thermal springs and geothermal wells.

In moving thermal energy from the interior to the exterior of the Earth, transfer by advection is more effective than transfer by conduction.

Thermal losses at the Earth's surface vary widely across the globe. They will be huge in areas of active volcanism. They will be high under oceans, higher than under continents because oceanic lithospheric plates are thinner and more conductive than continental plates. On average, for the globe as a whole, the *geothermal flux* is very low: $0.05 \, W/m^2$.

This flux is 7000 times weaker than the average annual solar flux at Central European latitudes. However, it is 100 times stronger than the mechanical energy dissipated within the Earth (in the form of earthquakes, for example).

4.4.1.2 Geothermal gradient

The geothermal gradient is defined as the temperature variation with depth in the Earth. Near the surface (to several kilometers of depth), the geothermal gradient is 3 degrees per 100 meters, on average. It is evident that this value varies locally in a significant way due to geological conditions (for example, in volcanic areas where it may reach more than 100°C per 100 m). At depth, the gradient diminishes progressively (Fig. 4.61). In the deepest borehole ever drilled on Earth, in the Kola Peninsula (NW Russia), the temperature at a depth of 12 km in the continental crust was 180 degrees; this corresponds to a mean gradient of 15 degrees/km. The high temperature prohibited the advance of the borehole to its anticipated depth of 15 km.

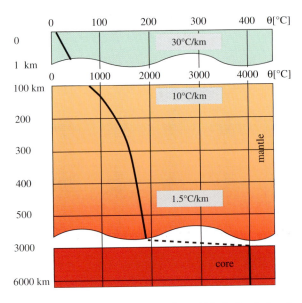

Fig. 4.61 Variation of the average geothermal gradient within the Earth. See also figure 6.18.

4.4.2 Exploitation of geothermal energy

Geothermal resources have been exploited for years. Long ago, steam turbine stations were constructed at sites of strong magmatic activity, in Italy (Larderello), Iceland and Tibet (Fig. 4.62), for example. To exploit geothermal energy, boreholes are installed to be in direct contact with steam or water is injected into a borehole and recovered as steam in a neighboring well.

Fig. 4.62 The Yangbajain geothermal field in Tibet, NW of Lhasa, which produces electricity from 150°C vapor collected by boreholes drilled to depths of 70 to 160 m into altered granites. This tectonically active zone has numerous thermal springs associated with faults. It is the highest geothermal field in the world (about 4500 m), which explains the paradoxical juxtaposition of subsurface vapor and perennial snows. GEOLEP Photo, A. Parriaux.

More recently, people have become interested in lower temperature energy resources that can be exploited in many regions. Two systems are currently in use: (1) deep well exploitation of hot groundwater, and (2) conductive geothermal probes, a technology that is already highly developed at present.

4.4.2.1 Exploitation by advection

This principle consists of finding a regional water-bearing layer with a normal geothermal gradient at a depth of more than 1500 m. The groundwater temperature should be between 50 and 60°C at a minimum. If the water-bearing layer is sufficiently permeable, the water can be pumped by a deep well and circulated through an exchanger to heat a group of buildings, without the need for a heat pump. The water in the primary circuit cools in the exchanger and is restored to the water-bearing layer through a second borehole, sufficiently far away to prevent it from cooling the extraction zone over the long term. It is generally not possible to dispose of it in a river because of its temperature and especially because of its chemical composition, which is often highly mineralized and as a result would harm the ecological balance of the river.

This system has numerous advantages, notably its low impact on the environment and its insensitivity to seasonal variations. Unfortunately, it has often failed. For example, in Switzerland strong financial encouragement has led to the installation of about ten boreholes. Only one exploration borehole that has proved very promising has led to the construction of a heating facility in the Riehen neighborhood near Basel (Fig. 4.63). The primary reason for the failures is the fact that permeability is generally too low at these depths. Karst terrains are often plugged by clay fillings, as was the case in a project for the city of Geneva: the Thônex borehole traversed upper Jurassic limestones to a depth of 2530 m. The temperature was as high as predicted, but production tests were able to extract only 20 m³/h. The high cost of the infrastructure, corrosion problems at the facilities, and precipitation in the primary circuit are also difficulties to be overcome.

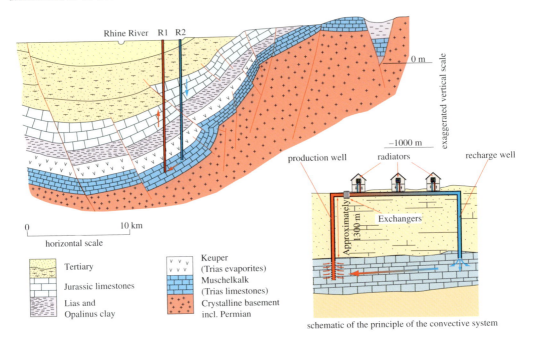

Fig. 4.63 Geothermal production by hot water extraction. Example of the Riehen facility, Swiss Jura (from [121]).

These shortcomings of natural permeability of rocks at depth and the progress of hydraulic fracturing techniques have given rise to a new concept: The Hot Dry Rock (also called Stimulated Geothermal Systems) developed at Soultz in the Rhine graben in Alsace. The principle consists of using recent techniques of pressurizing wells to widen existing fissures or even to fracture massive rocks at depths of several thousand meters in order to increase the permeability of the rock mass, and also the circulation flux. The pilot facility at Soultz is very promising in terms of energy exploitation. It consists of three wells 600 m apart and 5,000 m in depth into the granitic basement. The hydraulic stimulation phase, coupled with various chemical treatments to promote unplugging of fissures, is effectively completed. Cold water is injected into the central well. After circulation and heating in the fractured rock, the water is recovered by two peripheral wells. 1.5 MW of electrical power was obtained after the stimulation treatment. A project of the same type was begun in Basel under similar conditions under the name of Deep Heat Mining (Fig. 4.64). During the fracturing phase, earthquakes, some as large as magnitude 3.4 (Fig. 4.65) caused the project to be suspended.

Let us mention that deep boreholes represent a risk of uncontrolled extrusion of fluids. For example, an oil exploration borehole is supposed to be responsible of the huge mud flow devastating the region of Porong District (Java); since May 2006, hot and acidic mud is continuously invading rivers and villages in this region and no solution was found to stop it yet.

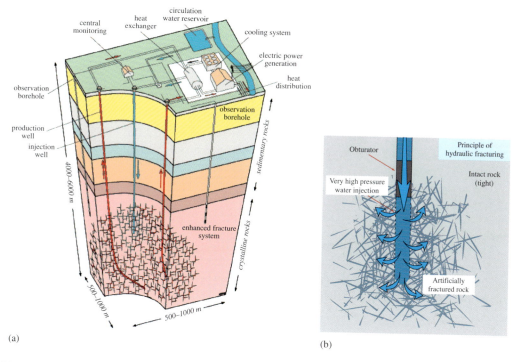

(a)   (b)

Fig. 4.64 Principle behind the Deep Heat Mining project. (a) General view. (b) Detail of the principle of hydraulic fracturing. From the "Deep Heat Mining" project, Federal Energy Office, Bern.

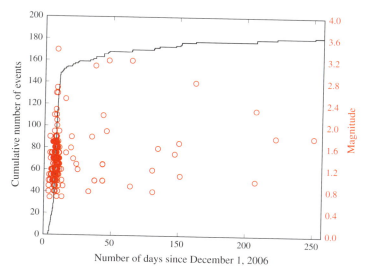

Fig. 4.65 Earthquakes caused by hydraulic fracturing conducted as part of the Deep Heat Mining project in Basel (red circles). Hydraulic fracturing began in December 2006. On December 8, an earthquake of magnitude 3.4 was clearly felt in the city, causing the fracturing to be stopped and the project to be suspended. From the Swiss Seismology Service ETHZ.

4.4.2.2   Exploitation by conduction

Because many of the deposits and rocks in the subsurface (between 0 and 100 m) have low permeability, the only reliable way to exploit geothermal energy is by conduction. Boreholes are installed in the subsurface to a depth of 40 to 300 m. A ***geothermal probe*** composed of two tubes connected at the base (Fig. 4.66) is placed in the hole. The probe is sealed in the borehole by a watertight grout. A fluid is circulated in the pipes (a mixture of water and ethylene glycol, for example). The fluid is slightly heated as it descends and rises in the pipe in contact with the wall of the borehole, which is naturally heated by the geothermal flux. This primary circuit arrives at the ***heat pump*** (which operates likes a refrigerator in reverse) which extracts about 5°C from the fluid to create water at 40°C to heat a house for example.

This system has been phenomenally successful. The cost for a single well-insulated modern home is about 30,000 dollars. This heating method is certain to become more common. However, it is important to avoid installing these probes too densely in order to avoid decreasing the yield in the long term due to the exploited flux being higher than natural recharge.

The use of geothermal probes in a mixed system (summer storage of air-conditioner wastes or of solar hot water – winter heat exploitation) promises great success in the future. This perspective is much more likely now that the ***energetic geostructures*** diversify the solutions available to the engineer. These structures are parts of underground construction (foundation piles,

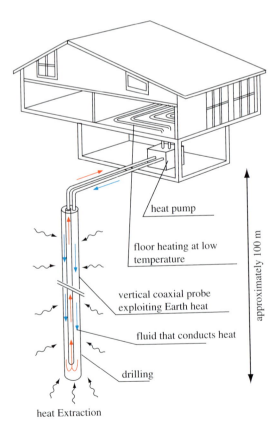

Fig. 4.66 Schematic of a conductive geothermal probe. Federal Energy Office, Bern.

slabs, slurry-trench walls, etc.) that contain heat exchanger tubes (Chapt. 14). It is obvious that the urban subsurface offers interesting opportunities for the sustainable development of the city.

Geothermal energy will never resolve the energy problems of the world by itself. But its development will increasingly contribute in the future to diversify energy sources and increase "environmentally friendly" energy. It is one of the major objectives of sustainable development.

[36, 238, 247, 266, 267, 268, 269]

Chapter review

Physics contributes crucial data to geology, on both a small and large scale. It helps us draw conclusions about the internal structure of the Earth and to recreate its history. Geophysical methods will be complemented in the remainder of the book by geochemical, mineralogical and petrological aspects to provide an overall view of geologic environments.

Geophysics is also useful for natural resource exploration (water, minerals, energy). Some methods also contribute to the work of the field geologist in the reconnaissance of sites for the construction of engineering projects and the study of natural hazards. The engineer will make use of various geophysical methods for these applications.

5   Rock Forming Minerals

Before studying all the varieties of rocks we should spend some time describing minerals, the units that make up rocks. ***Mineralogy*** is the science of minerals. It is related to inorganic chemistry and ***petrology***, the scientific study and description of rocks.

There is a wide variety of mineral species in nature. In this book designed for engineers, we will limit ourselves to minerals that play a preponderant role in the formation of the major rock types. We will see that the technical properties of these rocks are strongly influenced by the properties of their minerals and we will emphasize those that affect engineering projects. As for identification methods we will limit ourselves to field criteria that are easy to apply under all circumstances.

Why is mineralogy useful to geological engineering? For a long time engineers have considered geological material as a black box. Traditionally, engineers have conducted mechanical behavior tests and then assigned the results to a gross technical classification. For example, even today crushed rock is treated as a single geotechnical unit, which causes many difficulties in underground projects. Recent research shows that the rocks should be analyzed more completely and their microstructure and mineralogical composition taken into account in order to correctly characterize their strength (from [37, 38]).

Minerals are defined as natural solids formed by predominantly inorganic processes, but also in rare cases organic processes (aragonite in pearls or calcite in shells). They have a defined chemical composition or range of compositions: for example, the chemical formula of olivine ranges between two extremes, forsterite (Mg_2SiO_4) and fayalite (Fe_2SiO_4), and all proportions of Mg and Fe are possible mixtures. This compositional range is expressed in a chemical formula by enclosing the substituting elements in parenthesis: $(Fe,Mg)_2SiO_4$.

The internal atomic structure lets us distinguish two categories of minerals:

*   Crystallized minerals (or ***crystals***) whose orderly geometric arrangement of atoms reflects particular properties (classification according to seven crystal systems); for example, ***quartz*** = crystallized silica;
*   Amorphous minerals whose elements are randomly distributed; for example, some varieties of ***opal*** (hydrated silica).

Amorphous minerals will be discussed as we study the important role they play in the genesis of rocks (for example, in biogenic rocks) and in engineering (for example, the alkali-reaction of cement).

At the Earth's surface, under normal pressure and temperature conditions, minerals are generally present as solids, with several exceptions. Let us cite mercury, which exists as a liquid in its native state in the pores of ores (Fig. 5.1).

There are also synthetic minerals, made by man for industrial applications (for example, carborundum, SiC, used as an abrasive).

5.1   Crystallography

The crystalline arrangement of atoms depends primarily on their relative sizes (ionic radius) and their chemical bonds.

Fig. 5.1 Native mercury in the form of drops. The Almaden mine, Spain. GEOLEP Photo, L. Cortesi.

5.1.1 Internal structure of minerals

We will examine the elementary components of minerals and their arrangements.

5.1.1.1 Principal bonds

There are several types of bonds between two atoms, depending on how the electrons are arranged around the nucleus. This depends on their ***electronegativity***, or the capacity of an atom within a molecule to attract an electron to itself. Anions have a strong negativity compared to cations (Fig. 5.2).

		IA	IIA	VIIIB	IB	IIIA	IVA	VIA	VIIA
	1	H (2.1)							
	2						C (2.5)	O (3.5)	F (4.0)
Period	3	Na (0.9)	Mg (1.2)			Al (1.5)	Si (1.8)	S (2.5)	Cl (3.0)
	4	K (0.8)	Ca (1.0)	Fe (1.6)	Cu (1.9)				

Fig. 5.2 Electronegativity of some elements that make up common minerals (in electron-volts).

Minerals found in nature use two principal types of bonds (Fig. 5.3).

Ionic bonds

These are the bonds found in halite (NaCl) and anhydrite ($CaSO_4$), for example. The atoms in these minerals are highly ionized and are attracted to oppositely charged ions. One of the

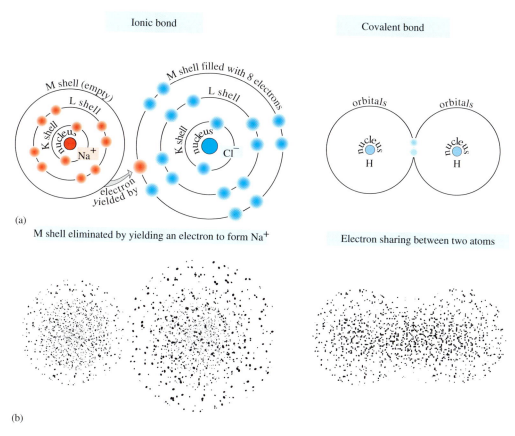

Fig. 5.3 (a) The two principal bonds in natural crystals. (b) probabilistic representation: the density of points is in each place proportional to the probability of finding an electron in that place.

two atoms has a strong electronic affinity (anion) and the other has a strong tendency to lose its outer electrons (cation). The cation and the anion are attracted to each other because of the electrostatic force between opposite charges (Coulomb's attraction). In ionic bonding, the structure tends toward maximal compactness. Minerals with this type of bond have moderate hardness and are brittle (they break cleanly).

Covalent bonds

When a substance contains atoms of identical or similar electronegativity, an ionic bond cannot occur. In this case, electrons are shared between atoms: this is a covalent bond, also called homopolar. The electrons in the outer layers of the two atoms orbit around the two atoms indiscriminately, simultaneously completing two external orbitals without any gain or loss of electrons. This is the case of the diamond, for example, in which the carbons atoms contain four electrons in the outer orbital. Minerals characteristic of this type of bond are very hard and brittle.

Many crystals have mixed bonds. For example, the bonds between Si^{4+} and O^{2-} in a large group of silicates are partially ionic and partially covalent. The difference in electronegativity is too small for the bond to be strongly ionic.

5.1.1.2   Crystal structures

Crystallography allows us to represent the crystalline structure of a mineral as a three-dimensional arrangement of atoms. Crystallography classifies crystals according to geometric rules. This detailed classification is beyond the scope of this book.

On the atomic scale, the crystal structure is made up of a ***motif*** (group of atoms or ions) repeated exactly in space (Fig. 5.4).

In the study of crystalline structure, it is generally easier to ignore the form of the motif and concentrate only on the geometry of the repetitions in space. The motifs are thus represented by points called ***nodes***.

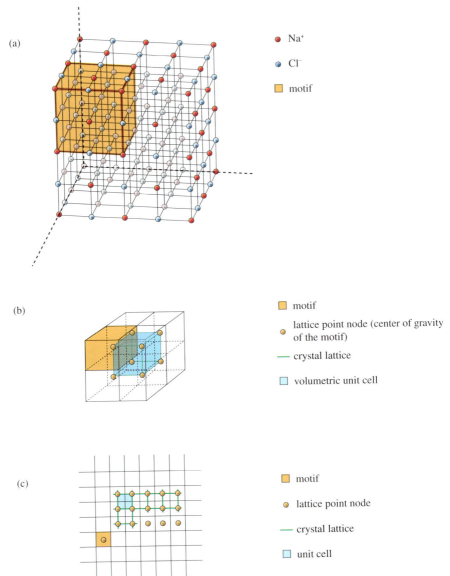

Fig. 5.4 Crystalline structure of halite (NaCl). (a) Atomic structure (ratio of atom size/length of bonds is decreased for a better visualization of the particular geometric properties); the orange cube represents the motif. (b) Relationship between motif–node–unit cell; each motif is replaced by a point, the node. (c) Two-dimensional visualization.

The nodes, repeated by translation, are linked to one another by straight lines defining an imaginary framework, the ***crystal lattice*** (Fig. 5.5). Each node is surrounded by an environment identical to that of every other node in the lattice. The distance between each node is the period of the translation. The figure obtained can be considered practically infinite since the periods are on the order of several angstroms. The ***lattice row*** (Fig. 5.5a) is the one-dimensional representation of the crystal lattice. There are a series of nodes spaced at equal intervals along a straight line.

In two dimensions, translation occurs along two axes. A plane passing through three nodes that are not in the same lattice row defines a ***lattice plane*** (Fig. 5.5b). The smallest unit that repeats infinitely by translation in the crystal lattice is a parallelogram (defined by two vectors a and b); this is the ***unit cell*** (case for the cubic system, Fig. 5.4 and Fig. 5.5b).

In three dimensions, the crystal structure is described by the translation repetition along three axes of the smallest three-dimensional geometric motif (parallelepiped along three vectors, a, b and c); this is the ***volumetric unit cell*** (Fig. 5.4b and Fig. 5.5c). A crystal structure contains many lattice planes. A halite crystal that measures $1\,cm^3$ contains about $2 \cdot 10^{21}$ unit cells.

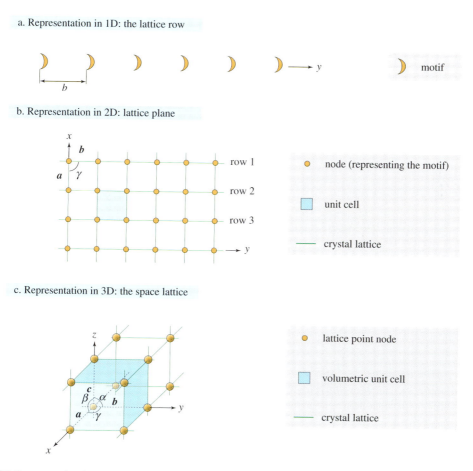

Fig. 5.5 Representation in space of a crystal lattice (here, the cubic system). The unit vectors ***a***, ***b***, and ***c*** are represented on axes x, y and z, each separated from the others by the angles of α, β and γ.

Lattice density

If we pass a straight line through the nodes in a crystal lattice, we obtain an infinity of possibilities: we can choose a line that passes through many nodes or just a few. The lattice density is the number of nodes per unit length of the line (1D); it is high in the case of line 1, low in the case of line 5 (example of the cubic system, Fig. 5.6). In two dimensions (2D) the *lattice density* (or *reticular density*) corresponds to the number of nodes per unit of surface area, and in three dimensions (3D) it is related to the volumetric unit.

The atomic arrangements along the various lattice planes give rise to an anisotropy of some properties (mathematically represented by tensors). The amplitude of these vector properties varies with the direction (for example hardness, electrical conductivity, light velocity).

The directions of the lattice planes play a fundamental role in crystal growth (crystal faces) and in the way crystals break when they are subjected to mechanical stress (cleavages).

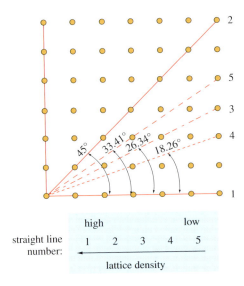

Fig. 5.6 The plane of a cubic crystal lattice in two dimensions. Several straight lines (lattice rows) of variable lattice density are possible. These straight lines (or these 3D planes) make variable angles with the major directions of the unit cell. The planes of maximum density are parallel to one of the axes (line 1). Those with a slightly lower density cut them at 45 degrees. Those of low density (line 5) can have almost any angle.

5.1.1.3 Symmetry elements

Crystal lattices can be characterized by three elements of symmetry (Fig. 5.7).

Center of symmetry

An object has a center of symmetry if it remains the same after inversion with respect to this point. This is the least degree of symmetry.

Symmetry planes (or mirror)

An object has a plane of symmetry if it remains the same after reflection through this plane. After reflection, the object and the image can no longer be superimposed. Almost all crystalline arrangements have at least one symmetry plane.

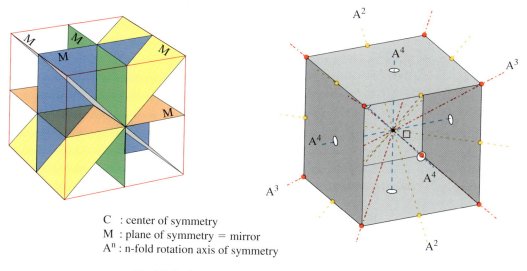

C : center of symmetry
M : plane of symmetry = mirror
A^n : n-fold rotation axis of symmetry

Fig. 5.7 Basic symmetry elements of crystalline structures.

Axes of symmetry

A crystal lattice has a n-fold axis of symmetry if it is identical after a rotation of $2\pi/n$. For example, a 4-fold axis rotates 90° four times and generates four images of the atoms of the lattice, three of which are new; the last ends up on the starting position. Thus the various objects obtained are superimposable in this example.

Crystal structures frequently have 2-fold axes. High symmetries have 3-fold axes (cubic and rhombohedral), 4-fold (cubic) and 6-fold (hexagonal prism).

5.1.1.4 Symmetry systems

The study of crystal structures has led to the establishment of seven systems of symmetry (Fig. 5.8). They go from unit cells with few elements of symmetry (triclinic) to more elaborate ones (cubic).

5.1.2 Crystal morphology

We have seen that in a crystalline arrangement it is possible to identify planes of different lattice density that cut the axes of the unit cell according to certain angles (Fig. 5.6). To these latter, it is possible to define parallel planes that have the same angle properties. These planes are possible faces that may appear in the growth of a crystal. Thus a crystal can be pictured in the following way: lattice planes correspond to crystal faces, lattice rows to edges and the nodes to summits.

In 1669 Steno noticed that the angles between equivalent faces of crystals of the same substance, measured at a given temperature, are constant (Steno's Law or law of the constancy of dihedral angles). In nature we observe that crystal faces and the most common cleavages are parallel to planes of the highest lattice density.

In the cubic system, the most abundant faces are parallel to faces of the unit cell or cut them at 45° C (Fig. 5.9) Other faces rarely occur. The image in figure 5.10 illustrates the morphology of the hexagonal system.

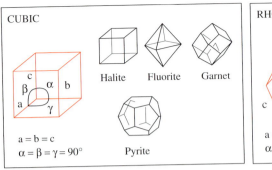

CUBIC

Halite   Fluorite   Garnet

Pyrite

$a = b = c$
$\alpha = \beta = \gamma = 90°$

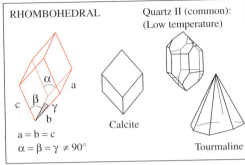

RHOMBOHEDRAL   Quartz II (common):
(Low temperature)

Calcite

Tourmaline

$a = b = c$
$\alpha = \beta = \gamma \neq 90°$

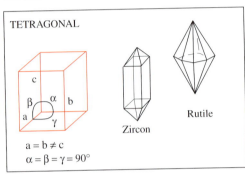

TETRAGONAL

Zircon   Rutile

$a = b \neq c$
$\alpha = \beta = \gamma = 90°$

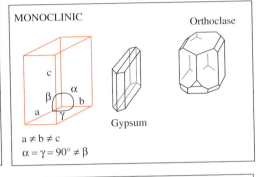

MONOCLINIC   Orthoclase

Gypsum

$a \neq b \neq c$
$\alpha = \gamma = 90° \neq \beta$

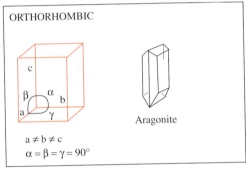

ORTHORHOMBIC

Aragonite

$a \neq b \neq c$
$\alpha = \beta = \gamma = 90°$

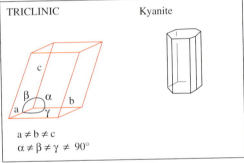

TRICLINIC   Kyanite

$a \neq b \neq c$
$\alpha \neq \beta \neq \gamma \neq 90°$

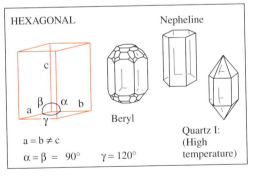

HEXAGONAL   Nepheline

Beryl

Quartz I:
(High
temperature)

$a = b \neq c$
$\alpha = \beta = 90°$ $\gamma = 120°$

Fig. 5.8 The seven systems of crystal symmetry, with shape and properties of the unit cell. Examples of typical crystal shapes with some corresponding minerals.

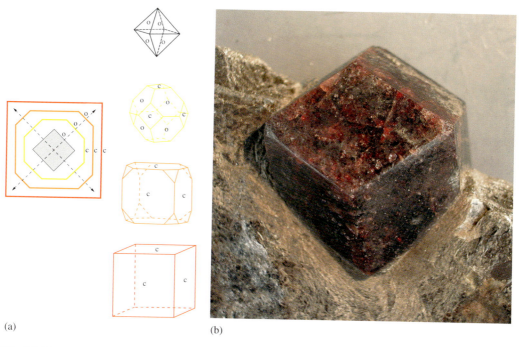

(a)   (b)

Fig. 5.9 Morphology of crystals in the cubic system: (a) Growth of a cubic mineral from an octahedral nucleus; the arrows indicate the direction of the most rapid growth of faces; the faces perpendicular to these arrows end up disappearing (from [149]). (b) Perfect habit of the cubic system illustrated by cuprite. Photo of the Cantonal Geological Museum of Lausanne, N. Meisser.

Fig. 5.10 Ice crystal showing hexagonal symmetry. The edges measure between 5 and 7 cm. GEOLEP Photo, L. Cortesi.

The angles that the faces form with each other are thus representative of the unit cell and of the symmetry properties. By measuring the angles of a crystal, it is possible to determine the symmetry system and obtain a first indication on the nature of the mineral. The distance between faces is in principle a multiple of the unit cell. On the macroscopic scale, since the size of these cells is on the order of a nanometer, there can be any distance separating the faces. These distances depend on the size of the crystal and the shape of the space available during its growth.

Crystal growth also depends on external factors such as pressure and temperature conditions as well as the nature of the fluid phase. In rocks, the form and also the perfection of a crystal depend on the degrees of freedom it has for its development (Fig. 5.11). *Automorphic* crystals (mineral with a perfect crystal shape) are formed in a very deformable environment: magma in a magma chamber, gas, water in geodes or open fissures. They have well-developed, regular faces, often of large size. But most of the time, crystallization takes place in an already solid environment. The mineral grows in a space limited by neighboring minerals. This creates *xenomorphic* crystals whose faces are hardly visible. This difference is shown by quartz: its crystals are magnificent in geodes, but altogether ordinary in granite where they are the last minerals to form (Fig. 6.13).

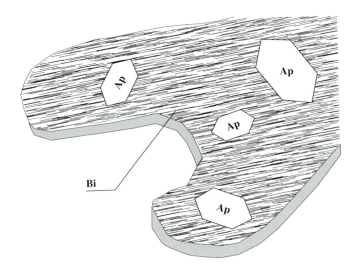

Fig. 5.11 Automorphic and xenomorphic minerals. Example of automorphic crystals of apatite (Ap) contained in a biotite xenomorph (Bi).

Identification of crystals is thus limited to rare automorphic crystals. For xenomorphic crystals, the measurement of angles and determination of the mineral structure requires X-ray diffraction techniques.

Twinning

Twins result from the symmetrical growth of two or more crystals of the same substance. The individuals are linked by an element of symmetry lacking in the single individual. One of these individuals is brought into the orientation of its partner, either by rotation (generally 180°) around a common axis of symmetry, or by reflection by a plane of symmetry or even by inversion around a center of symmetry. The twinning law governs which element of symmetry operates in a given species and establishes the crystallographic orientation of the axes or planes. The result is a bond between the two crystals (contact twins) or interpenetration (Fig. 5.12). Twins may be simple (two or a small number of individuals) or multiple, also called polysynthetic (a large number of individuals). Polysynthetic twinning of plagioclase (Fig. 5.20) is common but it is difficult to observe without a microscope.

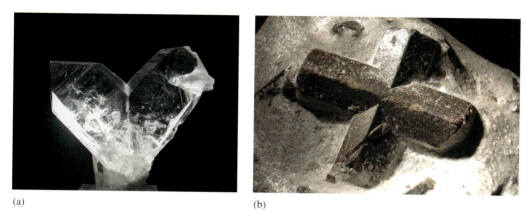

(a)   (b)

Fig. 5.12 Two famous twins: (a) contact twin "Japanese twin" typical of quartz, (b) penetration twin "cruciform twin" of staurolite. Photo Collection of the Cantonal Geologic Museum of Lausanne, N. Meisser.

5.1.3   Cohesion properties

These properties are responsible for the mechanical strength of a mineral. They depend on the mineral structure and the forces linking the atoms.

5.1.3.1   Hardness of minerals

In 1822, for the purpose of mineral identification, Mohs created a relative hardness scale using a series of 10 common minerals (Fig. 5.13). Mohs' scale is still used today.

1	2	3	4	5	6	7	8	9	10
Talc	Gypsum	Calcite	Fluorite	Apatite	Orthoclase	Quartz	Topaz	Corundum	Diamond

Fig. 5.13 Mohs' Hardness Scale.

The principle is simple: a mineral with a higher number scratches a mineral with a lower number when they are rubbed together. Boxes of minerals containing the ten specimens are used for identification. In addition, the following intermediate non-mineral hardnesses are useful: the fingernail (2.5), and a knife blade (slightly greater than 5), glass (5.5) and a steel file (6.5).

The hardness of minerals is also useful in judging the mechanical strength or the abrasiveness of rocks. However, the relationship between the hardness of a mineral and the hardness of a rock is complex. Rocks containing soft minerals are generally non-abrasive and have low mechanical strength. A rock made up of hard minerals will certainly be abrasive, but not necessarily have high mechanical strength. The latter property depends on the strength of the aggregation between the minerals. Let us look at two examples for illustration.

- A sandstone containing quartz grains cemented by quartz is abrasive and hard,
- A sandstone containing quartz grains weakly cemented by calcite will be abrasive but not mechanically strong (Chapt. 10).

The abrasiveness of a rock should be taken into account when choosing tools for working with the rock in order to prevent wear on cutters of tunneling boring machines (TBM) and drilling tools, for example. It is also useful for evaluating the effects of water-borne suspended mineral particles on pumps and turbines: turbidity from fine suspended quartz particles does much more damage than soft materials such as clays.

5.1.3.2   Fracture and cleavage

To study a rock, the geologist uses a hammer to take a sample. The fracture of minerals revealed by the sample can be random or planar and oriented parallel to families of lattice planes (cleavage).

Cleavages are particular fracture planes along which a crystal breaks preferentially when subjected to a mechanical stress. It can be compared to a brick wall damaged by an earthquake: cracks appear at places where the bricks are joined but few bricks are broken.

Four categories of fracture occur, as a function of the number of cleavages (Fig. 5.14).

- *Crystals without cleavage*: these break randomly because there is no lattice density plane that is particularly dominant. Example: Quartz has a ***conchoidal fracture*** that looks like circles around the point of impact (smooth wavy surface) similar to glass.
- *Crystals with one dominant cleavage*: the crystal breaks into fine flakes. The boundaries of the crystal in other directions are random. These are leaf-like minerals. Example: The phyllosilicates, and particularly, the micas.
- *Crystals with two dominant cleavages*: here the minerals are elongated prisms that may form fibers, sometimes even flexible fibers as in the case of asbestos (a variety of amphibole or serpentine).
- *Crystals with three dominant cleavages*: these have the shape of compact prisms, without noticeable elongation. Examples: orthoclase, calcite and halite.

The number of cleavages and the angles formed by the cleavages are excellent aids for identifying minerals with a hand lens or an optical microscope. It is not always easy to know whether the plane of a fracture is a cleavage or a crystal face, but this is not important since they are evidence of the same internal structure.

5.1.3.3   Elasticity and plasticity

Mechanical stress applied to a mineral will produce deformation ranging from elasticity (reversible deformation) to rupture with an intermediate plastic stage (irreversible deformation). Depending on the type of bond, minerals will respond in various ways when subjected to stress. The study of these mechanical properties is very important, in particular for the deformation of rocks (Chapt. 12).

5.1.3.4   Tenacity

Mineralogists use specific terms to describe the behavior of minerals subjected to various mechanical tests: fragile or brittle (easily broken or pulverizable), malleable (easily pounded

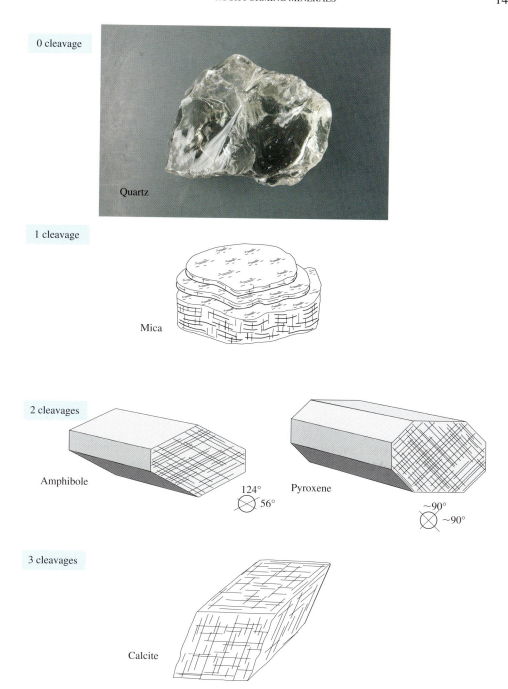

Fig. 5.14 Geometry of broken crystals as a function of the number of cleavages. Photo Collection of the Cantonal Geologic Museum of Lausanne, N. Meisser.

into sheets), sectile (easily sectioned with a knife), ductile (can be drawn into wire), and flexible (pliable, but without reverting to their initial shape).

5.1.4 Optical properties of minerals

The atomic structure or chemical composition of a mineral gives it specific optical properties that can be used for field identification. Characteristics visible to the naked eye or through the hand lens such as luster, transparency and color are used for identification.

5.1.4.1 Macroscopic characteristics (observable to the naked eye)

Reflectivity, luster

The reflectivity of a mineral is the ratio of reflected flux to incident flux. It directly influences the luster of a mineral. Metallic luster (as shown by the sulfides) is distinguished from nonmetallic luster. In the non-metallic category there is vitreous luster (quartz, feldspars), adamantine (diamond), resinous (sulfur, sphalerite), greasy (chalcedony), silky (gypsum), pearly (micas), porcelaneous and earthy (no luster because it is an aggregate of small variously-oriented grains).

Transparency

A mineral is considered transparent if the details of an object can be seen through the mineral. The absorption of a certain wavelength of incident white light causes various colors to appear. A mineral is called translucent if only a portion of the light passes through it (this is the case of the majority of rock forming minerals, particularly the silicates). If the light is totally absorbed or reflected from the surface of the crystal, it is called opaque (oxides and metallic sulfides, for the most part).

Color

The color of a crystal seen by transparency or by reflection is due to the attenuation of certain wavelengths of incident white light. The complex optical phenomenon of interaction between light and crystals can be summarized by the following types of coloration:

- Characteristic color: the color is directly associated with the presence of a coloring element in the chemical composition (Fe^{2+} generally causes a green or greenish-yellow color, whereas Fe^{3+} gives a reddish color), the crystal structure of the mineral, inclusions (very fine inclusions of rutile give quartz a pink or blue color) and physical phenomena – reflection, refraction, light interferences (for example, asterism is the reflection of a star-shaped arrangement of rutile inclusions in corundum) (Fig. 5.15).
- Coloration due to anomalies: some colorless minerals can occur as colored crystals. This coloration can be the result of the presence of imperfections in the crystal lattice, for example by replacement of atoms of different valence that causes an electronic disequilibrium (replacement of Si by Al in smoky quartz, by Fe in amethyst).
- Coloration due to impurities (trace chemical elements). This is the case of quartz, which is normally colorless but can also have various pastel colors: mauve (iron), pink (Ti^{4+}), red (hematite), etc.

In mineral identification, it is important to distinguish between a mineral that has its own characteristic color from one whose color results from secondary phenomena. For this reason, mineral identification uses transmitted color (characteristic color) rather than reflected color. The color is obtained by scraping the mineral on a non-glazed porcelain plate. The color of the powder is the characteristic color of the mineral; it is not influenced by secondary optical phenomena. In this way, black crystals of hematite can be distinguished from black crystals of

(a)   (b)

Fig. 5.15 Mineral colors due to coloring agents: (a) Manganese causes the coloration in rhodochrosite; (b) Uranium causes the yellow color of autunite. Photo Collection of the Cantonal Geological Museum of Lausanne, N. Meisser.

magnetite, for example: the former gives a red streak (from which it gets its name) while magnetite's streak is black.

5.1.4.2 Examination of thin sections

Optical properties difficult to determine with the naked eye or hand lens can be rigorously examined and measured under a polarized light microscope to resolve the identity of a mineral.

To identify a mineral under a polarized light microscope, a ***thin section*** is prepared. This involves gluing a small cube of rock to a glass plate and reducing its thickness to 30 μm (which is why it is called a thin section). An animation on the DVD shows minerals and rocks as they appear under polarized light microscope. The examination of the thin section under an optical microscope, with a light source under the thin section, allows opaque minerals to be distinguished from translucent minerals.

The translucent category can be subdivided into two groups:

- Optically isotropic minerals: These are minerals that are non-crystalline (opal for example), or crystals of the cubic system (halite, fluorite, etc. in which light is propagated in all directions with the same velocity; they have only one ***index of refraction*** $n = c/v$ (c = velocity of light in a void, v = velocity of light in the mineral); amorphous material in volcanic rocks behaves in the same way.

- Optically anisotropic minerals: these are minerals that belong to the other crystal systems; the light is propagated at different velocities depending on the angle of incidence; they have infinite indices of refraction ranging between the extreme values of n_o (ordinary ray perpendicular to optical axis) and n_e (extraordinary ray parallel to the optical axis). $n_o - n_e$ is called **birefringence** and is characteristic of the mineral.
- The majority of minerals are anisotropic to various degrees. Calcite is extremely anisotropic (Fig. 5.16), quartz is less so.

Fig. 5.16 Display of calcite's high birefringence. The object seen through the rhombohedron is doubled. One of the images comes from refraction according to the index of refraction n_o (1.658 maximum value) and the other one comes from n_e. (1.486 minimum value); thus calcite has a negative birefringence of -0.172. Photo Collection of the Cantonal Museum of Geology, Lausanne.

[29, 105, 172]   The geologist studies thin sections under the polarizing microscope (Fig. 5.17), which allows him to measure the anisotropy of the index of refraction. Many other characteristics can be determined by using this valuable tool. This is the meticulous world of optical mineralogy.

PROBLEM 5.1

Question 1

Determine the elements of symmetry of a hexagonal lattice. In particular, find the different values of n for the n-fold axis of symmetry. Draw them.

Question 2

Same question for the monoclinic lattice.

Look at the solution on the DVD

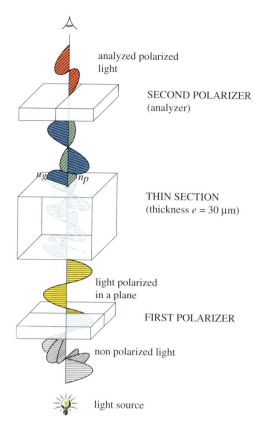

analyzed polarized
light

SECOND POLARIZER
(analyzer)

ng np

THIN SECTION
(thickness e = 30 μm)

light polarized
in a plane

FIRST POLARIZER

non polarized light

light source

Fig. 5.17 Trajectory of light waves through a polarizing microscope with the polarizers in perpendicular position. If the mineral is anisotropic, there is no extinction: the observer perceives a color that depends on the birefringence of the material.

[24]

5.2 Geochemistry

Before examining the principal minerals from a geochemical point of view, we shall define several essential notions linking the physical structure of a mineral to its constituent elements.

Substitution

Substitutions in a mineral are governed by strict rules regarding ionic radius and the electronegativity of the substituting ion. The principal atom of a structure can be replaced by another of similar size and electrochemical properties without a major change in the structure. For example the replacement of silica in aluminum silicates changes the ionic charge of the silicate group and thus allows the introduction of a monovalent cation.

This replacement may be accidental, for example, if it occurs only as rare replacements of the crystal structure. The "foreign" atom does not change the mineral properties of a mineral, except perhaps the color. For example, corundum (Al_2O_3) is colorless in its pure state. If Cr^{3+} replaces some aluminum atoms, the mineral becomes red (ruby); it would be blue as a result of substitution by Fe^{2+} or Ti^{4+} (sapphire).

If the replacement of one atom by another of a similar ionic radius becomes systematic, *solid solutions* occur. In nature there are continuous solid solutions (all proportions between the two extremes are possible), as in the plagioclase series. Other discontinuous series (only certain compositions are possible) occur, such as the alkali feldspars (§ 5.2.1.2). In the mineral formula, the two atoms are written in brackets.

Polymorphism

Some mineral substances can exist as two crystal forms that are stable at pressure and temperature conditions that are unique to them. This is the case of carbon, C, which can occur as diamond (cubic system, hardness of 10) or graphite (hexagonal system, hardness of 1). This drastic difference is the result of their structures having formed under completely different crystallization conditions (pressure and temperature). The diamond crystallized under very high-pressure conditions, whereas graphite is formed under more moderate pressure conditions. The boundary between these two forms of carbon occurs at a depth of about 50 km.

Let us now look at the minerals or principal groups of minerals that are fundamental to rock formation. We will classify them by geochemical families. In each case, we present the chemical composition (sometimes simplified), the most common appearance, physical properties, the environments where they are found in nature, and their technical uses.

5.2.1 Silicates

This group includes minerals made up of the tetrahedral structural element SiO_4: four atoms of oxygen surround a silicon center. The silicates are the principal class of minerals that make up the mantle and the lithosphere, in terms of their abundance and diversity of species (silicon is the most abundant metal on Earth – about 30% by weight of the Earth's crust). Many are aluminosilicates linked to many possible chemical elements. The latter can be major, minor or trace elements. Thus the chemical formula of a mineral can be very complex.

Silicates form in the endogenic cycle (within the Earth) and the exogenic (on the surface) cycle. Under endogenic conditions, they are essential and accessory constituents of metamorphic and plutonic rocks; in the latter case, they are formed during the orthomagmatic stage (first stage in the crystallization of magma) and the pegmatitic stage (second stage when residual magma enriched in dissolved volatiles becomes very fluid often resulting in the formation of very large crystals). At the surface, they occur as weathering products, particularly clays in certain metallic ores.

Economically, they are used for their own qualities (precious stones, geomaterials) or as raw materials for industrial products (porcelain, glass, refractory bricks, cement, etc.).

In this large geochemical group, there are sub-groups based on the way in the SiO_4 tetrahedra are linked. Some structures are composed of isolated tetrahedral, other of chains, other in layers.

5.2.1.1 Quartz: SiO_2

Some classifications place quartz in the oxide group because of its chemical formula. Others, including the present one, classify it in the silicate group because of the typical tetrahedral structure, in spite of the absence of a characteristic SiO_4^{4-} chemical group.

Quartz is a major mineral in crystalline and detrital rocks. It is the form of silica that is stable over the long term. It has various forms, blocky pyramidal hexagonal crystals

Fig. 5.18 Needle quartz. Photo from the collection of the Cantonal Museum of Geology, Lausanne, N. Meisser.

("rock crystal", Fig. 5.18) but most of the time, it has no easily identified crystal shape because it is the last mineral to form (See fractional crystallization, Chapt. 6).

It is an essential mineral in silica-rich plutonic rocks such as granite (Chapt. 6). It can be recognized by its variable shape, the absence of cleavage and its conchoidal fracture. Transparent gray in color, rarely colored, it has a hardness of 7 on the Mohs scale. It is chemically inert. Its density is $2.65 \cdot 10^3 \, \text{kg/m}^3$.

Quartz is the most common mineral in the Earth's crust (12% by volume), the environment in which it is most stable. It is found in magmatic rocks and in the sedimentary environment where its high mechanical strength and chemical stability make it the principal component of sands and silts (Chapt. 8). It can also be formed by the diagenesis of the skeletons of unicellular organisms (radiolarians, for example) and form rocks called radiolarites (Chapt. 9 and 10). It can be mobilized by hydrothermal circulation to form whitish veins in the fissures of rocks.

Quartz sand is melted to manufacture glass and insulation materials (rock "wool"). It is the raw material of fiber optics. In the solid state, it is used as an abrasive and, in the electronics industry, it is used for its piezoelectric properties. Artificial quartz is used to replace natural crystals in numerous technological applications.

Quartz is not the only mineral composed of silica: Opal is a practically amorphous variety of hydrated silica (sometimes crystallization as tiny spheres); chalcedony is slightly hydrated but here the silica is already in the shape of microcrystals of quartz (§ 10.3.2). Rocks containing these varieties of silica react with alkaline compounds of cement (alkali reaction). This leads to a loss of adhesion of the binder on the aggregate which decreases the strength of the concrete. Mineralogical examination of the aggregate can be used to detect this phenomenon.

5.2.1.2 Feldspar family: $(K,Na)AlSi_3O_8$; $CaAl_2Si_2O_8$

The feldspars are the most important family of rock-forming minerals (60% of magmatic rocks). They are a simple chemical group composed of a radical $AlSi_3O_8^-$ for the potassium and sodium species and $AlSi_3O_8^{2-}$ for the calcic one.

The feldspars are commonly represented on a ternary diagram whose extremities are the K, Na and Ca species: orthoclase, albite and anorthite, respectively (Fig. 5.19). Under ordinary conditions of magmatic crystallization, not all theoretically possible solid solutions actually occur. There is solid solution between the Na and Ca poles; these are the plagioclases. Another solid solution exists between Na and K but it is discontinuous: these are alkali feldspars. The center of the triangle and the Ca-K side does not exist as natural crystals.

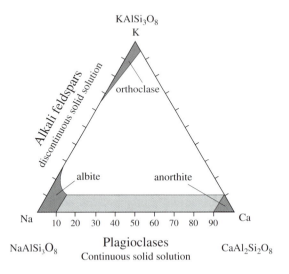

Fig. 5.19 Ternary diagram of the feldspar family at a temperature of 650°C and pressure of 100 MPa.

Macroscopically, feldspar crystals occur as short, blocky prismatic crystals with two or three cleavages, sometimes orthogonal (from which the name of orthoclase). They have a whitish color, perhaps tinted pink or green, due to the degree of alteration of iron-bearing mineral inclusions. The luster of the minerals of this family is vitreous to pearly. They have hardnesses of 5 to 6 on the Mohs scale and densities between $2.5 \cdot 10^3 \, \mathrm{kg/m^3}$ and $2.7 \cdot 10^3 \, \mathrm{kg/m^3}$; high densities correspond to the calcic varieties. The feldspars, with quartz, are the light-colored patches in magmatic rocks (for example, granite) and metamorphic rocks (for example, gneiss).

Twins are common and varied; they may be simple or multiple. The simple twin of alkali feldspars, called *Carlsbad*, is often visible with a hand lens or even to the naked eye (Fig. 5.20). Under the microscope, the multiple twins of plagioclase, called *polysynthetic*, result from adjacent growth of tabular crystals.

Feldspars are used in the manufacture of ceramics: the addition of feldspars reduces shrinkage and frost sensitivity; it increases resistance to thermal shock.

5.2.1.3 Ferromagnesian silicates

This category includes silicates that contain iron or magnesium as the principal cation. We will discuss only those necessary for an understanding of magmatism and rock formation. All of the iron-bearing ones are dark colored; the color varies with the degree of oxidation. They are generally hard, except for the phyllosilicates.

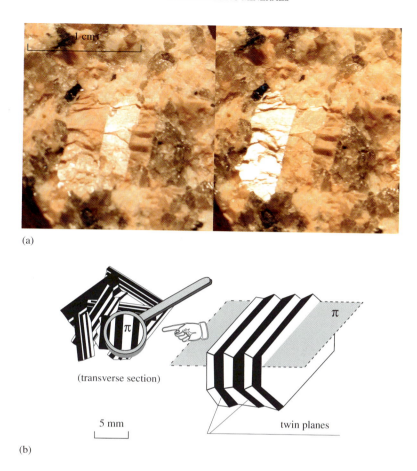

(a)

(b)

Fig. 5.20 (a) Alkali feldspar with Carlsbad twinning. GEOLEP Photo (b) Diagram of a polysynthetic twin in plagioclase as seen in cross section under the microscope.

Olivine: (Mg, Fe)$_2$SiO$_4$

This is the term for the continuous solid solution in all proportions between a purely magnesian and iron-bearing composition. The mineral often occurs as a mass or as rounded grains, more rarely in prismatic shape. Greenish yellow to olive green (from which the name), olivine is translucent and has vitreous luster. Its density $(3.27 \cdot 10^3 \text{kg/m}^3 – 4.37 \cdot 10^3 \text{kg/m}^3)$ increases with an increase in iron content. Its hardness is 6.5 to 7, and it has conchoidal fracture.

Olivine is a common mineral in silica-poor magmatic rocks (mafic to ultramafic) (Chapt. 6); it is the major constituent in plutonic rocks called peridotites; this name comes from peridot, a *gem* (precious and semiprecious stone) variety of olivine. It is also found as a crystalline inclusion (mineral mass) in basaltic lavas. Apart from its value as a gem, olivine is used as a refractive aggregate.

Pyroxenes: example of augite (Ca,Na)(Mg,Fe,Al)(Si,Al)$_2$O$_6$

The structure is simple chains of tetrahedra (Fig. 5.21). It includes various solid solutions as the chemical formula above suggests. The pyroxenes form stocky prisms, brown to dark green

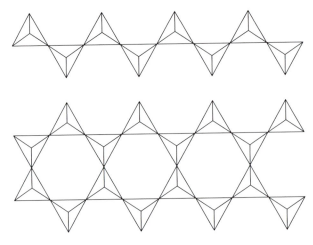

Fig. 5.21 Structure of pyroxenes (simple chain of tetrahedra) and of amphiboles (twinned chain).

or even black in color, with vitreous luster. Its density depends on its composition, ranging from $3.2 \cdot 10^3 \, kg/m^3$ to $3.6 \cdot 10^3 \, kg/m^3$. Its hardness ranges from 5 to 6. Pyroxenes have two cleavages at right angles that can be easily seen by the naked eye.

Pyroxenes are common minerals in mafic rocks and high temperature metamorphic rocks.

Amphiboles: example of the hornblende group $(Ca, Na)_{2-3}(Mg, Fe, Al)_5Si_6(Si, Al)_2O_{22}(OH)_2$

Members of the same family as the pyroxenes, amphiboles are composed of tetrahedra organized in bands of twinned chains (Fig. 5.21). All amphiboles are ferromagnesian, but they also contain various amounts of other elements, principally calcium and sodium. They are similar in appearance and physical properties to pyroxenes (hardness ranging from 5 to 6 and density between $3 \cdot 10^3 \, kg/m^3$ and $3.4 \cdot 10^3 \, kg/m^3$); luster can be vitreous to silky. Amphiboles have two characteristic cleavages at an angle of 124° which are easily visible.

These are common minerals in magmatic and metamorphic rocks. Amphiboles in metamorphic rocks form elongate, even fibrous, crystals often arranged in sheafs.

The ferromagnesian phyllosilicates

Phyllosilicates, in the broad sense, are one of the most important and complex mineral families. They are composed of a stack of tetrahedral (T) and octahedral (O) layers, extended to infinity in the lateral direction (for example, muscovite, Fig. 5.24). This repeated stack of layers from an initial motif produces minerals with very clear perfect cleavage. Phyllosilicates occur as very fine and flexible sheets and lamellae.

The ferromagnesian phyllosilicates, often dark-colored, get their color from certain coloring elements, green (Fe^{2+} and Ni), brown or black and violet (Cr); however, their streak is colorless. They have vitreous luster, low density ($2.76 \cdot 10^3 \, kg/m^3$–$3.2 \cdot 10^3 \, kg/m^3$) and a hardness between 2 and 3.

Ferromagnesian phyllosilicates are omnipresent. They are found in particular in silica-rich magmatic rocks, in metamorphic rocks (minerals such as serpentine, asbestos, chlorite, etc.) or in sediments (exogenic cycles).

Aside from the Ni-phyllosilicates, their economic importance as ores is limited. Some varieties of serpentines are used as asbestos because of their refractive and insulating properties (see below).

The family of phyllosilicates includes three groups of minerals: micas, serpentines and clays. Only the first two contain ferromagnesian minerals that are discussed below.

Ferromagnesian mica group

Biotite: $K(Mg, Fe)_3(AlSi_3O_{10})(OH)_2$

Biotite (or "black mica") is a ferromagnesian mica, black or smoky brown in color, with vitreous luster that is pearly on the cleavage (Fig. 5.22); its hardness ranges from 2.5 to 3 and its density from $2.8 \cdot 10^3 \, kg/m^3$ to $3.2 \cdot 10^3 \, kg/m^3$.

Biotite is found in many magmatic rocks, particularly granites, as well as in metamorphic rocks such as gneiss and micaschists. It is also a mineral present in detrital sedimentary rocks.

Chlorites: $(Mg, Fe)_3(Si, Al)_4 O_{10}(OH)_2 \cdot (Mg, Fe)_3(OH)_6$

Chlorites are a complex family of hydrated ferromagnesian phyllosilicates that includes Al, Fe and Mg-rich representatives. Here we will use the general formula and common characteristics for these diverse minerals.

Chlorites often occur as lamellar aggregates or masses of small particles. With hardness between 2 and 2.5 and density between $2.6 \cdot 10^3 \, kg/m^3$ and $3.3 \cdot 10^3 \, kg/m^3$, they are recognized by their color, which is various shades of green. They are translucent with pearly luster. Their cleavage is perfect.

Fig. 5.22 Large biotite crystal. GEOLEP Photo, G. Grosjean.

Chlorite is a mineral formed by hydrothermal alteration of other ferromagnesian silicates (biotite, pyroxenes, amphiboles). It is an essential constituent of certain low temperature and pressure metamorphic rocks (chlorite schists).

Serpentine group

We include in this group the principal serpentine-forming minerals (antigorite and chrysotile) and talc, a geochemically similar mineral. They result from primary weathering of magnesian silicates (olivine, pyroxenes and amphiboles) and they form a particular rock called serpentinite.

Antigorite-chrysotile: $Mg_3Si_2O_5(OH)_4$

These two minerals do not exist as well-developed crystals. Antigorite is characteristically massive and finely granular, while chrysotile is fibrous (Fig. 5.23). Their hardness ranges from about 3 to 5 and density from $2.5 \cdot 10^3\,kg/m^3$ to $2.6 \cdot 10^3\,kg/m^3$. The massive varieties of serpentine minerals have waxy luster and the fibrous varieties have silky luster. The dominant color of antigorite is green, from light to dark, with bluish nuances; chrysotile is white.

Fig. 5.23 Common appearance of chrysotile. GEOLEP Photo, G. Grosjean.

When chrysotile fibers become excessively long (length/diameter ratio > 100), the mineral is called **white asbestos**. The properties of the asbestos minerals, including their nonflammability and resistance to rot, flexibility, resistance to chemical products and high tensile strength means that they are very sought-after minerals for the manufacture of asbestos cement products (panels, walls and ceiling coatings, potable water pipes, etc.) and fireproof materials (brake shoes, flame retardant clothing, etc.). However, they have been increasingly abandoned because their dust creates a risk of respiratory cancer (asbestos disease). Many buildings that were built with asbestos cement panels have required remediation. Today, the only type of asbestos still being exploited is chrysotile. Its fibers are encapsulated in a cement or resin matrix that makes the material dense and non-friable, and thereby should not present a real health risk. However, the toxicity and a possible interdiction of chrysotile asbestos is a hotly debated topic on the political scene.

Talc: $Mg_3Si_4O_{10}(OH)_2$

Occurring as foliated aggregates or flexible pseudohexagonal lamellae, talc has a hardness of 1 on the Mohs scale. It is soapy to the touch and has a pearly luster. It is silvery white or apple green in color and has a density ranging from $2.7 \cdot 10^3\,kg/m^3$ to $2.8 \cdot 10^3\,kg/m^3$.

It forms by alteration of magnesian silicates and is found in slightly metamorphosed rocks, especially in serpentinites. It is used in numerous industries, notably for lubricants, paper and cosmetics.

5.2.1.4 Non-ferromagnesian phyllosilicates

There are numerous other types of non-ferromagnesian phyllosilicates. We will focus on muscovite because of its abundance and properties and the clays because of their importance for the engineer.

Muscovite: $KAl_2(AlSi_3O_{10})(OH)_2$

This aluminous mica (or "white mica") occurs as sheets with pseudo-hexagonal faces. Its internal structure is a combination of an octahedral layer sandwiched between two tetrahedral layers (Fig. 5.24). Its hardness is low, between 2 and 2.5; its density ranges from $2.76 \cdot 10^3 \, kg/m^3$ to $2.88 \cdot 10^3 \, kg/m^3$. Its luster is vitreous. The mineral is colorless and transparent if it occurs as thin sheets; it can have a whitish or yellowish tinge when it occurs in a thick stack.

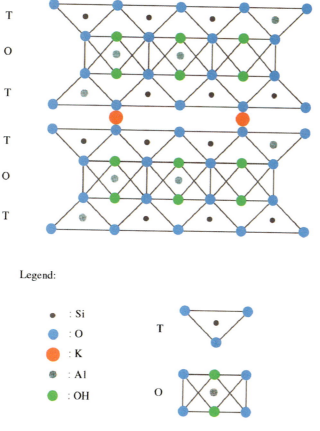

Fig. 5.24 The internal structure of muscovite is layers of tetrahedra and octahedra, typical of phyllosilicates (from [149]).

This mineral is common in silica-rich magmatic rocks and in metamorphic rocks such as micaschists. It forms crystals several meters in diameter and several decimeters thick in *pegmatite* veins (coarse-grained dikes that form in granitic massifs).

Because of its transparency and resistance to heat, white mica is used as an electrical and thermal insulator. Muscovite owes its name to its use as a glass substitute in ancient Russia.

The clay group

The name "clay" is used to designate both a family of minerals and unconsolidated sediments composed essentially of these minerals. The term is also used in the geotechnical field where it designates detrital grains smaller than 0.002 mm (Fig. 8.1). Mineralogical analyses show that many particles below this size are not clay minerals and that there exist clays larger than this size.

Clays are hydrated phyllosilicates that occur as very small plate-like crystals (Fig. 5.25), often hexagonal. They can be found in argillaceous sediments, in residues of altered rocks, in soils, etc.

253 (2-6)   00001   10μm

Fig. 5.25 Clay as it appears under an electron microscope. Upper Cretaceous "Red Beds", Swiss Prealps. University of Lausanne, photo M. Jaboyedoff.

The sheets are electrically charged anionic groups. Because of the charge, they have a tendency to repel each other, and as a result they have a very weak structure. Consequently, they have high specific surfaces, that is, a high ratio of the particle surface to mass. For kaolinite this ratio is $20 \, m^2/g$; for illite, $115 \, m^2/g$; and for montmorillonite $750 \, m^2/g$. They attract volunteer cations that are present in the interstitial spaces. This promotes their function as ion exchangers, which is fundamental in the soil geochemistry.

The observation and identification of these minerals is done primarily by electron microscope and X-ray diffraction. Mineral species are often identified, but there are also a series of structural hybrids called ***interstratifieds*** that will not be discussed here.

Illite: (K, H_3O)(Al, Mg, Fe)$_2$(Si, Al)$_4O_{10}$[(OH)$_2$, H_2O]

This is the most common clay mineral in temperate climates. It comes from the alteration of feldspars and, more rarely, micas. It dominates the fine fraction of detrital sedimentary rocks such as claystones.

Technically it forms aggregates with moderate plasticity that are not very susceptible to swelling by water absorption (Chapt. 13). Illite is used in the terra cotta industry, primarily in the area of construction materials (bricks, tiles, etc.).

Kaolinite: $Al_2Si_2O_5(OH)_4$

Kaolinite comes from the alteration of feldspar-rich magmatic rocks in tropical climates, or by ***hydrothermalism*** (underground circulation of high temperature fluids). Kaolinite has low plasticity. When it is pure, it occurs as a white powder.

It is not sensitive to modifications in the environment (dehydration, water chemistry). It does not swell. It is an essential mineral in porcelain manufacturing (because of its refractory qualities and its whiteness) and technical ceramics. It is also used in the paper industry (white coating to the surface of sheets that improves their surface properties).

Smectites: example of Montmorillonite: Na(Al$_3$, Mg) [Si$_4$, O$_{10}$(OH)$_2$]$_2$ · nH$_2$O

Smectites are a group of clay minerals whose principal property is the capacity to absorb enormous quantities of water molecules into their structure and to swell as a result. Montmorillonite is the main variety. Layers that contain montmorillonite cause significant deformation of building foundations. They have a very high plasticity.

Bentonite is fundamentally a rock composed of montmorillonite as a result of the weathering of volcanic ash; in the technical arena, bentonite has become synonymous with montmorillonite. It is exploited in numerous locales for all sorts of industrial uses, either in its natural state, or combined with sodium to increase its plasticity. Here are several uses:

- Drilling mud or for the slurry trench walls, making it possible to stabilize the walls during excavation without stabilization,
- Plasticizing agent in concrete,
- Addition to cement for injection grout,
- Confinement barrier near contaminated sites and storage of toxic wastes, particularly as a component in certain geotextiles.
- Domestic uses (cosmetics, cat litter).

5.2.1.5 Garnet group (Ca, Mn, Fe, Mg)$_3$, $Al_2Si_3O_{12}$

Among the many more non-ferromagnesian silicates, we will discuss only garnets, which are easily identifiable in common rocks. The garnet group contains many minerals of various colors (colorless, yellow, brown, red, green or black), with vitreous luster, compact shape (Fig. 5.26), hardness between 6.5 and 7.5, and density of $3.5 \cdot 10^3 \, kg/m^3$ to $4.3 \cdot 10^3 \, kg/m^3$. They play an important role in the study of metamorphism; they occur as secondary minerals in metamorphic rocks, more rarely in magmatic rocks. Well-crystallized garnets are used as gems.

Fig. 5.26 A mass of garnets. Photo from the collection of the Cantonal Museum of Geology, Lausanne, N. Meisser.

5.2.2 Carbonates

Two carbonates, calcite and dolomite, play an essential role in sedimentary rocks (limestone and dolomite) and metamorphic rocks (marble). They are economically important as Fe, Mn, Cu, Zn, and Pb ores, construction materials (dimension stone, lime, cement, etc.), and agriculture (acidic soil neutralization, Chapt. 14).

5.2.2.1 Calcite: $CaCO_3$

Calcite is the stable form of calcium carbonate. Calcite forms white or colorless translucent crystals when it is pure. It can also be beige or pink in color. Its luster is vitreous to pearly and can be iridescent. It has low hardness (3 on the Mohs scale) and has a density of $2.7 \cdot 10^3 \, kg/m^3$. It breaks along three perfect cleavages to form rhombohedra (see Fig. 5.16). It reacts with strong and even dilute acids.

Calcite is mainly a mineral of sedimentary origin. It forms gradually over time by the transformation of skeletons of mollusks and other marine organisms that were originally composed of *aragonite* (an orthorhombic polymorph of calcite, of biogenic origin). In its microcrystalline form, it is the main component of limestones. As CO_2 is released from water rich in calcium bicarbonate, chemical precipitates are formed, for example, calcite concretions (*stalactites* and *stalagmites*, *tufa*). Similar solutions can fill geodes and fissures with calcite veins (crystals as rhombohedra or elongate prisms) and cement the majority of detrital rocks (for example, alpine Tertiary sandstones). The calcite contained in rocks tends to dissolve in slightly acidic waters (Chapt. 13.) *Marble* is the result of the transformation of limestone by metamorphism: the fine particles of calcite recrystallize as large grains (Chapt. 11); it is exploited for ornamental stone.

Calcite is the essential mineral for the production of *lime* (calcination product of limestone). Mixed with clays, this calcination produces *Portland cement* (cement that can be used under water).

5.2.2.2 Dolomite CaMg(CO$_3$)$_2$

H.B. de Saussure named dolomite after D. Dolomieux, professor at the University of Grenoble, for his research on this mineral. Dolomite is the intermediate member of the solid solution between calcium carbonate and magnesium carbonate. Dolomite occurs as rhombohedral crystals. It is commonly found as a monomineralic rock either as a compact mass of microcrystals resembling fine limestone or as a poorly cohesive crystalline assemblage (saccharoidal dolomite). Like calcite, this carbonate has perfect rhombohedral cleavage. Colorless or whitish in color with a hint of pink or gray, its hardness ranges from 3.5 to 4 and it has a density of $2.85 \cdot 10^3$ kg/m^3. Its luster is vitreous to pearly. In contrast to calcite, it reacts almost imperceptibly with dilute strong acids. But when it is reduced to powder, it reacts visibly to acid. Rock composed essentially of dolomite is called dolomite. Mixed with calcite, it produces dolomitic limestone. These rocks can make up entire rocky massifs, the most famous of which is the Dolomites on the border between Italy and Austria. Dolomitic rocks are a bit less susceptible to dissolution by water than limestones (Chaps. 7 and 13).

Hydrothermal veins of dolomite are often associated with lead and zinc mineralization. Dolomites are exploited for construction or as rip-rap.

5.2.3 Sulfates

These minerals are not widespread on the Earth (they make up <5% of the Earth's crust). The two forms (hydrated and anhydrous) of calcium sulfate, gypsum and anhydrite, are of particular interest to the engineer.

5.2.3.1 Gypsum: CaSO$_4$ · 2H$_2$O

Gypsum is a transparent mineral that is usually colorless or white (also gray or yellow), tabular or elongate in shape, with vitreous or pearly luster. When elongated, it occurs as long fibers with two very visible principal cleavages. The tabular varieties can develop characteristic sword twinning (Fig. 5.27). Very soft (hardness of 2 on the Mohs scale) and with a density of $3.32 \cdot 10^3$ kg/m^3, it is the most soluble mineral after halite ($\sim$3g/l).

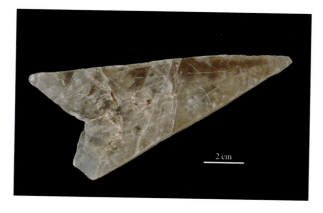

Fig. 5.27 Sword-like twinning typical of gypsum. GEOLEP photo, G. Grosjean.

Gypsum can also form a monomineralic hydrochemical sedimentary rock that precipitates as sea water evaporates. It is also formed by the surface hydration of anhydrite. One of the best-known forms of gypsum is the "**sand rose**" where aggregates of microcrystalline gypsum surround sand grains (Fig. 8.76).

Rocks formed essentially of gypsum cause numerous foundation problems because of their solubility (collapse cavities). **Gypsiferous** water attacks concrete (Chapts. 10 and 13).

Gypsum is widely exploited for **plaster** manufacturing: heated to 165°C, it loses some of its water of hydration; when it is mixed with water, it is rehydrated and progressively recrystallizes as gypsum. It is also commonly added to some cements as retarder of set. **Alabaster** is a variety of microcrystalline gypsum, of varied colors, that is used as an ornamental stone (vases, lamps).

5.2.3.2   Anhydrite: $CaSO_4$

Anhydrite is most often found as granular crystalline aggregates, rarely as tabular or stubby prismatic crystals. In well-crystallized specimens, its three orthorhombic cleavages can be seen. It is transparent and colorless (rarely white or reddish), with vitreous, greasy or pearly luster. Its hardness, 3 to 3.5, and its density of 3 are higher than gypsum.

Anhydrite is often associated with gypsum in sedimentary deposits. Anhydrite forms from the loss of a water molecule from gypsum as pressure and temperatures increase after gypsum is buried. The principal characteristic of anhydrite is that it can be retransformed into gypsum by hydration, accompanied by a significant increase in volume. This can create problems in underground constructions (Chapt. 13).

It is used primarily to retard the set of cement.

5.2.4   Phosphates

This geochemical group is not abundant. We will discuss only the most common phosphate, apatite.

Apatite: $Ca_5(PO_4)_3(Fe,Cl,OH)$
This calcium phosphate occurs as very elongate or stubby hexagonal prisms (Fig. 5.28).

Fig. 5.28 Apatite crystals. Photo from the Collection of the Cantonal Museum of Geology, Lausanne, N. Meisser.

This mineral has a vitreous luster. It is faintly colored in tones of yellow-green or pink-purple. Its hardness is 5 on the Mohs scale; its density is $3.2 \cdot 10^3 \, kg/m^3$. Its principal cleavage is poorly developed. Apatite is the most important ore of phosphorus; it is widely disseminated in all types of rocks, as an accessory mineral and in oxidation zones of some sulfide deposits.

Rocks rich in phosphates, mainly the sedimentary rock phosphorite which contains carbonate-rich varieties of apatite, are extensively exploited in Morocco and the United States. These deposits are accumulations of lithified animal excrement. They are used as fertilizers and for the laundry industry; this last use is being phased out by many countries because it contributes to eutrophication of lakes and rivers.

5.2.5 Halides

The principal representative of this family is "rock salt" or halite.

5.2.5.1 Halite: NaCl

Halite is the mineralogical name of table salt. It forms cubic crystals, colorless or white. In nature, it is also pink, gray, or purple because of inclusions of algae, organic matter or radioactive elements. It has vitreous luster and perfect cubic cleavage (Fig. 5.29).

Fig. 5.29 Customary appearance of halite. GEOLEP Photo, G. Grosjean.

Halite is a mineral with low hardness (2.5). Its density is $2.16 \cdot 10^3 \, kg/m^3$, very low, which explains why salt-bearing rocks form salt domes. It is extremely soluble in water (360 g/l). Its salty taste is the easiest way to identify it. Halite is formed by the evaporation of sea water in restricted lagoons or in closed seas with high evaporation rates (Dead Sea, Aral Sea, saline lakes of Turkey and Ethiopia). In arid regions, it also forms by capillary action: saline groundwater rises to the surface and forms a crust: this is the salinization that makes some irrigated regions unproductive.

The importance of salt in the development of humanity begins with its use in food. Salt commerce opened up the first long distance communication routes. Today salt is exploited in numerous countries around the world, either as geologic salt, or present day salterns. In the mines of Bex (Switzerland), for example, salt is extracted from deep mines dug into salt-bearing rocks. These were formed during the Triassic epoch by evaporation of ancient arms of the Tethys sea and then intensely folded during Alpine orogenesis. Salt is used as a raw material for the chemical industry, particularly for the manufacture of chlorine. It is widely used as a de-icing agent for roads and water demineralization on ion exchange columns.

5.2.5.2   Fluorite CaF$_2$

Fluorite forms beautiful cubic or octahedral crystals with perfect cubic cleavage (Fig. 5.30). It has a value of 4 on the hardness scale, and a density of $3.2 \cdot 10^3 \, kg/m^3$. Fluorite is transparent to translucent with vitreous luster. Its most common color is purple, but it can also be yellow, green or blue. It often has double coloring: green by reflection and blue by transparence. It occurs in hydrothermal veins or more rarely in limestones (important deposits in the USA). It is used for the manufacture of hydrofluoric acid and as a flux in metallurgy.

Another fluoride, **_cryolite_** (Na_3AlF_6), should be mentioned for its important environmental impact. It is one of the principal minerals used in electrolytic reduction for aluminum production. In the 1970s, fluorine emanating from aluminum factories had a catastrophic effect on alpine flora, decimating an extraordinary amount of woodland pines. Filters were subsequently installed on the factories to prevent this damage.

Fig. 5.30 Specimen of fluorite on smoky quartz. Photo Collection of the Cantonal Geological Museum of Lausanne, N. Meisser.

5.2.6 Sulfides

Sulfides are the most stable metal compounds in a reducing environment. These minerals are found in the endogenic cycle (especially volcanic eruptions) and in the exogenic cycle. They are also found in meteorites. In spite of their low abundance (0.15% of the Earth's crust) the sulfides have a significant economic importance for their role in metal deposits, such as Fe, Cu, Zn, Pb, etc. (and even Ag). Here we will limit ourselves to the most abundant sulfide, iron sulfide.

Pyrite: FeS_2

Pyrite occurs as cubic crystals, as prisms with cubic symmetry and pentagonal faces (Fig. 5.31) or as fine granular masses in nodules. The faces are striated, the luster is metallic and the color is golden yellow. Because of this characteristic, it is also called "fool's gold." Much harder (6 to 6.5) and less dense (5) than gold, its streak is black in color. Pyrite has a conchoidal fracture and no cleavage. It forms an interpenetration twin called an "iron cross."

Pyrite is a common accessory mineral in numerous rocks. It is not rare to find it in animal fossils (for example, pyritized ammonites). Pyrite is stable in a reducing environment, but it oxidizes easily in outcrop and produces sulfuric acid, which can attack metallic structures and concrete if it occurs in great abundance. In slag heaps and some tunnel excavation deposits, acidification of water and soils is observed. Pyrite is exploited for the production of sulfuric acid and as iron ore.

Fig. 5.31 Pyrite specimen. Photo Collection of the Cantonal Geological Museum of Lausanne, N. Meisser.

5.2.7 Oxides and hydroxides

Oxygen is the most abundant element on Earth (crust, upper mantle and hydrosphere). It makes up 49% of the weight of the Earth's crust and 92% of its volume. Like the sulfides, we will study only iron varieties. These are accessory minerals in rocks. Those that contain iron oxides in exploitable quantities are ores.

5.2.7.1 Hematite: Fe_2O_3

This mineral is often found as fine platelets (Fig. 5.32) and sometimes in a rose-shaped aggregate called an "iron rose." The rhombohedral shape is also common. Hematite is recognized by its opaqueness, its very obvious metallic luster, its black or reddish brown color, its high density ($5.3 \cdot 10^3$ kg/m^3) and hardness between 5.5 and 6.5. Rubbing it on an unglazed porcelain plate produces a blood red streak, from which the Greek root of its name.

It is a very widespread mineral (volcanic rocks, hydrothermal veins, masses and diverse rocks as a minor mineral) and it is the principal iron ore.

Fig. 5.32 Common appearance of hematite. Photo collection of the Cantonal Museum of Geology, Lausanne, N. Meisser.

5.2.7.2 Magnetite: Fe_3O_4

Magnetite occurs as perfect octahedra that are black in color with a slight bluish iridescence and metallic luster (Fig. 5.33). Its density is $5.18 \cdot 10^3$ kg/m^3 and its hardness is 6. The color of its streak is black; this is what distinguishes it from hematite. It can be identified primarily by its shape and by its high magnetism, which will move the needle of a compass or attract iron filings.

Geologists use its property to preserve the direction of the ambient magnetic field at the time of crystallization to conduct paleogeographic reconstructions (see § 4.3.4). Magnetite is one of the richest iron ores.

5.2.7.3 Goethite: FeO(OH)

This hydroxide commonly occurs as fibrous or radiating masses because of its perfect cleavage, or in concretionary masses, and also as vertically striated prisms, in tablets or as flakes. It has adamantine-metallic luster (sometimes dull) and a dark brown to black color, a density of $4.37 \cdot 10^3$ kg/m^3 and a hardness of 5 to 5.5. Goethite is typically formed by the hydration of iron ores. In sufficient quantity it can be exploited as iron ore. It is the cause of the red and ochre colors of some Mediterranean and equatorial soils.

[39, 123, 133, 144, 191]

Fig. 5.33 Magnetite specimen. Photo of the Collection of the Cantonal Geological Museum of Lausanne, N. Meisser.

5.3   Practical mineral identification

Mineral identification can be a complex operation requiring highly sophisticated equipment. For this book, we will confine ourselves to the methods used by the field geologist to which the engineer has easy access. In this way, the principal rock-forming minerals or their families can be identified.

These methods are based on observation with a hand lens and several simple tests. An animation on the DVD shows how these test are made in practice.

The following criteria are used:

*   Appearance (morphology): prisms (compact, elongate, tabular), sheets, fibers, random shapes,
*   Values of interfacial angles,
*   Color of the mineral and streak,
*   Luster (adamantine, vitreous, metallic, silky, flat, etc.),
*   Fracture (along planes, conchoidal or random),
*   Cleavage (number, angles)
*   Twinning,
*   Hardness test (fingernail, glass, steel),
*   Reaction of carbonate minerals with dilute HCl,
*   Other tests: estimate of specific gravity, magnetism, taste, touch, etc.

In the figure 5.34 we propose a flow chart for the identification of the principal minerals using field methods.

For a more refined identification, mineralogists use other methods (laboratory methods) and particularly measurements of density, thermal properties (specific heat, conductivity, fusibility), magnetic and electrical properties (pyroelectricity and piezoelectricity), radioactivity, surface properties (wettability), optical properties under the microscope (birefringence, etc.), X-ray diffraction, etc.

[70, 245]

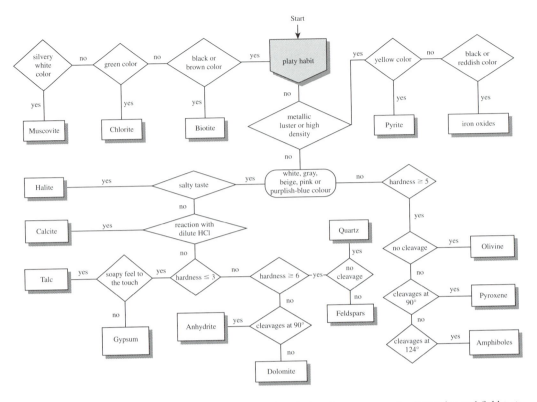

Fig. 5.34 Simple chart for the identification of principal minerals, based on macroscopic observation and field tests. For hardness, see § 5.1.3.1.

PROBLEM 5.2

Examine the photos of the three minerals (Fig. 5.35a to 5.35c) then determine them with the help of the identification chart from figure 5.34. As well as the photo, you receive the results of a few identification tests.

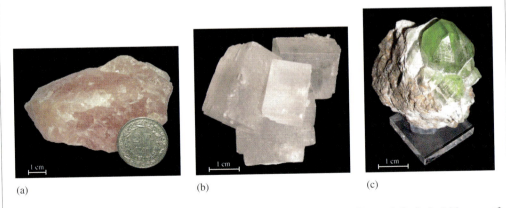

(a)   (b)   (c)

Fig. 5.35 Photos for problem 5.2. GEOLEP collection and collection of the Cantonal Geological Museum of Lausanne, photo GEOLEP D. Marques.

Question 1

- Mineral from photo 5.35a: The HCI test is negative. The mineral scratches glass. No particular taste.
- Mention one of the main uses of this mineral.
- What particular danger is present when excavating rocks containing a lot of this mineral?

Question 2

- Mineral from photo 5.35b: Cannot be scratched with a nail but is scratched by a steel file. No reaction to HCI and no particular taste.
- What precautions must be taken when excavating tunnels that cut through rocks containing a large amount of this mineral (see also Chapt. 13)?

Question 3

- Mineral from photo 5.35c: Scratches glass. No reaction to HCI and no particular taste.
- Which family of rocks is this mineral typical of (see also Chapt. 6)?

Look at the solution on the DVD

PROBLEM 5.3

Why do certain rocks contain iron in the form of pyrite and others in the form of oxides or hydroxides?

Look at the solution on the DVD

Chapter review

Rocks and sediments are composed of a wide variety of minerals. In practice, the common geological problems that the engineer faces are only related to a limited number of mineral species or mineral families. A thorough knowledge of these mineral families is necessary to understand rock-forming processes and to predict rock behavior.

In some cases, much greater mineralogical knowledge is needed, as for example in the aggregate industry. History has shown that ignorance of a process such as the alkali reaction has caused significant socioeconomic losses.

6 Magmatism and Magmatic Rocks

Having covered the rock-forming minerals in the previous chapter, we are now prepared to study rocks themselves. We will begin with the magmatic rocks because they were the first to form on the Earth. As they are eroded, they become the raw material of sedimentary rocks. Sedimentary and magmatic rocks together are the raw material of metamorphic rocks. Sedimentary and metamorphic rocks will be discussed in subsequent chapters.

Magmatism is a series of processes including the melting of the mantle or crust (magma formation) and the solidification of magma as magmatic rock.

Our study of the rocks will follow the fundamental principle of this book: we will understand the processes before focusing on the rocks they create. Because magmas originate deep in the Earth, we must supplement the general picture of the Earth's internal structure that we sketched in chapter 4, and discuss its composition. Then we will cover the theory of plate tectonics, which is related to magma generation and magmatic composition. Magmatism in turn plays an important role in the creation of the crust and mantle.

6.1   Composition of the Earth's layers

The model of the Earth as a series of concentric layers is the culmination of a long period of observation. Since the deepest well has reached only about twelve kilometers, the Earth's structure has been largely determined by various indirect methods, including the following:

- The analysis of outcropping rocks: crustal rocks occur at the surface as a result of plate collisions that created mountain chains. Rocks from the top of the upper mantle, part of the lower lithosphere, are also brought to the surface by the same process (for example, the peridotites of Finero in the Southern Alps and the diamond-bearing kimberlites of South Africa). Rocks come to us from the asthenosphere as fragments that are pulled out of deep chimneys involved in intraplate magmatism (§ 6.3.3.3).
- Laboratory tests: the composition of deeper layers can be deduced from melting tests on rocks.
- Various geophysical parameters measured at the surface: seismic wave velocity, density, magnetism (Chapt. 4).
- Analogy with other bodies of the solar system through the study of meteorites.

Figure 6.1 shows the layers of the Earth and their physical and chemical characteristics. The Earth is essentially composed of oxygen, silicon and iron, but in highly variable proportions depending on the layer. The layers are the core (internal and external), the mantle (lower and upper), and the crust. Matter exists as different states ranging from an elastic to a ductile solid (flowing slowly over the course of time) and finally to a liquid.

On the basis of these properties, we can distinguish two environments: the *lithosphere*, a combination of the crust and the outer part of the upper mantle, and the *asthenosphere*. The lithosphere can be likened to rigid plates (Fig. 6.8) that float like rafts on the asthenosphere and move as a result of *convection currents*, the force behind plate tectonics. The crust and the mantle are separated by the Mohorovicic discontinuity (the Moho), which marks a contrast in the velocity of seismic P waves.

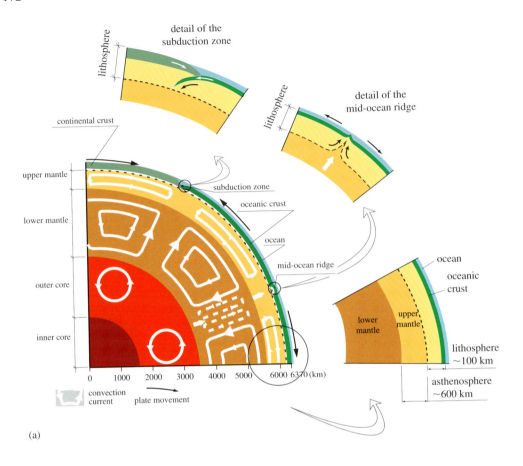

(a)

(b)

Layer	Physical properties	Geochemistry	Petrology
crust (upper lithosphere)	elastic solid	O, Si, Al, Fe	felsic, intermediate and mafic rocks
upper mantle – lower lithosphere	elastic solid	O, Mg, Si, Fe	ultramafic rocks
upper mantle – athenosphere	ductile solid and 3% of liquid (partial melting)	O, Mg, Si, Fe	ultramafic rocks
upper / lower mantle	ductile solid and 3% of liquid (partial melting)	O, Si, Mg, Fe	ultramafic rocks
outer core	liquid (residual magma)	Fe, Ni, (S)	amorphous mixture
inner core	ductile solid and 3% of liquid (partial melting)	Fe, Ni	metallic alloys

Fig. 6.1 Concentric layers that make up the Earth: (a) Geometry of the layers and thermal convection cells that occur in the layers, (b) Properties of the different layers.

6.2   Magmatism and plate tectonics

The coupling of mechanical (tectonic) phenomena and the generation of rocks (magmatism) helps us understand the diversity of magmatic rocks.

6.2.1   The history of plate tectonics

The history of plate tectonics began long ago. As early as 1620 the philosopher Bacon noted the similarity between the shapes of the African and South American coasts. Even so, the development of the theory of plate tectonics and its acceptance was a long and difficult process. In 1915 Wegener, a German meteorologist, gathered and summarized numerous existing observations and proposed the theory of *continental drift*. According to this theory, continents composed of light rocks (called SiAl because they are rich in **Si**licon and **Al**uminum) drift on a heavier layer (called SiMa, because it is rich in **Si**lica and **Ma**gnesium) that makes up the ocean bottoms. In the beginning there was a single continental mass that broke up into smaller continents that moved around with respect to each other. Wegener invoked various arguments to support his hypothesis (Fig. 6.2):

- *Shape of the continents*: The northeastern corner of South America and the Gulf of Guinea fit together almost perfectly. Moving to the north, the convex curve of the Sahara and western Morocco fits into the concave shape of the Florida coast of North America. Greenland fits into the space that separates Scandinavia from northern Canada. This coincidence of coasts is further emphasized if the continental slope of the plate boundaries is taken into account. The continental slope occurs at a depth of 200 m whereas the shores are often slightly inside the margins of the continental plates. For example, England and Ireland are part of the same plate as central Europe.
- *Continuity of geologic facies*. The ancient Saharan basement ends at the northern coast of the Gulf of Guinea and these rocks continue near the mouth of the Amazon River. Similarly, the old rocks that form the substratum of Gabon continue to the south of Recife, Brazil. These two examples show that the similarity is both geometrical and geological; radiometric dating methods confirm these observations.
- *Paleontological Continuity*: Certain climatic conditions are linked to particular sediments. For example, coal deposits of the Carboniferous, which are typical of an equatorial climate, are found today in eastern North America and northeastern Asia. These regions were formerly located at the paleo-equator.

In spite of these pertinent observations, the theory was refuted. It was difficult to imagine how the SiAl could move on the SiMa, given the solid nature of the SiMa. Geophysicists also rejected the forces Wegener used to explain the drift (tidal forces, precession forces and other isostatic forces) as being too weak to cause such movements. These challenges led to a total rejection of Wegener's theory.

But Wegener's idea resurfaced in the middle of the 20th century when research on oceanic plates revealed the process of lithosphere generation, particularly as a result of the study of paleomagnetism (see § 4.3.4). The theory of continental drift became the theory of continental plate movement. The plates no longer moved along the boundary of SiAl and SiMa, but much deeper in the asthenosphere due to its viscosity. Convection currents became a much more realistic engine. Wegener's successors advanced several arguments, notably that of large crustal structures. In 1960, Hess presented a process of oceanic expansion based on the presence of convection currents in the mantle below continents. He identified two types of primordial structures that emphasize the dynamic nature of ocean floor formation and support Wegener's theory.

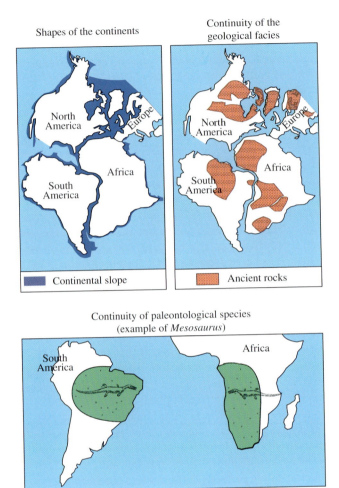

Fig. 6.2 Geologic indicators that imply that at one time, the continents were grouped together.

- *Mid-oceanic ridges*. This is where oceanic crust is produced. The process occurs symmetrically on both sides of the ridge, spreading the two lithospheric plates farther apart toward subduction zones. The oceanic ridges are true submarine mountain chains that extend tens of thousands of kilometers. They rise to an elevation of about 1500 m above the abyssal plains and have an average width of 1000 km. There is a gash at the center of the ridge that extends in the length of the rift. The rift has sharp walls, about 20 to 50 km apart, and the walls can be as deep as 1000 meters. Oceanographic exploration shows magmatic and hydrothermal activity at these ridges. The Mid-Atlantic Ridge emerges onto land in Iceland, making it possible to observe its topography directly (Fig. 6.3).
- *Oceanic trenches*: These are places where plates are consumed. Moving away from the oceanic ridge, the lithosphere progressively cools and thickens as a result of underplating of mantle material. When the plate reaches another less dense plate (such as a continental

Fig. 6.3 The Mid-Atlantic Ridge outcropping in Iceland: "Laki Fissure" (A 25-km long volcanic chain composed of a hundred volcanoes). Photo M. Marthaler in "Le Cervin est-il africain?" LEP Publications, Le Mont-sur-Lausanne, 2005.

plate), it sinks into the asthenosphere, disappears, and is thus recycled. The trenches are several thousand meters deep (11,502 m for the Marianas Trench) and several hundreds of kilometers long. The largest are located in the Pacific and occur near island chains (Japan, Philippines, etc.) or near mountain chains (Andes) when they occur on the margin of continents.

This is the basis of the second main principle of plate tectonics: the theory of *seafloor spreading*, which also introduces a new idea, that of the perpetual recycling of the crust. The ocean floor created at the ridges progresses to subduction zones where it sinks into the mantle. The continents also move but their lighter density prevents them from sinking into the mantle.

This new theory was not really taken into consideration until 1963 with the work of Vine and Matthews. They created the theory of oceanic floor spreading to explain the interpretation of magnetic anomalies.

In summary the theory of plate movement is related to magmatism in a cyclical way. Intense magmatic activity at the ocean ridges causes plate movement and oceanic expansion. Elsewhere in the cycle, plate movement causes magmatism in subduction zones.

6.2.2   Paleomagnetic reconstruction

Magnetic surveys show interesting linear anomalies on the oceanic crust on both sides of the Mid-Atlantic Ridge (Fig. 6.4). The striped pattern of magnetic bands signifies the inversion of the magnetic field. Positive anomalies indicate that the magnetic field at some time in the past was the same as the present-day field. Negative anomalies show an ancient magnetic field that was opposite the present day field. For example, measurements in Iceland show that the island is made up of adjacent bands indicating a series of SE-NW anomalies. When these measurements were done in the Atlantic Ocean, the image was repeated over the entire oceanic floor (Fig. 6.5). When the rocks were dated by radiometric methods, isochrons or lines of equal age of the basalt floor were traced, showing that lava near continental margins is older. The oldest lavas are about 140 Ma old.

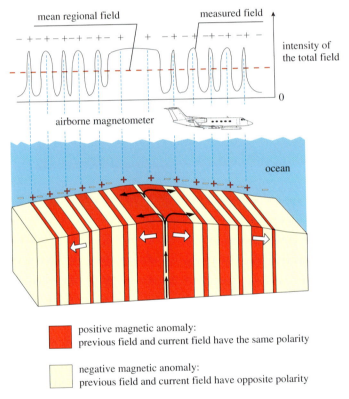

Fig. 6.4 Aerial measurement of paleomagnetic anomalies in bands in the Atlantic. Deviations from the regional average produce positive and negative anomalies.

This image of isochrons can be extended to other oceans. We see a similar ridge extruding modern-day basalts in the Pacific. The ridge passes between Australia and Antarctica and reappears in the Indian Ocean. The isochrons make it possible to break up the ocean floor into a mosaic of oceanic plates. The oldest ages were measured along the margins of subduction zones. These ages do not indicate the maximum age of the opening of the oceans, because a part of the plate has already been consumed. The shape and spacing of the isochrons allow to

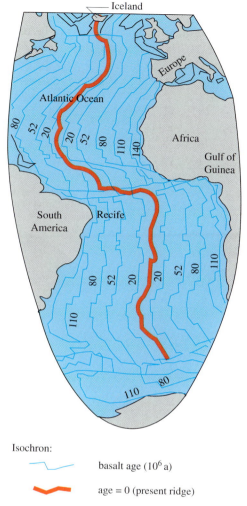

Iceland

Europe

Atlantic Ocean

Africa

Gulf of
Guinea

South
America

Recife

80   52   20   20 52 80 110 140

110   80   52 20   20   52   80   110

110   80

110

Isochron:

basalt age (10^6 a)

age = 0 (present ridge)

Fig. 6.5 Ages of basaltic rocks on the Atlantic seafloor (from [255]).

calculate an average velocity of plate movement over long periods, generally several centimeters per year. The isochron map also reveals another important tectonic element: ***transform faults***. These ruptures running more or less perpendicular to the ridges cut the latter into segments by horizontal shearing. They keep the direction of the ridge constant when plate boundaries are curved. This is particularly evident in the Mid-Atlantic Ridge along a line connecting the Gulf of Guinea and Recife, Brazil. A detailed study of the behavior of these transform faults shows that they behave differently from classical strike-slip faults (Chapt. 12). Assuming the velocity of contiguous plate movement to be essentially constant, we see that the shearing takes place only in the part of the fault located between two segments of a ridge (Fig. 6.6).

Paleomagnetic measurements confirm that the magnetic pole indicated by paleomagnetism on one continent does not coincide with the pole determined on another continent. Using these measurements, the original location of the continents can be reconstructed (Fig. 6.7).

These conclusions regarding plate movement based on the past are corroborated by current geodetic measurements collected by satellites (Fig. 6.8). The greatest velocities of spreading exceed 17 cm/year along the Pacific ridge.

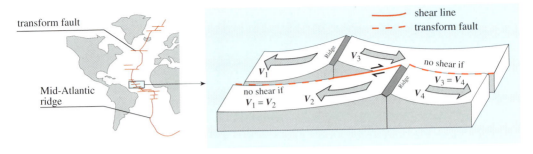

Fig. 6.6 Diagram of movement along a transform fault. An animation on the DVD shows how such fault is moving.

I Determination of the azimuth of the paleopoles and the angular distance between the sample location and the paleopole.

III Translation along parallels
→ Adjustment of the continents according to shape

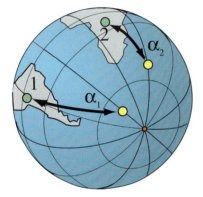

II Bring the paleopoles to the current pole:
  a. By rotation of the paleomeridians onto the current meridians,
  b. Then by translation of the continents along the meridian to make the paleopoles coincide with the current pole.

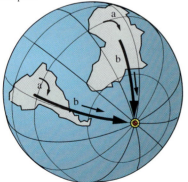

Legend:

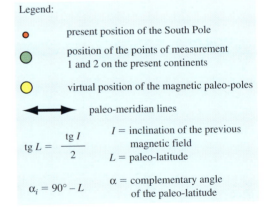

●   present position of the South Pole

◉   position of the points of measurement 1 and 2 on the present continents

○   virtual position of the magnetic paleo-poles

◄—►   paleo-meridian lines

$$\text{tg } L = \frac{\text{tg } I}{2}$$

$$\alpha_i = 90° - L$$

I = inclination of the previous magnetic field
L = paleo-latitude
α = complementary angle of the paleo-latitude

Fig. 6.7 Principle of paleomagnetic reconstruction of continent locations. The method consist of comparing two lavas of the same age, one is South America, the other in Africa. The previous magnetic field is characterized in terms of azimuth and inclination measured in magnetite contained in the lava. If the azimuth is not parallel to the present-day field, it means the continent has moved. The continent is then moved to align the previous field parallel to a present-day meridian. Using equation 4.28, the measurement of the paleoinclination gives the paleolatitude. The latitude of the continent is corrected by moving it along the meridian. The positions of the two continents are determined by orientation and latitude. Their mutual longitudinal distance must be determined by other criteria, particularly the age of lava samples compared to the age of the ocean floor separating the two continents. An animation on the DVD shows this reconstruction.

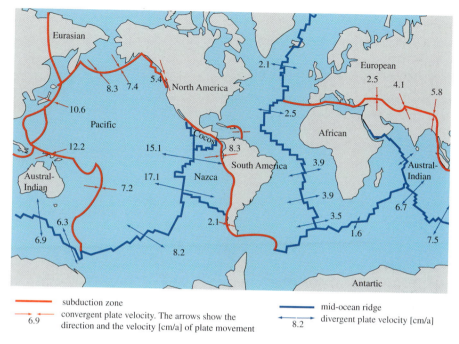

Fig. 6.8 Present-day relative movement of plates at the Earth's surface. Note: Antarctica is assumed to be immobile (from [187]).

6.2.3 Possible relationships between plates

Considering the convergent and divergent movements and the different types of plates, there are five principal relationships between plates (Fig. 6.8 and Fig. 6.9).

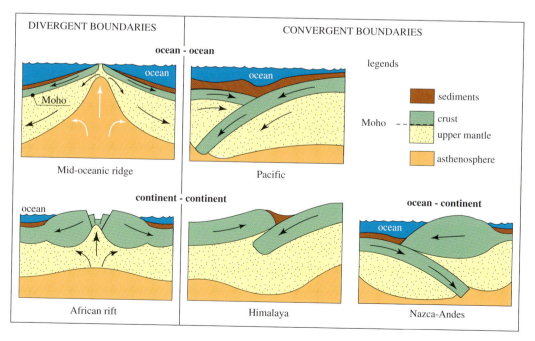

Fig 6.9 Principal relationships between plates (from [163]).

6.2.3.1   Divergence of two oceanic plates

This is the situation at the mid-oceanic ridges in the Atlantic, Pacific and Indian Oceans.

6.2.3.2   Divergence of two continental plates

This configuration is typical of intra-continental rift valleys (also called **graben**). One of the most extensive rifts links a series of large lakes in East Africa. In Europe, the Rhine Graben extends from Basel to Frankfurt (Fig. 6.10); the Rhine plain is sinking while the Vosges and the Black Forest are moving apart.

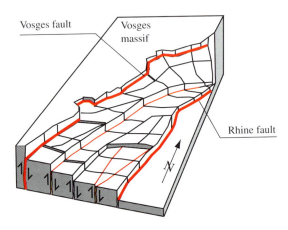

Fig. 6.10 Divergence of two continental plates. Example of the intra-continental opening of the Rhine Graben.

6.2.3.3   Convergence of two oceanic plates

This situation occurs in the Pacific, and the resulting subduction created the Aleutian Island archipelago.

6.2.3.4   Convergence of two continental plates

Let us consider a lithospheric plate made up of oceanic and continental parts connected by a **passive margin** (for example, India in the case of the formation of the Himalayas). If this plate encounters a continental plate (Eurasia is the chosen example), subduction of the oceanic part will occur first. Afterwards, the subduction will turn into a **collision** between two continental plates. Their low density prevents one of them from plunging into the mantle. Therefore, subduction ceases. The collision creates large mountain chains located at the heart of continental plates, like the Himalayas and the Alps. This relationship does not create volcanoes because there is no subduction.

6.2.3.5   Convergence of an oceanic plate and a continental plate

This is the most typical case of subduction. The Pacific "**Ring of Fire**" coincides with major subduction. It is found particularly at the convergence of the Nazca and South American plates. The continental margins are then called **active continental margins**.

[54, 242, 320]

6.3   From magma to magmatic rocks

Following this overview of lithospheric plate movement on Earth, we must understand how a liquid becomes a solid.

Magmatic rocks are formed as a result of the creation of a liquid phase (magma) followed by solidification. Recall that magmatism has been a fundamental process in the evolution of our planet and has created the various layers within the Earth. Although the mechanisms of magma creation have not been elucidated with certainty, it is recognized that three essential factors lead to melting: temperature increase, pressure decrease, and the possible presence of water, which lowers the melting temperature of the rock (Fig. 6.18).

Magmas are present in reservoirs located in the lithosphere. They also exist at great depth in the mantle (to 300 km) and at a few kilometers from the surface under the mid-oceanic ridges.

6.3.1   Process of magma generation

When a rock is subjected to a temperature increase, the rock melts gradually and selectively, beginning with the least to the most *refractory* (mineral having a high melting temperature). With increasing temperature, total melting occurs.

In general, during magma creation, if temperatures remain below this maximum temperature, *partial melting* occurs. The fact that melting is partial and selective implies that the composition of the magma that has already formed must be different from that of the initial rock. We will examine other processes of differentiation that modify the chemical composition of magma.

A wide geochemical variation of magma results from the different cycles of melting and crystallization that have occurred over geologic history. We will group them into four large families on the basis of their silica content:

- *Felsic magmas* (from **fel**dspar and **si**lica, formerly called acidic): rich in silica ($SiO_2 >$ 65%) and poor in iron;
- *Intermediate magmas*: average silica content (52% $< SiO_2 <$ 65%);
- *Mafic magmas* (from **ma**gnesium and **f** from the Latin word for iron, formerly called basic): low silica content (45% $< SiO_2 <$ 52%), abundant iron.
- *Ultramafic magmas*: very low silica content ($SiO_2 <$ 45%), magnesium and iron are very abundant.

It should be noted that the majority of magmas produce silicate lavas, but lavas containing sulfide and carbonate also exist.

6.3.2   Magma solidification processes

Let us imagine a region of the upper mantle that is particularly hot because of convection currents. A mass of molten rock has formed within other rocks that have remained solid; this is a *magma chamber* (Fig. 6.11).

To begin, let us make the simplifying assumption of a completely melted magma: this is a "soup" of a wide variety of atoms, mostly oxygen and silicon (Fig 6.12). In addition, aluminum, iron, calcium, potassium, sodium, and magnesium are present, along with other trace elements.

Let us examine the four phenomena that control the transformation of magma from liquid to solid, during slow cooling (plutonic conditions).

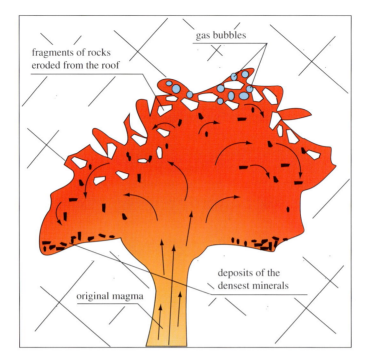

Fig. 6.11 Diagram of a magma chamber. The size of the magma chamber can range from several tens of meters to several kilometers.

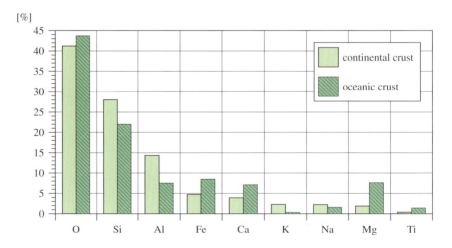

Fig. 6.12 Concentration of elements in rocks in the continental and oceanic crust.

6.3.2.1   Fractional crystallization

Unlike water, magma does not crystallize to a unique solid. Its chemical complexity is such that a single magma can produce different solids as a result of crystallization conditions.

The rocks that surround the magma chamber are always cooler than the magma; as heat dissipates toward the exterior, the temperature of the magma decreases very slowly. The molten mass starts to solidify. Minerals form gradually until the mass becomes entirely solid.

Let us look in more detail at how crystallization of a magmatic rock occurs. We will discuss theoretical conditions of a homogeneous and closed system, without the addition or loss of matter. Under these conditions, a given magma can produce only a single rock since it contains all its potential future minerals.

Different minerals have their own crystallization temperatures. As they crystallize from a completely melted magma, the first to grow are those that have the highest crystallization temperature (refractory minerals), then others gradually crystallize as the temperature decreases. According to this principle, finally all minerals will coexist at the end of the crystallization process, the first formed along with the last. In reality, things are more complicated because of reactions between already-formed minerals and the residual magma. These reactions result in the formation of new minerals that are more stable under the thermodynamic conditions of the lower temperature. The first crystals are thus partially or totally resorbed. Once the magmatic residue is used up, crystallization ceases and the rock is formed.

Two mineral families are particularly interesting with respect to fractional crystallization because they have different minerals at each temperature stage:

- Ferromagnesian silicates: the varieties that are poor in silica and rich in iron appear first and evolve toward those rich in Si and poor in Fe (olivines -> pyroxenes -> amphiboles -> biotite -> muscovite);
- Plagioclases: at high temperature a calcic plagioclase called anorthite is formed, then the sodium content progressively increases, to the detriment of calcium.

Orthoclase appears at the same low temperature as quartz.

Let us now examine how these general principles can be applied to our four magma types.

Felsic magma

The amount of silica is such that the first minerals formed leave an abundant liquid residue that allows fractional crystallization to go to completion (Fig. 6.13). There is too much silica for olivine to form. Pyroxene appears immediately but it reacts with the silica-rich magma to form amphibole via a solid-liquid reaction. Amphibole then reacts to form micas. The calcium plagioclase is unstable and react with the sodium-rich melt to form sodium-rich plagioclase. At this stage, there is still a residual magmatic melt from which potassium feldspar will form. The last liquid residue contains nothing more than the oxide of silica that will form quartz. The mica + sodium plagioclase + alkali feldspar + quartz mélange is typical of ***granite***, a felsic plutonic rock par excellence.

Intermediate magma

The first minerals to form are olivine and anorthite. The remaining liquid reacts with these minerals to form amphibole and a slightly more sodic plagioclase. The reaction stops here as a result of the complete consumption of the magma (Fig. 6.14). The composition of amphibole and intermediate plagioclase with less than 50% calcium is typical of ***diorites***. When the calcium and sodium concentrations are less than potassium, this kind of magma will form a rock called ***syenite*** in which potassium feldspar predominates.

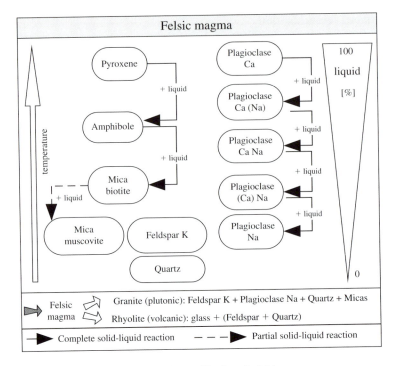

Fig. 6.13 Fractional crystallization of a felsic magma.

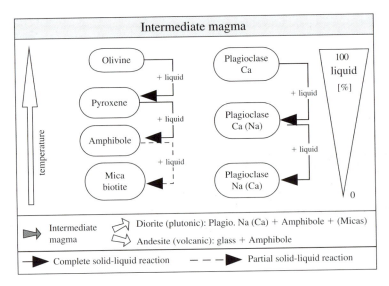

Fig. 6.14 Fractional crystallization of an intermediate magma.

Mafic magma

This series is even shorter because there is less silica available. After forming, olivine reacts totally or partially with the melt to form pyroxene (Fig. 6.15). The plagioclases remain very calcic (>50% Ca). This composition is typical of *gabbros*.

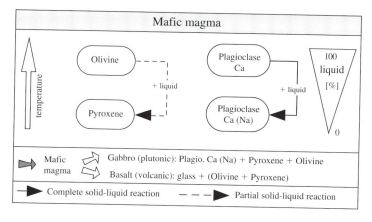

Fig. 6.15 Fractional crystallization of a mafic magma.

Ultramafic magma

Magmas rich in iron and magnesium come from the mantle. Ferromagnesian minerals with high melting points predominate, particularly olivine. They consume almost all of the available silica. The remainder is used to form calcic plagioclase or ***feldspathoids*** (minerals that are similar to feldspars but contain less silica). Peridotites, very rich in olivine, are the most common plutonic rocks of this category (Fig. 6.16).

Fig. 6.16 Core of green mafic rocks caught in red claystones in the oceanic scar at the center of the Himalayas, Ladakh. These rocks are remainders of the ocean that separated India from Asia and that are now located at an elevation of about five thousand meters, at the center of the mountain chain. GEOLEP Photo, A. Parriaux.

The case of a magma chamber evolving in a closed system is useful for understanding the basic mechanism of differentiation. However, in nature the system rarely evolves in this way.

6.3.2.2   Magmatic segregation

The ideal model of fractional crystallization in a homogenous and closed system is often disrupted in nature by the process of magmatic segregation, which modifies the physical and mineralogical evolution of the creation of magmatic rocks. Minerals that form first are rich in iron and calcium and are much denser than the residual melt. They thus have the tendency to settle to the bottom of the magma chamber while the fluid residue occupies the upper part. The concentration of these refractory minerals forms what are called *cumulates*; under certain cases they form metalliferous ore deposits. The fluid that is depleted of refractory minerals is called *evolved*. If the magma is expelled upwards, the liquid phase is separated from the solid phase. The two masses evolve separately. These final rocks are said to be *differentiated* from the source magma.

6.3.2.3   Assimilation

Assimilation can occur in two ways. First, as the magma cools, the dissipated heat increases the temperature of the surrounding rocks (margins of the magma chamber), which may then reach their melting points. A non-refractory liquid can be extracted from these rocks and incorporated into the magma. A second case is the immersion of fragments of the solid surrounding rocks into the magma; small and not very refractory fragments can thus melt completely and be assimilated into the magma. Larger elements will not necessarily be completely assimilated; they can form *inclusions* surrounded by a reaction rim in the magmatic rock. This process can locally change the composition of a magma.

6.3.2.4   Mixtures

If a new contribution of magma is mixed with the preceding magma in the conduits or the magma chamber, magmatic hybridization will result. Consequently, magmas are generally heterogeneous. A certain homogeneity must have existed in rocks that are called "primitive" (product of the solidification of original magma). However, this type of rock probably no longer exists today, because most of them were transformed by later magmatic cycles during the Archean.

6.3.3   Magmatic configurations

Let us now look at the Earth's magmatic environments where different rocks are created. The type of magmatic activity and the nature of the magma are linked to the tectonic setting. A profile across the Atlantic Ocean shows three main configurations linked to plate tectonics (Fig. 6.17): ocean ridges (called rifts on continents), subduction zones, and hot spots (magmatic eruptions through plates). To understand the conditions of magma production in each of these situations, we must understand the thermodynamic domains of rocks and temperature variation at depth. A temperature-pressure graph shows the transition between solid rock and magma as determined by two curves (Fig. 6.18), unlike a pure substance like water and ice. There are thus three areas:

- At low temperature, the rock is completely solid up to the *solidus* line;
- At an intermediate temperature, the rock is partially melted up to the *liquidus* line;
- At high temperature, the rock is entirely liquid (magma).

In the graph in figure 6.18, we can superimpose onto the pressure axis a graduated depth axis by dividing pressure by the average density of the rock (chosen at $3.2 \cdot 10^3 \, kg/m^3$). We can then add the geothermal gradient, or *geotherm*, to the same graph. This temperature-depth function is slightly different across continental (slightly insulating) and oceanic (slightly conductive) plates, as we saw in chapter 4. The geothermal gradient is a maximum at oceanic ridges because of the thinness of the lithosphere at this location.

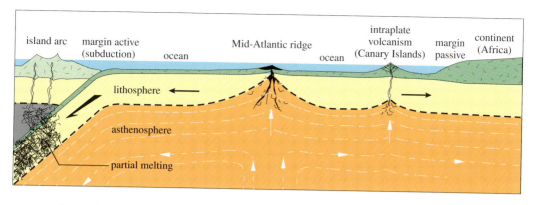

Fig. 6.17 Three magmatic configurations found in the example of the profile across the Atlantic.

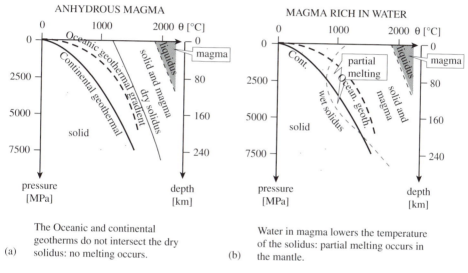

The Oceanic and continental geotherms do not intersect the dry solidus: no melting occurs.

(a)

Water in magma lowers the temperature of the solidus: partial melting occurs in the mantle.

(b)

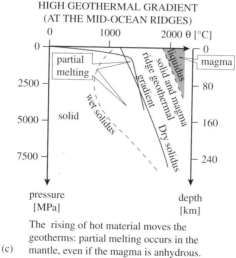

The rising of hot material moves the geotherms: partial melting occurs in the mantle, even if the magma is anhydrous.

(c)

Fig. 6.18 Areas of partial and total melting in mantle rocks compared with geothermal gradients in oceanic and continental plates as well as in mid oceanic ridges.

The intersection of the geotherms and solid curves (***dry solidus*** line) identifies the conditions of possible melting. It should be noted that, according to theoretical diagrams, in the first several hundreds of kilometers of depth, total melting never occurs (Fig. 6.18a). Partial melting takes place at a depth of 20 to 80 km only under ridges (Fig. 6.18c). But this means there should be no magma production in subduction zones, which does not correspond to reality. It was finally understood that oceanic plates carry water along as they are subducted. Water gets trapped as oceanic sediments gradually accumulate, and this water lowers the melting point of the rocks. A second solidus line, called the ***wet solidus*** is clearly displaced toward lower temperatures. These new thermodynamic conditions cause partial fusion in all three areas (Fig. 6.18b). Total melting remains impossible without abnormal geothermal gradients. Let us examine the three configurations in detail.

6.3.3.1   Ridge magmatism

Ridges are the places where oceanic lithosphere is created (Fig. 6.19). The oceanic crust is particularly thin (on the order of 5 to 10 km) and covers 55% of the surface of the Earth. Thus the lithosphere has an average thickness that can reach 100 km but it is only several kilometers at the ridges. The ridges have very strong ascending currents as shown by the rapid rise of material from the asthenosphere without a notable change in temperature. Due to decompression and the water present in faults, the mantle material is able to melt and then cool rapidly as it approaches the surface, creating rigid oceanic lithosphere.

At depth, magmas form mafic and ultramafic plutonic rocks at the base of the plate. Submarine volcanic flows form the ocean floor, creating pillow lava at the surface (Fig. 6.32) and columnar basalts below (Fig. 6.20).

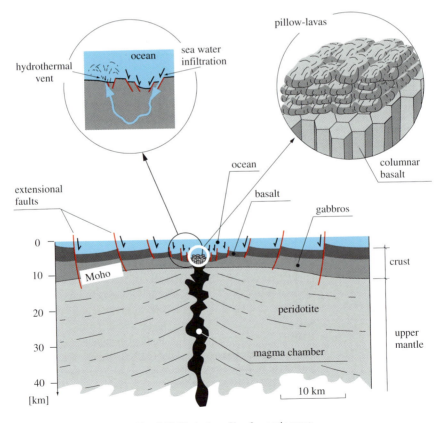

Fig. 6.19 Typical profile of oceanic crust.

Fig. 6.20 Basaltic rocks often occur in nature as ***columnar basalt***, parallel columns perpendicular to the flow. Cracks that separate the basalt columns result from shrinking of the basalt as it solidifies. This phenomenon gives the rock fissure porosity that makes it an excellent aquifer in volcanic terrains. Northern Tibetan Plateau. GEOLEP Photo, A. Parriaux.

6.3.3.2   Orogenic magmatism

The principal subduction zones that consume oceanic plates under continental plates are the places where continental lithosphere is created (Fig. 6.21).

The sinking plate, which is silica-poor in this example, partially melts at a depth of about 100 km because of the water content of the subducted plate. This magma reacts with the base of the continental lithosphere then rises gradually toward the surface through deep faults. The magma is thus enriched in silica.

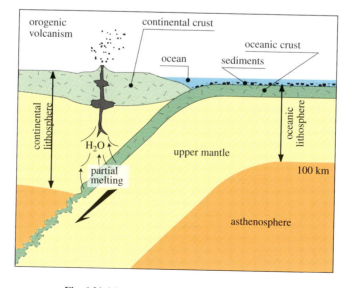

Fig. 6.21 Magma production in a subduction zone.

At depth this type of magma forms plutonic felsic rocks, granite in particular. These rocks occur in a complex association with metamorphic rocks (Chapt. 11). At the surface, volcanoes spread lava and ashes during violent eruptions, often accompanied by water vapor and gas.

Continental lithosphere is more complex than its oceanic counterpart in terms of emplacement processes. Two main phenomena explain its origin: Continental crust was first created from mantle material 4.2 Ga ago (§ 3.3.1), and it was later recycled (in subduction zones) during various orogenies. Continental crust is 100 to 150 km thick on average. Conceptually it can be divided into two parts (Fig. 6.22):

- *Upper crust*: 10 to 20 km thick, it contains sediments (the part called surface crust) and surface volcanic rocks, then granites and slightly metamorphosed rocks;
- *Lower crust*: on the same order of thickness, it is made up of denser rocks with intermediate to mafic composition and highly metamorphosed rocks.

The continental crust has a total thickness of 30 km on average, but it may be 70 km thick under orogenic zones, as revealed by isostatic anomalies (§ 4.2.5). Its density of $2.7 \cdot 10^3\,kg/m^3$ to $2.9 \cdot 10^3\,kg/m^3$ is characteristic of silica rich rocks.

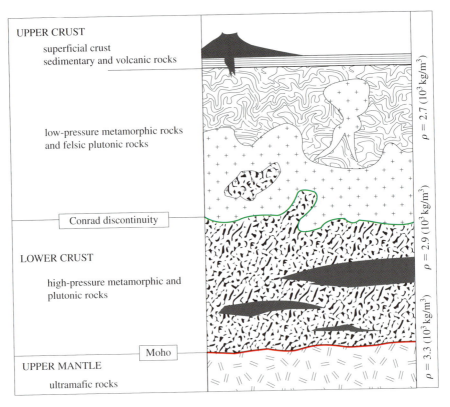

Fig. 6.22 Composition of the continental crust.

6.3.3.3   Intraplate magmatism

Hot spots are volcanic phenomena located far from plate boundaries. These are volcanic and plutonic phenomena related to a plume of solid but ductile hot material (Fig. 6.23) that rises from the depths of the mantle. According to some hypotheses the material may come from the boundary of the core. As the plume approaches the surface, pressure decreases and the mantle materials melt. The magma can then travel through the lithosphere and spread out on the surface. These columns of hot material are fixed and do not move with the lithospheric plates. The lithosphere is pierced like a steel plate moving above a torch.

Active volcanic regions like Yellowstone Park or the Hawaiian archipelago, for example (Fig. 6.24), are located at the middle of the North American continental plate and the Pacific oceanic plate, respectively. The drowned volcanoes of the Emperor Seamount chain extend northward from the Hawaiian Islands as an island chain. The southern islands are the most active; the volcanoes located to the north are all extinct.

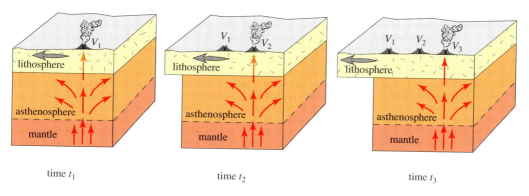

time t_1 time t_2 time t_3

Fig. 6.23 A hot spot above a moving lithospheric plate. V_1 = older volcano; V_3 = younger volcano. An animation on the DVD shows this phenomenon.

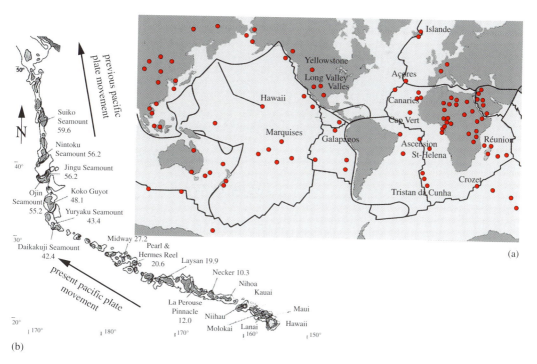

Fig. 6.24 (a) Hot spots active during the last ten million years (Antarctica has eleven hot spots, but they are not shown on this figure). (b) Detail of the Hawaiian archipelago with the age of basaltic rocks in millions of years.

6.4   Magmatic events

Now we shall discuss the geologic conditions of magma crystallization. Magmas that crystallize at depth or at the surface can have the same chemical composition but they will have a different mineralogical structure.

6.4.1   Plutonism and plutonic rocks

The ability of magma to move toward the surface depends on the presence of preferential pathways, or fractures. In general, the lithosphere is not very deformable, which explains why the great majority of magma remains trapped in the crust, never reaching the surface. Plutonism is the process of crystallization at great depth. Solidification takes thousands to several millions of years. As a result, the molten mass has plenty of time to form well-developed crystals. The texture of a plutonic rock is thus said to be **granular**, or **coarse-grained**, meaning that all the minerals are well crystallized. They are visible to the naked eye (Fig. 6.25).

Fig. 6.25 Texture of plutonic rocks, here a granite. GEOLEP Photo, G. Grosjean.

Granite, which is the most abundant plutonic rock in the Earth's crust, has a tendency to rise to the surface slowly after solidification because it is slightly less dense than the surrounding rocks ($\rho = 2.65 \cdot 10^3 \, kg/m^3$). It forms **batholiths** (Fig. 6.26), a type of large massif with a root that extends to depth. Batholiths can also rise along deep faults, as the granites of the Mont-Blanc Massif (Fig. 6.27).

In the final stage of crystallization of a pluton, low temperature (less than 100°C up to 500°C) hydrothermal fluids remain. They can either crystallize in the fractures of an already-solidified rock where they form veins of well-developed minerals, as in **pegmatites**, which are rich in large micas. Or the hydrothermal fluids can cause alteration of the rock mass through a process called hydrothermal alteration. The circulation of hydrothermal fluids in the rock causes chemical modifications that create new equilibria and new minerals. This type of alteration

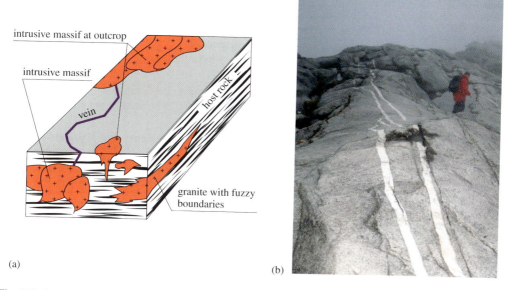

Fig. 6.26 Granites. (a) Geologic environment. (b) Late aplitic veins cutting the Mont Kinabalu granitic batholith, upper Neogene, northwest of Sabah, Malaysia. The **aplites** are microgranitic vein rocks from residual magmas deficient in ferromagnesian minerals. GEOLEP Photo, L. Cortesi.

Fig. 6.27 Massive granites today. Granites cover vast regions of the Earth. Their presence is easily determined by the shapes they display when eroded. (a) Rounded shapes when the uplift of the massif is not significant or if it is old. Example: the granites of the Hentiyn Muruu Massif east of Ulan-Bator, Outer Mongolia. GEOLEP Photo, A. Parriaux. (b) In recent and active orogenies, granites have much more angular relief similar to the summits of Mont Blanc (France). GEOLEP Photo, L. Cortesi.

is fundamental to the formation of ore deposits. Uranium is one of the minerals found in these deposits. Granites have a special relationship with radioactivity. They generate radioactivity, due the presence of uranium itself, and one of its decay products, **radon**. This radioactive gas is a serious public health concern that was unknown until several years ago. Although radon is one of the rare gases, it can occur in considerable concentrations in the subsoils of buildings. Inhalation of radon leads to a significant risk of lung cancer. Actually, it is not the radon itself that irradiates tissues but rather the daughter products of radon (polonium and bismuth isotopes, for example) that radon carries in suspension. The daughter elements, as solid microparticles, are adsorbed to the cells of the lungs where they continue to decay. Numerous countries have learned of this danger and have prepared risk maps based on geological substrata. Regions with granitic substrata are the most dangerous. But uranium can also be incorporated into sedimentary rocks, particularly shales. In places, houses that are in contact with faults or karst conduits are also vulnerable because radon can easily spread via these conduits. Once discovered, this danger can be prevented either by air-proofing the foundation or by "draining" the gas under the foundation and removing it. The risk is equally present in certain geomaterials that include uranium-rich aggregate. Granites are also important to another aspect of the nuclear industry since they are suitable rocks for the disposal of power plant waste (Chapt. 14).

Plutonic rocks are massive and mechanically strong, and as a result they are often used as dimension and ornamental stone. However, they can be weathered at surface conditions, so it is important to monitor this phenomenon (Chapt. 13).

6.4.2   Volcanism and volcanic rocks

Like earthquakes, volcanoes are one of the most spectacular examples of geologic activity. They are also one of the major geologic risks that have threatened humanity since the beginning of time.

6.4.2.1   Processes, volcanic events and rocks

Volcanism includes a whole series of eruptive phenomena.

Lava eruptions
Lava is magma that flows to the surface continuously and forms massive rocks as it cools. If these rocks have vacuoles that result from degassing they are called *scoriated*. The temperature of a magma when it arrives at the surface depends on its composition; the majority of silicate magmas have temperatures between 900 and 1200°C. Other more rare types of magma have much lower temperatures: for example, a *carbonatitic* magma solidifies below 500°C. This magma is the product of a very advanced stage of differentiation and is very rare (the only occurrence today is at the Oldoinyo Lengai volcano in Tanzania).

Pyroclastic eruptions
Pyroclastic deposits (Fig. 6.28) result from viscous magmas ejected during explosions. They are classified according to grain size (d = diameter):

- *Blocks* (already solid at the time of the eruption) and *bombs* (masses of viscous lava at the time of the eruption) ($d > 64$ mm);
- *Lapilli* (2 mm $< d < 64$ mm);
- *Ash* ($d < 2$ mm).

Fig. 6.28 Lateral pyroclastic deposit rich in blocks and bombs. This deposit, which is less massive than lava, is highly eroded. Caldera de Los Frailes, Spain. GEOLEP Photo, L. Cortesi.

The accumulation of pyroclastic particles on land or in water generally creates unconsolidated formations. However, if the particles have not completely solidified by the time they reach the Earth, they can form a welded aggregation with a vitreous matrix, a *"welded tuff."*

Depending on the mode of transport, different types of deposits can form:

- *Pyroclastic falls*: These result from vertical eruptions that fall back onto the cone. As the particles fall, grain size sorting occurs. Large and non-porous particles reach the ground first, followed by the finer material, creating stratified deposits. These deposits occur everywhere on the cone independent of its topography.
- *Pyroclastic flows*: This is pyroclastic material transported by very high temperature gases. The gases support the particles in the flow (*"nuées ardentes"* or "burning clouds"), which can travel down valleys on the slopes of the volcano and spread out over considerable areas (20,000 km^2 or more). Their speed and heat make them particularly lethal; they kill primarily by suffocation (examples are Pompeii, Mount Pelée, and Soufrière). The famous volcanologists Katia and Maurice Kraft were killed by a nuée ardente at Mount Unzen in 1991. Grain size sorting occurs laterally as a result of a pyroclastic flow; the non- stratified sediments are composed essentially of ashes. Pyroclastic deposits can also be transported beyond the cone when they fall on lava flows that are still moving. *Surges* are pyroclastic flows with a much lighter particle load. They move even more rapidly than nuées ardentes, reaching speeds of 200 to 250 m/s. They form thin deposits of finely stratified beds.
- *Atmospheric suspensions*: Extremely fine particles can spread over the entire Earth due to high altitude currents and can cause significant climatic disturbances. The powerful Pinatubo explosion in 1991 sent ash as high as 40 km. Some authors claim it caused a global average drop in temperature of 0.5°C (over a period of 5 years, the estimated time necessary for the particles to fall back to Earth). These deposits are almost imperceptible because they are mixed with other eolian sediments.

There are post-volcanic phenomena such as *lahars* (Indonesian term). These devastating mudflows result from the erosion of freshly deposited volcanic ashes that are not stabilized by vegetation. They occur after abundant rainfall, snow, or ice melt, or even the failure of the walls of a crater lake. Because of their density, lahars flatten everything in their path (forests, houses, etc.) and can move enormous blocks over tens of kilometers. These flows caused extensive damage during the eruption of Pinatubo in 1991.

Through diagenesis, pyroclastic masses that were originally unconsolidated can become lithified (Chapt. 10) and produce volcano-sedimentary rocks, also called *volcanic tuff* (or *cinerites* when the size of particles is inferior to 2 mm).

Gas eruptions

Volcanic activity is generally accompanied by gas eruptions (exhalative activity) that can take the following form:

- *Exhalations* during a paroxysmal activity: They accompany an eruption and occur in the magma chimney. The gases are the major force behind the rising magma; after the magma rises, decompression occurs, gas bubbles form and reduce the density of the magma. In certain cases, the abundance of these exhalations contributes to the explosive nature of the eruption.
- *Fumaroles*: These are gentle, persistent and diffuse eruptions. They can occur on the lower or exterior slopes of the crater, sometimes even at great distances from the chimney. Depending on their temperature, different minerals can be associated with them (for example, "sulfur flowers").
- *Geysers:* These are hot springs of steam and gas that may reach a height of several tens of meters. These are cyclic eruptions. The water comes primarily from groundwater of meteoric origin that comes into contact with shallow magmatic reservoirs. Sometimes the water comes solely from the magma. These events are limited to regions of active volcanism and strong geothermal anomalies. They are found primarily in Iceland, Yellowstone Park (USA) and northern New Zealand.

The effects of eruption conditions on the texture of volcanic rocks

Generally when magma erupts at the surface, it solidifies rapidly in contact with air or water (it takes a few seconds to several days). In extreme cases, the result is an amorphous matrix, similar to glass, where the atoms are randomly disseminated. But generally the texture is microcrystalline: only the embryos of minerals can form, because there is not enough time to allow them to develop completely. In many volcanic rocks, magnificent well-formed crystals can be found in the matrix (Fig. 6.29); they come from fractional crystallization in the magma chamber, well before the eruption. Olivine basalts are an example.

When lava contains lots of gas, the rock texture is called vesicular (*scoria, pumice*). The variety called *liparite* (from the island of Lipari in northern Sicily) has abundant vesicles that are not interconnected, which allows them to float. *Pozzolans* are scoriaceous volcanic ashes rich in silica, aluminum and iron oxide.

In contrast to plutonic rocks, volcanic rocks are not widely used in industry, except for basalt, which is traditionally used as building stone, and some lavas used for roofing slabs. Today, volcanic rocks are used primarily for light aggregate. Because they contain amorphous silica, the risk of alkali reaction must be taken into account (see § 5.2.1.1). Pozzolans are important in the cement industry because they react with low-temperature calcium hydroxide in the presence of water to create very stable hydrated compounds. The resulting cement is very resistant over the long term.

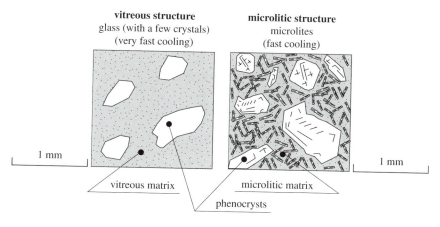

Fig. 6.29 Two textures of volcanic rocks (viewed under a polarizing microscope).

6.4.2.2   Types of eruptions

There are a variety of volcanic behaviors on Earth, from quiet eruption of magma (lava) to highly explosive pyroclastic eruptions that eject magma fragments. An animation on the DVD shows different forms of volcanic activity.

Volcanic eruption dynamics and the types of rock they form depend essentially on two factors: magma viscosity and gas content. These factors are related to the silica content of the magma. Mafic magmas are generally fluid and gas-poor compared to more felsic magmas.

A classification based on magma type shows four principal types of eruptions on land (Fig. 6.30). This is an older classification that is useful for a general presentation of eruption dynamics and the volcanic landforms created. The types are listed by silica content (viscosity) and gas content (explosivity).

- *Hawaiian* type: The eruption of hot and fluid lava spreads without significant seismicity (this type of activity is called extrusive). The lavas flow quickly and the cones they form have gentle slopes and are laterally extensive. Example: Kilauea (Hawaii, USA).
- *Strombolian* type: This type of eruption can be explosive. It alternates between fluid lava flows and ejecta. The cone is taller. Example: Stromboli (Eolian Islands, Italy).
- *Vulcanian* type: the viscous material and high gas pressure cause an explosive eruption consisting of repeated eruptions of ash and consolidated magma into the atmosphere. The cone is made up of ash layers accumulated on the steep slopes, which are prone to erosion. Example: Vulcano (Eolian Islands, Italy). When pressure is particularly high, it causes extraordinary ejecta called *plinian* activity (in honor of Pliny the Younger who described the phenomenon). Plinian activity is characterized by an imposing vertical plume of magma mixed with gas.
- *Pelean* type: In this type of volcano, the magma is so viscous that it forms a plug that stops up the chimney, causing a pressure buildup in the abundant gas. The resulting explosion often causes nuée ardentes that can destroy the entire volcano. We can cite the examples of the famous eruption of Montagne Pelée in Martinique in 1902, and more recently, the eruption of Mount St. Helens (USA) in 1980.

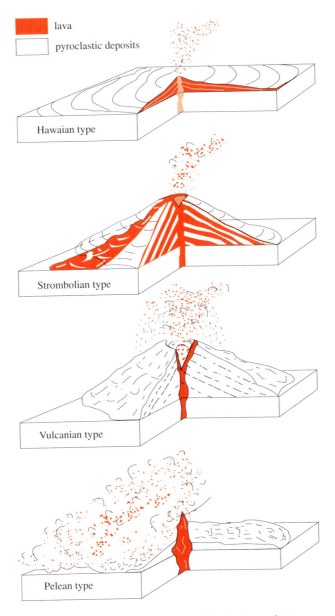

lava

pyroclastic deposits

Hawaian type

Strombolian type

Vulcanian type

Pelean type

Fig. 6.30 The four types of terrestrial volcanic eruptions.

In addition to these four types, *phreatomagmatic* eruptions, which are generally highly explosive, involve the interaction of magma and water. They can be submarine or sub-lacustrine (Fig. 6.31) eruptions or can result from the interaction of magma and an aquifer. Submarine eruptions that take place mainly along the mid-oceanic ridges are similar to the Hawaiian type since the magma solidifies as *pillow lava* (Fig 6.32). This shape is due to the fact that the outside of the lava mass solidifies in contact with water while the center is still melted and continues to flow.

Fig. 6.31 Crater lake at the Poás volcano, Costa Rica. The diameter of the lake ranges from 200 to 400 m and its water is highly acidic (pH of 1) and sulfurous. Fumarolic activity is continuous and is accompanied by phreatomagmatic (magma and water) and phreatic (water and gas) explosions. GEOLEP photo, L. Cortesi.

Fig. 6.32 Pillow lava photographed at the northern end of the Lau ridge, Lau Basin, Polynesia. The photo shows an area five meters wide. Photo courtesy of C. Langmuir, Harvard University and D. Fornar, Woods Hole Oceanographic Institution.

6.4.2.3 Structure and evolution of volcanoes

The term "volcano" includes external and internal structures produced by an eruption.

The principal morphologies of eruption on land are summarized in figure 6.33. Their variability depends on the proportion of pyroclastic deposits and the fluidity of the lava. Depending on the fluidity of the lava, the morphology ranges from ***basaltic shields*** to ***domes*** (extrusion of very viscous lava, predominantly solid eruptions) including lava ***cones*** (with almost

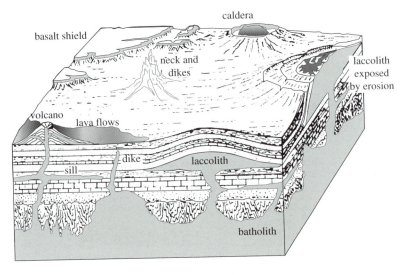

Fig. 6.33 Various volcanic structures.

45° slopes). Some volcanoes are mixed types, formed by the alternation of flows and pyroclastic layers: these are stratovolcanoes (examples are Etna and Vesuvius). The crater at the top is a flared volcanic chimney. In some particularly violent eruptions, the magma chamber partially empties, causing the circular collapse of the cone. This large depression is called a ***caldera***. Sometimes a lake fills in the caldera (Fig. 6.34). Calderas vary widely in size; for example, the La Garita caldera in Colorado is 85 km long and 40 km wide.

Inside the volcano, there are several unusual structures that produce surprising landscapes as a result of erosion. These are very typical of volcanic regions:

- ***Dikes***: these are rocks that form when massive lava fills open subvertical fractures. Often the surrounding volcanic rocks are rapidly eroded, leaving only a system of dike fields

Fig. 6.34 The filled caldera of Crater Lake (diameter of 8 to 10 km), Oregon, USA. Earth Sciences Photographic Archive, U.S. Geological Survey.

that resembles fortifications (Fig. 6.35). The dike rocks are very resistant and can serve as hydrogeological screens between aquifer compartments. They control the distribution of water on volcanic islands, as in Tenerife for example (Fig. 6.36).

- *Sills* and *laccoliths*: These are injections of magma from the magmatic chimney, but they intrude horizontally between preexisting layers of rock that offer a low resistance to magma invasion.

Fig. 6.35 Example of a dike, Massif Central, France. GEOLEP Photo, J.-H. Gabus.

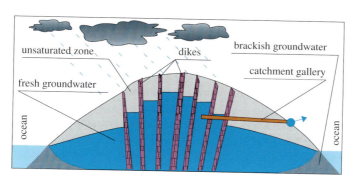

Fig. 6.36 Compartmentalization of groundwater resources by dikes on the island of Tenerife. Unscaled diagram.

- *Necks*: These are formed when a group of angular blocks solidifies in the volcanic chimney. Often this chimney is more massive than the cone. This neck structure is more resistant to erosion. Once the cone erodes, the neck is a spectacular pinnacle that is the relic of the volcano that has disappeared (Fig. 6.37).

Fig. 6.37 Example of a neck. Massif Central, France. Geolep photo, J.-H. Gabus.

6.4.2.4   Volcanism and man

The legend of Atlantis is thought to relate to the explosion of the volcanic island of Santorini in the Aegean Sea. Today all that remains is a partially submerged caldera. In Pompeii, nuées ardentes swallowed up the city and its inhabitants in 79 A.D. The images of people immobilized by asphyxiation are unforgettable.

The relationship between man and volcano has always existed. Gods, fertile soils, and economic factors continue to attract humans to live near volcanoes regardless of the risks they pose, including destruction of land and loss of life.

The favorable soil structure and mineral properties (potassium for example) of pyroclastic deposits make them ideal for agriculture. Metal deposits associated with these systems can be highly prized resources. The geology of volcanic areas may include significant aquifers and geothermal resources, which are of more and more interest today.

But volcanism can also cause catastrophes and massive destruction. Lava and ejecta cover the areas around volcanoes. Nuées ardentes and lahars are generally the most devastating. Cities such as Naples or Mexico City where millions of people live are particularly vulnerable. For example, Mexico City uses continuous video camera and seismic measuring equipment to monitor the volcano Popocatépetl. Another potential catastrophe is the tidal wave that can accompany the eruptions of coastal, island or submarine volcanoes. Table 6.1 shows several important events that have occurred in the past centuries.

[49, 98]

Table 6.1 Catastrophic volcanoes that have an important place in history.

Date	volcano	place	fatalities
1586	Kelut	Indonesia	10 000
1783	Laki	Iceland	10 000
1792	Mont-Unzen	Japan	15 000
1815	Tambora	Indonesia	92 000
1883	Krakatoa	Indonesia	36 000
1902	Montagne Pelée	Martinique	30 000
1912	Mont-Katmai	Alaska	0
1951	Le Lamington	New Guinea	3000
1980	Mont Saint-Helen	USA	57
1982	El Chichon	Mexico	2000
1985	Nevado del Ruiz	Columbia	23 000
1991	Pinatubo	Philippines	350

PROBLEM 6.1

Your task is to study the volcanic risk posed by two different volcanoes:

– Vesuvius (Naples region, Italy)
– Mauna Loa (Pacific)

For each volcano, you receive a map showing the extension of the volcanic flows (Fig. 6.38a and b). This data is completed by a chemical analysis, which is representative of the volcanic rocks (Tab. 6.2).

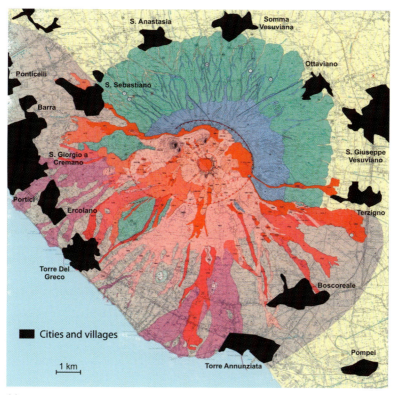

(a)

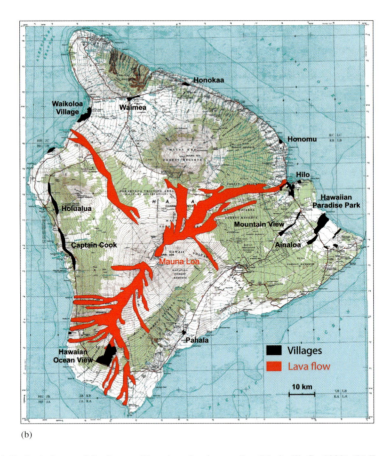

(b)

Fig. 6.38 (a) Geological map of the Somma Vesuvio volcanic complex (Marinelli, G., 1988). (b) Topographic map of Hawaii (USA) with lava flow (source USGS).

Table 6.2 Average geochemical composition of volcanic rocks from the two sites.

% Weight	Vesuvius (I)	Mauna Loa (USA)
SiO_2	56.19	49.16
TiO_2	0.62	1.9
Al_2O_3	18.07	10.7
Fe_2O_3	2.79	–
FeO	1.9	11.2
MnO	0.17	–
MgO	0.87	16.5
CaO	3.6	8.55
Na_2O	4.89	1.66
K_2O	7.02	0.3
$H_2O + CO_2$	1.57	0.08

Questions

- What can you say about the behavior of these two volcanoes and how do they differ?
- What types of risks threaten the different populations?
- What protective measures would you propose to set up in each case?

Look at the solution on the DVD

6.5   Major magmatic rocks

Here we give a classification summarizing the principal rocks based on field methods.

6.5.1   Classification

The classification of these rocks rests on their mode of emplacement and their composition. Magmatic rocks are formed by the solidification of magma at depth (plutonic rocks) or under subaerial/subaqueous conditions (volcanic rocks). Their texture (mineralogical arrangement on a microscopic scale) ranges from coarse to vitreous. Their structure (appearance in outcrop) is a result of their mode of emplacement. Their ***mineralogical composition*** is the first descriptive step in the identification of well-crystallized plutonic rocks. Additional analyses may be needed to determine the ***chemical composition***; this step is essential for microlitic and vitreous rocks (volcanic rocks).

6.5.1.1   Criteria for classification

There are various types of classification. In figure 6.39 we present a block diagram classification of the principal rocks. It has the advantage of describing the rocks in a simple way, although the method is rather approximate because it is based on mineralogy observed in the field. The rock name depends on:

- The relative abundance of different major minerals, represented on an axis showing the silica content of the rock. In this way, ***felsic, intermediate, mafic*** and ***ultramafic*** rocks can be distinguished.
- The geodynamic context (depth of rock solidification): the front of the block diagram shows plutonic rocks; the volcanic equivalents are shown on the back. In nature there are rocks intermediate in texture (microgranular), considered to be formed at intermediate depths.

6.5.1.2   Nomenclature

There are four principal families. Let us remember that their constituent mineral associations are the result of the fractional crystallization process (Fig. 6.13, 6.14, 6.15). We discuss the rock families beginning with the most felsic members and progressing to the most mafic:

- ***Granite*** family (felsic rock): They have lots of quartz, alkali feldspar, plagioclase, and black and/or white mica (Fig. 6.25 and 6.40a and b). The volcanic equivalent is ***rhyolite***, made of a light amorphous matrix often containing ***phenocrysts*** (large well-developed crystals) of feldspar and quartz (Fig. 6.41a). Particularly rapid cooling results in glass such as ***obsidian*** (Fig. 6.41b) and ***pumice*** if the lava was rich in gas (Fig. 6.41c) Microcrystalline dike equivalents are aplites (Fig. 6.26b).
- ***Diorite*** family (intermediate rock): These are essentially composed of plagioclase and amphibole. They do not have enough silica to form quartz in any significant quantity. Their plagioclase still contains sodium. The volcanic term is ***andesite*** (Fig. 6.41d and e). The ***granodiorites*** are an intermediate between granites and diorites. They differ from the diorites because they have a generally higher silica and potassium content, and sometimes amphibole or pyroxene replace biotite. ***Dacites*** are the volcanic equivalent.

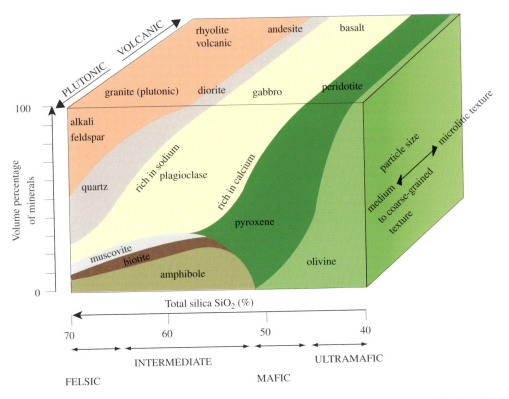

Fig. 6.39 Schematic diagram of the principal magmatic rocks based on their mineralogical composition (from [174]). Plutonic rocks, completely crystallized, are represented on the front of the block-diagram. Volcanic rocks, not very or not at all crystallized are represented on the back face of the block. There is, in fact, a continuum between these two classes of rock. Each mineral occupies a range that corresponds to the silica content necessary for its existence. By convention, silica decreases toward the right side of the diagram. Theoretically, for a given silica content of a rock, there is a mineral association that cuts the vertical line corresponding to this silica content. For example, a rock with a total silica content of 70% (felsic rocks) should lead to the creation of a rock containing approximately 22% quartz, 50% alkali feldspar, 14% plagioclase and 14% mixture of micas and amphiboles. This rock is called a granite. On the right part of the diagram, a mafic rock containing 50% total silica should contain 24% plagioclase, 72% pyroxene, and 4% olivine, and would thus be called a gabbro. The identification of volcanic rocks is not really possible using such a representation because only part of the minerals are formed. Geochemical analysis is thus necessary.

(a)   (b)

Fig. 6.40 Some plutonic rocks. (a) Porphyritic granite with phenocrysts of orthoclase. (b) Thin section of granite under a polarizing microscope, a typically plutonic coarse texture. (c) Gabbro. (d) Peridotite. Photos GEOLEP, P. Christe.

Fig. 6.40 (Continued).

- *Gabbro* family (mafic rock): Lacking quartz, they are composed mainly of pyroxene and calcic plagioclase, and sometimes amphibole and olivine (Fig. 6.40c). The volcanic equivalent, **basalt**, is made up of a dark gray microlitic matrix often containing phenocrysts of pyroxene or olivine (Fig. 6.41f).
- *Peridotite* family (ultramafic rocks): dark-colored magmatic rocks (Fig. 6.40d), they are composed almost exclusively of ferromagnesian silicates (90% olivine and pyroxenes); volcanic equivalents are rare.

Rocks other than those cited above exist in nature, but they are not common.

6.5.2   Field identification methods

The examination of a rock's texture, which is generally isotropic, is done prior to any other analysis. If the rock is entirely composed of minerals visible to the naked eye or through a hand lens, it is a plutonic rock (Fig. 6.40). If it is composed of a vitreous, microcrystalline or vesicular matrix, it is a volcanic rock (Fig. 6.29 and 6.41).

6.5.2.1   Plutonic rocks

These are massive rocks that are entirely crystallized and have negligible interstitial porosity. They are identified on the basis of the principal minerals present in the rock. Geologists use the same field methods that are used in mineral identification (§ 5.3). Thus, quartz, feldspars in general, and the principal ferromagnesian minerals (micas, amphiboles or pyroxenes, olivines) can be identified. The identification of the mineral assemblage makes it possible to determine the rock type (Fig. 6.39).

Fig. 6.41 Volcanic rocks. (a) Rhyolite with phenocrysts of quartz and altered feldspars (sawn surface). (b) Obsidian. (c) Pumice lava (liparite) with occluded pores, floating on water. (d) Andesite with phenocrysts of amphibole. (e) The same andesite in thin section under analyzed light, microlitic matrix containing plagioclase. (f) Basalt with olivine. GEOLEP Photos, P. Christe.

6.5.2.2 Volcanic rocks

Unlike their plutonic equivalents, these rocks often macroscopically display stratification related to their method of aerial deposition. The identification of volcanic rocks is more difficult due to the nature of the matrix, which generally lacks crystallized minerals (Fig. 6.41). Chemical analysis in the laboratory is the only way to identify them. In practice, some typical rocks can be identified using the following rules:

- The identification of minerals crystallized prior to the eruption (phenocrysts): They indicate the approximate silica content of the magma. For example, the presence of olivine or pyroxene is characteristic of a lava similar to basalt; mica is typical of an intermediate or felsic lava; the obvious presence of quartz is indicative of rhyolite.
- The color of the matrix: for poorly crystallized or uncrystallized rocks, the color of the matrix can be an indicator. Depending on their iron and magnesium content, the color can range from light (predominantly felsic) to dark (predominantly mafic).

Obsidian, a vitreous intermediate lava, is an exception (Fig. 6.41b).

This method is also used in the general flow chart for the identification of rocks in chapter 11 (Fig. 11.26 and 11.27).

[20,
59,
84,
296]

PROBLEM 6.2

Magma located 60 km down in an oceanic ridge is supposed to have the following chemical composition:

SiO$_2$	Al$_2$O$_3$	Fe$_2$O$_3$	MgO	CaO	Other
49%	15%	10%	8%	10%	8%

Questions

- What rock should form if this magma crystallizes deep underground? What minerals would it be composed of?
- If on the other hand this magma came to the surface, what rock would expect to form?

Look at the solution on the DVD

PROBLEM 6.3

A rock has a plagioclase which contains 85% Na, 8% Ca and 7% K. This rock also contains orthoclase.

Questions

- Draw the location of the plagioclase on the triangular feldspar diagram (Fig. 5.19).
- What plutonic rock is this?

Look at the solution on the DVD

Chapter review

Magmatic rocks result from the solidification of magma either at depth (plutonic rocks) or at the surface (volcanic rocks). Plutonic rocks are completely crystallized, massive and solid. Volcanic rocks are microcrystalline or amorphous, massive and hard (lavas) or powdery (pyroclastic rocks) depending on the type of eruption. Magmatic rocks form at oceanic ridges, in subduction zones or through lithospheric plates. Their composition is directly influenced by the tectonic environment in which they occur.

7   The Water Cycle

After discussing magmatism and the great internal cycle of the Earth, we will now move on to continental and oceanic sedimentary environments. We lack one element for discussing these topics: the water cycle. It is one of the driving forces for the transfer of material on the surface of the globe: erosion, transport, and sedimentation.

Water is also a necessary element for human life. It is essential to our survival: we need several liters per day. In developed countries, comfort is defined as a daily consumption of about 400 liters per day per person (Fig. 7.1).

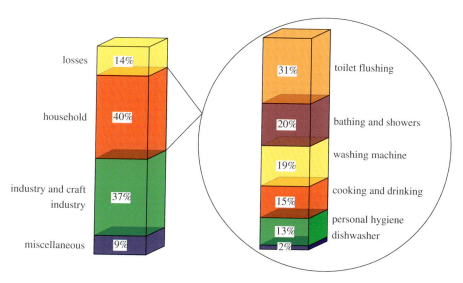

Fig. 7.1 Water consumption in Switzerland, an example of an industrialized country. Data from the Swiss Gas and Water Industry Association.

Water is also indirectly a food commodity, and it is a major factor in agricultural consumption. The following numbers give some idea of its importance:

- It takes 1,000 liters of water to produce 1 kg of bread;
- It takes 10,000 liters of water to produce 1 kg of meat.

A vegetarian diet is thus much more economical in terms of water: 1000 kcal/day requires 300 liters/day, compared to about ten times more for a carnivorous diet.

Figure 7.2 shows the anticipated evolution of world consumption depending on whether humans continue to be carnivores or adopt a vegetarian diet (from [319]).

Since countries that today have limited water supplies also have the highest birth rates, serious conflicts between the north and south appear to be inevitable in the near future if nothing is done to limit population growth and distribute resources more equitably.

Water is thus one of the keys to sustainable development. Engineers are closely involved in the use and management of this natural resource and its preservation: there are many reasons to focus on this subject.

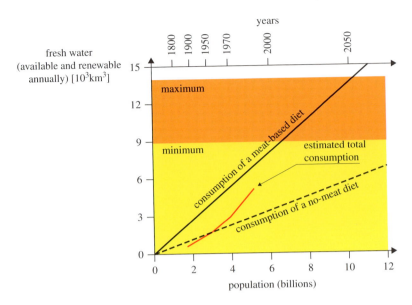

Fig. 7.2 World consumption of fresh water for recent centuries and forecasts for the future. There is an obvious difference in consumption between a vegetarian and a carnivorous diet (with 20% meat) (Values shown are for a consumption of 2500 kcal per person per day). Assuming that the global population doubles between now and 2050, we will consume all the renewable fresh water resources in the coming decades. A vegetarian diet can push back this date almost a century. The horizontal lines at 9,000 to 14,000 km³ represent the range of uncertainty on fresh water resources that are available and renewable each year (from [319]).

7.1   Water reserves and their exchanges

What we call the hydrosphere has intrigued humanity since the dawn of time. In the Old Testament, for example, the various forms of water were presented allegorically (Fig. 7.3).

Today we break down the water cycle into a series of reservoirs connected by mass transfer vectors. Each reservoir is characterized in terms of volume, the physical and chemical form of the water, the input it receives from other reservoirs, and the losses that it passes on.

If we agree that the volume of accumulated water V does not vary (which is realistic over several decades) we can calculate a mean *transit time* T_t of the water molecules in the reservoir (often expressed in years); Q is the flux of water entering, which is equal to the flux leaving

$$T_t = V/Q \qquad\qquad (7.1)$$

The inverse of the mean transit time is the mean *renewal rate* of water; it is the percentage of new water that enters the reservoir each year. In reality, water molecules have extremely variable transit times, which means that the average value itself does not reflect reality very well. For example, the water of Lake Geneva takes thirty years to travel from one end of the lake to the other on average. But the water that remains near the surface of the lake due to stratification travels the distance in just several years. This difference is even more noticeable in other reservoirs, particularly in the subsurface.

Over long periods, the water cycle has undergone significant modifications, particularly during the Quaternary glaciations (Chapt. 3). Today it is probably necessary to add short-term

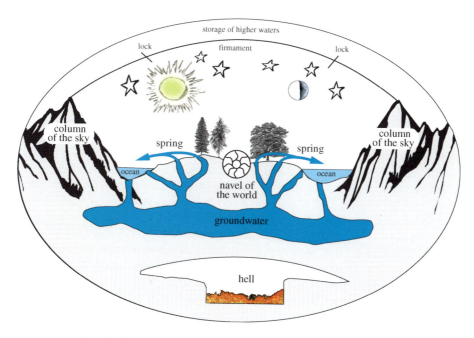

Fig. 7.3 Waters of the Earth according to the Old Testament (from [80]).

global climate changes due to man's activities to these natural modifications. We shall discuss the mechanisms a bit farther on.

We will now discuss the water cycle beginning with a description of the atmosphere. We will then move on to surface water (rivers, lakes, and glaciers) followed by groundwater. Finally, the role of the ocean will be presented. Table 7.1 gives the relative volumes of these reservoirs. A different weight will be given to each reservoir depending on its role in the processes of material transfer and the creation of new rocks.

Table 7.1 Volume of the principal water "reservoirs" on Earth.

ocean water	$1\,362\,200\,000\,\text{km}^3$	97.3%	
fresh water	$37\,800\,000\,\text{km}^3$	2.7%	100%
Total	$1\,400\,000\,000\,\text{km}^3$		

	continental ice sheets	$29\,181\,600\,\text{km}^3$	77.2%	
	groundwater	$8\,467\,200\,\text{km}^3$	22.4%	
Fresh water	lakes & marshes	$132\,300\,\text{km}^3$	0.35%	100%
	atmospheric water vapor	$15\,120\,\text{km}^3$	0.04%	
	rivers	$3780\,\text{km}^3$	0.01%	

7.2   The atmosphere

We will first discuss the atmosphere in general as a water reservoir. We will then see how the atmosphere transfers its water to the soil.

7.2.1   Atmospheric reservoir

The atmosphere is the most mobile water reservoir in space and time. The volume of water accumulated as vapor, droplets and ice crystals is small compared to other reservoirs (Tab. 7.1). Exchanges between the atmosphere and the ocean and between the atmosphere and the continents are very important; water molecules have transit times of only about ten days.

The composition of the atmosphere has evolved significantly over geologic time (Chapt. 3). Today the atmosphere is clearly oxygen-rich, or oxidizing. It is composed principally of nitrogen and oxygen (Tab. 7.2). However, a few minor components play an important role in the thermal balance of the atmosphere. These are the greenhouse components that man has introduced (for example: chlorofluorocarbons, aerosols and refrigerants) or those components whose concentrations he has greatly increased, such as carbon dioxide.

Table 7.2 Principal components of the atmosphere.

Elements	Percent volume
N_2	78
O_2	21
Ar	0.9
CO_2	0.03
H_2O	<1

7.2.2   The greenhouse effect

The greenhouse effect is a natural phenomenon that maintains a mild temperature at the Earth's surface. It results from the fact that visible rays reaching the Earth are reflected back into the atmosphere and are enriched in the infrared range in the process. The lower atmosphere does not allow this radiation to escape and in turn sends it back to the Earth, where it causes heating similar to what happens inside a greenhouse (Fig. 7.4). Water vapor, methane and carbon dioxide are primarily responsible for the screening role. Anthropogenic increases of methane and carbon dioxide lead to warming.

The increase of carbon dioxide results from the return of large quantities of geologic carbon to the atmosphere. This carbon was previously isolated from surface geochemical cycles and it has been released by the use of fossil fuels: coal, petroleum, and natural gas (Chapt. 10). The beginning of industrialization is clearly marked by an increase of carbon dioxide in the air, and this increase is no longer balanced by photosynthesis (Fig. 7.5). Scientists forecast that by 2050 the concentration of carbon dioxide will be double the natural concentration if we do not drastically change our consumption habits.

Methane also plays an important but less recognized role. Methane has increased significantly since the introduction of rice farming. Underwater agriculture creates reducing conditions

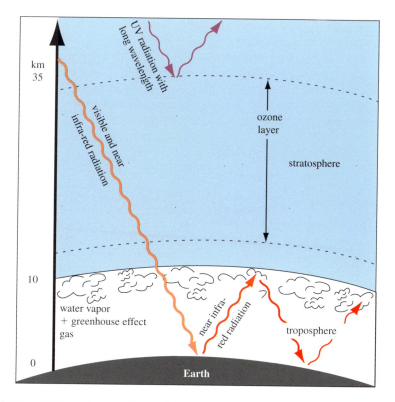

Fig. 7.4 Solar radiation in the atmosphere and the greenhouse effect. See also the animation on the DVD.

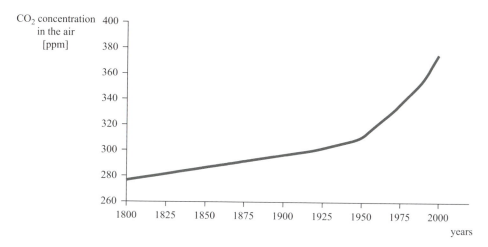

Fig. 7.5 Evolution of the carbon dioxide content in the atmosphere since the beginning of industrialization (smooth curve). Ancient data analyzed from Antarctic ice, data since 1958 in the atmosphere of Mauna Loa. University of California, La Jolla [297].

in the soil that mineralize organic matter as methane. Waste produced by intensive cattle production is also a significant source of methane.

Scientists use planetary-scale simulation models to attempt to estimate the scope of these
changes. These simulations involve numerous uncertainties due to the complexities of the
phenomena involved. Some anthropogenic factors promote warming while others have the
opposite effect. Aerosols, for example, tend to lower the temperature because of their screening
effect; these small particles are released when fossil fuels are used (Fig. 7.6). The connection

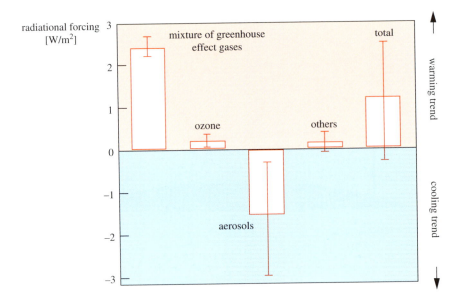

Fig. 7.6 Influence of greenhouse effect gases, ozone and aerosols on global warming (rectangles). The segments indicate the uncertainties regarding their role, which are very large in the case of aerosols (from [253]).

between the atmosphere and the hydrosphere is not well understood at this time. Great prudence is thus necessary in the interpretation of this relationship. In chapter 14 we will present
more details on the issues linked to climate change and its implications for the engineer and for
society (see § 14.2.3).

To complement the simulations on the current and future climate status, scientists study
paleoclimatic data to understand naturally possible variations and correlations between
the climate and the dominant factors (Problem 3.3). Drilling into polar ice sheets and analyzing old ice and air bubbles trapped in the ice lets scientists to "go back in time" several
hundred thousand years (Fig. 7.7). They analyze greenhouse gases and oxygen isotope
18, which is an indicator of paleotemperatures; when the climate is warm, the air contains
more heavy oxygen than it does during a cool period due to isotopic fractionation during
evaporation.

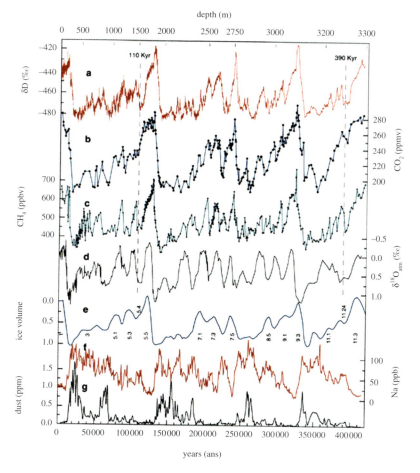

Fig. 7.7 Analyses of ice from the Antarctic ice sheet at Vostok. From top to bottom: Deuterium in ice, carbon dioxide, methane, ^{18}O in air bubbles, volume of ice estimated on the basis of isotopic variations in the oceans (used for time calibrations of the curves) marine aerosols, terrigenous aerosols (from [214]).

7.2.3   Atmospheric precipitation

The distribution of precipitation on Earth is one of the major components of climatology. We will discuss the two principal factors that control this distribution, and greatly simplify them for this discussion.

7.2.3.1   Climate zones

If we cross the northern hemisphere from the Congo to the North Pole, we encounter several large climatic zones based primarily on latitude L (Fig. 7.8). There are six principal zones:

Humid tropical zone ($0° < L < 15°$): The equatorial zones which have little seasonal variation. Precipitation is very abundant (up to more than 5000 mm/yr), soil humidity is very high, and the temperature does not vary by much. Runoff and transpiration by vegetation are significant. Soils are leached by the high rainfall, leading to the formation of thick alteration crusts (laterites).

Arid tropical zone ($15° < L < 35°$): Desert and sub-desert areas with large temperature deviations. Annual precipitation is less than 22 mm/y. Evaporation exceeds input, which can lead to the creation of closed basins with saline lakes.

Subtropical zone ($35° < L < 40°$): A transition zone between tropical regions and temperate regions. The average temperature is still very high. This zone can be dry, as in the case of the Mediterranean basin. But it can also be very humid when exposed to warm ocean winds that contain abundant water vapor; it has intense monsoons (for example: Southeast Asia) that bring annual rainfall of several meters per year.

Temperate zone ($40° < L < 55°$): These are regions with well-identified seasons and moderate temperature deviations. Precipitation is distributed throughout the year (on average 1000 mm/yr); at high elevations, some of it falls as snow. There is an equal distribution between evapotranspiration, runoff, and infiltration.

Boreal zone ($55° < L < 70°$): This is the cold zone that transitions to the polar zone. Precipitation is less significant and a good part of it occurs in solid form. Evaporation is low due to the weak solar flux.

Polar zone ($70° < L < 90°$): These are cold and dry regions (<300 mm/yr) where precipitation occurs primarily as snow. Since precipitation is low, losses by evaporation and infiltration are also low; there is a short summer period, and a flow of solid water in the form of glaciers.

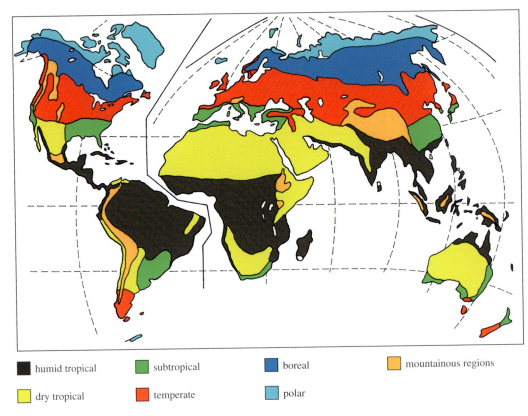

| ■ humid tropical | ■ subtropical | ■ boreal | ■ mountainous regions |
| □ dry tropical | ■ temperate | ■ polar | |

Fig. 7.8 Major climate zones of the Earth.

7.2.3.2   Orographic phenomena

On the surface of the Earth, mountain chains play a particular role in climate. They are an exception to the latitudinal distribution of climatic zones (Fig. 7.8). By examining rainfall distribution, it can be seen that mountains often separate dry zones from wet zones. These topographic phenomena are known as the *orographic effect*.

Rainfall amounts and altitude

Topography forces air to rise, which in turn causes it to cool off and condense. The turbulence created by the topography is also favorable to rainfall production. Thus the volume of rainfall generally increases with altitude (average rate of about 100 mm/y for 100 m altitude in the northern alpine region). But there are important variations depending on the position of the mountain range in relation to the main atmospheric currents (Fig. 7.9).

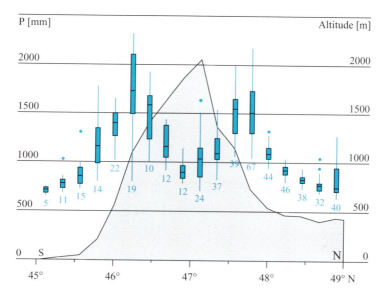

Fig. 7.9 Rainfall variation related to altitude throughout the Alps between the Gulf of Venezia (Italy) and Bayerischer Wald (Germany). The gradient is highly variable: steep on the northern flank, a bit less on the southern side. In the center of the range (the Tyrol), the gradient becomes negative. This distribution is more or less symmetrical because the Alps receive humid air from both sides. Statistics (maximum, 75%, median, 25%, minimum) on annual rainfall between 1971 and 1990. Topography is integrated on ribbons of 0.22 degrees of latitude and 2.5 degrees of longitude. The numbers indicate the number of stations considered in each ribbon. After Schwarb M., Frei C., Schär C., Daly C., Hydrological Atlas of Switzerland, sheet 2.6, 2001.

The foehn effect

The *foehn* (Swiss word for this type of wind) effect, a warm wind that flows down hillsides under the wind, is closely related to the temperature gradient as a function of altitude.

When an air mass encounters a mountainous barrier that it cannot skirt around, the air mass rises and as it does so its temperature gradually decreases. If the air is initially dry, its temperature drops by about 1.2°C/100 m. After passing the crest, the air warms as it descends according along the same thermal gradient. The wind temperature is thus the same at equal altitudes on both sides of the mountain. When the rising air is humid, the air-water vapor mixture reaches the dew point and condensation begins; the windward side receives abundant rainfall as a result. Each molecule of water that goes from gas to liquid adds energy equivalent to the latent heat of vaporization to the environment. As a result, the air mass and water droplets cool less quickly than the dry air. Thus the temperature gradient is no more than about 0.6°C/100 m. The rising of the air causes a less significant temperature drop than in the case of initially dry air. On the other side of the mountain, the drier air will follow the "normal" gradient of 1.2°C/100 m and the air temperature on this side will thus be higher than the temperature at the same elevation on the windward side (Fig. 7.10). This unsymmetrical distribution is typical of ranges that have one dominant humid wind direction.

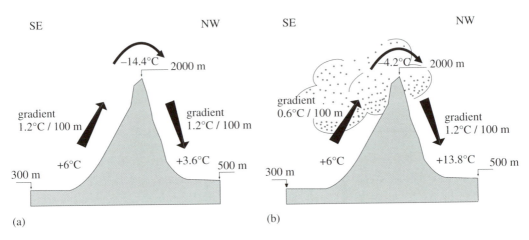

Fig. 7.10 Climatic effect of atmospheric foehn currents crossing a mountain chain. (a) Case of dry air, without the foehn effect. (b) Case of humid air, with the foehn effect; example of the situation on 20 February 1966 in the Alps of western Switzerland. The temperature south of the Alps was 6°C. It was 14°C in Montreux and at the same time, it was 6°C in Geneva and 5°C in Neuchâtel (from [32]).

There are numerous examples of asymmetrical distribution of rainfall on the Earth. Let us cite the following:

- The Himalayas: the monsoons flood the southern slope of the mountains while Ladakh is practically a desert.
- The Northern Andes: the eastern currents that are full of humidity on the Atlantic side drop rainfall on the Amazon slope; on the other side are the coastal deserts of Peru.
- The Southern Andes: contrary to case of the Northern Andes because the dominant winds are from the west, bringing humidity from the Pacific; to the east of the mountains are the Argentinian deserts.

7.2.4   Rainfall fractionation on the soil

The water from liquid precipitation P is distributed into several components when it reaches the ground, and these components determine the water balance of a unit soil surface.

$$P = ET + I + R \tag{7.2}$$

ET = *evapotranspiration*: it is the sum of physical evaporation and the transpiration of the biomass.

I = *infiltration*: this is the quantity of water that infiltrates the soil.

R = *runoff*: this is the excess rainwater that exceeds the capacity of evaporation and infiltration; the water runs off along the surface of the soil.

All these components are conventionally expressed in millimeters, like precipitation (a 1 mm rainfall is equivalent to an input of $1\,1/m^2$). Their relative contribution varies:

- In space, at all scales, as a result of climate, slope of the ground, composition of the ground, soil use, presence of drainage, urbanization, permeability of the soil and subsurface;
- In time, according to the seasons (plant activity, degree of saturation, presence of snow, frozen ground).

The mean annual balance on the scale of all emergent land is approximately the following:

- Input – rainfall and snow: $110,000\,km^3$,
- Return to the atmosphere: $70,000\,km^3$,
- Water remaining at and in the ground: $40,000\,km^3$.

In cold regions, the water on the ground occurs as snow and ice. In warmer regions, the distribution between surface water and groundwater essentially depends on the nature of the geological substrate. There are two extreme cases (Fig. 7.11):

- Impermeable substrate: infiltration is practically nonexistent, very significant runoff with well-developed river system (Fig. 7.12). For example, argillaceous substrata (claystones, marls, Quaternary clays) or non-fissured magmatic rocks.
- Permeable substrate: all the water infiltrates, there is no runoff, thus no river system (Fig. 7.13). For example, karst rocks (limestone, gypsum) and gravelly alluvium.

[4, 6, 32, 95, 99, 153, 154, 197, 215, 217, 226]

Rocks with average permeability such as marly limestones or poorly cemented sandstones are intermediate cases.

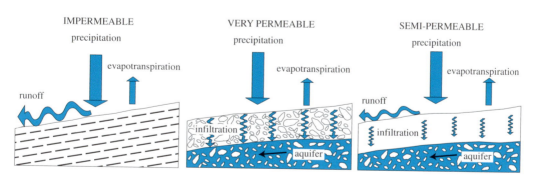

Fig. 7.11 Distribution of surface water–groundwater as a function of the nature of the geological substrate.

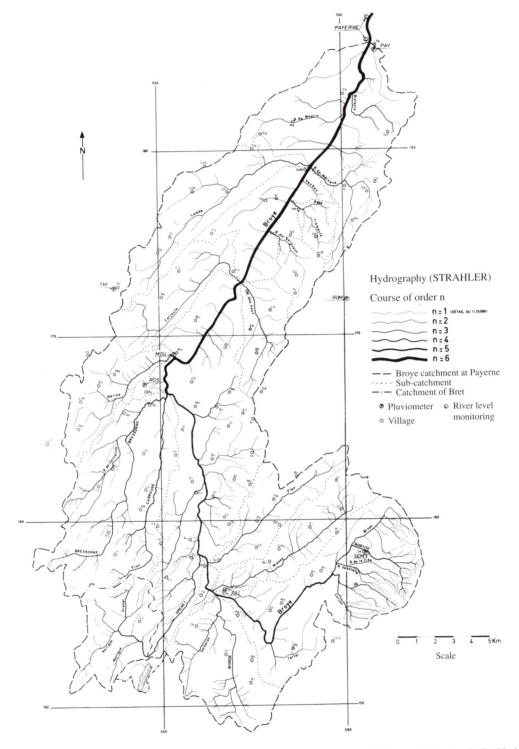

Fig. 7.12 River system on top of an impermeable substrate. Case of the Broye basin (Western Switzerland). Strahler's classification, see § 7.3.1.3 (from [202])

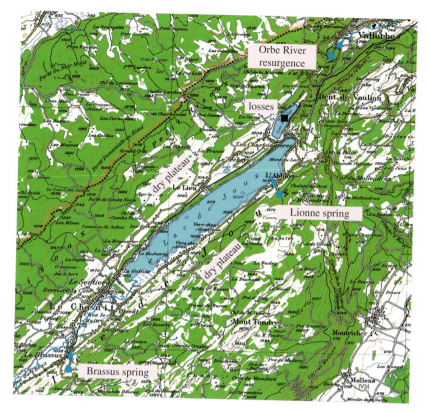

Fig. 7.13 River system in very permeable terrain. Example of the Orbe karst basin (Swiss Jura). Lake Joux and Lake Brenet are supplied by the Orbe river, which comes from Les Rousses (France) and by two large karst springs. The outlet from the lakes is underground. The Orbe reappears upstream of Vallorbe after flowing through caves. Aside from the Orbe alluvial plane upstream of the lake, the area is almost completely devoid of streams. Reproduced with authorization of Swisstopo (BA056911).

7.3 Surface water

Continental surface water occurs in a series of reservoirs in parallel and in series that communicate with each other in space and time. We first discuss the rivers, most dynamic reservoir, then the more static ones, lakes and glaciers.

7.3.1 Rivers and streams

Moving water in streams and rivers is called *flow*. It is the sum of several hydrologic components:

- Water that runs off the surface of the topographic basin,
- Water from springs that appear at the surface in this basin,
- Water that is contributed artificially to the streams (wastewater, groundwater pumping stations, hydroelectric and other diversions).

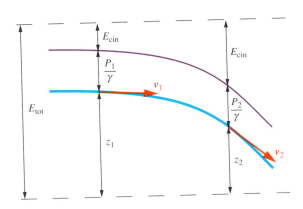

Thus the water in a river comes not only from the surface, but also from groundwater. The groundwater input maintains the flow in the river when there is no surface runoff. It is called *base flow*.

7.3.1.1 Basics of river hydraulics

It will be useful to define some of the physical principles that govern flow conditions in rivers. They are used to calculate the magnitudes of the velocities and head losses in natural beds. These principles will be useful in the discussion of erosion, transport and sedimentation conditions in rivers (Chapt. 8).

The **Bernoulli equation**, or the principle of energy conservation, is the basis for calculating the various forms of energy that play a role in the permanent flow of a perfect fluid (flow without friction) under gravitational conditions. The hydraulic head H is constant at all points along the flow path. It is the sum of the potential energy of position, the potential energy of pressure, and kinetic energy (Fig. 7.14). It can be expressed as:

$$H = z + \frac{P}{\gamma} + \frac{v^2}{2g} \tag{7.3}$$

where
  z = elevation of the point along the path
  P = pressure at the same point
  γ = specific weight of the fluid
  v = fluid velocity
  g = gravitational acceleration

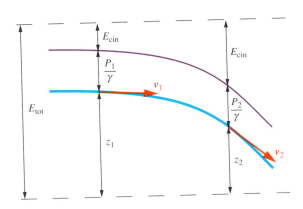

Fig. 7.14 The three energy terms in the flow of a perfect fluid.

In reality, because of its viscosity, water does not behave like a perfect fluid. The sides and the bottom of the river have a frictional effect on flow. This frictional effect causes erosion of the streambed. The passage from point 1 to point 2 along the path causes a loss of head ΔH (**generalized Bernouilli law**). This head loss, which is equivalent to an increase of internal energy of the media, will depend on numerous factors.

One of these factors is the type of flow in the channel. When flow is slow and the velocity vectors are parallel, flow is *laminar* and the head losses are proportional to the velocity (Fig. 7.15). When flow becomes rapid, the pathways are turbulent, which gives rise to the term *turbulent flow*; the head losses are equivalent to the square of the velocity.

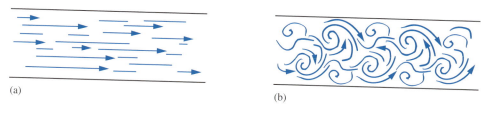

(a)   (b)

Fig. 7.15 Laminar flow (a) and turbulent flow (b).

The geometry of the bed and roughness of the walls play a significant role. Numerous empirical expressions have been proposed on the basis of artificial channel tests and mathematical models. One of them that is often used is Strickler's law, which describes the head loss ΔH as:

$$\Delta H = \frac{v^2 \cdot L}{K_s^2 \cdot R_H^{4/3}}$$

(7.4)

where

v = water velocity (m/s)
L = length of the channel over which the head loss ΔH is being measured (m)
K_S = Strickler's head loss coefficient ($m^{1/3}$/s) (Tab. 7.3)
R_H = hydraulic radius (m)

The hydraulic radius is the ratio between the cross sectional area of flow S_m and the wetted perimeter P_m (Fig. 7.16).

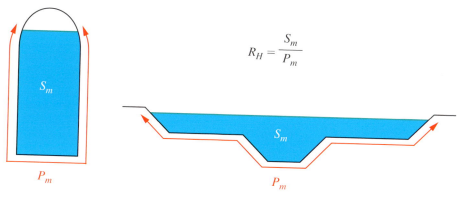

$$R_H = \frac{S_m}{P_m}$$

Fig. 7.16 Definition of hydraulic radius of a river.

Table 7.3 Order of magnitude of the Stickler's head loss coefficient (from [228]).

Type of watercourse or conduit	Type of bed	K_S [m$^{1/3}$s^{-1}]
Mountain stream	Bed with blocks and small falls	10 to 15
River	Irregular bed containing boulders	20 to 30
Canal	Coarse gravel bed	30 to 40
Gallery	Regular fine gravel bed	40 to 45

Strickler's equation can also express the water velocity v as a function of the hydraulic gradient i:

$$v = K_s \cdot i^{1/2} \cdot R_H^{2/3} \tag{7.5}$$

with

$$i = \frac{\Delta H}{L} \tag{7.6}$$

It is thus obvious that it is possible hydraulically to distinguish moderate velocity flows with small head losses (rivers) and violent flows in which the head losses are considerable (mountain streams). We can develop this distinction, which will be useful in determining the type of water-course we are dealing with in a given region. Let us consider an experimental channel with a rectangular cross-section of width B that must evacuate a constant discharge Q. The slope i of the channel is modified and the water depth measured for each value of slope (the hydraulic gradient i is approximated by the slope of the bed, simplification that is valid if the slope is low).

- Either discharge Q flows along a low slope and the water depth in the channel will be significant,
- Or the slope is increased, and thus the velocity increases, and the water depth will be reduced.

Let us examine the relationship between slope i and water depth d. For this we will need a term that expresses the *specific head* H_S over the entire cross section (Fig. 7.17):

$$H_s = d + \frac{v^2}{2g} \tag{7.7}$$

This head is effectively constant across the entire cross-sectional area of the channel since flow is permanent and the water surface is parallel to the bottom.

If we apply a loss of head law to this channel, as for example Strickler's law, we can express the relationship between the slope i and the depth d by introducing the width B:

$$i = \frac{v^2}{K_s^2 R_H^{4/3}} = \frac{Q^2}{B^2 d^2 K_s^2 \left(\dfrac{Bd}{B+2d}\right)^{4/3}} \tag{7.8}$$

$$i = \frac{Q^2}{K_s^2 B^{10/3}} \cdot \frac{(B+2d)^{4/3}}{d^{10/3}} \tag{7.9}$$

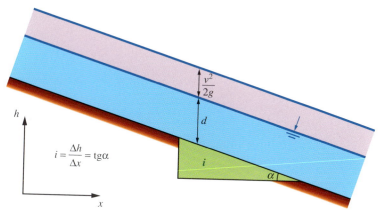

Fig. 7.17 Specific head along the entire section of the channel.

This non-linear relationship $i = f(d)$ is shown in the left part of figure 7.18.

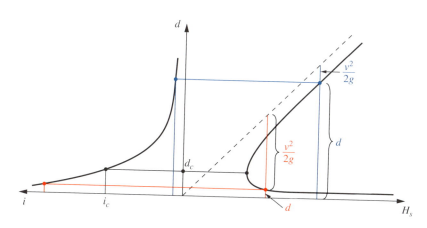

Fig. 7.18 Relationship between slope, water depth and specific head. Distinction of fluvial and torrential flow on both sides of the critical depth. Case of a channel with a rectangular cross-section and constant discharge in space and time; v is the velocity in the channel.

Now we will express the relationship between the specific head H_s and the depth d:

$$H_s = d + \frac{v^2}{2g} = d + \frac{Q^2}{2gB^2d^2} \tag{7.10}$$

This function $H_s = f(d)$, which is also non-linear, is represented graphically on the right part of figure 7.18. The derivative of the specific head with respect to depth is expressed as follows:

$$\frac{\partial H_s}{\partial d} = \frac{\partial}{\partial d}d + \frac{Q^2}{2gB^2} \cdot \frac{\partial}{\partial d}d^{-2} = 1 + \frac{Q^2}{2gB^2}(-2) \cdot d^{-3} = 1 - \frac{Q^2}{gB^2d^3} \tag{7.11}$$

The curve shows that the specific head is minimal for a ***critical depth*** d_c, which corresponds to the condition:

$$\frac{\partial H_s}{\partial d} = 0 = 1 - \frac{Q^2}{gB^2 d_c^3} \tag{7.12}$$

$$d_c = \sqrt[3]{\frac{Q^2}{gB^2}} \tag{7.13}$$

The specific head at this point is:

$$H_s = d_c + \frac{d_c^3}{2d_c^2} = \frac{3}{2} d_c \tag{7.14}$$

since

$$\frac{Q^2}{gB^2} = d_c^3 \tag{7.15}$$

This critical depth value separates the two areas of the function $H_s = f(d)$ (Fig. 7.18):

* When $d > d_c$, the kinetic energy term $v^2/2g$ is low; flow is essentially potential. It is ***fluvial flow***.
* When $d < d_c$, the kinetic energy term predominates; flow is essentially kinetic. This is ***torrential flow***.

7.3.1.2   Rivers over time

Rivers have variable discharges over time. During rainy periods, discharge often increases abruptly and then slowly diminishes, according to a negative exponential function (Fig. 7.22); these two phases constitute a ***flood*** event. Floods are sometimes catastrophic, particularly when buildings have been constructed within the flood corridor, when protective dikes are unable to contain exceptional floods or when they fail (Fig. 7.19). An animation on the DVD shows a flood in a torrential flow condition.

The engineer must ensure that the structures are correctly designed to withstand high intensity floods on the basis of probabilities established for each type of project, in other words, a risk category. Dams must be able to resist severe floods that have high ***return periods***, for example a thousand years (a flood corresponding to a thousand-year return period would occur on average every 1000 years). In contrast, a small bridge used to cross a river should be able to withstand only a one-hundred year return period flood (Fig. 7.20). The determination of flood discharges equivalent to very low probabilities is determined by extrapolation using statistical adjustment functions based on historic flood data (Fig. 7.21).

Fig. 7.19 Exceptional flood in Vaison-la-Romaine on 22 September 1992. Remains of a house built in the flood corridor. GEOLEP Photo, A. Parriaux.

Fig. 7.20 The rivers of central Asia react intensely to precipitation when the soil has not thawed sufficiently. Building bridges across these wide alluvial plains is a problem for engineers. In poor countries the construction of large metal or concrete bridges is not possible; wood, a local resource, is used to construct large works such as this bridge over the Tuul Gol River in Outer Mongolia, SE of Ulan-Bator. Its piles are protected by upstream cutwaters made of wood to protect against the impact of floating debris, especially blocks of ice transported during the spring thaws. GEOLEP Photo, A. Parriaux.

In contrast to floods, after long droughts, the discharge is very low; this situation is called *low-water level* (Fig. 7.22). In hot countries, the low water level in rivers can be so low that the rivers dry up. Although the low water situation is less spectacular than floods, it is no less worrisome. Irrigation has traditionally caused the diversion of river water, resulting in significant ecological damage. Many dams have been built for irrigation. Recent legislation generally guarantees a minimal discharge into rivers downstream of these dams.

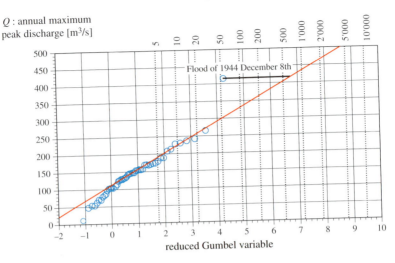

Fig. 7.21 Graph showing the adjustment of the annual maximum peak discharge Q (the maximum instantaneous discharge per year) measured for 52 years at the hydrometric station on the Broye in Payerne (Western Swiss Plateau). The measured discharges (points) are reported on graph probability paper (here Gumbel's Law) showing the reduced variable on the x axis and the variable Q on the y axis. Each point is determined by its experimental frequency, obtained by organizing the data of the 52 years in ascending order. If the point discharges followed the selected law of probability, they would line up along a straight line. It is obvious that the discharges of the Broye river correspond very well to the straight line indicated on the graph. However, the point that corresponds to the exceptional flood of 1944 and its 400 m³/s is found far above this adjustment. According to Gumbel's law, this event would be a flood with a return period T equal to 700 years; that is, in the course of a half century of observation, an event occurred that should occur on average only every seven centuries. This operation is scientifically questionable because it assumes a continuity of flood-control processes that in reality have not been verified for such discharges. The method presupposes a continuity of basin conditions, which is evidently not the case in urbanized areas or areas of intensive agriculture (from [202]).

The normal variations in river discharge, variations that respond to typical hydrometeorological conditions of an average year define its *flow regime*. There are three principal types:

- *Pluvial regime*: flow responds only to rain events, with a slight time offset. This regime has the highest variability of discharge, and as a result, the most violent floods. Pluvial regime-type basins occur at low elevations and at low latitudes. They are the most common on the Earth.
- *Nival regime*: water supply comes primarily from snowmelt. These are boreal or high altitude basins. In reality, many basins in and around mountainous areas are of a mixed pluvial-nival regime.
- *Glacial regime*: the summer ice melt of glaciers provides the principal input to flow. This regime is characteristic of the polar zone and high mountain chains; there is a lot of water in the rivers during the dry season at the foot of the mountains, which is intensively used for irrigation. In winter, by contrast, the discharge is practically nonexistent. Many mountains basins have a nival-glacial regime.

As a result of climate warming, the regime of rivers should tend to change from glacial to nival and from nival to pluvial (see § 14.2.3).

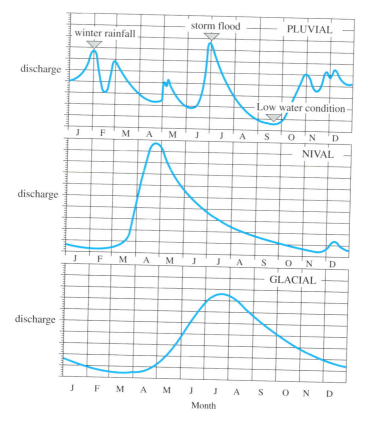

Fig. 7.22 Typical hydrographs of the three principal flow regimes.

7.3.1.3 Rivers in space

Flow is distributed into geographic units called **watersheds** or **catchment basins**. These are surfaces defined by the geometric place of points in an area within which water reaches a particular river.

For the part of water that comes from runoff, the basin can be easily identified. It is the ridgeline that surrounds the basin. This is the **topographic basin**.

For the groundwater that reaches a river, it is more difficult to understand since this is no longer a problem of a surface, but of volume. Groundwater circulates in a very complex way, flowing from one layer to another. It is mainly the structure of the subsurface that determines this second type of basin: the **groundwater supply basin**.

Where the subsurface is quite homogeneous and not very permeable, the two types of basins are almost identical. In areas of high permeability, they are very different (Fig. 7.23). We will discuss groundwater supply basins in section 7.4. For this chapter on surface water, we will use the topographic criterion.

Streams generally coalesce, with the exception of some rivers that separate temporarily on alluvial plains or around alluvial fans, and they generally form a tree-shaped **river system**. Moving from upstream to downstream, a river system begins with small ravines several decimeters wide, where flow is often temporary. These are first-order segments. When two stretches meet, they form a second-order stream, a bit wider. This process continues, creating

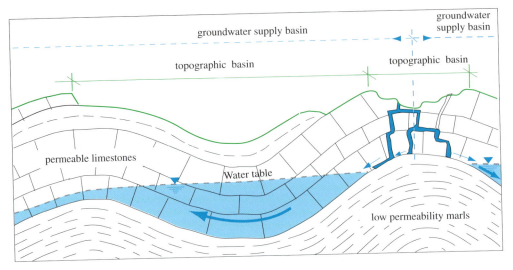

Fig. 7.23 Difference between the topographic basin and the groundwater supply basin in karst rocks.

the hierarchy of the flow system. Strahler [277] proposed the system of numbering river basins on the basis of the order of tributaries, their number and their length (Fig. 7.24), in order to facilitate a comparison of basin morphologies.

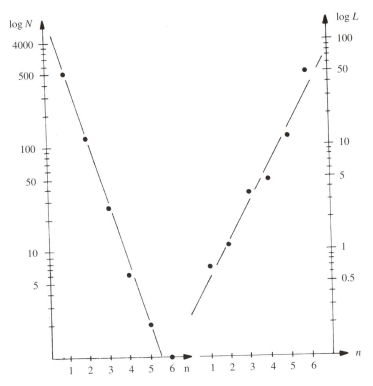

Fig. 7.24 Strahler's classification of the n-orders of tributaries for the Broye basin (Western Swiss plateau). N = number of tributaries, L = average length. See also Fig. 7.12.

The geometric shape of the system often reveals deep geologic structures, such as faults or folds in the geological substrate (Fig. 7.25).

In chapter 8, we will return to the functions of flow in rivers when we discuss the continental phenomena of erosion and sedimentation. We will see in particular how rivers adopt a meandering path when the slope is low and the geological substrate is homogeneous.

[33, 235, 240, 260, 261, 274, 275]

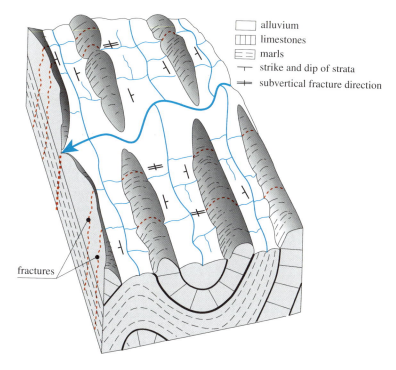

- alluvium
- limestones
- marls
- strike and dip of strata
- subvertical fracture direction

fractures

Fig. 7.25 The geometry of the system shows a network of fractures in the rocky substratum. These zones are equivalent to rocks that are less resistant to weathering and mechanical erosion by the river. Water follows zones of weakness. Water does the same with weak stratified and folded beds as compared to harder beds. NB: the dip determines the orientation of the beds. The long dash shows the orientation of the horizontal of the bed, the small dash shows the direction of maximum dip (§ 12.4.1).

PROBLEM 7.1

We are looking at the Broye watershed in Payerne (Fig. 7.12). We know:

- The surface area of the basin, closed at the Payerne federal gauging station: $393\,km^2$,
- The spatial distribution of average annual rainfall in the basin: *isohyet* map (lines of equal rainfall) (Fig. 7.26),
- The mean inter-annual regime of the Broye watershed at the "Payerne" station (Fig. 7.27).

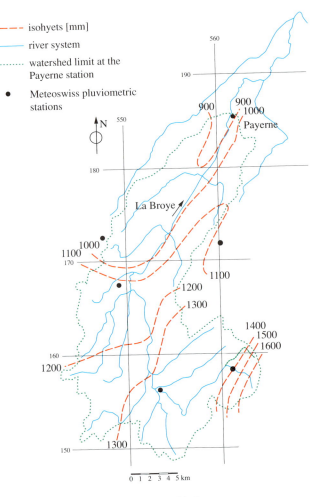

Fig. 7.26 Mean annual isohyet map.

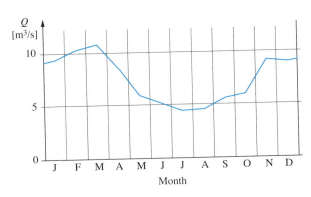

Fig. 7.27 Mean inter-annual regime of the Broye basin at the "Payerne" station. Monthly discharges

Questions

- Calculate the order of magnitude difference in water volume at the entrance to and the exit from the basin.
- To what can we attribute the difference?
- We are assuming that the discharge of all the groundwater outlets from the basin are included in the discharge of the river at the "Payerne" station. Given that the station is located on an alluvial plain, evaluate this assumption.

Look at the solution on the DVD

7.3.2 Lakes

Lakes play the role of a hydraulic buffer in a river system. A lake can regulate the discharge of the river that leaves the lake.

7.3.2.1 Geological classification

Along the profile of a river, a lake occurs upstream of an obstacle that dams up the flow. The land-water contact line thus marks a concavity before the threshold. The origin of the obstacle and the concavity may be related to very different phenomena. There are six main phenomena, the two major ones linked to the presence of glaciers in the past.

Morainal dam lakes

We will see in chapter 9 that valley glaciers form a ridge of debris called a frontal moraine in advance of their front. When the glacier retreats, and if the moraine resists erosion, a lake can form behind the moraine (Fig. 7.28).

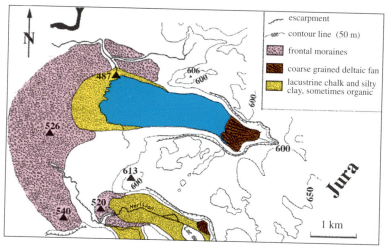

Fig. 7.28 Morainal dam lake behind the frontal moraine formed by the Jurassian ice cap during the last glaciation. Lake Chalain, Franche-Comté, France (from [41]).

Lakes formed by glacial overdeepening

The profile along a glacial valley does not resemble that of a valley of a mountain stream or river. The bottom has a series of thresholds (a convex profile) where rocks are hard and over-deepenings, also called ombilics (a concave profile) that occur where rocks are soft. As a solid material, ice can move over these thresholds due to the pressure from the upstream mass of glacier. When a valley glacier melts, it leaves in front of it large lakes that are the over-deepened segments upstream of the obstacle. This phenomenon creates the majority of mountain and piedmont lakes.

Lake Geneva is one of the best examples of this phenomenon (Fig. 7.29). Let us study its longitudinal profile and the course of the Rhone downstream of Geneva in detail. The rocky substrate under the Quaternary deposits at the bottom of the lake is below sea level; this shows that glacial erosion cut very deeply into the soft rocks of the Plateau. Downstream of this trough, the glacier had to cross the Jura Mountains at that time. In the transverse gorge of Fort-L'Ecluse, the level of the rock under the present-day alluvium (the lowest point of the threshold) is higher than sea level because of the greater resistance of the Jura rocks. Such a profile is possible only with a solid erosion vector: the Rhone glacier. It was able to cross the threshold because of the pressure transmitted by the ice in the upstream part of the basin. Not far from Geneva, the lakes of Neuchâtel, Morat and Biel all have the same origin. They formed one large lake at the time the glacier retreated (see also Fig. 8.52).

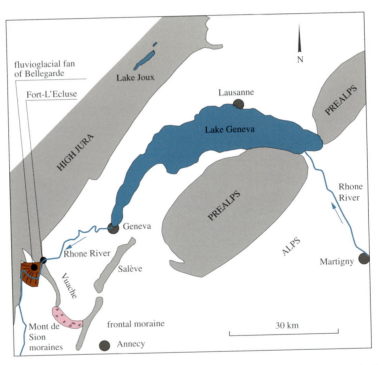

Fig. 7.29 One of the most beautiful lakes resulting from glacial over-deepening: Lake Geneva (or Leman lake).

Lakes dammed by alluvial fans, landslides or rock-mass falls

As valley slopes erode (Chapt. 9), obstructions of the major channel can result in breakup deposits from a tributary. They form an alluvial fan that the main river cannot clear away. Water accumulates upstream, creating a shallow lake. The same phenomenon can result from a landslide or a rock-mass fall. This situation occurred in the Alps north of Venice in the 1960s, and is known as the Longarone catastrophe. This large landslide filled the reservoir of the Vajont dam and created a small lake upstream (Fig. 7.30). Lake Sarez in Tajikistan is another example. The lake formed behind the enormous Usoy rock-mass fall in 1911 following an earthquake. The possible rupture of this natural barrier threatens the whole of the River Murghab basin.

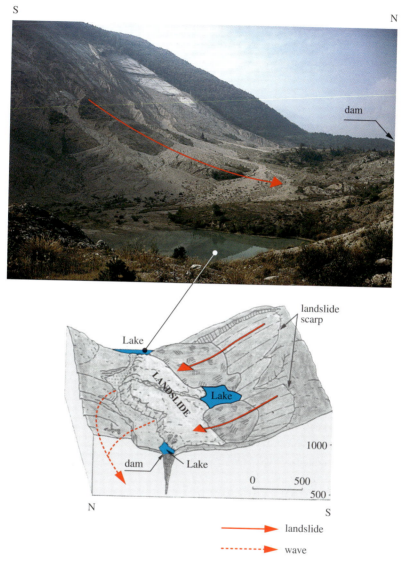

Fig. 7.30 Lake formed upstream of the Vajont slide (Northern Italy) on 9 October 1963 at 22:39. From the southern slope of the valley, a mass of 300 million m³ slid at high speed (about 60 km/h) into the reservoir behind the Vajont dam. An enormous wave flowed over the crest of the dam and destroyed the city of Landslide Longarone situated on the Piave alluvial plain downstream of the dam. The death toll was 1900. GEOLEP Photo, J.-H. Gabus.

Lava-dam lakes

Fluid lava from a volcanic structure located on the side of a basin can flow along the axis of the main valley and permanently obstruct the river when the lava solidifies. This situation is typical of old volcanoes in the Chaîne des Puys in the Massif Central of France (Fig. 7.31).

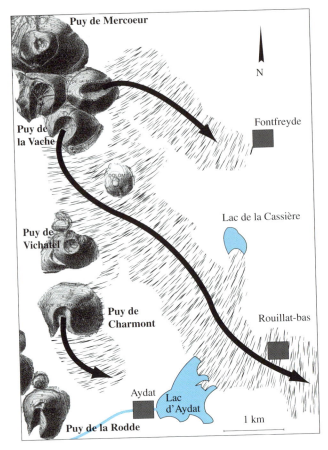

Fig. 7.31 Dam resulting from a lava flow (dashed areas). The case of the lake of Aydat in the Chaîne des Puys, Massif Central.

Crater and caldera lakes

At the end of an active phase, the crater of a volcano or a caldera may accumulate rainwater much like a large rain-gauge, if the walls are intact and the bottom is not permeable. This creates a circular lake (Fig. 7.32).

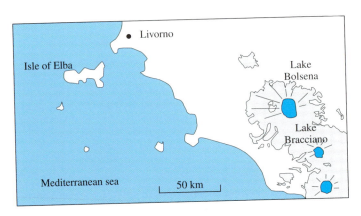

Fig. 7.32 Series of Roman lakes with Lake Bolsena, the largest volcanic lake in Europe.

Tectonic lakes

Lakes of tectonic origin are generally large. They occupy the bottom of rifts in divergent plate zones at the center of the continents (Chapt. 6). They are bordered by large normal faults.

This is the case of the great lakes of East Africa located from south to north along the rift. Lake Baïkal in southern Siberia has occupied a tectonic depression north of the Himalaya for 25 million years; it is the largest liquid fresh water reservoir on the planet, with a maximum depth of more than 1600 meters. In the southern part, the lake is now threatened by significant industrial pollution.

7.3.2.2   Thermal classification

The thermal behavior of lakes is the reason for their hydrodynamic behavior and ecological consequences. The temperature of the water has a direct influence on its density. We know that the maximum density of fresh water occurs at 4°C. The ability of water to mix depends on density contrasts, particularly related to temperature. The phenomenon led to a definition of the thermal regime for lakes to illustrate how they function. In the nineteenth century, Forel, the illustrious limnologist of Lake Geneva, proposed a classification of three types (Fig. 7.33).

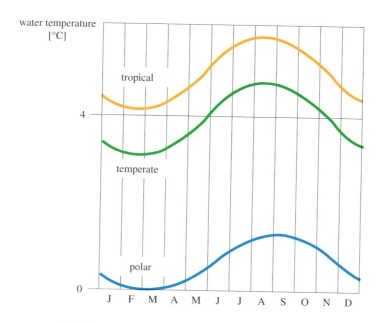

Fig. 7.33 Thermal classification of lakes according to Forel.

Polar lakes

The temperature of polar lakes always remains between 0 and 4°C. The coldest part is close to the surface; this is why lakes freeze in winter. Polar lakes are located on the perimeter of glacial regions and at high elevations in mountain chains.

Temperate lakes

These are lakes typical of medium latitudes, when they don't have a large mass of water. The temperature curve near the surface crosses the 4°C line, while the bottom water is stable at about 4°C.

Tropical lakes

These lakes have a surface temperature that does not go below 4°C. As their name indicates, they are located in regions of low latitude. However, there is one exception: Lake Geneva. In fact, because of its large volume, great depth, its high thermal inertia, and temperate climate that surrounds it, the temperature of the surface water does not generally go below 4°C. This implies a particular thermo-mechanical behavior to which Lake Geneva owes its good health. The diagram in Fig. 7.34 shows its temperature and density variations in winter and summer.

[3, 90, 91, 97, 290]

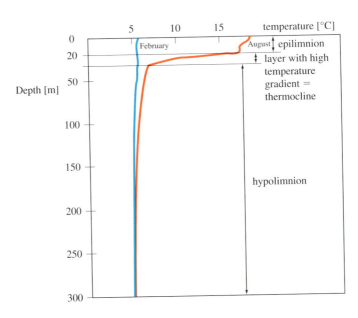

Fig. 7.34 The "tropical" behavior of Lake Geneva. In summer, the surface water (***epilimnion***) heats up. The water is light and floats on the heavy water of the main mass (***hypolimnion***), which is at about 5°C. The density contrast is such that no mixing takes place between the surface and bottom water. In winter, especially during very cold winters, the surface water reaches the same temperature as the hypolimnion. The well-oxygenated surface water can thus mix with water in the large mass, a phenomenon amplified by wind action. This oxygen input from the surface is facilitated by input from sub-lacustrine currents. If this mixing ceased to occur, as a result of climatic change, for example, organic matter in the mud at the bottom of the lake would soon consume the free oxygen dissolved in the water for its minerali- zation. Once the oxygen supply is used up, anaerobic mineralization would occur, which is not favorable to lacustrine life. The production of reduced forms of sulfur (hydrogen sulfide), nitrogen (ammonia) and carbon (methane) would degrade water quality (from [3]).

Today, limnologists use more detailed classifications particularly to describe the frequency of water mixing phases.

7.3.3   Glaciers

Glaciers in polar regions or high mountain chains at lower latitudes represent an important reservoir of surface water. Their contribution to the water cycle is welcome since they produce abundant water during warm periods when rainfall is often deficient. Mountain dwellers have used the significant summer discharges of mountain streams fed by snowmelt and glaciers to create irrigation channels (called by local names, for example "bisse" in the Valais Alps) (Fig. 7.35). These artificial channels take water from mountain streams and divert it along a very low slope to fields for agriculture. They are generally earthen ditches but also sometimes rock tunnels or wooden gutters suspended from cliffs.

Fig 7.35 Ancestral use of melt waters by "bisses". Along this steep section of the Bisse du Ro (Valais, Switzerland), one of the oldest channels, built in the 15th century, and one of the best preserved. It has not been used since 1947. Exceptionally steep, the bed of the bisse followed Quaternary deposits and rock. To cross particularly steep areas, holes were made into the rocky wall to set beams that supported the channel and planks to enable the caretaker to conduct daily monitoring. GEOLEP Photo, L. Cortesi.

Reservoirs are constructed to make the best use of "white coal" the significant energy produced by hydroelectric dams during the winter period. The Grande-Dixence dam in the Alps (Valais, Switzerland) is an impressive example, with a height comparable to the Eiffel Tower, and 400 million cubic meters of water accumulated behind it and multiple intakes and collection conduits from the valleys on the southern side of the Rhone basin (Fig. 7.36).

The glacier is thus a component of the water cycle that cannot be ignored.

7.3.3.1   Morphological classification of glaciers

Before we look more in detail at the hydrological function of glaciers, we must break them down into four types on the basis of morphology and size.

Cirque glaciers

These are small glacial structures that form in a concavity between two rocky ridges converging to a summit (Fig. 7.37), usually on slopes protected from the Sun. They have a

Fig. 7.36 Grande-Dixence Dam and a part of its glacial reservoir. Photo of Alfred Stucky, former professor at the Federal Institute of Technology Lausanne and one of the pioneer dam builders of the world.

Fig. 7.37 Cirque glaciers on the eastern flank of Piz Morteratsch, Bernina Massif (Graubünden Alps), seen from the Boval hut. On the right side is a regenerated glacier made up of ice that has fallen off the ice cap covering the summit. GEOLEP Photo, A. Parriaux.

compact shape and a small mass. They are in contact with *firn* (new snow accumulated against and at the foot of slopes) and ice from the walls but separated by a fracture called a *bergschrund*. Because of their low inertia, many of these glaciers have rapidly disappeared during the warming observed in the Alps beginning in the 19th century. These disappearances have intensified in recent decades.

Valley glaciers

The cirque glacier can be followed by a glacier tongue that flows down along the bottom of the valley (Fig. 7.38). This tongue may be very long: The Bering glacier holds the world record at 200 km and in the Alps the Aletsch glacier is the longest, at 23 km. The thickness of these glaciers may also be very significant (around 900 m at Konkordiaplatz on the Aletsch Glacier). The ice breaks up over rocky thresholds to form *seracs* or *ice cones* (Fig. 7.39). Valley glaciers contain a complex aquifer in which the water flows through crevasses, faults or channels before draining into the sub-glacial torrent and coming out into the open air through a tunnel at the end of the tongue. Channels or pockets of water can empty abruptly when the ice moves, creating catastrophic flows (Fig. 7.40).

(a)   (b)

Fig. 7.38 Valley glaciers. Example of the Rhone Glacier at Gletsch (Central Switzerland). The two photos illustrate the spectacular retreat of the glacier from the beginning of the twentieth century (a) until the present day (b). In 1830, at the end of the Little Ice Age, the tongue was 100 m from the hotel. Photo (a) Gabler, (b) GEOLEP G. Franciosi.

When a glacier ends at the top of a cliff, it is called a *hanging glacier*. It often feeds the main valley as massive ice falls. The Giétroz glacier for example caused a major catastrophe in the Val de Bagnes (Valais Alps) in 1818 before the existence of the Mauvoisin hydroelectric dam. A large ice fall from the hanging glacier formed a lake that threatened the entire valley. To empty the lake gradually (Fig. 7.41), engineers constructed a tunnel in the ice that had accumulated at the foot of the cliff. They were able to drain a significant portion of the water. Unfortunately, this structure could not withstand the violence of the flows and the ice dam was washed away while there were still 20 million m^3 of water in the lake. This volume of water drained in half an hour, causing a devastating flood in the entire Val de Bagnes. At Martigny, at the mouth of the Drance in the Rhone Valley, the wave was so high that it carried away several people. The incident killed a total of 150 people.

Fig. 7.39 Seracs in the "Mer de Glace" at Mont Blanc. Crossing a convexity in the rocky bottom that was thought to occur at the foot of the peak caused the glacier to break up along tension fractures, very visible downstream the threshold. These ice columns are very fragile and can collapse catastrophically. Photo Copyright Mario-colonel.com.

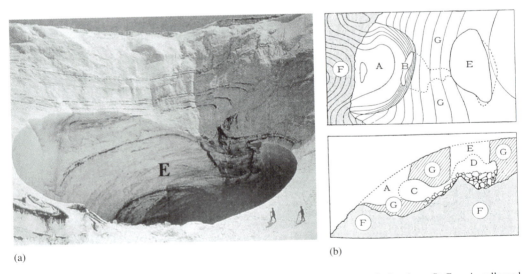

(a)   (b)

Fig. 7.40 Rupture of a water pocket at Mont Blanc in 1892. (a) The Tête-Rousse glacier above St-Gervais collapsed after a water pocket 20 m by 38 m and 85 m deep suddenly emptied. The water and mud flows killed 130 people in the village of St-Gervais and destroyed some buildings. The wave was felt in the Arve River as far away as Geneva. Diagram and photo taken from [239]. (b) A: scarp; B: lower opening; C: lower cave: D: upper cave; E: collapsed part; F: crystalline schists; G: glaciers.

Fig. 7.41 Slide of the Giétroz Glacier at the bottom of the Val de Bagnes on 16 May 1818 (Valais Alps). Aquatint, sepia from «Course à l'éboulement du glacier de Giétroz et au lac de Mauvoisin» (attributed to Philippe-Sirice Bridel, pastor of Vevey, Lortscher and Sons).

The same type of event occurred much more recently (1965) at the worksite of the Mattmark earthen dam in Valais. During the night, seracs in the Allalin Glacier collapsed on the workers and engineers housed at the worksite, killing 88 people (Fig. 7.42).

Fig. 7.42 The Mattmark catastrophe in the Val de Saas (Valais). Photo Kraftwerke Mattmark AG.

A glacier that forms at the foot of the wall as a result of falls from the upper glacier is called a ***regenerated glacier***.

Glaciers move at various velocities. Those in alpine valleys advance between 40 and 200 meters per year on average, whereas those in large polar valleys move almost 10 times more rapidly because they generally flow into the sea and the terminal part of the glacier is able to float. Sometimes intense accelerations, called *surges*, occur. They are due to a disequilibrium situation resulting from a thickening in the accumulation zone and a simultaneous depression in the ablation zone. There is thus an increase in the driving force and a decrease in the resistant force, as in a landslide. Velocity accelerations of up to 11 km in three months have been observed in the Himalayas. During the surge of 1982–1983, advances of the Variegated Glacier (Alaska) reached peaks of 65 meters per day. An animation on the DVD shows the relationship between the ice movement and the fluctuation of the glacier front.

Piedmont glaciers

These are valley glaciers that discharge from a mountain onto a vast flat area and spread out in the shape of a large foot. Piedmont glaciers were very common in the Southern Alps during the Quaternary glaciations. They are now found only in polar regions.

Ice sheets and ice caps

The accumulation of snow in the center of mountainous areas in cold regions forms a continuous ice cover, with a regular convex shape (Fig. 7.43). Large extensions (> 50,000 km^2) are called *ice sheets*. If they are smaller, they correspond to *ice caps*. Sometimes rocky peaks, *nunataks*, poke through the top of them. Ice sheets and ice caps give rise to numerous valley glaciers that export the ice toward the periphery of the mountains. Polar ice sheets are famous for their thousands of meters of ice (Fig. 7.44) and the extraordinary information they have produced on paleoclimatic evolution during the Quaternary.

Rock glaciers, which are richer in mineral debris than ice, will be treated in chapter 8 in the discussion of the features of polar regions.

Fig. 7.43 Small ice cap on the border of the Tibetan Plateau. In places rocky spurs pierce the mantle of ice (nunataks). Melted snow persists on the surface of the plateau because the soil is frozen. GEOLEP Photo, A. Parriaux.

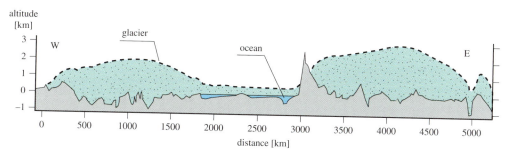

Fig. 7.44 Polar ice sheet. Example of the Antarctic ice sheet.

7.3.3.2 Hydrologic balance of glaciers

All glaciers have two distinct areas (Fig. 7.45):

- *Accumulation zone*: The upper part where snow deposits are more abundant than losses from melting and sublimation. In valley glaciers, these are often large plateaus and the beginning of the tongue.
- *Ablation zone*: The lower part of the glacier where losses exceed snowfall deposits. They are the lower part of the tongue of valley glaciers.

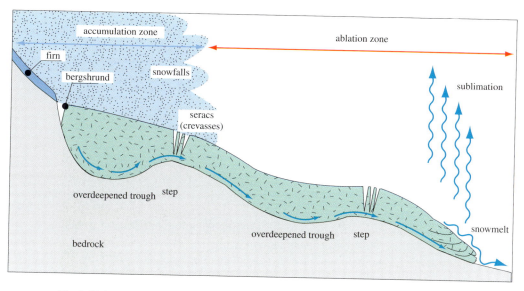

Fig. 7.45 Accumulation and ablation zones in a valley glacier (schematic longitudinal profile).

The hydrological balance of an entire glacier is obtained by computing the algebraic sum of the volumes accumulated and the volumes exported over a given period (Fig. 7.46).

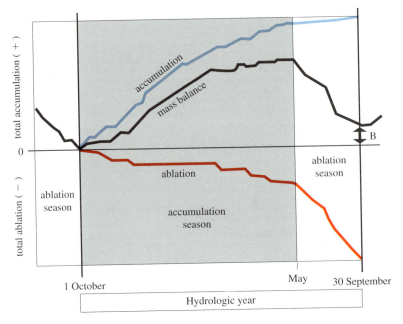

Fig. 7.46 Water balance of an entire glacier in the northern hemisphere. B = yearly balance (from [262]).

During years when the balance is positive, the surface of the glacier in the accumulation zone rises because of the added snow. It will probably be several years before the tongue will move forward, due to the inertia of the glacier.

At the present, we are seeing a continuation of the retreat of the glaciers that began in the middle of the 19th century (Fig. 7.38). Now that global climate change is confirmed, this tendency will increase substantially and soon in the Alps there will be only a few rare glaciers such as the Aletsch. That is, if the natural glacial cycles of Milankovitch (Chapt. 3) prevail, we will move into another large cycle of glaciation …

[22, 112, 114]

7.4   Groundwater

Groundwater is the principal source of drinking water on Earth. In Switzerland, for example, groundwater provides 80% of drinking water. Paradoxically, groundwater resources are also the least well known. It is difficult to find and observe them. Their circulation mechanisms are strongly influenced by complex and variable geologic conditions. In this book, we will first present the reservoir properties of natural geological materials, and then the ways that groundwater circulates. We will then progress to practical problems of exploiting groundwater resources, the effects of drainage and irrigation, and the management and protection of groundwater, an area in which the engineer is directly involved.

7.4.1   Porosity of geological materials

How do these water-storing voids occur in geological formations? There are three basic types of porosity based on the shapes of the pores (Fig. 7.47). This classification of rocks is supplemented by the information in table 7.4.

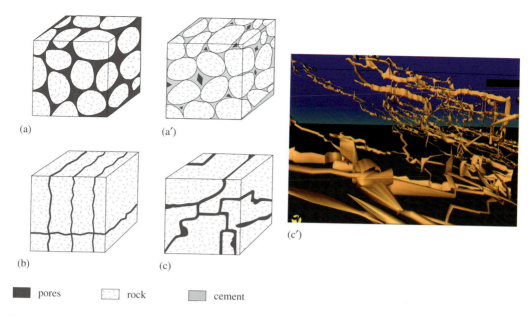

| pores | rock | cement |

Fig. 7.47 Three types of porosity. (a) Interstitial pores in sand. (a′) Interstitial porosity in sandstone. (b) Fissure poros-ity. (c) Karst porosity (from [168]). (c′) 3D model of a karst network of the Siebenhengste in Swiss Alps (document Swiss Institute of karstology). Look at the animation on the DVD.

7.4.1.1 Interstitial porosity

In recent unconsolidated sediments and in some poorly aggregated rocks, there are intersti-tial spaces between particles in the structure. If these pores are small in size (<0.1 mm), they can contain water so strongly attached to the mineral matrix that it cannot be removed by grav-ity, as in the case of clays. If the pores are large and there is communication between them, water can circulate in the voids as in sand, gravel and poorly cemented sandstones. For exam-ple, a cubic meter of gravel can contain about two hundred liters of water that can drain by gravity (free water). If the pores are not interconnected, such as vesicles in lava, they do not contribute to underground flow because the fluids remain trapped in pores (Tab. 7.4).

7.4.1.2 Fissure porosity

Indurated rocks can develop fractures (planar pores) when they are deformed by mechanical forces. If they are slightly open, water can infiltrate into them, move around and accumulate. The volume of water stored in these rocks is generally low (several percent of the volume of the rock) because of the small amount of space between the sides of the fissures.

7.4.1.3 Karst porosity

Water that infiltrates into fissures in soluble rocks has a tendency to enlarge the fissures. In the case of limestone, the acidity of the percolating water plays a major role in the dissolution. It comes from rain (strong acids, usually anthropogenic) and soils (weak acids). The Karst area of Slovenia has given its name to this type of weathering; this geologic region is composed of limestone that is full of very large voids; channels and caves that are more than 10 m in

Table 7.4 Summary of porosity in different rock families. The columns show three geometric shapes of pores. Cellular porosity is associated with interstitial porosity, planar porosity with fissure, and cylindrical porosity generally with karst. The top half of the table gives primary or syngenetic porosity, that is, porosity created at the same time as the rock. The bottom half of the table groups together all the phenomena that promote secondary or postgenetic porosity: weathering, tectonic activity, etc. We distinguish between rocks with open, semi-confined and confined porosity (such as vesicles in volcanic rock). The figures cited after the names refer to illustrations of the rocks in question.

genesis	pore shape			
	cellular		planar	cylindrical
Syngenetic (primary porosity)	**intergranular pores of detrital rocks** *puddingstone* M (Fig. 10.7a) *sedimentary breccia* M (Fig. 10.7b) *sandstone* m (Fig. 10.12) *siltstones* m *claystone* m (Fig. 10.13)	**intercrystalline pores of magmatic and metamorphic rocks** *granites* m (Fig. 6.25) *gneiss* m (Fig. 11.5) *marbles* m (Fig. 11.20) *etc.*	bedding plane M contraction joints of lavas (*columnar basalt*) M (Fig. 6.20) desiccation cracks (*clay*) M (Fig. 13.19)	traces of burrowing organisms M+m
	intergranular pores of biogenic rocks *corallian rocks, coral masses* M+m (Fig. 10.15) *oolites* m (Fig. 9.22) *tufa* M+m (Fig. 13.12) *travertines* M+m (Fig. 13.13) traces of burrowing organisms	*Vesicles in volcanic rocks* pumice M+m (Fig. 6.41c)		
Postgenetic (secondary porosity)	**physical weathering** Rocks highly deformed by tectonics *tectonic breccia* M+m *kakirites* m (Fig. 11.13) frost sensitive rocks – *marls* m (Fig. 10.19) – *mudstone* m (Fig. 10.13) – *marly sandstone* m Lithophage borings	**chemical weathering** granular rocks with heterogeneous solubility – *cargneule* M+m (Fig. 10.34) – *calcareous sandstone* m – *shelly limestone* M+m (Fig. 1.5) – *limestone and dolomite with heterogeneous matrix (fossils and microfossils)* M+m (Fig. 10.16) – *impure gypsum* M+m (Fig. 10.31) – *heterogeneous salt rocks* M+m *granites* m (Fig. 13.4)	**joints** faults (*with or without karstification*) M (Fig. 12.18) overthrust surface (with or without karstification) M sliding surface M topple M exfoliation surface M	**conduits of karst rocks** *limestones* M (Fig. 13.11) *dolomites* M (Fig. 10.20) *gypsum* M (Fig. 10.31) mineralization of plant fossils *wood trunks in Molasse* M

Legend: M = macroporosity: > 5 mm   m = microporosity: <5 mm
Relationship porosity <--> permeability
not underlined: open porosity ____ semi-closed porosity _____: closed porosity

diameter (cylindrical pores). All limestone mountains in the world are affected by this process to variable degrees. These rocks are so permeable that the regions where they outcrop are devoid of rivers. They can convey large quantities of water over long distances. This is the field area of speleologists (Fig. 7.47c'). The Jura Mountains are a perfect example of this phenomenon (Fig. 7.13). Rocks made of gypsum ($CaSO_4 \cdot 2H_2O$) are even more soluble in water and are also susceptible to these processes.

7.4.2 Water flow in geologic media

First let us say in a very general way that water that infiltrates into permeable ground will seep more or less vertically through unsaturated porous ground called the **unsaturated zone** or *vadose zone*. When the water encounters a impermeable layer, the vertical seepage ceases and water accumulates in the underground pores to form a *saturated zone*. Together, saturated and unsaturated zones of a permeable medium constitute a **water-bearing formation** (or **aquifer**, § 7.4.4). The geometric and hydrogeologic conditions of these aquifers are diverse (Fig. 7.48). Water in the aquifer is not immobile; it flows along gentle slopes to natural outlets (springs) or artificial outlets (pumping wells).

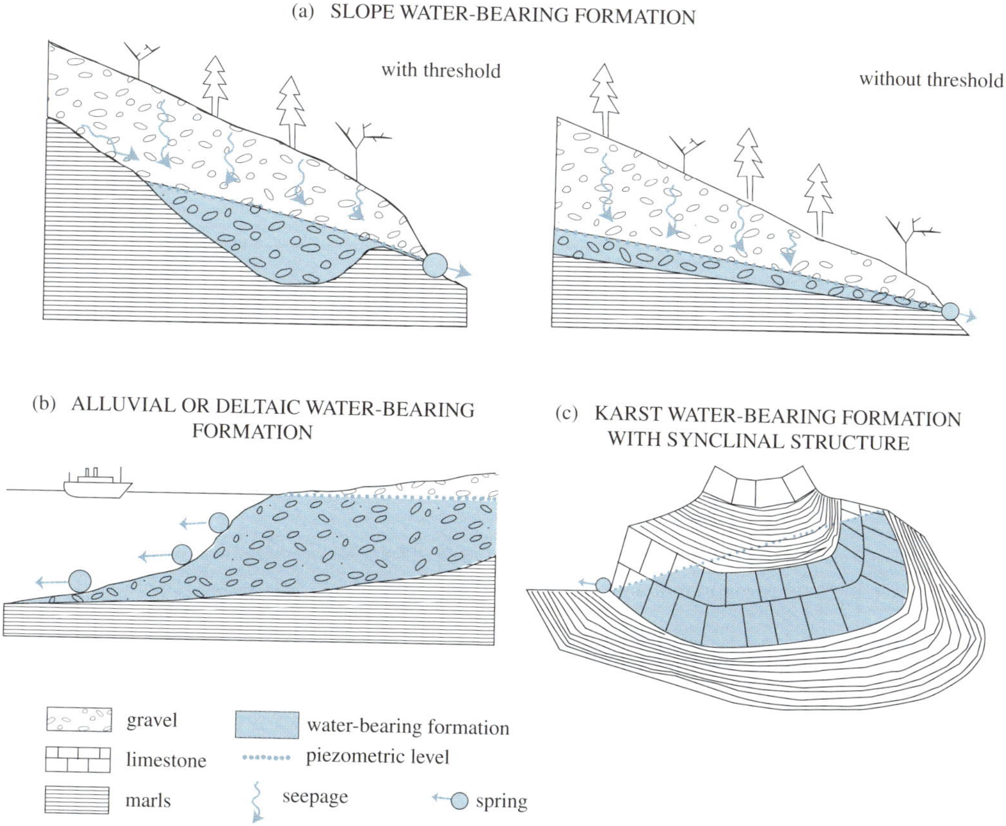

(a) SLOPE WATER-BEARING FORMATION

with threshold

without threshold

(b) ALLUVIAL OR DELTAIC WATER-BEARING FORMATION

(c) KARST WATER-BEARING FORMATION WITH SYNCLINAL STRUCTURE

gravel | water-bearing formation
limestone | piezometric level
marls | seepage | spring

Fig. 7.48 Various types of aquifers. (a) A water-bearing gravel formation on a slope situated on an impermeable substrate: a spring occurs on the slope at the contact between the two formations; the possible presence of a concavity in the lower bed promotes more stable behavior of the spring than in the case of a linear gravel veneer. (b) A deltaic water-bearing layer: the base level is the level of the lake or sea; the outlets are sub-lacustrine or submarine. The formation may have a considerable thickness. (c) A synclinal karst formation; a layer of karstified limestone is folded in a concave way and rests on a horizon of impermeable marls; the spring appears at the contact between the limestone and the impermeable layer.

7.4.3 Basics of hydrodynamics

Some physics basics on flow in porous media are useful for discussing the notion of permeability. Because of the great complexity of biphasic flows in the unsaturated zone (air and water) we limit ourselves here to the saturated zone.

Although underground flow is analogous with surface flow in the application of the Bernouilli equation (§ 7.3.1.1) it is also very different because groundwater has a reduced section through which to move. This reduced section consists only of porosity between grains and rock fragments. Whereas losses due to friction play a relatively minor role in river flow, they are fundamental in underground flow. The generalized form of Bernouilli's law must be used, because of the significant head losses and wide variations of these losses depending on the geologic setting. For example, given identical hydraulic and geometric conditions, water will flow one billion times faster in a gravel stratum than in a clay stratum. This relationship between hydrodynamic boundary conditions and the nature of the geologic material through which the groundwater travels is thus particularly important. To understand it, let us examine the experiment that led to Darcy's law.

Let us fill a cylinder of section S and length l with a granular material, for example, sand (Fig. 7.49). This sand rests up against the two ends of the cylinder on two porous walls that let water pass but hold back the sand grains. The cylinder is connected to a supply pipe on one side and has an outlet on the other end that is equipped with a controllable valve. The sand is saturated and put under a slight pressure by injecting water through the two conduits. Two manometers M_1 and M_2, two simple open tubes, allow the pressure to be measured at the two ends of the cylinder depending on the water level inside the tube. At static conditions the water heights h_1 and h_2 in the two manometers is equal to a horizontal reference point. With the cylinder in any position, the manometer M_1 is at elevation z_1, which is different from the other manometer M_2 at elevation z_2. In each tube, the water height h' above the cylinder is equal to:

$$h' = \frac{P}{\gamma_w} \tag{7.16}$$

where

P = hydrostatic pressure
γ_w = specific weight of water
Since $z_1 \neq z_2$, $P_1 \neq P_2$.

By applying the Bernouilli equation, we can say that the hydraulic potential at the two ends is equal. The potential energy of pressure compensates for the difference in potential energy of position.

$$z_1 + \frac{P_1}{\gamma_w} = z_2 + \frac{P_2}{\gamma_w} \tag{7.17}$$

If we now open the outlet valve a little and provide a constant discharge Q to the cylinder, we create permanent flow conditions. The friction of the water against the grains of sand causes a loss of head that appears between M_1 and M_2. The potential at M_2 is lower than at M_1. If the position of the cylinder does not change, the pressure term shows a decrease, visible as the reduced head in manometer M_2.

The head loss is expressed as a relationship to the flow path, here the length l. We obtain the hydraulic gradient i:

$$i = \frac{(h_1 - h_2)}{l} \tag{7.18}$$

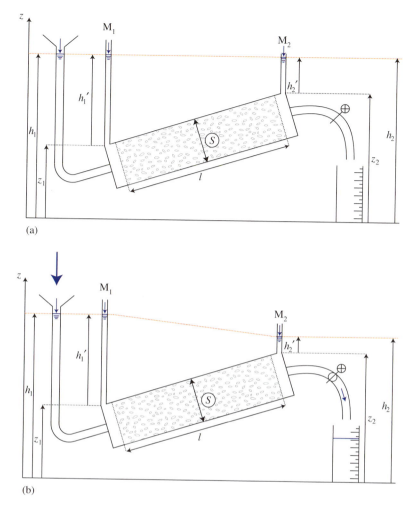

Fig. 7.49 Flow in porous saturated media and Darcy's law: (a) static (closed valve); (b) permanent flow (open valve).

Darcy observed experimentally that if the velocities are low and flow thus approximates laminar flow, for a given material, there is a linear relationship between the discharge Q and the hydraulic gradient i:

$$\frac{Q}{i} = -K \tag{7.19}$$

The proportionality constant K is called the ***Darcy permeability coefficient*** or ***hydraulic conductivity***. It has dimensions of velocity [m/s]. The negative sign before the K shows that the hydraulic gradient is negative. Writing this relationship in terms of velocity, we obtain:

$$Q = v_D \cdot S = -K \cdot i \cdot S = -K \frac{h_1 - h_2}{l} S \tag{7.20}$$

This velocity v_D called the Darcy velocity is not a real velocity. It is the velocity of the water if porosity n were equal to 1. The real velocity v_R of the water molecules that travels through the pores is given by:

$$v_R = \frac{v_D}{n} \tag{7.21}$$

Since the porosity is closer to 0 than to 1, we see that the real velocity is generally a multiple of the Darcy velocity.

The coefficient of permeability K is not a characteristic intrinsic to the material since it depends on the density and the viscosity of the material that passes through it. In practice, for groundwater flows at ordinary temperatures (about 10°C) it is used as a de facto parameter appropriate for the medium. In petroleum geology or geothermics, this simplification is no longer valid (see § 4.4.2.1); in this case, *the intrinsic* or *specific permeability* k must be used. The intrinsic permeability does not depend on the fluid viscosity or the value of gravity g at the measurement point. It is calculated from the following equation:

$$k = K \frac{\eta}{g} \, [\text{m}^2] \tag{7.22}$$

where

η = kinematic viscosity of the fluid (for water it is more than two times less at 30°C than at 0°C)

The generalized Bernouilli equation can be simplified in environments of interstitial porosity because of the slow flow velocity. The kinetic energy term $v^2/2g$ can be neglected in most cases. For rapid flow in fissures or in karst, this term must be taken into account, as in the case of flow in pipes.

7.4.4 Application of permeability to the subsurface

An *aquifer* is a geologic formation that contains interconnected pores large enough to conduct significant quantities of water; an aquifer is not necessarily saturated. An *aquitard* is a semi-permeable unit that does not produce significant quantities of groundwater. An *aquiclude* is a layer that is so impermeable that it behaves as a watertight barrier.

If the piezometric surface of an aquifer is in contact with an aquiclude, the aquifer is considered to be *confined*. An *unconfined aquifer* has a free water table (Fig. 7.50). An *artesian*

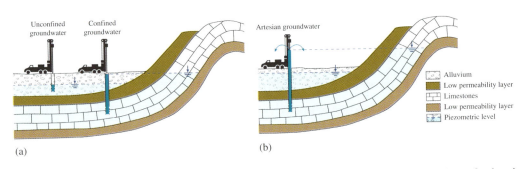

Fig. 7.50 Principal types of aquifers. (a) Unconfined aquifer with a water table within the aquifer and a confined aquifer with a piezometric surface under pressure under the confining layer. (b) Artesian aquifer with a piezometric surface above the ground surface. An animation on the DVD shows the boring phases and their effect on the groundwater.

aquifer is a confined aquifer whose head is higher than the land surface; a well drilled into this layer is a flowing well (Fig. 7.51).

Fig. 7.51 Artesian fluvioglacial aquifers confined below lodgement till. a) Artesian venue during boring operation (Solalex, Swiss Alps). b) Water flowing to a height of several meters from a piezometer (Reculanne aquifer, Lake of Geneva Basin). GEOLEP photos, A. Parriaux.

7.4.5   Hydrodynamic application of flow toward a well

An interesting example of Darcy's law is the calculation of flow toward a pumping well. We will present flow to the well in the case of an unconfined aquifer and a confined aquifer, based on *Dupuit's equations*.

7.4.5.1   Dupuit's theory

For both cases, we will use the following assumptions:

- The aquifer has interstitial porosity, and is homogeneous and isotropic;
- Geologic contacts are planar and horizontal;
- The aquifer is horizontal and extends infinitely in all directions (at rest, there is thus no flow);
- Flow toward a pumping well is permanent (water pumped from wells is renewed indefinitely);
- The well is perforated across the entire aquifer;
- The discharge pumped from the well is such that the groundwater drawdown is small compared to its thickness.

The following variables are defined:

K = coefficient of permeability (or hydraulic conductivity) of the aquifer [m/s]
H = Head at rest with respect to the impermeable base of the aquifer [m]
e = thickness of the aquifer when the aquifer is confined [m]
r = radius of the well bore [m]
R = radius of influence [m]
h = thickness of the aquifer in the pumping well [m]
Δh = drawdown in the well [m]
v = Darcy velocity [m/s]
S = cross-sectional flow [m²]
x = horizontal radial distance from the center of the well [m]
z = elevation [m]

In both cases, flow is symmetrical around the well center. Figure 7.52 shows the vertical radial profile.

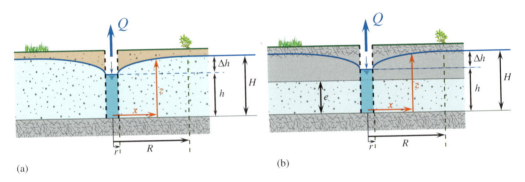

(a)   (b)

Fig. 7.52 Diagram of the Dupuit equations. (a) Unconfined aquifer. (b) Confined aquifer.

Case of an unconfined aquifer

The piezometric surface that produces the head H is within the aquifer.

We can write that the discharge Q through a cylindrical surface at distance x is:

$$Q = v \cdot S = v \cdot 2 \cdot \pi \cdot x \cdot y \tag{7.23}$$

From Darcy's law we express velocity as a function of the coefficient of permeability K and the gradient i:

$$v = Ki = K\frac{dz}{ds} \tag{7.24}$$

ds being the length of the path along the flow line equivalent to the loss of head dz (Fig. 7.52). Since we have assumed that drawdown Δh is low with respect to H, the gradient i is also low. We can thus approximate the exact expression for the gradient by writing:

$$i \cong \frac{dz}{dx} \tag{7.25}$$

The discharge Q thus becomes:

$$Q = K\frac{dz}{dx}2\pi xz \quad \text{or} \quad Q = \frac{x}{dx}2\pi z \, dz \tag{7.26}$$

$$\frac{1}{x}dx = \frac{2\pi K}{Q}z \, dz \tag{7.27}$$

By integrating this equation, we obtain:

$$\ln x = \frac{2\pi K}{Q}\frac{1}{2}z^2 + C \tag{7.28}$$

The extreme boundary conditions are those in contact with the well and those at infinity. On the subject of this latter limit, we must introduce the notion of the radius of influence R which extends to the distance where drawdown due to pumping can still be measured in practice (about 1 cm).

For $x = r$, $z = h$, we have:

$$\ln r = \frac{\pi K}{Q}h^2 + C \tag{7.29}$$

For $x = R$, $z = H$, we have:

$$\ln R = \frac{\pi K}{Q}H^2 + C \tag{7.30}$$

Subtracting equation 7.29 from equation 7.30, the constant C disappears:

$$\ln R - \ln r = \frac{\pi K}{Q}(H^2 - h^2) \tag{7.31}$$

where

$$Q = \frac{\pi K(H^2 - h^2)}{\ln(R/r)} \tag{7.32}$$

By introducing drawdown in the well Δh, the ratio can be written as:

$$Q = \frac{\pi K(2H - \Delta h)\Delta h}{\ln(R/r)} \tag{7.33}$$

The well discharge is proportional to the coefficient of permeability. However, it is linked to drawdown by a parabolic function; discharge increases less quickly than drawdown. This is characteristic of an unconfined aquifer (Fig. 7.53).

Case of a confined aquifer

This case is similar to that of an unconfined aquifer except for the fact that the thickness e is constant, the cross section is proportional to the distance x:

$$S = 2\pi xe \tag{7.34}$$

We finally obtain the ratio:

$$Q = \frac{2\pi Ke\Delta h}{\ln(R/r)} \tag{7.35}$$

The discharge is always proportional to the coefficient of permeability but this time it is related to drawdown in a linear way (Fig. 7.53). This makes it possible to distinguish a confined layer from an unconfined layer during a pump test on the basis of the ***characteristic curve*** of the well.

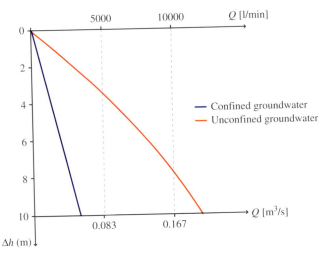

Fig. 7.53 Characteristic curve of a pumping well in a confined aquifer and an unconfined aquifer (case of the numerical example of § 7.4.5.2).

7.4.5.2   Numerical application

The numerical example below makes the difference between aquifers very clear. The data common to both cases is the following: a well with a radius of 1 m drilled into an aquifer with a permeability of 10^{-3} m/s. The radius of influence is approximately 100 m.

Case of an unconfined aquifer

The aquifer thickness is 20 m. To calculate the constant C_u of the function $Q = f(\Delta h)$:

$$C_u = \frac{K}{\ln(R/r)} = 0.683 \cdot 10^{-3} \tag{7.36}$$

$$Q = C_u(2H - \Delta h)\Delta h \tag{7.37}$$

For 1 m of drawdown, we obtain $Q = 2.66 \cdot 10^{-2}\,\mathrm{m^3/s}$ or $1600\,\mathrm{l/min}$. Other drawdown values allow us to plot a characteristic parabolic curve of the unconfined aquifer (Fig. 7.53).

Case of a confined aquifer

Head remains constant at 20 m on the impermeable floor, but the actual thickness of the confined aquifer is only 5 m.

The constant C_c of the function $Q = f(\Delta h)$:

$$C_c = \frac{Ke}{\ln(R/r)} = 6.83 \cdot 10^{-3} \qquad (7.38)$$

$$Q = C_c \Delta h \qquad (7.39)$$

For 1 m of drawdown as in the unconfined aquifer, we obtain a discharge $Q = 6.83 \cdot 10^{-3}\,\mathrm{m^3/s}$ or only $410\,\mathrm{l/min}$. This point establishes the slope of the characteristic straight line of the well (Fig. 7.53).

7.4.6 Hydrologic balance of aquifers

The aquifer has a buffering effect on input and output. In a ***transient regime***, that is, when flow varies over time, the variation of the volume of the aquifer V between time t_2 and t_1 is the algebraic sum of the integral of the entering discharges Q_e and of the integral of the exiting discharges Q_s during this period of time, or:

$$\Delta V = \int_{t_1}^{t_2} Q_e \, dt - \int_{t_1}^{t_2} Q_s \, dt \qquad (7.40)$$

In fact, aquifers are fed by numerous natural and artificial inputs (Fig. 7.54). Several outputs are also possible.

7.4.7 Groundwater intake

In this part, we will focus principally on the use for drinking water supply. Thermal water was specifically treated with geothermics (see § 4.4.2.1). The collection or extraction of groundwater is within the domain of engineering. In collaboration with a geologist, the engineer should adapt the groundwater intake system to the aquifer structure. Two types of capture sites should be considered.

7.4.7.1 Springs

Groundwater can be collected at the natural outlets of the aquifers. However, the water should be captured upstream of the spring area, properly speaking, to avoid mixing the spring water with surface water that may not be suitable for human consumption. To do this, trenches or galleries (Fig. 7.55) are constructed. The groundwater is conveyed by gravity into a collection chamber where it is monitored for quality control, flow measurement, and distribution according to the rules of water supply regulation.

7.4.7.2 Production from a pumping well

When there is an aquifer at depth, but the water does not naturally flow to the surface, the water can be extracted by a pumping well. The well artificially penetrates the aquifer. A slotted

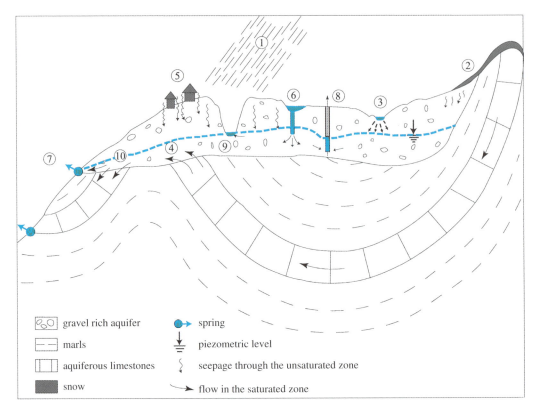

Fig. 7.54 Diagram of inputs and outputs from an aquifer. Supply. 1: rainwater infiltration. 2: infiltration of snowmelt. 3: infiltration of streams and rivers. 4: contributions from other aquifers. 5: artificial infiltration of slightly polluted urban water (roofs, squares, access roads). 6: artificial recharge facilities. Losses: 7: springs. 8: pumping wells. 9: base flow of rivers. 10: losses to other aquifers.

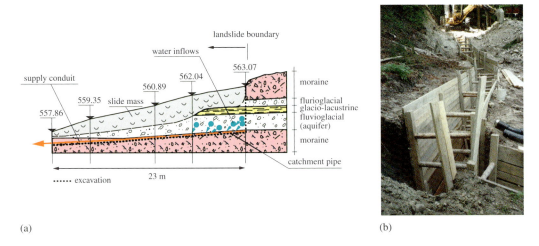

(a)   (b)

Fig. 7.55 Use of a trench with a perforated pipe to capture a spring. (a) Plan for capture of the Biscou Spring in Begnins (Lake Geneva Basin). (b) Worksite showing the excavation of a capture trench in water-bearing gravels, with vertical support walls shored up by timber. GEOLEP Photo, A. Parriaux.

screen is placed in the well. The screen is composed of small slits whose size is selected on the basis of the geologic formation in order to allow water to flow through while holding back the aquifer material (Fig. 7.56). A pump is installed to bring water to the ground surface. Discharge from the well is supplied by aquifer production through the slits in the pipe.

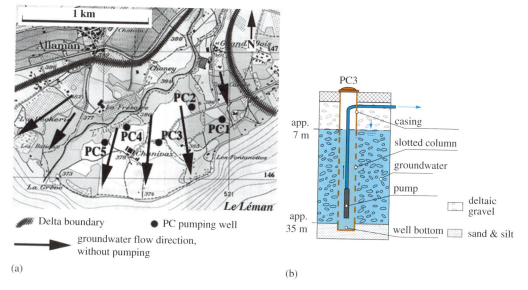

Fig. 7.56 Groundwater production wells. (a) Case of a curtain of five wells producing water from the delta aquifer of the Aubonne (Lake Geneva Basin). Discharge 10,000 l/min (drinking water for 30,000 people). Reproduced with authorization of Swisstopo (BA056911). (b) Cross section of one of the wells through the delta sediments.

7.4.8   Effects of drainage and irrigation

The improvement of swampland by drainage will be described in chapter 8. It involves a lowering of the groundwater in surface aquifers. It can cause two effects:

- A decrease of the natural recharge of the deeper groundwater. The bottom of the swamp may allow water to percolate into a deeper layer, but reducing the pressure of the surface water decreases this supply flux (Fig. 7.57a).
- A new outlet for groundwater that previously seeped upwards into the swamp: the drainage causes water to leave the aquifer and thus causes competition with springs (Fig. 7.57b).

Irrigation produces the opposite effect. Quantitatively the possible excess of water applied to the fields recharges the groundwater if the base of the soil is permeable. Qualitatively, the effect is often detrimental to water; for example nitrates and micropollutants are leached from the soil (they are not desirable in drinking water), and salts can rise to the surface through capillary action in desert areas.

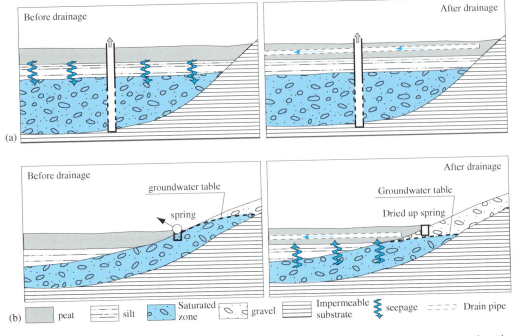

Fig. 7.57 Possible hydrogeological effects of drainage. (a) Decreased surface recharge. (b) A drainage system lowering a deep aquifer and causing a spring to dry up.

7.4.9   Groundwater management

Groundwater management can be approached in two ways:

- A water reservoir to be exploited like ore in the mining industry;
- A renewable resource that is equivalent to the average annual recharge of the aquifer; this recharge can be natural or artificial.

In terms of sustainable development, groundwater should be considered a renewable resource. This is possible in countries with temperate climates. In arid regions, exploration has revealed deep aquifers that have abundant fresh water, such as the Continental Intercalary Aquifer of the Sahara and the deep aquifers of the Dead Sea in Israel that make it possible to grow tomatoes in the desert. These ancient water reservoirs recharge very slowly and groundwater mining is occurring. The consequences of overexploitation are numerous:

- Gradual decrease of the available reservoir by lowering the water level (increasingly deeper wells must be dug);
- Subsidence of the soil may extend to several meters;
- Salt intrusion into coastal aquifers (Fig. 7.58)

The long-term maintenance of populations in these regions poses significant ethical problems.

In numerous countries it has been necessary to fight against aquifer overexploitation (see also Chapt. 14). In Geneva, the aquifer of the paleo-valley of the Arve River was so overexploited until the 1970s that an ***artificial recharge*** facility was constructed (Fig. 7.59). This facility collects raw water from the Arve River, filters it to make the water potable, and then injects it into the aquifer. This water then flows several kilometers through the ground, unpiped, toward pumping wells.

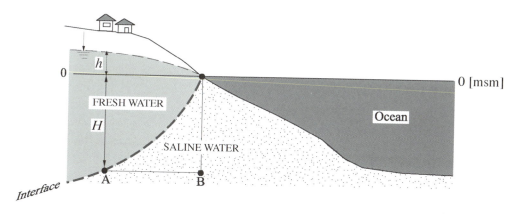

Fig. 7.58 Hydrogeology of coastal areas. Under the hydrostatic conditions of the aquifer, fresh water forms a lens that rests on heavier saline water. The Ghijben-Herzberg equations make it possible to calculate the thickness of this lens at point A in the coastal zone. The pressure P at point A is equal to that at point B, at the same level but at the extremity of the lens. H is the thickness of the fresh water below sea level, whereas h is the elevation of the water table at point A. The specific weight of the fresh water γ_{wf} and saline water γ_{ws} are known. Hydrostatic equilibrium gives $P_A = \gamma_{wf} (H + h) = P_B = \gamma_{ws} \cdot H$ or $H = h\gamma_{wf}/(\gamma_{ws} + \gamma_{wf})$. With numerical values of 10 and 10.25 kN/m³ for the specific weight, we obtain $H \cong 40\,h$. In other words, if the piezometric level is 1 m above sea level, we can expect to find a fresh water thickness of approximately 40 m. In practice the shapes of the contact are much more complex because of the simple fact that equilibrium is hydrodynamic and not static. However, heterogeneities of the aquifer structure are usually the principal source of complication.

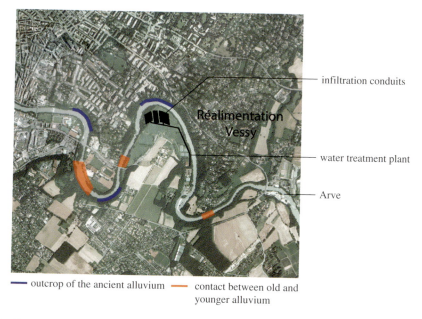

Fig. 7.59 Artificial recharge of the aquifer of the ancient Arve valley in Geneva. The areas colored blue and orange are areas of natural recharge. An animation on the DVD shows how recharge water is diffusing around the infiltration zone.

7.4.10 Groundwater protection

Groundwater protection is based on mapping *aquifer vulnerability*, which varies greatly depending on hydrogeological conditions. For example, an open karst aquifer is much more vulnerable than a fluvioglacial aquifer covered by a moraine (Fig. 7.60); the former does not have a protective filtration bed and the water flows rapidly into its large voids without effective filtration; the latter, on the contrary, is well protected by the low permeability moraine and the water flows slowly into the pores of the aquifer is filtered by the granular matrix.

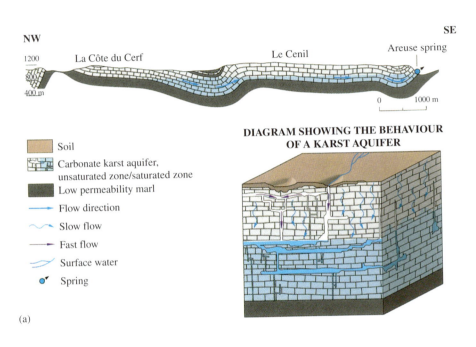

DIAGRAM SHOWING THE BEHAVIOUR OF A KARST AQUIFER

Soil

Carbonate karst aquifer, unsaturated zone/saturated zone

Low permeability marl

Flow direction

Slow flow

Fast flow

Surface water

Spring

(a)

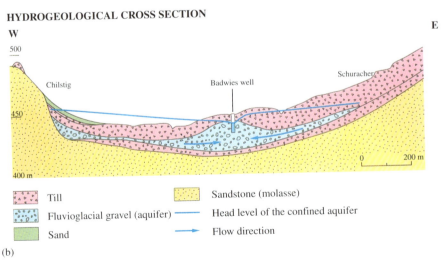

HYDROGEOLOGICAL CROSS SECTION

Till

Fluvioglacial gravel (aquifer)

Sand

Sandstone (molasse)

Head level of the confined aquifer

Flow direction

(b)

Fig. 7.60 Vulnerability of aquifers. (a) An uncovered karst aquifer is very vulnerable. (b) A fluvioglacial aquifer covered by a moraine is naturally protected (from [67]).

Pollution prevention involves the use of two tools (Fig. 7.61):

- Land development measures to prohibit construction that threatens groundwater in high vulnerability areas;
- Technical measures to prevent contact between pollutants and water in areas where pollution is likely to occur.

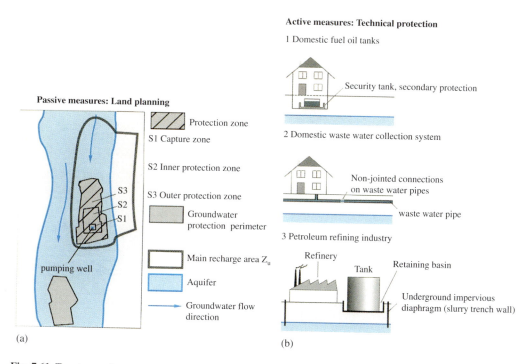

Fig. 7.61 Two types of measures to protect groundwater resources. (a) Land development measures, example of the different kinds of protective areas used in Switzerland. (b) Technical pollution protection measures (from [79]).

PROBLEM 7.2

A drinking water production well in the alluvial plain of the Rhine in Germany is near the river (Fig. 7.62a). It extracts groundwater by pumping the alluvial aquifer. Although a groundwater protection area was established and the recommendations were followed, the well's water quality has sometimes deteriorated. The authority responsible for the groundwater supply has decided to monitor it for two years and make systematic analyses to determine the cause of the problem. Your engineering consulting office has the job of interpreting the analytical data and proposing solutions.

Analyses show that the chemical quality of the water has degraded between year i and year $i + 1$. The second year is characterized by a rainfall deficit during the spring. Analyses of oxygen isotope 18 and deuterium were also done in conjunction with the chemical

analyses in the well and in a borehole installed in the aquifer 100 m upstream of the well
(Tab. 7.5). The purpose of this investigation was to determine the hydrogeologic supply conditions of the alluvial aquifer and of the well. In problem 3.3 we saw how isotopic fractionation during evaporation is used to reconstruct climatic variations. This process is also
involved in the condensation of water vapor in clouds: when the clouds are high, they are
poor in heavy isotopes. This is valid for oxygen 18 and deuterium. If we plot these data
pairs, we observe that rainwater from the same climatic region plots along a straight line
called the local meteoric line.

Figure 7.62b shows the isotopic signatures of local rainfall and of the Rhine water sampled near the well.

Table 7.5 Isotopic analyses used in Problem 7.2.

Date	Sampling point	$\delta^{18}O$ [‰]	δ^2H [‰]
Jan. year i	well	−9.6	−68.5
Jan year i	borehole	−8.8	−62.0
June year $i+1$	well	−10.7	−78.0
June year $i+1$	borehole	−9.2	−63.5

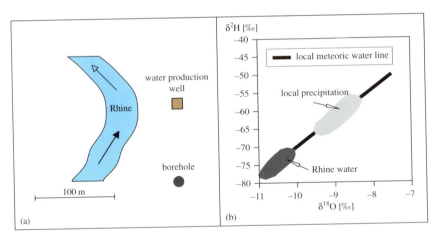

(a) (b)

Fig. 7.62 (a) Location map. (b) Isotopic composition of water from the Rhine and local precipitation.

Questions

- Why is the water of the Rhine low in heavy isotopes, compared to local rain water?
- What can we say about the results of the isotopic analyses?
- What solution can be suggested to the producer of drinking water?
- How can isotopic methods be used in the case of investigations on the domain of
  potable water?

Look at the solution on the DVD

7.4.11   Groundwater and civil engineering projects

We cannot end this section on groundwater without discussing engineering projects. We will see practical examples that show problems caused by these interactions and errors that can be made if they are not understood and analyzed during their planning and resolved with the necessary confidence. The relationship between water and construction functions in two ways: groundwater's effect on construction and the effect of a structure on groundwater.

Groundwater has long been a problem for the engineer because of the difficulties it causes during underground construction (Fig. 7.63).

Fig. 7.63 The construction of the Mont-d'Or tunnel between Vallorbe and Frasne (French-Swiss Jura) encountered major difficulties because of the presence of karst zones. On 23 December 1912, the tunnel excavation had penetrated 4366 m upstream of the Vallorbe portal when a karst conduit burst violently, producing a 3000 l/s current of water that carried off the tunnel access platform and flooded the plain below Vallorbe. Several springs in the area dried up (from [271]).

The effects of construction projects on water were ignored for a long time in large civil engineering projects although systematic spring measurements were often done before the beginning of construction (springs inventory). The reason for these observations was more legal than technical, because its goal was to handle complaints for damages in the event that springs dried up. Impacts were "repaired" by financial compensation. At that time, only water quantity was considered. Rarely were there concerns about possible qualitative impacts.

This idea began to change in the 1970s when people began to understand the environment and the protection of natural resources. This was the time of the first water protection legislation. At that time, the position of builders was much stronger than that of defenders of drinking water resources, so that the engineer who proposed projects did not really have to consider these resources as a priority.

Since then, this idea has evolved significantly. New legal tools have come into existence, such as environmental protection laws. In the majority of developed countries, complete environmental impact studies are required for engineering projects of a certain size. This has greatly changed the mentality of the civil engineer who must now place a priority on the location of the project in its natural setting. The relationship between Man's activities and the geologic environment will be developed in chapter 14 with regard to the major issues in society.

7.4.11.1 General problems related to groundwater during the construction of engineering projects

In a quasi-systematic way, the presence of groundwater in the subsurface tends to deteriorate the stability conditions of geological materials and thus complicate construction operations. Figure 7.64 summarizes the various influences that we will discuss below.

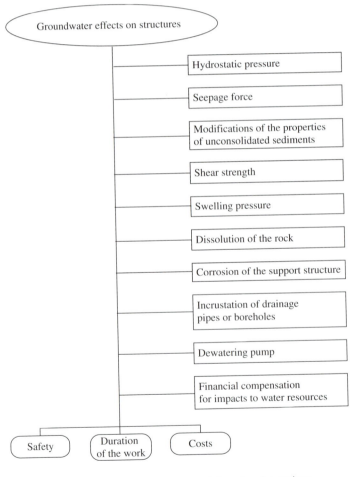

Fig. 7.64 Effects of groundwater on civil engineering projects.

Hydrostatic pressure

Pascal's barrel experiment (a barrel equipped with a vertical tube of small diameter) showed how a small addition of water into the pipe could explode the barrel. A similar experiment that is more realistic for a geologist can be done using a fissure in a rock mass (Fig. 7.65). This phenomenon was well known and used in the past at quarries where blocks were extracted from the front of the quarry by infiltrating large quantities of water at the top of the exploitation. It plays an important role at the working face of tunnels in fractured rock masses if the mass has not been drained in advance (Fig. 11.14), and in rockslides.

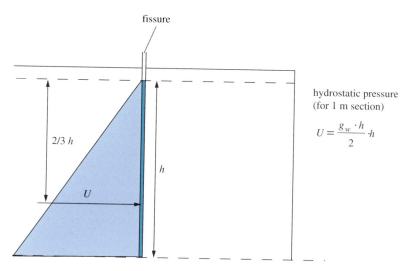

hydrostatic pressure
(for 1 m section)

$$U = \frac{g_w \cdot h}{2} \cdot h$$

Fig. 7.65 Pressure exerted by simple hydrostatic pressure in a fissure. The pressure is equal to half of the water height in the fissure multiplied by the specific weight of water and by the surface in contact with it. Its point of application is at 1/3 of the saturated height.

Seepage forces

Water was static in the previous case, but in this case it is moving. Head losses that occur as water travels through the media generate a force called the **seepage force**. This force acts on the particles in the ground that "brake" the water. Because this force is proportional to the hydraulic gradient, it is particularly noticeable in impermeable formations where head losses are significant. The most unfavorable case is that of fine-grained cohesionless sediments (sands, non-argillaceous silts) whose grains are easily eroded, creating quicksand (when the weight of grains is compensated by an upward seepage force corresponding to the critical gradient), internal erosion and piping which is very dangerous in excavations (Fig. 7.66). Deep holes in water-bearing aquifers are very subject to this risk, as well as landslides. In the case of earthen dams, this phenomenon can be prevented by placing a drainage facility at the toe of the dam. This drainage bed makes the flow lines inside the dike quasi-vertical, allowing the seepage force to stabilize the dam.

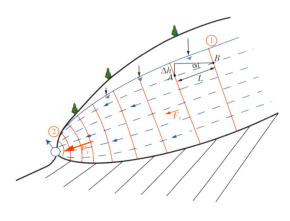

- Choice of two sections:

  S_1 far from the outlet ①

  S_2 close to the outlet ②

- Constancy of flow Q

  $$Q_1 = Q_2 = v_1 \cdot S_1 = v_2 \cdot S_2$$

  As $S_2 \ll S_1$

  $$\Rightarrow v_2 \gg v_1$$

- $v = K \cdot i = K \cdot \dfrac{\Delta h}{L} = K \cdot \sin\alpha$ K = Darcy permeability

  With

  $$\Delta h = z_B - z_A$$

  $$\Rightarrow i_2 \gg i_1$$

- Seepage force F_s on a volume of x, y, z dimension

  $$F_s = x \cdot y \cdot z \cdot \gamma_w \cdot i$$

- At the outlet: $i \nearrow$ then $F_s \nearrow \Rightarrow$ erosion of solid particles

Fig. 7.66 The effect of seepage forces in a landslide. The seepage force F_s is proportional to the hydraulic gradient i. When the cross sectional flow S is restricted, in a homogeneous material, a constant discharge Q implies an increase in gradient. In areas like this the seepage force can become high enough to destabilize a slope and can even carry away soil particles (underground erosion). A typical case is the front of landslides where the flow section is very reduced.

In tunnels the sudden entry of mud is a result of seepage forces. The tunnel is under atmospheric pressure, and the aquifers that they encounter are under very high hydraulic gradients. These flows of mud occur in tunnels, in unconsolidated sediments, in fissured rocks or in karst filled with granular material (Problem 13.1).

Modification of properties in unconsolidated sediments

The amount of groundwater present in unconsolidated sediments can significantly influence their mechanical behavior. In the dry state, soils with a fine matrix (silt and clay) behave as solids; potters clay is typical in this regard. When the ground gets wet, it proceeds to the "plastic" stage, where it deforms under pressure and does not return to the initial state, but keeps its solid consistency (Fig. 7.67). If the water content increases further, the soil can liquefy: the grains start to flow with the pore water. The water contents that delineate these three areas are called the *plasticity limit* and the *liquidity limit*. The difference between the two contents is the *plasticity index*. These limits are determined on the basis of simple standardized laboratory tests.

We understand that the water content of the ground during construction will thus have a very direct effect on excavations. The appearance of plasticity will be relatively easy to monitor. In contrast, exceeding the liquidity limit generally leads to non-controllable situations. There is thus good reason to prevent this condition by reducing the water content prior to excavation, using well points, for example (Fig. 7.68).

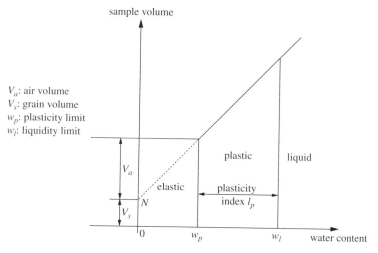

Fig. 7.67 Change in behavior as a function of water content in fine-matrix. Swelling and shrinking of the sample are visible on the volume axis.

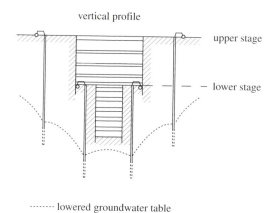

Fig. 7.68 Reduction of water content in fine-grained unconsolidated sediments using the well point technique. The points are tubes with a perforated lower part. They are installed by air injection through their lower extremity or by pounding. Each point is tightly connected to a collector pump that creates a void. Thus it is not necessary to put a pump in each well point. Drawdown in the well point is limited to several meters. To lower the water level to a deeper position, several stages of well points are used. The points must be very close together if the permeability of the ground is low (small radius of influence). For permeabilities higher than 10^{-4} m/s, the aquifer is dewatered by the use of larger-diameter water wells equipped with pumps. They are more widely spaced in that case.

Shear strength

The change in state also means that the shearing strength of clay sediments diminishes with an increase in water content. This change affects cohesion in particular (Fig. 7.69).

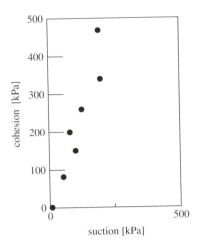

Fig. 7.69 Increase in cohesion with suction, and with desaturation of the ground. This example shows the case of a silt from a gravel washing treatment facility (from [103]).

Swelling

Water can hydrate minerals that are potentially susceptible to swelling, such as anhydrite and the swelling clays (§ 13.1.2.2). Preventing this hydration requires powerful support structures to balance this stress. This situation is often encountered in tunnels where decompression of the rock mass along walls and humidity in the air promote swelling (Problem 13.1).

Rock dissolution

Salt- and gypsum-bearing rocks are easily dissolved, which can cause the formation of cavities along the walls of engineering projects (Problem 13.1).

Corrosion of support structures

Groundwater can be acidic as a result of the oxidation of H_2S into sulfuric acid and it can corrode concrete and metal supports. Sulfate-rich waters attack traditional concrete (Chapt. 13).

Incrustation of drainage systems

Drainage systems installed around projects can have a limited lifespan due to the precipitation of mineral salts and the associated growth of bacterial gels. The most common case is the plugging of drainage boreholes and pipes with tufa (Fig. 7.70). Masses of tufa can form around springs by the release of CO_2 (Equation 9.6). Drains should thus be designed to be easy to inspect and clean out if necessary.

Dewatering wells

Excavations into aquifers cannot always be drained by gravity. It is then necessary to pump them permanently to keep water out of the project, and this entails significant long term costs and a low exploitation security.

Fig. 7.70 Precipitation of calcium carbonate covers roots ("foxtails") that keep their shape while plugging a drain. The efficiency of the drain is obviously reduced. GEOLEP Photo, N. Bulliard.

Compensation for threats to water resources

The partial or total drying up of existing resources, particularly springs, often entails compensation to water rights owners, which may often be significant.

All these difficulties linked to engineering projects lead to a decrease in construction safety, an extension of the project's duration and increased expenses.

7.4.11.2   Hydrogeologic impacts of engineering projects

We will discuss the quantitative and qualitative impacts of each type of project.

We are most familiar with the quantitative impacts of engineering projects. There are numerous examples from the history of civil engineering. Most commonly the discharge of a spring is reduced and in certain cases it can dry up completely. In rare cases, engineering projects can lead to an increase of spring flow or the appearance of totally new springs. The processes that govern these quantitative impacts are often related to artificial drainage phenomena that modify flow systems. Modifications of geomechanical constraints can lead to permeability variations in joints, which can impact flow distribution. These disturbances are especially common in underground projects and dams.

There are two types of qualitative impacts: 1) direct impacts due to the leaching of substances produced during construction that end up in the groundwater, and 2) indirect impacts such as a threat to the natural protection of the aquifer from the project or a modification of the flow system. In the latter case, mixing processes are disturbed and elevated concentrations of undesirable components can reach the aquifer outlet.

Surface projects

This classification includes all projects that only slightly penetrate the subsurface. They are usually related to communications and buildings. Underground excavations to a depth of several meters are also included in this category.

Roads on slopes

In terms of water resources, road construction on slopes often involves extensive excavations. In regions where shallow aquifers are present, numerous water inflows may be encountered. Drainage trenches are used to evacuate these inflows and divert this water from its underground pathway (Fig. 7.71). This phenomenon often occurs in hillside aquifers in morainal covers and in medium permeability rocks (weathered limestones and marls, flysch, sandstone, igneous and metamorphic rocks). Karst rocks are rarely affected by this phenomenon, with the exception of excavations near natural aquifer outlets.

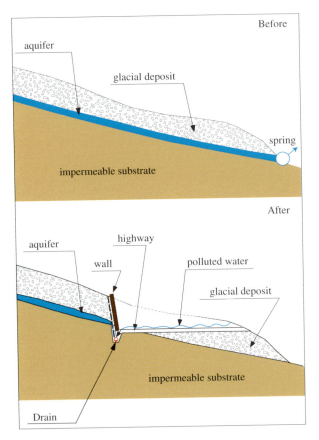

Fig. 7.71 Truncation of a slope aquifer by the excavation for a road. Groundwater flows into drainage ditches along the retaining wall and mixes with pavement runoff. The downstream spring dries up. A possible solution would be to capture the spring upstream of the road to prevent any deleterious effects of the road on water quality.

Roads on alluvial plains

On alluvial plains, the construction of subsurface linear projects necessitates either a permanent drawdown of the aquifer or the construction of a watertight structure (Fig.7.72). The first solution is to be avoided because it may damage other aquifer outlets, particularly drinking water wells. The construction of a tight structure also requires a drawdown of the aquifer, but it is temporary; in permeable sediments, the radius of influence can be wide enough to cause serious impacts over a wide area. It is preferable to construct a sheet pile or slurry trench wall enclosure prior to the pumping operation in order to draw down groundwater only from the excavation area.

The structures in the saturated zone also have a negative impact on flow in the aquifer because they can be a significant barrier if the flow section of the aquifer is reduced. In this case, artificial bypasses should be constructed to compensate for the loss of section, either as beds of calibrated gravel, or as conduits distributed all along the section that creates the barrier (Fig. 7.73).

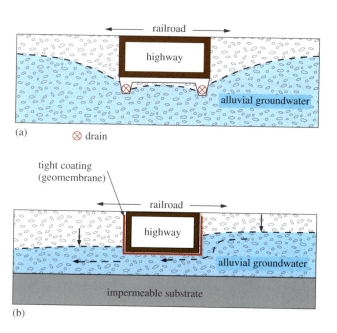

Fig. 7.72 Underpass beneath a railroad line on an alluvial plain. (a) Keeping water out of the underpass by permanently lowering the aquifer by pumping. (b) Construction of a water-tight structure that is immersed in the groundwater; a barrier effect can be observed if the flow cross-section is significantly reduced.

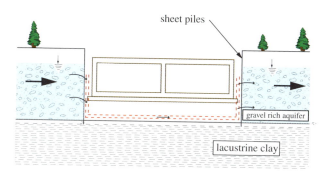

Fig. 7.73 Construction of a bypass using conduits under the foundation of the highway near Neuchâtel. The construction completely truncates the lacustrine gravel aquifer. Once the caisson of the road was constructed, the sheet piles were removed and the bypass began to function.

Infiltration of pavement runoff

Our discussion of water quality must include pavements that receive polluting substances from vehicular traffic. These are various hydrocarbons (mono-aromatics of the BTEX series, aliphatic and polycyclic aromatic hydrocarbons-PAHs) and also trace minerals, particularly heavy metals Br and B and salt used for snow removal ($NaCl$ and $CaCl_2$). Observation of aquifers shows that groundwater resources are rarely contaminated as a result of road runoff; where it occurs, it is generally of low intensity. This is because the majority of road pollutants do not travel far, with the exception of chlorides, which can be found over large areas (Fig. 7.74).

Along with the risk of non-point source pollution, roads may generate accidental risks. They are rare, but may have significant consequences (Fig. 7.75).

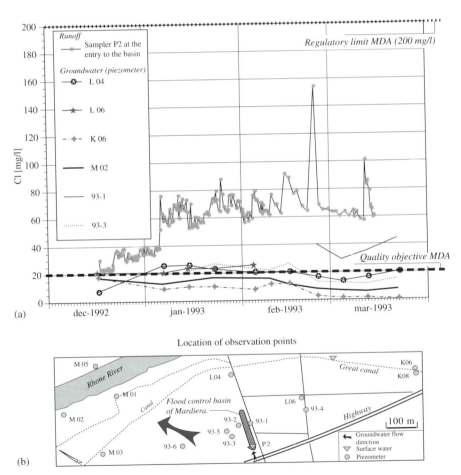

Fig. 7.74 Distribution of chlorides used for road deicing in the vicinity of the Mardiera flood control basin (Rhone plain in Valais, Switzerland). (a) Salt concentrations in winter are very high in the effluents reaching the head of the basin. In the piezometers installed around this infiltration basin, the salt contents are abnormal, but they approach the quality standards defined for drinking water, thanks to their dilution by the aquifer upstream in the alluvial plain. Other road contaminants have not been detected in the piezometers (MDA: Manual of food commodities, Switzerland). (b) Location of observation points (from [210]).

Schematic hydrogeological cross section

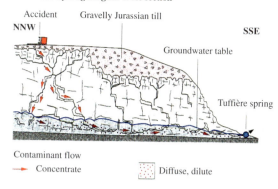

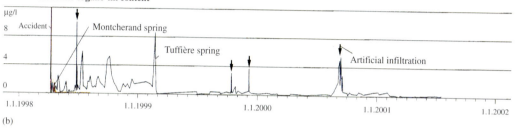

Fig. 7.75 Accidental point source pollution caused by road traffic. In 1998 a truck accidentally left the Vallorbe-Orbe highway (at the foot of the Jura Mountains). Its load of organic tin by-products spread over the ground in an area that is in direct hydrogeologic connection with the La Tuffière karst spring, one of the main resources of the city of Orbe (a). The contaminant reached the spring quickly, making it unusable for several years (b). To speed up the rehabilitation of the aquifer, massive infiltration (vertical arrows) of water were done, leading to the removal of large quantities of the contaminant (from [79]).

Today, our understanding of road runoff management is changing. For decades, water that ran off roads was collected, directed into basins where it was decanted and oil separated; then it was released to surface water. Some countries avoid gross hydrologic disturbances by favoring micro-cycles of water. This concept calls for an approach to the situation as if the project did not exist. It is preferable to allow road runoff from public squares and water collected from roofs to infiltrate near these structures. The subsurface can then play its role as a hydraulic buffer to smooth out the discharge peaks and to allow geochemical retention. This concept has been studied on an experimental road (Fig. 7.76): the slope of the road embankment is covered with a double soil horizon that ensures the effective retention of pollutants before they enter into the subsurface and contact the groundwater.

The railroad

As a transportation link on the surface, railroads are similar to roads in their impacts, especially because they use the same type of structures (viaducts, trenches), the transport of hazardous material, etc. Railroad differ from roads in terms of the following:

- Railroads are not as wide as roads, which limits earthworks on slopes;
- Many lines are old and constructed on embankments, thus do not have any subsurface extent;

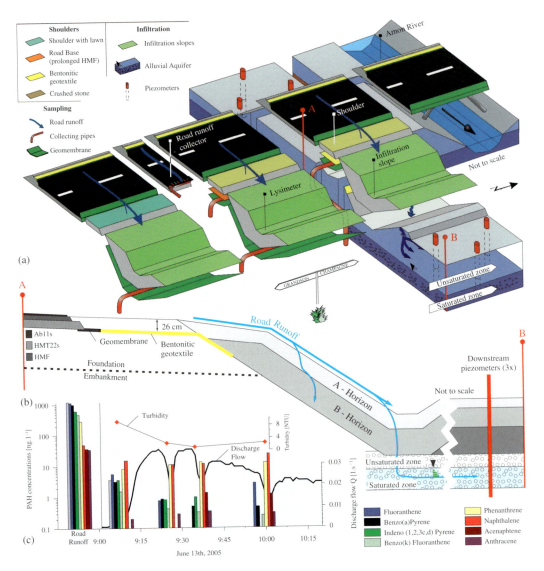

Fig. 7.76 Grandson experimental station (Swiss Plateau) testing the infiltration of road runoff. (a) Diagram of the test segments (an infiltration slope and two lysimeters). Two shoulders treated to be impermeable (with a bitumen concrete and with bentonite geotextile) are compared to a standard shoulder of grass covered gravel. The geomembrane at the base of these shoulders is there only to measure the performance of the two impermeable designs. The bentonite shoulder provides the best results in conducting the maximum amount of water onto the infiltration slope and by preventing the loss of water into the road foundation. (b) The AB profile shows the detail of the shoulder with the impermeable bentonite and the infiltration slope with its two soil horizons, as they should be constructed in practice. (c) An artificial infiltration test was conducted with rain water in the summer of 2005. The difference in concentration between gross runoff and water that leaves the lysimeter, after significant retention in the soil of the slope; example of polycyclic aromatic hydrocarbons (PAH), which are particularly toxic (from [216]).

- Substances released as a result of non-point source pollution are very different: herbicides, toilet wastewater;
- Accidents involving large quantities of pollutants because of the capacity of cisterns and the number of railcars involved; locomotives can release highly toxic transformer oil (Fig. 7.77).

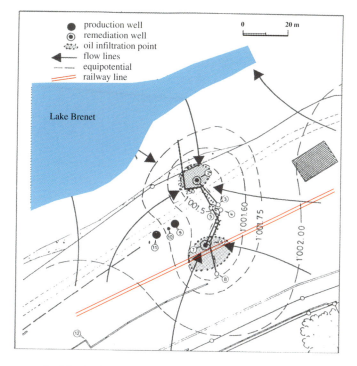

Fig. 7.77 Pollution as a result of the derailment of a locomotive near a drinking water well at the Le Pont Station, Jura Mountains. This event caused significant quantities of transformer oil to enter the fluvioglacial terrace aquifer bordering Lake Brenet. The well was immediately taken out of service. After removal of the contaminated sediments in the unsaturated zone, two recovery wells were immediately installed at the points of preferential infiltration. The water resource was remediated after three months of treatment at the site (pumping with oil separation, artificial rain with nutrients for bacterial development). Remarkably, this accident happened within three days of a pollution event caused by massive infiltration of liquid manure that took the other essential resource of this basin out of service. The return period of the occurrence of these two events together is estimated to be about 3000 years; even though improbable, this conjunction of events did actually take place in July 1989 (from [206]).

Deep building foundations

Buildings with deep foundations into the subsurface can cause serious disturbances in the urban underground environment as a result of their damming effect. There are various types of foundations (Fig. 7.78):

- Piles: Generally these are far enough apart so that the barrier effect is minimal (pile diameter << spacing); the problem is different when piles are juxtaposed to form a curtain.
- Slurry trench walls: the barrier effect is complete if they reach the base of the aquifer.

We will come back to this subject in chapter 14 where we will place it in the broader context of underground urban development (§ 14.1.3).

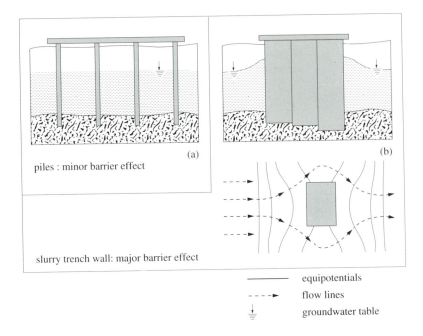

piles : minor barrier effect

slurry trench wall: major barrier effect

——————— equipotentials

- - - -► flow lines

groundwater table

Fig. 7.78 Barrier effect on groundwater flow. (a) Foundations on piles. (b) Slurry trench walls

Reservoir dams

Reservoirs affect the flow of water from the rock massif due to a sharp rise in the base level of aquifers, which can cause modifications of flow systems. However, this effect often is not important, because dam sites have been chosen for their location in impermeable rocks. It is important not to underestimate flow in fissures, which may play a major role in rocks that are thought to be impermeable. The basins most susceptible to these disturbances are in karst regions (for example, in the Dinarides Mountains, according to [186]). To minimize this detrimental phenomenon, both for the profitability of the reservoir and for groundwater resource protection, discontinuities of the rock masses in the basin must be very carefully mapped and the penetration of water into permeable zones must be prevented by means of waterproofing at the surface or by grouting at depth through boreholes. The location of the dam is an area of a steep head loss, which depends on the water level in the reservoir (Fig. 7.79).

Grouting operations in the rocky foundation under the upstream portion of the dam tend to cause the principal head loss to occur at this location. Downstream, in contrast, the rock mass is drained from the bottom to prevent any destabilizing hydraulic gradient.

Hydroelectric dams in the mountains impact the groundwater beyond the reservoir. This action is due to the indirect effects of water intakes in mountain streams in the same valley and in adjacent valleys; these mountain streams are part of the aquifer recharge, especially when they are in flood stage. Harnessing them prevents them from serving this function.

Run-of-river dams

Because alluvium is often very permeable, dams in rivers in alluvial plains often have a major problem of controlling the flow in the alluvial aquifer. Side canals (Fig. 7.80) are used for this purpose. These structures are used to prevent a large buildup of groundwater behind the dam by draining the water and to prevent a large depression below the dam by recharging the aquifer. Failure to regulate these levels effectively causes multiple problems, for example,

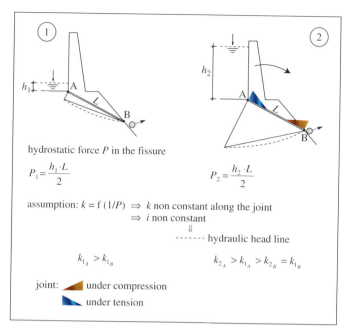

hydrostatic force P in the fissure

$$P_1 = \frac{h_1 \cdot L}{2}$$

$$P_2 = \frac{h_2 \cdot L}{2}$$

assumption: $k = f\,(1/P) \Rightarrow k$ non constant along the joint
$\Rightarrow i$ non constant
$\Downarrow$
------ hydraulic head line

$k_{1_A} > k_{1_B}$

$k_{2_A} > k_{1_A} > k_{2_B} = k_{1_B}$

joint: ◣ under compression
◤ under tension

Fig. 7.79 Fissure flow through rock joints under a dam. Case 1: Low reservoir level. Case 2: Full reservoir. The permeability of the joints varies according to the pressure: if the hydrostatic pressure in a joint is elevated, it dilates, facilitating the passage of water. The rotation momentum resulting from the force of the water against the dam tends to open up upstream discontinuities and close downstream discontinuities. This has the effect of reducing the permeability near the outlets and as a result creating hydraulic gradients dangerous to the stability of the foot of the structure.

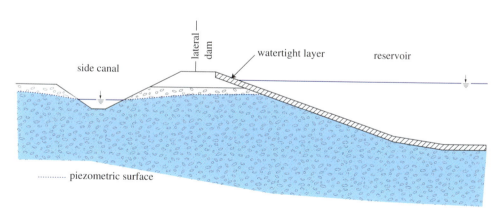

Fig. 7.80 Control of groundwater levels by side canals. Situation of a dam where the side canal prevents the groundwater from rising.

clogging of the side canals due to lack of maintenance can cause the following types of problems: in the dam area, flooding of the foundations of buildings, loss of soil fertility, drowning of old contaminant releases; downstream, drinking water wells may dry up. In addition, the smoothing of peak discharges tends to plug up riverbeds, inhibiting the recharge of aquifers that are in contact with the river.

Quarries

Although they are not engineering structures, properly speaking, quarries have various impacts on the environment in general and on groundwater in particular. Under the general term of quarry, we also include sand and gravel pits as well as rock quarries. There are four types of quarries (Fig. 7.81):

- "Alluvial" type: The interaction with groundwater is significant, especially if the deposit is exploited beneath the water table. If the aquifer outcrops after exploitation, it can be exposed to surface pollution and to undesirable thermal variations that affect the bacteriology of the water. If the trenches are filled in with the non-polluting mineral materials, which are generally impermeable, the groundwater has to detour around these obstacles. This may cause the groundwater level to rise. When larger areas are restored in this way, groundwater flow is so inhibited that the alluvial plain may become a swamp. Drainage systems are then installed to replace the natural flow within the aquifer. Numerous old excavations have been filled in with various wastes, which have created large contaminant plumes downstream from these sites. Many of the world's gravel pits and sand quarries are exploitations of this type.

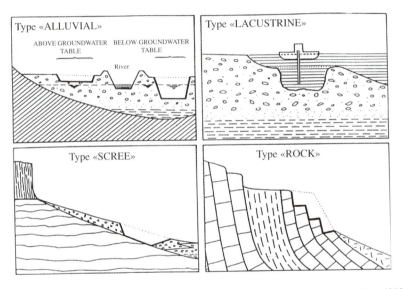

Fig. 7.81 Types of quarry exploitation based on their relationship to the environment (from [203]).

- "Lacustrine or marine" type: exploitation in water by dredging or sucking of gravels and sands affects surface water primarily (basically, turbidity and bank erosion, etc.) Underground water present in lacustrine deltas are generally not exploited in practice, although in the future it is conceivable that they may be exploited for naturally filtered potable water.
- "Scree" type: The exploitation of blocks and gravel from hillsides presents few problems for groundwater because the perched position of these deposits does not favor the accumulation of important quantities of groundwater.
- "Rock" type: In regions that lack alluvial gravel, rock is exploited and crushed to produce various aggregates. Quarries for dimension stone or ornamental stone are the minority of quarries. Because these rocks are hard, they generally have fissures that may contain groundwater. In the same way, carbonate rocks are generally karstic. Depending on the individual case, groundwater can be threatened for several reasons: loss of the

filtering cover on the rock, artificial fracturing of the rock by the use of explosives to break up the rock, and the massive entry of surface water by concentrated infiltration, hydrocarbon pollution as a result of the use of machines, plugging of fissures by the penetration of mud from washing, etc.

In all these cases, the quality of the material emplaced after exploitation plays a determin–ing role.

Underground construction projects

Before discussing tunnels and galleries, we must include borings as underground construc-tion projects, although they are performed from the surface. Boreholes can have impacts on deep formations.

Boreholes

Reconnaissance boreholes installed for the construction of structures and prospecting for various types of deposits (mineral, hydrocarbon, geothermal) involve two principal dangers (Fig. 7.82):

- Deterioration of the natural cover of aquifers: penetration of contaminated surface water directly into aquifers.
- Short-circuit between stacked aquifers: transfers of water between these horizons may cause springs to disappear and new outlets to appear; qualitative impacts are also

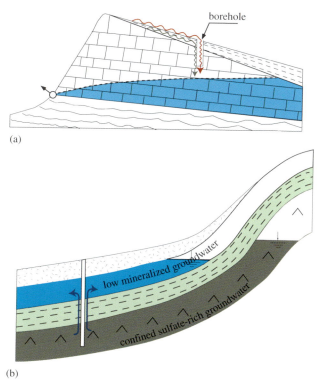

Fig. 7.82 Adverse hydrogeological effects from poorly-grouted mechanical borings or by piezometric tubes that have deteriorated by aging, crushing, or shearing. (a) Pollution by surface water. (b) Bypass between aquifers of different hydraulic potentials: in this example, gypsiferous water under pressure contaminates a fresh water aquifer when sul-fates rise up the annular space between the piezometric pipe and the rock.

possible as a result of the mixing of waters of different chemical properties and bacterio-logical contents.

Underground galleries

This term includes tunnels, galleries, and underground excavations more than about ten meters deep. Aside from numerous impacts common to this type of structure, their unique function will also cause their own effects. The safest technical method and the most economical way to manage water in an underground structure is to eliminate it by drainage into the foundation of the structure. If the ground is not permeable, this drainage will have little effect on the hydrogeology of the region. However, if the rock is permeable and the tunnel is located deep underground, there will be a considerable lowering of piezometric surface; the radius of influence of the disturbance will be large. As a result, numerous regional springs at high elevations could dry up, and these might be very significant for water supplies in mountain villages (Fig. 7.83).

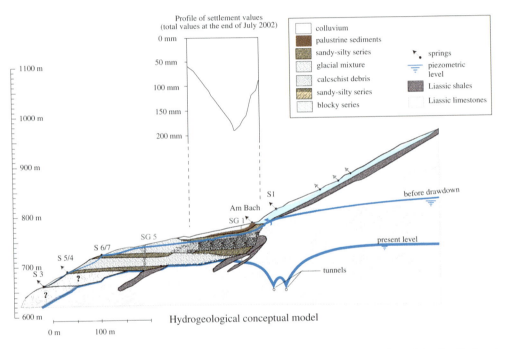

Fig. 7.83 Drying up of springs as a result of tunnels that draw down hillslope aquifers. Case of the Am Bach spring that supplied the village of St. German, Haut-Valais. Digging the base tunnel of Loetschberg caused a steep drawdown of the groundwater on the right slope of the Rhone Valley as it crossed the Liassic limestone aquifers, which were abruptly drained by the tunnel. The drawdown also affected the rockslide debris on which the village was built. The Am Bach spring, which had flowed naturally above St-German, disappeared in the summer of 2001. This phenomenon was accompanied by the suppression of the Archimedes pressure between the grains and the subsidence of buildings constructed on the palustrine spring deposits, which were highly compressible (a mixture of peat and tufa), causing significant damage (from [302]).

Let us recall that a decrease in interstitial pressure can cause subsidence that often damages structures, especially when the amplitude of subsidence is heterogeneous (see the example of the Tseusier dam, Fig. 4.12).

The different types of disturbance to flow systems that can be caused by tunnel construction are summarized in figure 7.84.

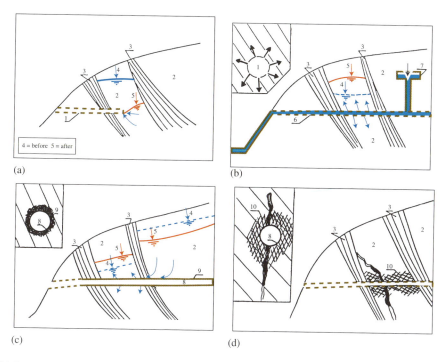

Fig. 7.84 Summary of disturbances to groundwater flow caused by a tunnel in a compartmentalized rock aquifer. (a) Excavation of tunnel 1 in aquifer panel 2 caught between two aquitards 3; lowering of the natural piezometric level 4 to position 5. (b) Case of a hydroelectric facility whose water supply gallery 6 is not watertight, with a surge tank 7; the momentary over-pressuring may cause water to penetrate into the aquifer layer causing a modification of its level (from 4 to 5) and the water quality. (c) Bypass between aquifer layers 2 through the artificially fractured zone 9 at the tunnel wall behind the lining 8; the natural levels 4 of the aquifers tend to equilibrate artificially 5. (d) Local barrier effect caused by the waterproofing of the natural permeability of the rock 10 following grouting near the project whose spreading is difficult to control (from [207]).

These effects can be prevented by waterproofing the gallery. There are two possible scenarios. The simplest is to temporarily draw down the aquifer for the duration of the construction; once the watertight lining is in place, the drains are plugged up so that the groundwater gets re-pressured and returns to its natural level. In certain critical situations, the temporary lowering is not possible, as in the case of tunneling through the Jungfraukeil for the Loetschberg base tunnel in the Swiss Alps; thermal water occurs in this area of the sedimentary rock located between crystalline formations that is suspected of being in communication with the thermal springs at Leukerbad. In this type of situation, dewatering of the aquifer layer is done by creating a cone that isolates the tunnel from the remainder of the rock mass (Fig. 7.85).

In mountainous areas, the hydrostatic heads that must be restored at construction projects often exceed several hundred meters and can be more than a thousand meters in alpine base tunnels. Pressures such as these cannot be tolerated by a concrete lining. The rock mass itself functions as the resistant ring. Grout injection is done not only to make the rock impermeable, but also to solidify the rock mass.

The construction of waterproof tunnels, as opposed to simple drainage in tunnels, entails significant additional construction costs. There is thus ample reason to construct them only when absolutely necessary. This requires a good understanding of the hydrogeological conditions of the rock mass in order to evaluate the risk of interactions between the structure and the aquifers.

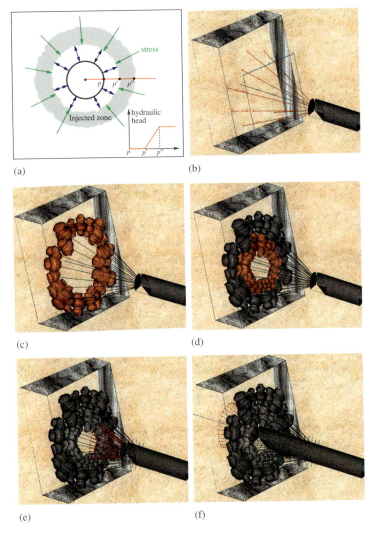

(a)   (b)

(c)   (d)

(e)   (f)

Fig. 7.85 Making the rock impermeable prior to construction of the sedimentary shell of the Jungfraukeil through the base tunnel of Loetschberg. In the southern part of the Aar Massif, the tunnel was supposed to penetrate several sedimentary layers of the Lias that was pinched between crystalline rocks. These argillaceous limestones, sandstones and shales were aquifers with pressures of more than 1000 m of water at the elevation of the tunnel. The water sampled in a deep borehole was similar to the thermal waters of Leukerbad, and engineers had received instructions not to lower the aquifer level, even during the construction of the project. This constraint required the rock to be made impermeable in the shape of a ring, at the center of which the tunnel would then be excavated. This ring was constructed by injecting grout into fissures in the rock. Permeability in the ring became so low that the entire head loss occurred in this area once the tunnel was excavated. (a) Between the ring and the tunnel, the rock mass was drained and functioned as a mechanical support to compensate the huge stress applied on the grouted cone. Figures b to f describe the succession of steps for grouting works and tunnel construction. (b) After stopping the tunnel 30 m from the aquifer layer, a series of boreholes was installed from the front in the shape of an open cone, while water pressure was maintained in the permeable rock layer. (c) Grout was injected under pressure into the boreholes; if it was insufficient, a second ring would be injected inside the first to ensure impermeability. (d) A boring was drilled along the axis of the injected cone to monitor the effectiveness of the injection; if it was insufficient, a second ring would be injected inside the first to ensure impermeability. (e) After checking the water-tightness of the ring by borings along the axis of the tunnel, the attack front was advanced to a distance of about ten meters from the layer; a cone of drainage boreholes within the injected zone ensured that atmospheric pressure was present in the future excavation. (f) The excavation was able to cross the "protected" aquifer; the drainage borehole will extend beyond the aquifer layer to continue the excavation of the tunnel (from [243]).

These studies require geologic knowledge of the massif, borehole data, and perhaps even exploration galleries. Geochemical and isotopic analyses are also helpful.

Case of water supply galleries

Many galleries constructed to supply water to hydroelectric dams pose particular risks to groundwater resources; in fact, their impermeability is often perfunctory. Two types of impacts are particularly important. One is groundwater recharge from surface water; the tunnels convey surface water and bacteriological impurities into rocky strata that may be permeable and thus may contaminate deep aquifers through artificial recharge. The other impact is dissolution; in some soluble rocks such as gypsum and anhydrite, flow in the tunnel can cause dissolution that may affect the permeability of the surrounding rock, thus facilitating infiltration. This usually weakens the tunnel (Problem 13.1).

Case of railroad tunnels

Railroad tunnels often date from the 19th century or the early 20th century. Thus they are often not watertight, particularly the base under the rails. Also, wastewater from the trains is discharged directly under the rails. The same is true of the liquids transported in the event of an accident. In unsaturated permeable ground such as limestone for example, pollutants can infiltrate into the groundwater (Fig. 7.86). Recent railroad tunnels do not have the same defect since they are equipped with high performance geomembranes, which considerably reduce the risk of pollution to the groundwater.

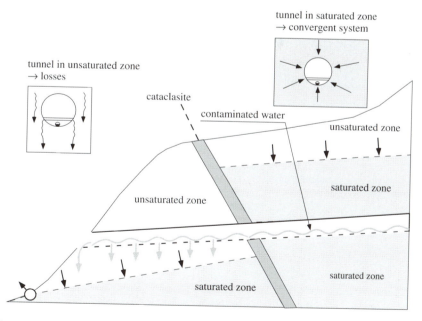

Fig. 7.86 A non-watertight railroad tunnel in a rock massif, which is unsaturated in places and saturated in other places. The path across saturated ground implies no problems, since the foundation of the tunnel is a point of convergence of all polluted fluxes. This is valid only if the foundation does not cross unsaturated rock along its path toward the exit; this last condition is rarely met in practice.

Case of highway tunnels

[46, 74,
96, 158,
192, 249,
286, 321]

Highway tunnels are generally more modern than railroad tunnels. Their wastewater is systematically collected in a pipe and evacuated from the area.

7.5   Water in the seas and oceans

Oceanography is a completely separate discipline. In this geology book we will limit ourselves to the broad outlines of oceanography, that is, the main currents that circulate in the world's oceans. We have already had an opportunity to discuss various aspects of oceanography in chapter 6 and we will examine other aspects related to the creation of marine sediments in chapter 9.

In terms of water movement, it is useful to describe the large convection cells of the oceans. These cells are controlled by complex interactions between the atmosphere and hydrosphere. Some basic data will help explain the principal mechanisms:

- The areal distribution of climates and the dominant wind directions have a direct influence on marine currents. In a very simplified way, the currents go from east to west between the tropics and then turn toward the east from there (Fig. 7.87.).

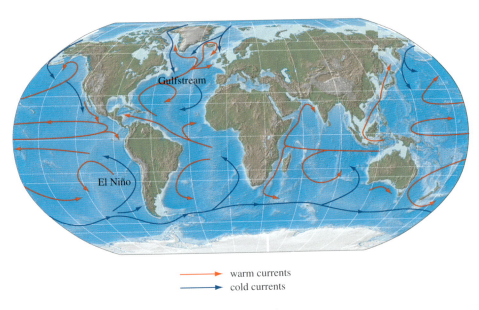

warm currents
cold currents

Fig. 7.87 The great oceanic currents.

- The *Coriolis* force causes ocean currents to rotate in a clockwise manner in the northern hemisphere and a counter-clockwise direction in the southern hemisphere.
- Water in the inter-tropical zone is warm, light, and rich in plankton (to a depth of about 200 m); this water has a high evaporation rate.
- Deep water and water in polar regions is cold, rich in oxygen, and poor in organic matter.

• As a result, equatorial "heat pumps" (Fig. 7.88) cause warm water to flow toward the poles where it cools. At the poles, water density is further increased by a high salt concentration due to the segregation that occurs during the formation of ice floes (Fig. 7.89).

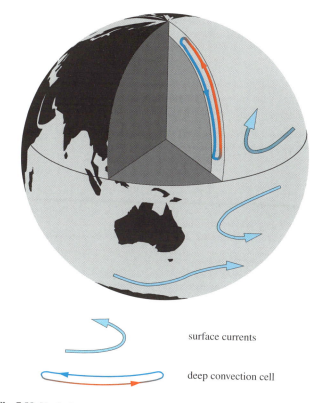

surface currents

deep convection cell

Fig. 7.88 Vertical movement of oceanic water, from the equator to the poles.

Fig. 7.89 Oceanic ice floes from the Glacial Arctic Ocean melting in the Ienissei Estuary at the beginning of the summer. GEOLEP Photo, A. Parriaux.

The difference in the land-water ratio in the two hemispheres causes north-south currents. Here is the reason: Oceans predominate in the southern hemisphere, thus favoring evaporation as compared to condensation. On the contrary, a larger portion of the northern hemisphere is occupied by land and, as a result, precipitation exceeds evaporation. This water disequilibrium leads to atmospheric currents that move from south to north, and compensating oceanic currents that flow from north to south.

The great marine currents are well known. Let us discuss two examples, one in each hemisphere.

The gulfstream

This is a warm surface current that brings thermal energy from the Atlantic tropical zone to far northern Canada and Europe (Fig. 7.87). It causes the climate of the eastern coast of the ocean to be mild and makes it possible for a good part of the Baltic Sea to remain free of ice. It is a stable current, at least up until now.

El niño

The El Niño current (Fig. 7.87) is often discussed because of the climate disturbances that it causes in the Pacific (Fig. 7.90).

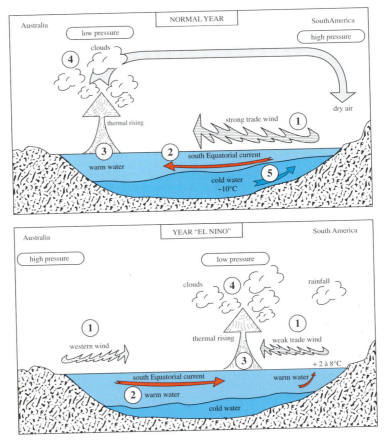

Fig. 7.90 Atmospheric circulation in the equatorial Pacific during a normal year (a) and during a year dominated by El Niño (b). Principal stages of the phenomenon: (1) direction of atmospheric currents, (2) movement of oceanic currents in response to winds, (3) atmospheric evaporation, source of storms and rainfall (4), (5) rise of oxygen- and nutrient-rich cold water along the South American coast.

In a normal year (Fig. 7.90 a), the part of the Pacific situated between the Equator and the Tropic of Capricorn is swept by trade winds that blow from the SE to the NW (1) pulling along with them an ocean current that contains warm water from the surface trench of the Pacific on the west side of the ocean (New Zealand, east of Australia and the Philippines) (2) thus creating a reservoir of warm water there; the water temperature is higher than in the east. Significant evaporation in the area of warm water causes tropical storms (3) and winds that are full of humidity after crossing the ocean produce violent rainfall in the west (4). In contrast, the eastern coast is sheltered from this intense precipitation and has a dry climate. As for the ocean, the thermocline is inclined toward the west and increases toward the American coasts, which are poor in warm water, allowing deeper colder water to rise toward the surface (*upwelling*) (5); this phenomenon is fundamental for marine life, which receives a supply of water particularly rich in oxygen and nutrients.

This process seems to be immutable, but according to a periodicity varying between two and seven years it becomes disequilibrated (Fig. 7.90 b) and gives rise to a phenomenon called El Niño ("Child Jesus" because it culminates around Christmas). The American high pressure and the Australian low pressure change places; this is called the **Southern Oscillation**. The trade winds become weaker or even disappear with the arrival of headwinds (1) that push warm surface water toward the center of the Pacific (2). The areas of high evaporation move with these warm waters (3) and tropical storms cause catastrophic rainfall in Peru and along the Equator (4). The thermocline decreases (it is almost horizontal) and does not allow cold currents from the deep water to rise, which reinforces the heating of the American coast (from 2 to 8°C); this causes a decrease of nutrients and oxygen content in the water, with significant consequences for the marine biomass. This phenomenon, which lasts about a year, has been recognized since the Spanish occupation of 1525. A particularly severe event in 1982–1983 and another in 1997–1998 led to thoughts of climate change. Nothing at the present time can prove a connection to climate change, since the reasons for this disequilibrium and the relationship between the atmosphere and hydrosphere are not precisely known on a global scale.

The **La Niña** phenomenon, in contrast to El Niño, is an unusual cooling of the Pacific equatorial waters. It is rather like the climate conditions of a normal year (the thermocline rises toward the South American coast) but in a way that is exacerbated by a reinforcement of the trade winds. There is an intensification of upwelling and of rainfall in Indonesia, and the climate along the South American coasts is particularly dry.

[52, 53, 66]

Chapter review

Water is the primary condition for life on Earth. Water is also the substance that directly controls human life, since it is our first necessity. However, water can be threatened by the effects of pollution and bad management.

Water exists in various forms, liquid, solid and gas. The external dynamics of the planet are in large part determined by exchanges between these phases.

This exchange cycle prepares the reader to understand the various environments and processes that create sediments and sedimentary rocks (Chapts. 9 and 10):

- slope deposits,
- river sediments,
- lake sediments,
- marine sediments.

Water can also have a significant influence on engineering projects, often by impeding the progress of operations because of time delays, but also in terms of project safety and public safety. For this reason, the engineer must take water into account and make the necessary and appropriate decisions for each situation.

8   The Continental Sedimentary Environment

Our study of the water cycle has introduced continental and oceanic reservoirs and their hydrologic exchanges. We can now make the connection between the transport of water and geologic materials, first on continents, then in the ocean environment (Chapt. 9).

The transport of geologic materials starts with **erosion**. We will examine the rapid processes that control the removal of material. The **weathering** of rocks, which prepares them for erosion, will be discussed in chapter 13 where we will cover the engineering problems encountered in rocks of degraded technical quality.

Transported geological materials form **sediments**. These are generally unconsolidated masses of particles of various shapes and characteristics that are deposited after a transport phase, which may be long or short. Three classes of sediments are considered:

- **Detrital** sediments: solid particles that come from the disintegration of existing rocks, classified according to several criteria, particularly grain size (Fig. 8.1);
- **Biogenic** sediments: material that consists of an accumulation of plant and animal organisms;
- **Evaporitic** sediments: precipitates deposited from supersaturated waters as mineral salts.

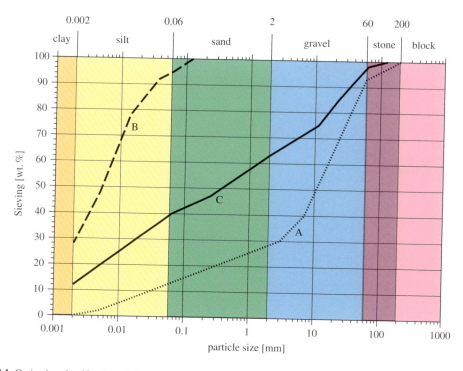

Fig. 8.1 Grain size classification of detrital sediments. The respective weight percent of the grain size fractions can be measured by sieving and sedimentation. The curve shows the percentage (by weight) of particles smaller than a given size. The boundaries between the classes are those of the Unified Soil Classification System used in the geotechnical field. In pedology, the upper limit of gravel is 20 mm and the sand-silt boundary is 0.05 mm. Examples of sediments: A = alluvial sandy gravel; B = lacustrine argillaceous silt; C = sandy and silty morainal gravel.

If these deposits are not eroded, that is, if they are covered by other sediments, they can be slowly transformed into rock. This evolution will be seen in detail in chapter 10 on diagenesis, after an overview of the various continental and marine sediments.

In this chapter, after a brief general introduction, we will describe typical continental environments (Fig. 8.2).

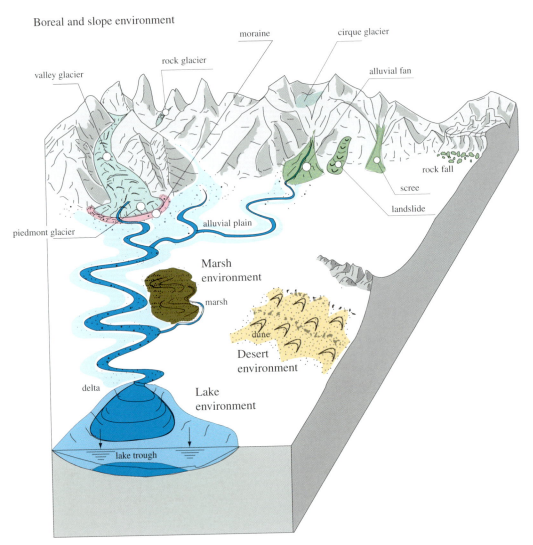

Fig. 8.2 Typical continental environments.

8.1 Generalities

We will identify the principal processes involved in erosion, transport, and deposition for the typical continental environments. Note that the transitional environments between continents and sea, such as coasts, estuaries, and marine deltas will be considered part of the oceanic environment (Chapt. 9).

8.1.1   Continental erosion

Continental erosion destroys areas of high relief created by various geologic processes. Erosion consists of in-place weathering of geologic material, the partial breakup of this material, and its removal and transport. The composition and structure of the rock, its topographic position, and climate are all key factors in erosion.

8.1.1.1   In-situ alteration and mechanical fragmentation

The dominant factors are the following: *pedogenesis* (soil formation), which involves the weathering of minerals and their fragmentation by roots, temperature variation, water content and finally freezing. These factors will be studied in detail in chapter 13.

8.1.1.2   Removal and transport

Insoluble particles are transported essentially by runoff and wind, either discretely or en masse during slope movements (Section 8.2). Soluble particles are transported as ions in continental waters.

[5, 7, 31, 75, 76, 164, 167]

Orogenesis and erosion are in perpetual combat on Earth. As soon as topography appears, erosion counteracts its development. A phase of very active erosion will take place in newly formed positive elements, creating a sharp profile (Fig. 8.3, 8.4 and 8.5). Active erosion also attacks freshly deposited sediments (Fig. 8.6).

Fig. 8.4 Differential erosion is a result of the weatherability of various rocks. This photo shows a detrital series in vertical position: hard rocks (conglomerates) stand out as blades in a matrix of intensely frost-shattered claystones. Stock La, Zascar, Himalaya. GEOLEP Photo, A. Parriaux.

Fig. 8.3 Intensive torrential erosion of morainic deposits, Ladakh. The sparse vegetation cover facilitates this type of erosion. GEOLEP Photo, T. Bussard.

Fig. 8.5 Active erosion in karst limestones in Southern China (Li River region, Guilin). The fissured zones have been eroded more actively than the others, which remain in high relief. GEOLEP Photo, A. Parriaux (diagram taken from popular Chinese images).

(a)   (b)

Fig. 8.6 Erosion of recent poorly-consolidated sediments produces surprising landforms. (a) A "tower" sculpted by surficial erosion of a glaciolacustrine terrace at Kamloops in the Canadian Rockies. The sedimentary strata that compose the rings of the tower can be clearly observed. (b) Effect of underground erosion (see § 7.4.11.1): piping in the silts of the glaciolacustrine terrace at Kamloops, Canadian Rockies, GEOLEP Photos, A. Parriaux.

The final stage of erosion is ***peneplanation*** (Fig. 8.7), a process that produces a region that has no relief. For example, Brittany is a peneplain resulting from the erosion of the ancient Hercynian mountain chain of the Armorican Massif. The Alps will undergo a similar fate.

smooth relief

fully weathered rocks

rock outcrop

Fig. 8.7 The Gondwana peneplain photographed in the Pampas region of Southern Brazil. The old continent has very low relief although it is rocky in nature. The old rock, intensely eroded and weathered, crops out in places where it is neither covered by weathering residues, nor by runoff or eolian deposits. GEOLEP Photo, A. Parriaux.

The quantity of material removed from a given region is defined as ***denudation***. It is expressed as the average thickness of material carried away per year. This factor makes it possible to compare different regions of the world. It is calculated indirectly by establishing a mass balance of material transported by rivers.

8.1.2 Continental sedimentation

Particles are deposited when the velocity of the transporting fluid decreases. Water deposition occurs in the following environments:

- Internal parts of meanders in alluvial plains,
- Flood zones during flooding,
- Alluvial fans,
- Lakes.

In the case of eolian transport, the most significant accumulations are located in areas protected by obstacles and in depressions on the ground. Fine particles can also be deposited thousands of kilometers from the erosional zones, due to atmospheric currents (§ 8.7.2).

Earth history shows that large continental sedimentary packages have occurred as a result of orogenesis. For example, thousands of meters of sediments from the dismantling of the Tertiary

Alpine mountains have accumulated in the molassic basin between Chambery and Vienna. Paradoxically this thick series accumulated at sea level in a shallow depression, as shown by the alternation of fresh water, brackish water, and completely marine fossils (Fig. 8.8). This is explained by the fact that these deposits gradually accumulated while the area subsided, and as a result the shallow sedimentary continental environment was maintained (Fig. 8.9).

Fig. 8.8 In the upper marine molasses of the Swiss Plateau, a shell-bearing sandstone formation provides a very rich marine fauna of lamellibranches with occasional perfectly preserved sharks teeth. La Moliere fossiliferous sandstone. GEOLEP Photo, GG. Franciosi.

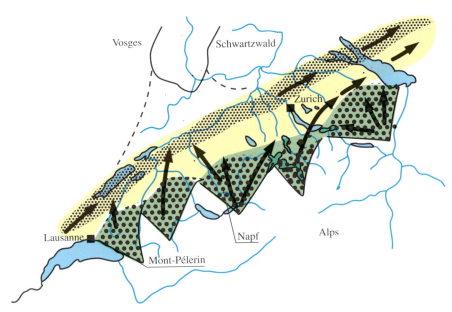

Fig. 8.9 Continental sedimentary complex of the mid-Tertiary molassic basin. Three principal depositional facies can be distinguished, from the proximal to the distal: in green, the delta widening at the mouth of ancient rivers north of the Alps (conglomerates); in yellow, the alluvial and lacustrine facies (sandstones and marls); in yellow with large dots, the brackish facies (sandstones and gypsiferous marls) (from [125]).

We could even say that sands that have later been transformed into the red sandstones of the European Permian (for example, in the Vosges) are the "detritus" of Hercynian mountain chains (Carboniferous).

Let us now examine the mechanisms of erosion, transport and deposition of sediments in various typical continental environments. We begin with the ones that are familiar to us and will end with two exotic settings.

8.2   The slope environments

From a physical point of view, the rocks and unconsolidate sediments that make up the sides of valleys are subject to gravity. The component of their weight that is parallel to the slope is the mechanical force of instability. The shear strength of the material tends to oppose the downward movement. If the driving force exceeds the maximum shear strength, there is disequilibrium and the mass moves downward until it reaches a new equilibrium position, which often is the base of the slope. Soil and rock mechanics have developed methods to calculate the forces involved in various cases: shape of the slope, material characteristics, presence of water, weight of a structure, etc (Chapt. 12).

From a geologic point of view, movement along slopes can occur in various ways:

- Vertical falls of blocks from a cliff, generally followed by sliding, rolling or saltating down a slope;
- Slide along a shearing surface;
- Removal of particles by non-point runoff along the hillside;
- Removal of rocky material by a snow or ice avalanche.

Surface water and groundwater play major roles in the mechanics of slope instability. This is the reason why many destabilizing events occur during rainy periods.

From the point of view of analyzing geologic hazards, it is useful to make a distinction between gravity phenomena that affect individual particles and mass movements.

8.2.1   Particle transport on slopes

Slope-shaping phenomena take place continuously over time. They affects isolated particles and does not constitute a significant geologic hazard. There are two types of processes, depending on the grain size of the moving particles:

8.2.1.1   Scree formation

Torrential erosion of valleys and artificial road-cuts in mountainous areas create cliffs in hard rocks. Fissures that cut through these rocks and the joints between layers form isolated parallelepiped-shaped blocks. When these joints open due to water pressure or dilatation from freezing, the blocks can become disconnected and fall to the foot of the cliff and roll down the slope. They accumulate in the shape of a scree apron or a series of fans whose summits correspond to particularly unstable areas of the cliff (Fig. 8.10). Scree aprons are sorted by grain size; large blocks are not slowed down by transport, and as a result they travel farther down the slope than smaller blocks.

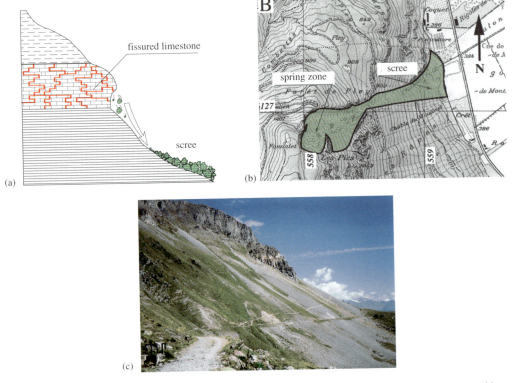

fissured limestone

spring zone

scree

scree

N

(a)

(b)

(c)

Fig. 8.10 (a) The principle of scree formation at the foot of a cliff. (b) Example of a very active scree corridor at Châble-Croix south of Vionnaz (Valais Alps). The material is used for the manufacture of aggregate. Reproduced with authorization of Swisstopo (BA056911). (c) Scree apron under a limestone wall. Active zones can be distinguished from older zones, which are covered with vegetation. Giffre Basin, Savoie France. GEOLEP Photo, A. Parriaux.

Although scree forms throughout the year, its frequency increases greatly during the freeze-thaw cycle. The engineer can protect structures by installing special netting to retain blocks, as long as they are relatively small in size (Fig. 8.11). An animation on the DVD shows tests of nets in heavy stress conditions (block of 9.6 tons falling at 25 m/s).

Technically speaking, a scree fan is near its stability limit. In the case where construction truncates this type of formation, the engineer must plan to support the talus up-gradient of the excavation.

These formations are highly permeable. They often hide a spring from a rocky aquifer; the water flows from the base of the scree as if the scree were the principal aquifer, even though it is in fact a secondary reservoir (Fig. 8.12).

8.2.1.2   Formation of colluvium

Soft rocks are susceptible to weathering and loose sediments form colluvium when slopes are wetted with strong rains or significant snowmelt. Fine particles are eroded or removed by the shear force of runoff water. They accumulate in places where the slope has an inflection or

Fig. 8.11 Very flexible net to retain blocks. Protection for the road, railroad and inhabitants of the Chillon Castle area. Swiss Prealps. GEOLEP Photo, A. Parriaux.

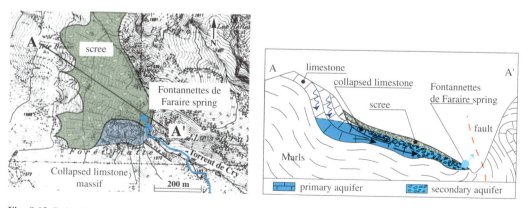

Fig. 8.12 Role of scree as a secondary aquifer, upgradient of the Fontanettes de Faraire spring, Chamoson (Valais Alps) at the foot of the Hauts-de-Cry Massif. Cross-section of the system, Copyright 2005 Canton du Valais.

at the foot of the slope (Fig. 8.13). Colluvium is composed primarily of sands and silts, with some organic material. It is very loose, often waterlogged and fluid, and unstable in excavations. The engineer must often contend with colluvium along road cuts (Fig. 8.14).

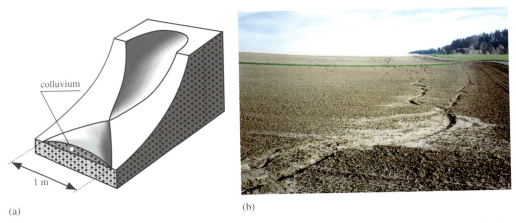

(a)   (b)

Fig. 8.13 (a) Colluvium formation (from [203]). (b) Phenomenon of active colluvium formation in the bare agricultural land as a result of storms. Soil of a Rhodanian moraine, north of Lausanne, Switzerland. GEOLEP Photo, A. Parriaux.

Fig. 8.14 This unusual erosion protection for road cuts resembles a natural landform more than a wall structure. Bioengineering can provide elegant solutions, as along this slope in South Africa where the thick residual soils are stabilized by plantings in small gouges dug horizontally along the slope. GEOLEP Photo, A. Parriaux.

8.2.2 Mass transport

Mass transport refers to geological materials that become disconnected from slopes in more or less homogeneous packages. With the exception of some regular landslides, these phenomena are often discontinuous in time and sometimes occur as natural catastrophes. We will describe the principal types, proceeding from solid rocks to semi-liquid materials.

8.2.2.1 Rock falls

The scree-formation process can occur in a more massive manner, with the rupture of an entire section of a cliff at one time. This occurrence is likely if the foot of the cliff is notched, grooved, or undercut by erosion, causing the cliff to overhang. This is the case when a soft easily-eroded layer occurs under a massive rock in the cliff. A typical case occurs in the limestones of the Upper Jurassic in the Jura chain, which rest on Argovian marls (Fig. 8.15). This is also the case of the conglomerates in the wine-growing region of Lavaux along the shores of Lake Geneva which required significant stabilization works; the instability results from the presence of marls sensitive to variations of water content and to freezing, which are present between benches of conglomerates.

Fig. 8.15 Potential rockfall situation on the northwestern flank of the Chasseron near Ste-Croix in the Jura chain. The overhang is obvious and the direction of fissures is favorable to a rupture. GEOLEP Photo, A. Parriaux.

Rock falls are highly variable in size. During recent centuries, the Alps have been the location of several large events (called ***rock-mass fall***), such as the famous Derborence rock fall, at the foot of the Diablerets Massif, which was described by C.F. Ramuz in a novel of the same name. Very recently, three events have upset Alpine valleys by creating significant plugs of material that have formed lakes: Val Pola in Valtelline in 1987, Randa in the Mattertal in 1991 and Sandalp in the Linthtal in 1996 (Fig. 8.16). An animation on the DVD shows a video on the rock-mass fall of St-Niklaus (Valais Alps).

(a)   (b)

Fig. 8.16 Rock-mass fall of Sandalp (Glärner Alps). It occurred as two main events, January 24th and March 3rd 1996. (a) A large part of the limestone massif of the Zuetribistock (2.2 millions m³) slid along a series of very smooth joints dipping toward the valley. (b) Due to its high kinetic energy, the mass collided with the foot of the opposite slope of the valley, creating an obstacle for the river, and inundating the hydroelectric basin and the Hinter Sand farm. The Vorder Sand farm built near the debris of a past rock mass fall was destroyed. GEOLEP Photo, A. Parriaux.

During some equally large events, the kinetic energy has been so great that debris has spread over many kilometers. The moving mass seems to flow on a cushion of dust that prevents any braking. These are called *rock avalanches*. Since the retreat of the Würm ice sheet, several events have occurred in the Alps, for example the collapse of Mt-Granier in the Chartreuse Massif and that of Arth-Goldau in the Canton of Schwyz (Fig 1.7).

8.2.2.2   Landslides

Landslides can affect both loose overburden and the rocky part of the slope.

The first case involves unconsolidated masses that are plastic and saturated with groundwater. The slip surface is often the contact between loose sediment and rock; it is thus very obvious. When this surface is within the overburden sediments themselves, it is *circular* in shape (Fig. 8.17). It is particularly plastic and may have slide striations. Loose sediment slides can also occur in artificial embankments, for example, mine spoils (Fig. 8.18).

In the second case, movement takes place along a slide surface that corresponds to a direction of weakness in the rock mass (stratification joint, fractures, or a combination of the two). The shear surface has particularly weak resistance properties, due to beds of plastic clays. A typical case is that of *planar slides* that affect argillaceous sedimentary rocks (Fig. 8.19) such as molasse, for example. When the slide is entirely composed of weathered rocks, the mass becomes highly plastic. Mobile slides form, which flow according to topographic irregularities (Fig. 8.20).

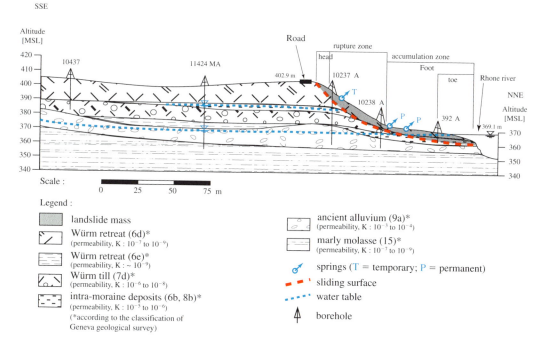

Fig. 8.17 Peney landslide, on the right bank of the Rhone downstream of Geneva. The moraine slides on clays, threatening the stability of the terrace and the Verbois dam. Permeabilities are given in m/s (from [294]).

Fig. 8.18 Mine spoils often pose significant environmental problems. In addition to the effects of acidification of water and soils, their stability can be compromised by the weathering of the rocky debris. Witwatersrand gold mines, South Africa. GEOLEP Photo, A. Parriaux.

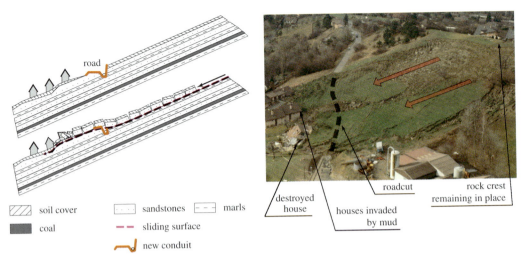

soil cover   sandstones   marls   destroyed house   houses invaded by mud   roadcut   rock crest remaining in place

coal   sliding surface

new conduit

Fig. 8.19 Planar sliding in the Belmont coal molasses (Lake Geneva Basin) on 14 February 1990. The notch excavated into the sandstone bench to enlarge the road is probably the cause of the rapid movement over the entire slope up to the crest. The landslide destroyed a house and invaded villas up to the level of the second storey. GEOLEP Photo. J. H. Gabus.

Fig. 8.20 Case of the Pavilion slide, Canadian Rockies. The original material is an altered Cretaceous claystone that contains a significant proportion of smectites. The slide comes from a large zone of ablation covered by a forest; it then flows through a sort of canyon, before spreading out as a vast accumulation lobe. This type of sliding does not generally reach a dangerous velocity (average velocity of several decimeters per year in active zones). GEOLEP Photo, A. Parriaux.

High relief slopes are also subject to a particular instability mechanism: the ***deep rotational slide***, also called ***sagging***. It is characterized by a steeply inclined circular rupture surface; the down-dropped mass preserves its structure rather well, with a slight amount of breakup and a slight tilting. In the mountains these slides can affect an entire slope and be several hundred meters thick (Fig. 8.21). In these cases, they can damage underground structures that cross the shearing surface and occasionally lead to the abandonment or destruction of important structures, if the movements are particularly rapid (Fig. 8.22).

Fig. 8.21 Deep landslide along the bank of Lake Lugano, Ticino, Switzerland. The limestone rocks slide slowly along the fissures. Note the perfect shape of the landslide scarp. J.F. Kauffmann Photo.

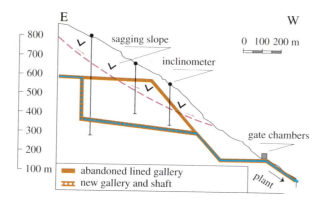

Fig. 8.22 Broad subsidence on the left bank of the Fraser River, British Colombia. The deep movement required the reconstruction of a part of the supply gallery and shaft of the Wahleach hydro-electric dam (from [189]).

8.2.2.3 Mudflows

When the water content of the unstable material is particularly high, "flows" can occur: this material is more mobile than the material involved in landslides. The tongue is more elongated and its thickness does not exceed several meters. There are two types, depending on the type of material.

- Cohesive material with an argillaceous component: these are plastic flows called "*earth flows*," whose propagation velocities are clearly higher than normal slides, although they generally do not exceed several meters per minute. Much higher velocities are encountered in the case of *"quick clays"*. These are clays with a highly cohesive appearance until a mechanical solicitation makes them unstable and causes them to demonstrate liquid behavior. Flows caused by this loss of equilibrium may reach several tens of kilometers per hour. This phenomenon occurs especially in northern countries (Scandinavia, Canada) because these are old marine clays that have been elevated by isostatic rebound in areas that were highly glaciated during the Quaternary. Exposed to the rainfall, the sodium cations ensuring their strong cohesion have been gradually replaced by hydrogen protons, which cannot maintain a strong bond. The structural arrangement of the clay layers persists but in an unstable equilibrium (Fig. 8.23). The addition of kitchen salt can make these clays cohesive again.

Fig. 8.23 Quick clays of the Champlain in the accumulation zone of the Lemieux landslide, Ontario. The original strata are completely deformed by the movement. These slide materials completely obstructed the flow of the South Nation River. GEOLEP Photo, A. Parriaux.

- Non-cohesive material: these are more liquid flows that move much faster as a result of their consistency. Their velocities can exceed 10 m/s. The material must be poor in clay and contain a significant proportion of silt or sand. Large size blocks and tree trunks are easily carried by this type of transport. **Debris flows** occur within river beds, triggered by surface water (Fig. 8.24). They generally terminate in an alluvial fan (Fig. 8.25). For example, the city of Brig (Valais Alps) was the location of such a flow caused by the sudden flooding of the Saltina, a mountain river that descends from the Simplon Massif (Fig. 8.26). The city, which was built on an alluvial fan, was invaded by a layer of mud and blocks several meters thick. Two people died and there were 300 million dollars of damage. An animation on the DVD shows a debris flow monitored on the Illgraben fan (Valais Alps). When the flow occurs on a slope, outside the river bed, the phenomenon is called *"debris avalanche"*. It is triggered by groundwater pressure, usually where a permeable water-bearing layer at depth is covered by less permeable sediments. The village of Gondo on the southern slope of the Valais Alps was cut in two by such a flow during the heavy rains of October 2000 (Fig. 8.27).

(a)   (b)

Fig. 8.24 Snapshot of the arrival of a debris flow at Zarvragia, tributary of the Rhine in Graubünden, Switzerland. (a) 18 July 1987 at 16:00. (b) 15 minutes later. Discharge was estimated at 600 m³/s. Velocity of the front: 8 m/s. Toni Venzin Photos, Trun.

Fig. 8.25 Debris flow that crossed the trans-Himalayan road between the Gange Plain and Ladakh. The flow demolished a house. The grain size of the flow shows its bimodal nature: silts at the surface of blocks. The edges of the blocks are angular. GEOLEP Photo. A. Parriaux.

There is no net boundary between cohesive and non-cohesive flows.

Sometimes an obstacle blocks the riverbed prior to a flow, for example, a landslide from a slope or an alluvial fan. It is generally followed by an abrupt dam break that causes an excessive increase in discharge and liquefies the material. This is an example of a coupling of phenomena, such as is often observed in catastrophes in mountain regions.

8.2.2.4 Geographic information systems and landslides

Several recent techniques of numerical land representation have very useful applications in the screening and monitoring of landslides. Geographical information systems (GIS) make it possible to graft onto numerical elevation models a whole series of geological parameters involved in risk evaluation:

- Active or past instability;
- Type of terrain (grain size distribution, lithology, weathering, fractured rock, etc.);
- Suspected and confirmed groundwater inflows.

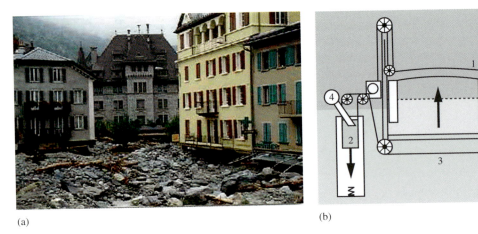

(a)   (b)

Fig. 8.26 Debris flow from the Saltina river in Brig on 24 September 1993. (a) Photo of the main street which was built on the old bed of an alluvial fan and performed that function again when the river flowed out of its artificial bed, upstream of the bridge at the top of the fan. (b) A new bridge was built to prevent a repeat of the congestion at this point; the bridge rises in response to the force of the water. A water intake in the channel upstream at a level reached only by a high flood supplies a reservoir that controls the raising of the bridge: 1 = foundation of the bridge, 2 = reservoir, 3 = pulley system, 4 = inlet from the water intake. G. Escher Photo (diagram from [182]).

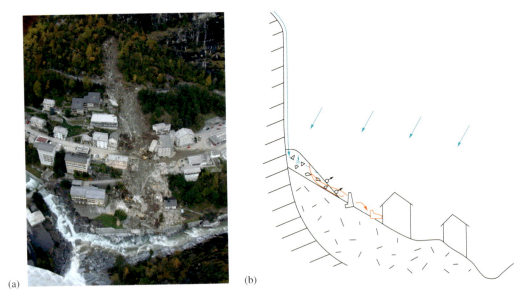

(a)   (b)

Fig. 8.27 The Gondo Catastrophe. In October 2000, the southern slope of the Swiss Alps experienced exceptional rainfall. A sort of mini-monsoon coming from North Africa and crossing the Mediterranean dropped 700 to 900 mm of water in five days. The village of Gondo (a) is built at the foot of a steep cliff on a moraine covered by a debris fan (b). A retaining wall had been constructed up-gradient to retain blocks that fall from the cliff. During the rain event, a large quantity of water infiltrated into the fan. It formed a temporary water-bearing layer in contact with the debris and the moraine. Exceptional seepage forces destabilized the foot of the fan and caused a rapid debris avalanche. Material accumulated behind the retaining wall, which had not been designed for such a load. After several minutes the wall collapsed causing a flow that crossed the village, demolished houses, and killed fourteen people. Photo www. crealp.ch.

[257, 299, 318]

GIS makes it possible to add this geological information to topographic criteria, in particular, slope and spatial variations (curvature, orientation …) the objects at risk, etc. The geologist is thus in a position to evaluate the situation by integrating all the data useful to the stability problem. The advent of the first laser aerial coverage by LIDAR systems (light detection and ranging) provides higher performance to this in-vitro treatment, a valuable complement to land surveying. LIDAR sweeps the territory with a laser bundle and measures the distance between the apparatus and the soil, making it possible to construct a precise three-dimensional model. The morphological details detected by this coverage (even in the presence of dense vegetative cover) may be indicators of the nature of the subsurface and potentially unstable areas (Fig. 8.28).

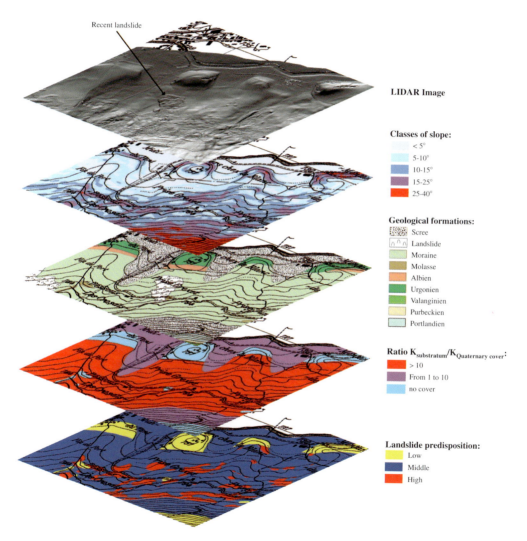

Fig. 8.28 Example of a GIS representation of a slide area with superposition of 5 layers of information. 1: LIDAR coverage describing the topography with very high resolution; a recent slide is easily identifiable. 2: automatic calculation of slope classes. 3: geologic map showing the distribution of the formations present. 4: relationship between aquifer permeability and permeability of the cover, an important factor in the formation of slides and flows; an impermeable cover on a permeable substratum is very unfavorable to stability. 5: synthesis of information of the first four layers as a map showing landslide predisposition. Case of the Travers slide, Jura, Neuchâtel. From Master thesis Severine Bilgot 2007.

8.3   The alluvial environment

Before discussing rivers in their natural context, it is useful to discuss some physical basics on particle transport in rivers.

8.3.1   Solid transport in rivers

Except for transport in solution, there are three forms of material transport in rivers (Fig. 8.29):

- Transport by *traction* or *rotation*: large diameter particles remain in contact with the streambed. The shearing force of water moves components by causing them to roll or slide.
- Transport by *saltation*: smaller particles temporarily lose contact with the bottom due to locally rising forces caused by water velocity or following a collision with bottom particles.
- Transport by *suspension*: the smallest particles, especially those of a foliated shape such as micas and clays, remain permanently in the water mass and follow the various current streams. Their weight is negligible compared to the forces that keep them in suspension (Fig. 8.30).

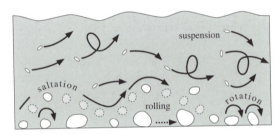

Fig. 8.29 Mechanisms of solid transport in rivers (from [248]).

Fig. 8.30 Solid transport is very important in a country with high relief and where vegetative cover is absent. This is the case in the Himalaya along one of the tributaries to the Indus, the Zascar. The solid transport is visible as suspended material that gives the water its color. GEOLEP Photo. A. Parriaux.

In a river, the mass transported per unit of time based on these three processes is the *solid discharge*.

Tests in experimental channels have made it possible to determine the velocities at which particles in the streambed are carried away and the velocities at which they are re-deposited (Fig. 8.31).

The longitudinal profile of a river (Fig. 8.32), from the source to the mouth, has a parabolic shape if the river is in equilibrium and if the geological substrate is homogeneous. In the ideal case the highest section is the location of erosion, and the lower part is the location of deposition. The mouth will be treated in the section on the lacustrine environment (Section 8.4) and in the chapter on ocean coasts (Chapt. 9).

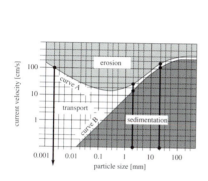

Fig. 8.31 Functions of the erosion and sedimentation of particles in a river (from [126]). Curve A: erosion function of the bed. Curve B: sedimentation function. Example of three particles: a particle with a diameter of 20 mm needs a current of 150 cm/s to pick it up from the bed and it is re-deposited if the current falls below 110 cm/s; 40 cm/s is sufficient to lift a 2 mm particle and it is re-deposited at 15 cm/s; in contrast, a clay particle of 2 μm has such cohesion that it takes 100 cm/s to move it from the bed; velocity must drop to practically zero for it to be re-deposited.

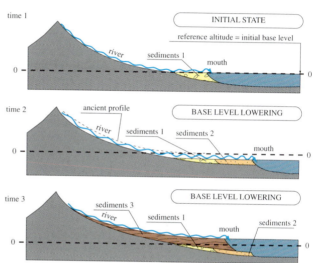

Fig. 8.32 Longitudinal profile of a river. The effect of modifications of the base level. Time 1: initial conditions; the river is in equilibrium; it erodes the higher part of its reach and deposits in the lower part. Time 2: lowering of the base level (for example, by rupture of a natural dam downstream of the lake); the river erodes its course more actively from downstream than from upstream (regressive erosion); sediments 1 are eroded and they form a new lower alluvial plain (sediments 2). Time 3: rise of base level (for example, the valley is again blocked downstream of the lake); the river bed fills with sediments 3; a higher alluvial plain is constructed.

8.3.2   Areas of stream and river erosion

In the upper part of basins, torrential erosion sculpts valleys into V-shapes (Fig. 8.33). This process depends on the strength of rocks: hard rocks produce *cliffs* relieved by *ledges* (narrow debris slope) in soft rocks. Slope debris are transported down the mountain stream by the various processes described in the section on slope environment.

The engineer tries to reduce erosion in the streambed by the use of energy-breaking structures, as thresholds to create obstacles to flow (Fig. 8.34). Erosion control along banks requires retaining walls (Fig. 8.35).

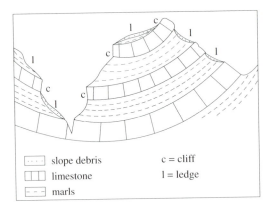

slope debris   c = cliff
limestone   l = ledge
marls

Fig. 8.33 V-shaped profile created by torrential erosion.
The hard and massive rocks create cliffs and soft easily-
weathered rocks make ledges.

Fig. 8.34 Breakwater made of blocks of reinforced concrete. River crossing the "Swiss Road" Lamosangu-Jiri,
Nepal. The river eroded the base of a large landslide that
covered the road built with Swiss technical coopera-
tion. This case of a high degree of stabilization is exceptional along this road where more moderate techniques
have performed very well since the inauguration of the
road in 1985.

Fig. 8.35 Gabions can be used to protect river banks in non-industrial areas. These are parallelopipeds of metal mesh
filled with blocks and stacked on top of each other. These walls can be made without machinery. They deform flexibly,
without abrupt breakage. The problem in the river is often erosion of the bed under the gabions, which ends up carrying
away the entire structure. Sun Kosi River, Nepal. GEOLEP Photo. A. Parriaux.

8.3.3   Alluvial zones

The lower part of a basin is an alluvial plain; it is bordered by alluvial fans. The slope
diminishes as velocity decreases, and as a result sedimentation takes place. The plain has a
tendency to advance slowly into the lake or the sea and the level of the plain rises progressively.

8.3.3.1   Alluvial plains

In alluvial plains that are densely populated by humans, rivers are generally rectilinear with some large regular curves. These sections are largely artificial, having been created during canal and barrier construction to provide flood safety and allow crops to be grown and sometimes for navigation to occur. In their natural state, rivers have few straight sections. Rivers flowing on their own alluvium have a tendency to make new channels and as a result riverbeds have highly varied shapes made up of a succession of curves. Maps prepared at various times also show that these curves move over time. The mechanism that governs the formation of many alluvial plains is called *meandering*. In its natural state, a river moves constantly over the alluvial plain due to the continuous evolution of *meanders* (Fig. 8.36).

Fig. 8.36 Meanders, a natural landform in low-slope alluvial plains. Case of Chesapeake Bay (York River, USA). Active meanders and oxbow lakes formed from captures can be seen. NOAA National Estuarine Research Reserve Collection Photo, A. Bahen.

Let us look in more detail at what happens in one of these meanders. There is a heterogeneous distribution of water velocity (Fig. 8.37). On the outside of the bend (the *cut bank*), water flows rapidly; water erodes the bank and digs up the bottom of the river; the flow is within the laminar regime, and the water velocity does not decrease. On the inside of the curve (the *point bar*), there are turbulences that reduce the average velocity and cause the bed load to be deposited; the section is thus smaller and the friction between the water and the river bottom causes braking to be more significant. This phenomenon is reinforced by a rotational flow of water, which creates a centrifugal component on the surface and a centripetal one on the river bottom. The latter gives rise to an oblique shearing stress on the particles in the streambed (Fig. 8.37). It is obvious that erosion and a bit of sedimentation tend to reinforce the velocity differences. This is why a meander can evolve only by continuing to dig outward and downstream and by new aggradation on the inside of the turn. The river moves continuously over time, taking away sediments from one side and creating new ones on the other. This is how the majority of alluvial plains evolve. The position of meanders is also influenced by the presence of alluvial fans.

Because the meanders move, two loops can end up by joining. The river chooses the most direct path. This is called **capture** (Fig. 8.36). The meander is abandoned (*oxbow*); sometimes the water is motionless at the bottom and is in contact with groundwater. The channelization works have also disconnected numerous meanders from the active riverbed. These depressions are often filled to promote agriculture, and are sometimes filled with waste. Those that are spared may be preserved as a wet biotope.

Alluvial sediments formed by meandering are a complex of discontinuous gravelly and sandy deposits (Fig 8.38). They are channelized, that is, they follow the shape of the river's migration. In a cross section perpendicular to the axis of the valley, they appear as lenses with oblique stratification. Between these gravelly deposits that have accumulated in the old bed of the river, there are more continuous beds of fine sand and silt that correspond to episodes when the plain was flooded during high water.

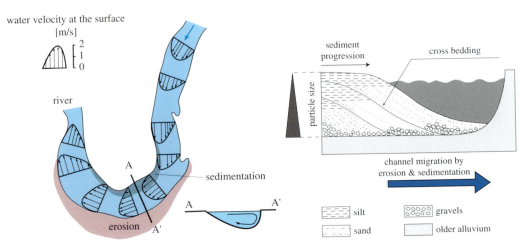

Fig. 8.37 Distribution of water velocity in a meander. Cross section A–A' shows the effect of the rotational flow.

Fig. 8.38 Schematic cross section of alluvial sediments generated by a fluvial migration.

In some high-elevation alluvial plains that have a slope of about one percent and are fed by coarse alluvium, the trace of the river is divided into multiple anastomosing branches that form a "***braided river***" (Fig. 8.39). The rambling is more active than in meandering zones. Sedimentation is intensely channelized.

The engineer must be suspicious of channelized sediments whose technical qualities may vary widely over small distances. This is also true for agronomy and for the foundation of structures. Gravelly and sandy alluvial sediments contain abundant groundwater, often highly exploited for drinking water supplies. They also are an excellent resource for the aggregate industry; the low content of fine particles and the rounding of gravel pebbles are very favorable for the manufacture of concrete.

(a)   (b)

Fig. 8.39 Alluvial Plains: Two Visions. (a) "Natural alluvium". It is necessary to go to inhospitable regions of the Earth to find alluvial plains that are still in their natural condition. Fluvial migration produces very aesthetic images along its path, causing continuous change in the water on the surface of the plain. A "braided river", region of Medicine Lake, Canadian Rockies. (b) "Anthropogenic alluvium". Some cities in the third world are not able to manage their household waste. The present day alluvium of rivers in these cities shows the detrimental effects of human habitation. Katmandu, Nepal. GEOLEP Photos, A. Parriaux.

8.3.3.2 Alluvial fans

As a tributary approaches an alluvial plain in a valley, the water velocity decreases rapidly because the slope is reduced. The transported particles are deposited in a wide zone called an ***alluvial fan*** (Fig. 8.40). These sedimentary systems are often complex because of base level variations in the alluvial plain (Fig. 8.41). This sedimentary environment is much more discontinuous than the environment in the river. It is often ephemeral. On the other hand, intense flood events upset the surface of the fan, creating new channels, and covering the land with sediments several decimeters thick in several hours. The material that forms the alluvial fans is poorly sorted and it occurs in discontinuous and oblique beds. The sediments are less rounded than the sediments in alluvial plains.

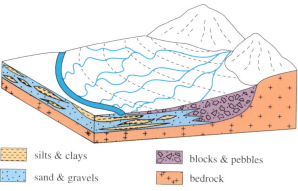

silts & clays   blocks & pebbles

sand & gravels   bedrock

Fig. 8.40 The process of alluvial fan formation (from [41]).

Fig. 8.41 Terraced alluvial fan. After the lowering of the base level in the main valley, the river has shaped its upper fan into several terraces showing five levels of lowering (numbers 1 to 5). Himalayas, Rothang La Ladakh, GEOLEP Photo, A. Parriaux.

The presence of fine particles in alluvial fan sediments makes them less favorable than those of the alluvial plain for the use of groundwater (lower permeability) or use in the gravel industry (need for more intensive material washing operations).

[165, 233, 3

PROBLEM 8.1

Figure 8.42 shows two sections of the valley, one representing "cut-in-bedrock" terraces and the other "nested fill" terraces. In both cases, determine the respective elevation and lowering of the base level over time by referring to the longitudinal profile of a river (Fig. 8.32). Indicate the phases of erosion with respect to sedimentation.

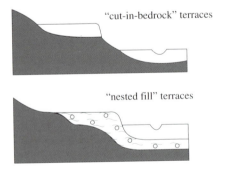

Fig. 8.42 Sedimentary structure of "cut-in-bedrock" and "nested fill" terraces. Schematic cross-sections.

Look at the solution on the DVD

8.4   The lacustrine environment

Sediments that are deposited in lakes come from all the rivers that flow into them. Depending on the distance from the mouth of the river, the sediments can vary widely.

8.4.1   Deltas

The characteristic type of sedimentation is the delta. It is the result of the massive accumulation of river sediments that occurs as the water velocity decreases when entering a lake. Outside of the delta area, the lake receives very few sediment (Fig. 8.43).

Fig. 8.43 Characteristic shape of a lacustine delta. Silvaplana, Graubünden. GEOLEP Photo, L. Cortesi.

The density relationship between river water and lake water determines how they mix. Density is influenced by water temperature, salinity, and suspended particle content.

8.4.1.1   Delta types

Depending on density relationships, three cases are possible (Fig. 8.44).

- River density = lake density

The river water collides at full force with the lake water. There is an abrupt drop in velocity and deposition occurs over a short distance. The flanks of the delta are steep and the particles are poorly sorted. An example of this is the modern delta of the Aar at the mouth of the Hageneck canal into Lake Biel (Swiss Plateau), dug in 1863 (see Fig. 8.52); the water has a light sediment load and the river temperature is highly buffered by the lakes of Brienz and Thun, well up-gradient in the basin.

- River density > lake density

In this case, the river waters sink under the lake water at the delta contact (Fig. 8.45). There can be an actual under-lake channel that conducts the water, essentially unmixed, over

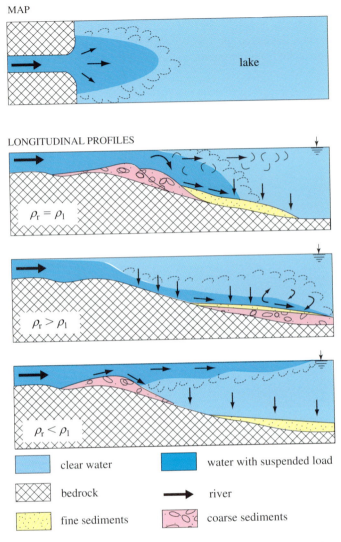

MAP

LONGITUDINAL PROFILES

$\rho_r = \rho_l$

$\rho_r > \rho_l$

$\rho_r < \rho_l$

clear water   water with suspended load

bedrock   river

fine sediments   coarse sediments

Fig. 8.44 The three types of deltas as a function of density relationships. ρ_r – river density; ρ_l – lake density

kilometers along the coast. The particles are more segregated than when the two water bodies collide directly. This is the case of Rhone river at its mouth in Lake of Geneva (Fig. 8.46). The river water is colder and has a higher sediment load than the lake. The slope of the delta front is gentler than in the previous case.

• River density < lake density

The situation prevents a collision between the waters. In this instance, the river water flows onto the surface of the receiving basin. As the water gradually slows down, the particles are deposited over a large area in front of the delta. Fine particles in suspension may be dispersed very far before being deposited. In fact, this case is the most typical of marine deltas, where the salinity of the water increases the density of the receiving medium (Chapt. 9).

Fig. 8.45 Lacustrine delta in the Saskatchewan River in Alberta. The water carries a high solid material load and enters the lake rapidly, forming a short turbid plume at the water surface. Several arms form in parallel. Note that in a narrow valley, the delta has a linear front. GEOLEP Photo, A. Parriaux.

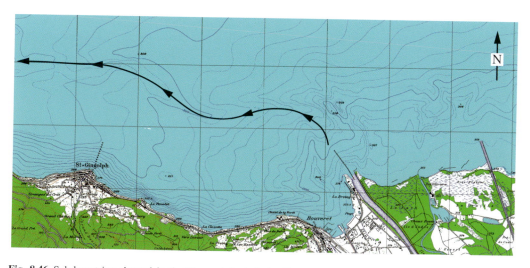

Fig. 8.46 Sub-lacustrine channel in the Rhone delta in Lake Geneva. Reproduced with authorization of Swisstopo (BA056911).

8.4.1.2 Sedimentary structures in deltas

If we look closely at the sediments that make up a delta, for example the first type described above, we note a very particular arrangement of strata, in which three areas can be distinguished (Fig. 8.47):

- *Topset beds*: these are beds of coarse material, generally gravels and sands, approximately horizontal; they correspond to the elongation of the riverbed onto the body of the delta.

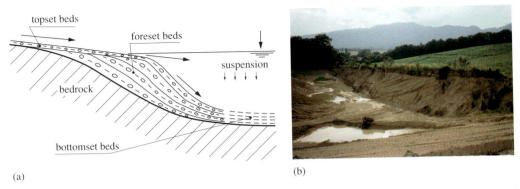

Fig. 8.47 Delta Structure. (a) The three sedimentary bodies of a lacustrine delta. (b) Progradational delta series of an old glaciolacustrine delta exposed in a gravel pit on the margin of the Lake Geneva depression, upstream of the village of Coinsins. The high dip of the beds shows that deposition occurred on a delta front. GEOLEP Photo, A. Parriaux.

- *Foreset beds*: beneath the topset series, the gravel and sand beds are highly inclined toward the lake; the angle of the strata corresponds to the angle of maximum stability of alluvium in water (about 35 degrees in the case of gravels, a bit less in the case of sands). At the base, the layers curve to follow the shape of the bottom of the lake depression.
- *Bottomset beds*: in front of the delta body, these are suspended particles (fine sands, silts, clays) that are deposited in thinner beds; they grade progressively to trough sediments.

8.4.2 Lacustrine basins

Far from the mouths of rivers, lacustrine basins receive only small quantities of sediments from two different sources:

- *Detrital sediments*: extension of bottom-set beds at the foot of deltas that cover the depression; average particle size decreases with distance from the mouth; the grains are arranged in sedimentary layers that form a series of fine millimeter-thick sand-silt-clay cycles which correspond to the succession of flood and drought periods of the lake tributaries. Many lacustrine sediments occupy perched positions today as a result of the lowering of the base level (Fig. 8.48). These sediments are occasionally used for raw material for the terra cotta industry; for example, the city of Katmandu is built on an ancient lake, and its deposits are used for construction.
- *Biogenic sediments*: in lakes whose temperatures are not too low and when the watershed contains carbonate rocks, biogenic sediments are deposited on the lake bottom. They are very loose calcareous silts called *lacustrine chalk*. These sediments come from the deposition of living organisms encrusted with calcite that precipitates from supersaturated water (Chapt. 9). Biogenic sediments also contain small carbonate shell debris.

(a)

(b)

Fig. 8.48 Horizontally stratified sediments of old Lake Lamayuru, Ladakh. (a) General view. (b) Detail of sediments eroded by the river, that causes their slide. GEOLEP Photo, A. Parriaux.

Both detrital and biogenic sediments can contain a significant proportion of organic matter as a result of vegetative growth in the watershed, but the organic matter can also be anthropogenic, as when wastewaters are not sufficiently purified. Lacustrine clays with a high content of fine organic matter that is disseminated throughout the sediment are called *gyttja*.

These lacustrine bottom series form layers that have very poor foundation qualities. They are highly compressible and they can even become plastic if the clay fraction is significant. They require special foundations. Federal Institute of Technology of Lausanne is partially built on this type of ground, which explains why the buildings are supported on a multitude of piles. Fine lacustrine deposits cause landslides, as shown by the highway project between Lausanne and the Franche-Comté, on the left bank of the Orbe gorges.

Large over-deepenings eroded by Alpine glaciers during the last glaciation are today mostly filled by fine detrital deposits. For example, the Rhone plain in the Valais, upstream of the rock threshold of St-Maurice, is made up of almost 1,000 m of loose sediments, of which a good portion are lacustrine sediments.

[41, 162, 252, 259]

8.5   The palustrine environment

A region is called swampy when it is permanently or very often flooded, to the point where a specific type of vegetation becomes established. Peat is the main palustrine deposit. It is generally formed from plant remains that are poorly decomposed due to anaerobic conditions. Two very different causes can lead to the formation of swamps.

8.5.1   Swamps at the bottom of depressions

When surface water converges in an impermeable depression where the slope is not sufficient to drain it, a swamp that is essentially horizontal to the bottom of the depression can form (Fig. 8.49). This case is very common in areas where the retreat of glaciers has created landscapes of small hills separated by numerous depressions. The till cover implies that the bottom of the depression is rather impermeable.

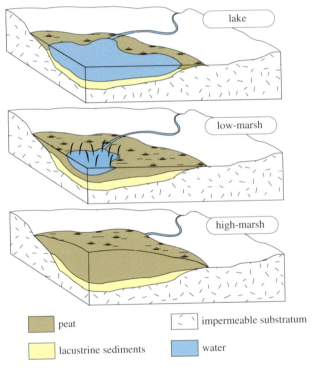

Fig. 8.49 Genesis and evolution of a swamp at the bottom of a depression (from [281]).

Often the swamp is the final phase of the filling of a lacustrine depression. When water depth is reduced to several decimeters, **hygrophilic** plants begin to colonize the water table (carex, reeds, birch, willows, etc.). When a plant dies, its remains contribute to peat formation and the filling of the lake. If the process stops at that point, a flat-bottom depression is formed (**low marsh**) (Fig. 8.49 and Fig. 8.50). If conditions are favorable, marsh vegetation may grow and form a convex mass of sphagnum, very loose and compressible, called a **raised bog**.

Fig. 8.50 Swamp on the margin of the Orbe River, Vallée de Joux, Jura. Carex and birch are characteristic of this environment. GEOLEP Photo. A. Parriaux.

8.5.2   Swamps created by springs

Down-gradient from a perennial spring, a swamp may occur on a slope. It is called a ***slope marsh***. Here a veneer of peat develops permanently in the flooded zone. The peat is associated with calcareous tufa when the groundwater contains calcium carbonate. A magnificent example is the slope marsh of Bois-de-Chênes, in the south-facing side of Lake Geneva near the city of Gland (Fig. 8.51). That kind of marsh was forgotten by the engineers of Loetschberg alpine tunnel and led to the subsidence of the village of St-German (Fig. 7.83).

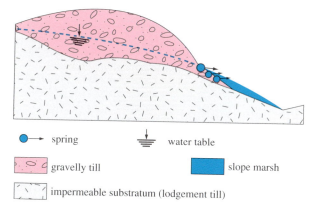

● spring	water table
gravelly till	slope marsh
impermeable substratum (lodgement till)	

Fig. 8.51 Swamp from springs. Case of the Bois de Chênes slope marsh, La Côte Region, Lake Geneva Basin. Schematic profile not to scale.

8.5.3   Marsh drainage

Man has always tried to recover land from swampy areas. Peat regions are very fertile when drained. It has also been considered important to drain swamps because of the risk of illnesses caused by insects that they harbor.

These operations were conducted in a minor way until the 18th century. The 19th century was the time of large hydraulic development projects on rivers. In Switzerland for example, they confined the principal rivers and made major corrections of their natural courses. The sub-Jurassian area was the location of two large operations called "Correction of the Jura water." One of the goals of this project was to lower the lakes of Biel, Neuchâtel and Morat by several meters. It thus became possible to drain the great swamp of Seeland, the Orbe, and the Lower-Broye plain (Fig. 8.52).

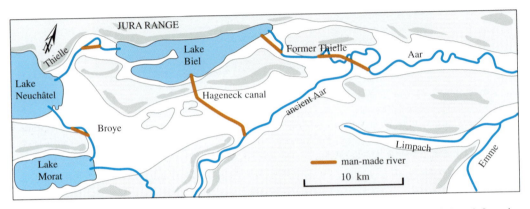

Fig. 8.52 The first water correction project in the Jura (second half of 19th century). The level of the sub-Jurassian lakes and the adjacent swampy plains was controlled by the mouth of the Aar River in the Thielle River downstream of Lake Biel. The Aar floods provided a great deal of alluvium to this region, which induced upstream inundations that extended very far. The most effective solution was to divert the Aar into Lake Biel. The Hagneck channel was created by cutting through a molassic hill just before the lake. The course of the Thielle, which became the Aar downstream of Lake Biel, and the course of the Broye were excavated and channelized, which lowered the water level about 3 meters. Ditches were dug to collect water from thousands of drainage pipes that had been buried under the old swamp surfaces. The result over the following hundred years was an undeniable success in agricultural productivity and food production for the population. A second correction was made in the 20th century, involving the deepening and widening of some channels. Hydro-electric plants were constructed along the river downstream of Lake Biel. In addition to energy production, they provide better regulation of the water in the channels.

Over the long term, swamp drainage causes damaging effects of several types. In addition to the suppression of a large, very specific ecological niche that may contain protected species, it has been observed that the soil in drained regions continues to subside. This has progressed to such a point today that certain areas of the Orbe plain, at the southern foot of the Jura chain, have become **polders** (depressions below the gravitational level of outlets), thus requiring water to be pumped out of them and disposed in the river. It is interesting to study the reasons for this subsidence.

Subsidence
Peat fibers in the saturated zone are subject to Archimedes principle (Fig. 8.53). The supporting force of a fiber on another is very weak. As a result the fibrous system can be particularly loose. By draining the peat, Archimedes force is eliminated (Problem 8.2). The result is compaction and the ground surface subsides.

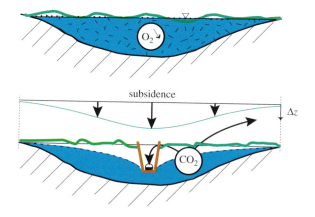

Fig. 8.53 Double effect of draining a swamp (topography in green).

PROBLEM 8.2

Considering a peat fiber with a volume of $1\,cm^3$ and a specific weight of $12\,kN/m^3$, calculate the supporting force that it exerts on the fiber that supports it, in a drained environment and in its natural undrained condition. Compare the results and draw conclusions based on these results. To simplify the problem, assume that the specific weight of the fiber is constant.

Look at the solution on the DVD

Mineralization of peat

The installation of drainage systems lowers the level of groundwater that saturates the peat. The pores of the peat that were initially occupied by anoxic water now also contain air. Conditions have changed from an anaerobic environment that preserves peat to an aerobic environment that promotes biochemical processes of mineralization. Organic chains such as lignin are decomposed into gas, and ions such as CO_2, NO_3, and PO_4 are released into the air or drainage water. In this way, peat is transformed from a solid to a gas or a liquid gas or liquid, which tend to escape.

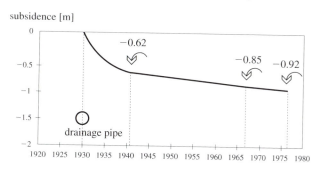

Fig. 8.54 Ground subsidence in the Orbe Plain, as a result of swamp drainage.

The combined result of these phenomena can be spectacular, as shown in figure 8.54, which shows the evolution of the soil elevation for a point on the Orbe plain. Often the drainage systems become so close to the surface that they are destroyed by plows.

In swampy areas, it is advisable to construct building foundations on piles. If embankments are necessary, they must be installed several years in advance so that the subsidence of the peat has occurred by the time the project is constructed.

8.5.4 Exploitation of peat

For centuries, peat has been the only household fuel in many regions. Thus a large number of peat bogs have been partially or totally destroyed by individual exploitation. Today, in many regions the last slightly-exploited or undestroyed peat bogs are totally protected. In countries rich in peat such as Poland or Russia, it continues to be exploited, particularly for export as garden soil.

[278]

8.6 Boreal and polar environments

This section focuses on the Earth's surface located above 55° latitude. In this environment we include the high mountains of lower latitude, which are very analogous to near-polar regions.

We divide this vast terrestrial environment into areas occupied by ice and areas that lack glaciers.

8.6.1 Glacial regions

In chapter 7 we examined the principal types of glaciers. In this section, we will discuss the deposits they form, focusing on the present day glacial environment, and also referring to deposits from the large Quaternary glaciations, which have left their mark on many countries.

Glacial deposits can be subdivided into two groups: glacial formations, that is, those in contact with ice, and periglacial formations defined here as the deposits that occur on the periphery of glaciers. Figure 8.55 is an illustration showing the respective locations of these different formations.

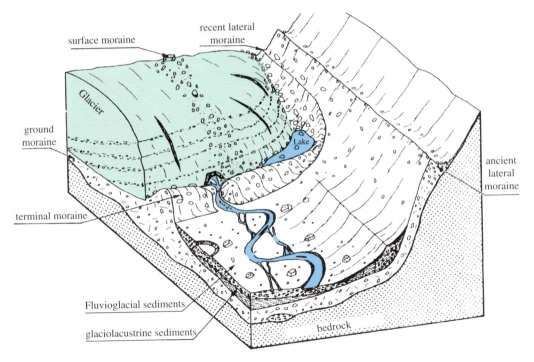

Fig. 8.55 The principal glacial and periglacial formations of a valley glacier.

8.6.1.1 Glacial formations

Deposits formed directly from the movement of ice are called *moraines*. The shape and composition of moraines vary widely according to their processes of formation and the rocky material associated with them.

The fundamental difference between a deposit of glacial origin and an alluvial detrital deposit is texture: the glacial deposit contains components of all sizes and shapes, generally angular, and the sedimentary structure is poorly organized.

Very generally, we can distinguish the following types:

- A **surface moraine** (Fig. 8.56) is composed of blocks and debris accumulated on the surface of the glacier; **erratic blocks** (Fig. 8.57) are the largest of them.
- A **lateral moraine** (Fig. 8.58) and a **terminal moraine** (Fig. 8.59) are ridge-shaped accumulations that contain blocks mixed with a fine matrix (abundant for the first moraine, less present for the second).
- The **ground moraine** or **lodgement till** (Fig. 8.60) is a layer of finely crushed debris at the base of the glacier, containing pebbles and blocks that are often striated. This moraine is intensely compacted by the weight of the ice; it rests on rock that has been polished and striated by blocks sliding over it (Fig. 8.61) ("sheepback" facies).

Moraines deposited by the retreat of the large glaciations are generally good-quality foundation material. However, it is necessary to pay attention to their clay content and to the presence of groundwater.

Fig. 8.56 Surface moraine of the Oberaargletscher, Grimsel, Switzerland. GEOLEP Photo, A. Parriaux.

Fig. 8.57 Erratic blocks are the most visible evidence of widespread glaciation. Interpreted by the Ancients as remnants of the Deluge, erratic blocks were understood only after research on moraines led to an understanding of the glacial invasions. This is an example of the Okotoks block on the eastern slope of the Canadian Rockies. This block of granite was 41 m long and weighed 18,000 tons before it broke up. The block was carried by the Athabascan glacier, and was deposited 580 km from the glacier's present location. It was transported during the last Quaternary glaciation, called the Wisconsinian in America, equivalent to the Würm in Europe. GEOLEP Photo, A. Parriaux.

Fig. 8.58 Lateral moraine. Verra Glacier (Italy), southern slope of Breithorn. GEOLEP Photo. L. Cortesi.

Fig. 8.59 Small present-day terminal moraine with lake formation upstream. The outlet from the lake erodes the non-cohesive material of the moraine. Ferpecle Glacier, central Valais. GEOLEP Photo, A. Parriaux.

Fig. 8.60 Upper ground moraine on the Cossonay-Dizy Plateau, western Swiss Plateau. GEOLEP Photo, A. Parriaux.

Fig. 8.61 Glacial polish with striations. Slab of shelly sandstone, La Molière Region, western Swiss Plateau. GEOLEP Photo, A. Parriaux.

8.6.1.2 Periglacial formations

Two types of geologic formations are linked to water that exits the glacier, depending on whether the environment is alluvial or lacustrine. Two other types are related to frozen sediments on slopes.

Fluvioglacial alluvium

In front of valley glaciers, the glacial torrent builds an alluvial plain that is affected by the same mechanisms described for ordinary alluvial plains (Section 8.3). Two major differences must be pointed out, however:

- The water-transported components have not had time to be weathered and rounded like alluvium;
- Sediments are often deposited on **dead ice** (immobile masses of ice that are no more in contact with the active glacier); as the ice melts the sediments move. Bed tilt, faults and folds occur, etc. (Fig. 8.62).

Fluvioglacial formations left by glaciers are good foundation grounds and good aquifers.

Fig. 8.62 Fluvioglacial sediments. An old gravel pit west of Lavigny. Western Swiss Plateau. The curved and faulted beds are obvious. GEOLEP Photo, A. Parriaux.

Glaciolacustrine deposits

It often happens that the tongue of a glacier ends in a lake (Fig. 8.63), where it forms a glacier front delta. The lake depression receives fine particles only, like the bottom-set beds we described in the lacustrine environment. There are two notable differences between the glaciolacustrine and the lacustrine sediments:

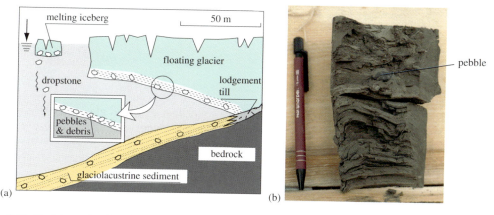

Fig. 8.63 Glaciolacustrine deposits with fine parallel laminations. (a) Diagram showing how the deposit forms and the origin of dropstones. (b) Drill core, Claie-aux-Moines region, western Swiss Plateau. GEOLEP Photo, G. Franciosi.

- The presence of pebbles caught in fine cyclical stratifications; these are **dropstones**, that is fallen pebbles, dispersed by icebergs that float on the lake with a veneer of moraine stuck to the ice;
- Absence or rarity of fauna or flora remains because of the very cold climate.

The fine glaciolacustrine sediments are very poor foundation material, for the same reasons as lacustrine sediments, except that the former may have been compacted by the ice load.

Rock glaciers

There is geological material permanently frozen in high mountains in the middle latitudes (see permafrost § 8.6.2). In the Alps, for example, frozen soils are found on north-facing slopes beginning at an elevation of 2,500 m. Mountain permafrost is frozen rock that has cracks filled with ice, as seen in the construction of the "Snow Metro" tunnel in Saas-Fee (Valais Alps), or masses of blocks and debris surrounded by ice. The latter gives the mass a plasticity that causes it to flow down a slope like a glacial tongue. These masses are called **rock glaciers** (Fig. 8.64). They move much more slowly than real glaciers, on the order of several decimeters per year. Many of them date from the Little Ice Age (1600 to 1820).

Fig. 8.64 Rock glacier of Muragl, Engadine, Switzerland. GEOLEP Photo, A. Parriaux.

Solifluction

If the surface soil thaws while the bottom of the soil is still frozen, the melt water cannot drain by infiltration. It waterlogs the soil, turning it into a liquid paste that begins to flow down the slope as little tongues about a meter in size: this is **solifluction**. The **tundra soil** is a type of "woven" envelope that contains and holds the fluid mass like a pocket (Fig. 8.65).

These particular slope instabilities must be taken into account in civil engineering projects in ski areas, pipes, and electric lines.

8.6.2   Non-glaciated regions

In all areas of the Earth where seasonal variations occur, the temperature of the subsurface is affected by the surface temperature. These variations decrease with depth. If we draw a

Fig. 8.65 Solifluction tongues. Réchy Valley, Valais. GEOLEP Photo, A. Parriaux.

vertical profile showing the temperature variation as a function of depth for the hottest and for the coldest season, we have curves that are very far apart at the surface but almost join at a certain depth (Fig. 8.66). This surface layer is on the order of ten meters and is called the

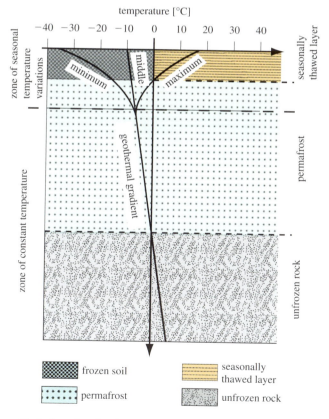

Fig. 8.66 Temperature profile in the subsurface in polar regions. The seasonally thawed layer occurs at the surface, dominating the permafrost.

zone of seasonal temperature variations. Below this, the temperature is not sensitive to seasonal variations. This is the *zone of constant temperature*. In fact, temperature in this zone is controlled by the geothermal gradient.

Temperature variations at depth in boreal and polar areas follow the same pattern. However, these zones are unusual in that their temperatures are below 0°C. Thus these regions may have no glaciers at the surface but ice is present. Instead of forming perennial accumulations at the surface, the ice occupies a part of the subsurface: it is the *permafrost*, a stratum where groundwater is permanently frozen. The permafrost can be several hundred meters thick, as in the very cold regions of Siberia, for example. In more temperate zones, it thins and becomes discontinuous. Permafrost sometimes occurs close to the surface of the soil, as ice bombs, called *pingo* (Eskimo term meaning "pregnant woman"), surrounded by a cap of loose sediments. They can reach a considerable size, up to several hundred meters in diameter and tens of meters high (Fig. 8.67).

deformed
sediments

pingo ice

Fig. 8.67 A small pingo at an altitude of 4700 m on the northern part of the Tibetan Plateau, Kunlun Pass. GEOLEP Photo, A. Parriaux.

The upper part of the zone of seasonal temperature variation melts in the summer. As the frozen pores make the permafrost impermeable, the melt water in the upper part of the soil cannot infiltrate. This seasonally thawed layer is thus completely saturated. This particularity causes these regions to have very characteristic landforms such as polygonal soils (Fig. 8.68).

Civil engineering projects are very delicate in areas such as these. The thawed layer has practically no bearing capacity (Fig. 8.69). Roads built on this layer can crack. During refreezing there are upheavals due to the formation of ice lenses. All pipes must be constructed above ground to monitor possible ruptures due to differential movement of the soil. Buildings must be constructed on piles that extend to the permafrost. A space is left under a house to guarantee that the top of permafrost does not become fluid in the foundation zone (Fig. 8.70).

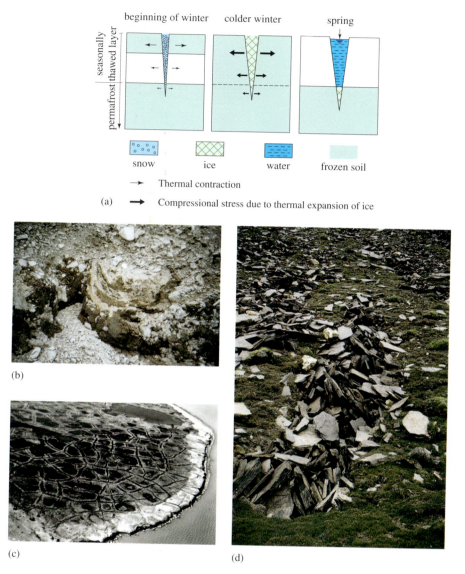

Fig. 8.68 Polar soils. (a) Diagram of the origin of *ice wedges*. (b) Deformations of the subsurface due to freezing; the sand strata are deformed to an almost vertical position. Ulan-Bator, Mongolia. GEOLEP Photo, A. Parriaux. (c) *Polygonal soils* along the coast of the glacial Arctic Sea, near Barrow Alaska; the diameters of the polygons are 8 to 15 meters. Earth Sciences Photographic Archive Photo, U.S. Geological Survey. (d) Alignment of blocks standing on edge. Ladakh, GEOLEP Photo, T. Bussard.

Climate warming also affects these regions by reducing the extent and the thickness of the permafrost. There is a concern that increased warming will result from methane produced during the melting of the palustrine series, which dominates the permafrost. This occurs when lakes form as the ice thaws and the top of permafrost is lowered. Let us recall that methane is about twenty times more active than CO_2 in terms of the greenhouse effect. For example, the region of western Siberia has warmed about 3°C in 40 years.

Fig. 8.69 Melting of the permafrost under building foundations can cause significant differential subsidence that leads to destroy solid structures. Historic buildings in Siberia are particularly affected by this phenomenon. The royal alley from the Tsarist period in Yenisseisk. GEOLEP Photo, A. Parriaux.

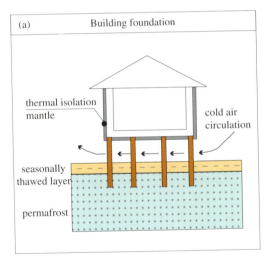
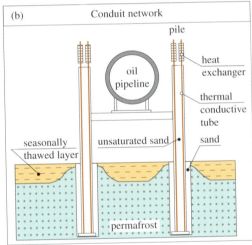

Fig. 8.70 Examples of construction methods in permafrost regions. (a) Buildings should be constructed on piles secured in the permafrost; an open layer of air is left between the soil and the base of the house to prevent the heating of the foundation. (b) Thermal piles; the pile is made of a metal cylinder filled with sand and equipped with a heat exchanger that can be disconnected. In the winter, the exchanger becomes very cold and cools the environment around the pile by conduction. In the summer, the exchanger is disconnected; the very low temperature around the pile causes this pile to remain in a frozen envelope throughout the year, which guarantees better stability. This system is used for the Trans-Alaska Pipeline.

[73, 82, 83, 85, 93, 113, 115, 138, 263, 306]

We will come back to certain aspects of freezing in our discussion about frost sensitivity of sediments and weathering of rocks (Chapt. 13).

8.7 The desert environment

By deserts, we mean regions of the world where vegetative cover is absent. Glacial and periglacial regions and high mountain environments that also meet this definition have already been discussed. We will focus our interest here on dry and usually hot deserts. We have seen in chapter 7 that low rainfall conditions are linked not only to tropical climate zones (for example, the Sahara) but also to the orographic effects of mountain chains (for example, the Gobi desert, or the deserts of South America).

8.7.1 Eolian erosion and transport close to the soil

The naked soil plays a fundamental role in the erosion and transport of sediments. Air is a more significant factor in particle transport than is water. In section 8.2 we saw the various phenomena of river transport (Fig. 8.28). The same processes operate in the fluid "air." The essential difference is fluid density, which affects bearing capacity. Even high velocities can transport only particles smaller than several millimeters. The largest particles remain near the ground. We can confirm the power of sand-bearing air layers on the erosion of rocks (Fig. 8.71). The impact of sand against rocks creates a polish typical of eolian erosion.

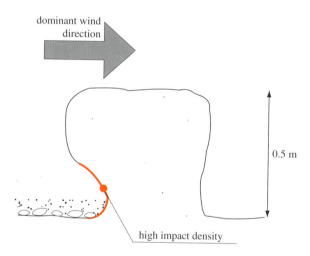

Fig. 8.71 Eolian erosion by traction and saltation. The rocks are eroded primarily at the base.

Transport near the soil causes the formation of ***dunes*** of various shapes depending on the changing direction of the wind. When it is stable, ***barchan*** (crescent-shaped dunes) form (Fig. 8.72). The dunes move with each sandstorm if the topography allows it. Their sedimentary structure gives us information about their formation (Fig. 8.73). The windward face has a low slope, and air flows in a laminar manner at great speed. The shearing force carries the grains and causes them to climb to the crest. Once past the crest, the abrupt lowering of the surface causes turbulent flow. Velocities decrease and particles are deposited at the down-gradient foot of the dune, at their angle of static stability in air (about 30 degrees). Thus the dune moves with the wind. The strata that make up the dune are nothing more than

sedimentation

Fig. 8.72 Barchans, Kunlun Pass, northern slope of the Tibetan Plateau. GEOLEP Photo, A. Parriaux.

Fig. 8.73 Internal structure of an eolian dune. See also the animation on the DVD.

a succession of inclined beds that are relics of old leeward slopes. This typical structure of dunes makes it possible to identify desert paleo-environments in geologic formations. (Fig. 8.74).

Fig. 8.74 The oblique sedimentation of ancient dunes provides evidence for the continental phase of Gondwana during the Pangean period. Sanga do Cabral (Permian and Triassic), Mata region, southern Brazil. GEOLEP Photo. A. Parriaux.

Eolian ablation (or ***deflation***) of a soil made of a mixture of pebbles and sand causes segregation that results in a continuous pebbly cover that protects the finer components at depth (Fig. 8.75) and creates ***pebble deserts*** (or ***regs***).

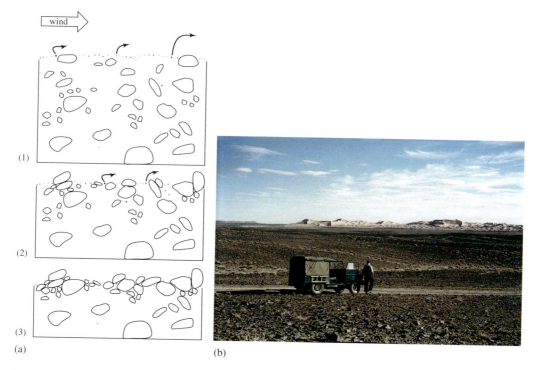

Fig. 8.75 (a) Genesis of pebble deserts in three stages. (b) Reg in the Erg Chebbi Region, Morocco. GEOLEP Photo, J.-H Gabus.

8.7.2 Transport of suspended particles

Particles in suspension can reach considerable altitudes. When the winds blow hard and the surface soil is very hot, particles measuring several microns can reach altitudes of 10 to 15 km. They can then be transported over more than 1000 km before descending in masses of colder air. For example, traces of Saharan dust can be observed in temperate areas of Europe during storms; when dust-laden water droplets evaporate they leave a beige or reddish residue.

During glacial periods, the climate was unfavorable to plant growth. Eolian erosion was very effective in these areas that lacked soil. As a result, sands and silts were deposited on their margins; these deposits are called *loess* (Fig. 8.76).

8.7.3 Water in deserts

In deserts, rainfall is scarce and evaporation is significant. However, there are rare but violent rains that are effective agents of material transport. They cause temporary rivers (*oueds*) that produce formidable floods. Often these rivers infiltrate into the ground before reaching the sea.

Groundwater is present in most deserts, as shown by the presence of oases. When the aquifer is shallow, intense desiccation causes powerful upward capillary action. Water carrying a

Fig. 8.76 (a) Loess area, Xining, northern China. These deposits accumulated to a thickness of several hundred meters north of the Himalayas, essentially during glacial periods. Easily eroded, they form a hilly landscape today. These sediments are used to reconstruct the paleoclimatic history of the Himalayas. (b) Detail of the structures and composition of loess; absence of stratification, concretions (called "***loess dolls***"), desiccation cracks, Tibetan Plateau. GEOLEP Photos, A. Parriaux.

high mineral salt load evaporates just as it arrives at the soil surface, and these salts precipitate. When the water contains sulfates, gypsum can form crystals shaped like flowers called "***sand roses***" (Fig. 8.77). When the water contains chlorides, the soil becomes salinated. These processes can also occur when desert areas are irrigated with groundwater to grow crops. There is a progressive increase in the salt content of the soils, which can cause the operation to fail over the long term.

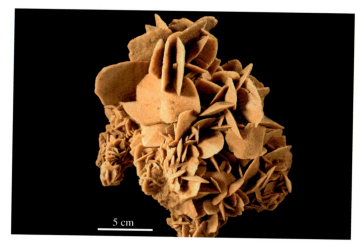

Fig. 8.77 "Sand roses" form as a result of gypsum precipitation at the surface after ground water has risen to the surface through capillary action. GEOLEP Photo, G. Grosjean.

8.7.4 Protective measures

It is often necessary to construct infrastructure in deserts: roads, railroads, pipelines, petroleum or mining facilities, etc. The engineer must be inventive in planning projects to ensure that the structures are not constantly invaded by eolian sands. Several fundamental rules are required (Fig. 8.78), for example:

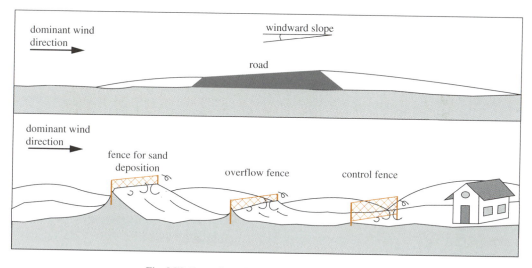

Fig. 8.78 Protection of structures against sand invasion.

- To stop the advance of a dune, the windward surface must be stabilized by a cover (vegetation, blacktop, etc.).
- A large dune can be broken up by covering some segments of its windward face, as described above; the big dune will evolve into a series of small dunes, which can be handled separately.

- To deposit sands before they reach the site, the same procedure is used as with snow: construct fences that create turbulence behind the obstacle and cause deposition.
- Roads can be banked slightly against the dominant wind direction to prevent the deposition of sand on the road.

[68, 150, 171, 265]

These measures become more complicated when the winds come from variable directions.

PROBLEM 8.3

You have to analyse the morphology of a territory on the base of a topographic map. The problem concerns a region in the Swiss Alps: the Goms valley, in the upper part of the Rhone basin (Fig. 8.79). You have to your disposition on the DVD (part Animations) the national topographic map at the scale 1/25000 sheet CN 1250 Ulrichen.

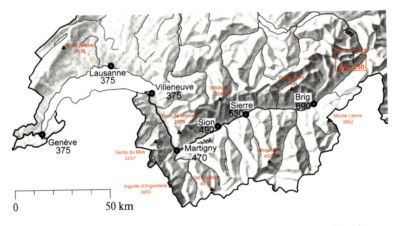

Fig. 8.79 Physical map of the Rhone basin with the situation of the study region.

Questions A

- Identify the following morphological elements on the map:
  1. An alluvial plain
  2. An alluvial fan
  3. An active meander
  4. An abandoned meander
  5. A terrace
  6. A swampy zone
  7. A scree apron
  8. A cirque glacier
  9. A valley glacier
  10. An area of seracs
  11. A lake created by glacial overdeepening
  12. A lateral moraine
  13. A cirque created by torrential erosion and its debris fan
  14. A protective barrier against debris flows

Some indications on the legend of the map are given on the DVD

Questions B

- How is it that the plain between Ulrichen and Geschinen is perfectly horizontal at such an altitude?

You have at your disposition on the DVD (part Animations) an extract of the sheet CN 1270, which shows the downward continuation of the plain.

Figure 8.80 describes the geological setting of the study area.

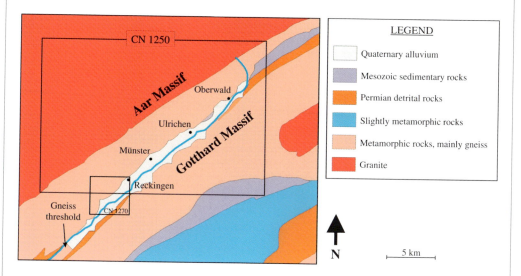

Fig. 8.80 Simplified geological map of the Goms valley

Questions C

- What type of foundation should be anticipated for the construction of an airport control tower?

Look at the solution on the DVD

Chapter review

Continental sediments appear to be simple since they are generally classified into four groups on the basis of grain size, ranging from clays to gravels. However, we have seen that the continental environment has an enormous variety of possible sedimentation sites, which are related to different erosion and transport phenomena. Thus, the resulting deposits are structurally complex of sedimentary origin often discontinuous. We will review these sediments in chapter 10 when we discuss rocks of sedimentary origin that form from this material and the processes that cause these transformations.

9 The Oceanic Sedimentary Environment

The oceanic sedimentary environment is not unknown to us. In chapter 6 we saw magmatic processes in detail, particularly those that take place at the bottom of the ocean: magmatic fluxes in the mid-oceanic ridges, the composition of the oceanic crust, islands above hot spots, etc.

Let us now consider the oceans as sediment "factories." A large part of the sedimentary rocks found on Earth come from marine sediments; we will now focus on this environment. Knowledge of marine sediments has increased in recent decades through the use of modern methods: remote sensing, high-resolution bathymetric methods, oceanic soundings, and reconnaissance by submarines. These observations on the present-day oceanic environment have made it possible to interpret the structures of rocks of marine origin that are now present on land.

The oceanic environment is the general receptacle for all materials that come from terrestrial erosion and are carried off by water. Material transport of various rocks toward the ocean can be summarized under two different processes (Fig. 9.1):

- Transport of material dissolved in water: these are mainly the ions Ca^{2+}, Mg^{2+}, Na^+, K^+ and HCO_3^-, SO_4^{2-} and Cl^-.
- Transport of detrital material in solid form: this is insoluble rock debris, particularly silicates.

Oceans exhibit a wide diversity of environments, processes, and interactions between the living world and the mineral world. We will see here how the supply of continental material is used throughout the entire ocean.

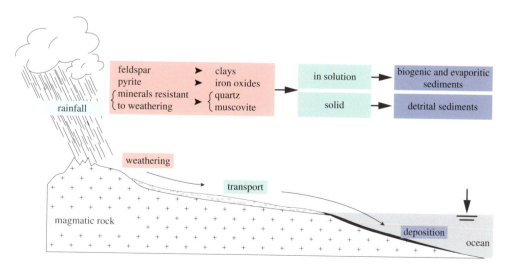

Fig. 9.1 Material flux from the continents into the oceans. Example of the weathering of a magmatic rock.

There are three types of oceanic sediments that form sedimentary rocks:

- Detrital sediments, similar to the continental deltaic sediments covered in chapter 8,
- Biogenic sediments, in which a living organism is the source of the sediment,
- Evaporitic sediments resulting from physico-chemical precipitation of salts contained in seawater.

As in the chapter on continental sedimentation, we will discuss the formation of recent sediments. The transformation of marine and continental sediments into rock through the process of diagenesis will be treated in chapter 10.

The oceanic environment is not homogeneous. Its variability is governed by the distance to coastlines and water depth. We will divide it into two broad areas: (1) the continental margin and (2) the continental rise and abyssal plains. In addition, there is good reason to distinguish two very different environments in terms of the profile from the continent to the ocean depth:

- Regions with large detrital supply in which a continental shelf made up of loose debris spread over a great distance offshore and is followed by a noticeable slope that extends to the continental rise (Fig. 9.2);
- Regions with low detrital supply, but with intense biogenic activity where the continental margin gradually meets the continental rise offshore of the reef zones (Fig. 9.3).

9.1   Continental margin

This is the transition zone between the coast and the continental rise. Here the height of the water column increases to about 2000 m. The continental margin extends tens to hundreds of kilometers offshore. It is more highly developed in areas where rivers empty into the ocean. It is a place of both coastal erosion and the final receptacle of river sediments that will be intensely reworked. It is also an ecosystem preferred by aquatic flora and fauna, particularly the littoral portion, the *photic zone*, (the zone where sunlight still penetrates into the water).

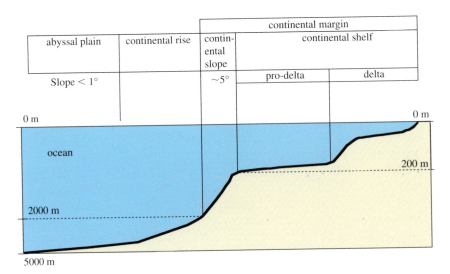

Fig. 9.2 Broad morphological units of the oceanic environment. Example of dominant detrital sedimentation. Not to scale.

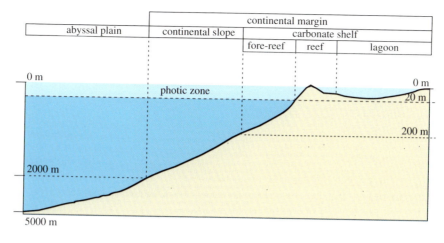

Fig. 9.3 Broad morphological units of the oceanic environment. Example of dominant biogenic sedimentation. Not to scale.

9.1.1 Detrital sedimentation

Rock debris from erosion of the coasts or fluvial sediment input undergoes the combined effects of the principal water movements that affect the ocean shoreline: waves, coastal currents, tides, and currents in areas where rivers flow into the sea. These movements abrade the particles, sort them by grain size and density, and also deposit them as sedimentary structures that are representative of local conditions.

9.1.1.1 Wave effects

Waves are created by the movement of water molecules, primarily as a result of the shearing effect of the wind on the water surface. This movement can be represented as trajectories of water molecules.

Far from the shore, where the water is tens of meters deep, the trajectories are almost circular (Fig. 9.4). The trajectories occupy a vertical plane parallel to the wind. The line of wave crests is perpendicular to the wind direction. The amplitude of the water's oscillation diminishes rapidly with depth. At the bottom, there is no shear effect.

Closer to the coast, the oscillation of the water reaches the bottom. The trajectories become ellipses, elongated along the horizontal axis. Sediments become involved in the back and forth movement and result in the creation of dunes parallel to the waves.

When the water reaches the beach, the waves turn into breakers. The trajectories are no longer elliptical. The shearing force carries away some of the bottom particles from the beach and then re-deposits them. This repeated effect causes abrasion of the grains and a segregation based on resistance to abrasion. A gradual enrichment in quartz (very hard and chemically stable) occurs at the expense of softer grains, for example shell debris made of aragonite. Wave action continuously reshapes the configuration of beaches. In order to maintain littoral infrastructure, significant constructions must be built to protect them from this erosion (Fig. 9.5).

Wave impact also attacks rocky coasts. Cliffs form as a result of the digging action of waves, which almost always operates at the same level to create a linear notch. The rock above

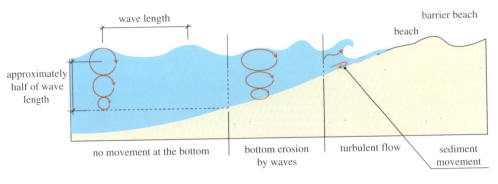

Fig. 9.4 Diagram showing wave motion as a function of distance from the coast (from [262]).

Fig. 9.5 Protection of shores from wave erosion makes use of natural rock blocks, except in regions that lack massive rock. In such places, artificial tetrahedral-shape blocks made of reinforced concrete can be used as protection. This example shows the Pacific shoreline south of Tokyo, Surugu Bay. GEOLEP Photo. A. Parriaux.

this erosional horizon begins to overhang. Periodically the wall collapses along vertical fissures, and the cliff recedes as a result (Fig. 9.6).

The result is a beach that is almost flat with a rocky bottom, called a ***wave-cut terrace***. The collapsed blocks are moved rapidly during storms. They are then broken up and mixed with particles carried by the waves. The force of the waves on oceanic coasts is spectacular; one example is the North Sea.

The French coast of the English Channel has receded up to 2 km since Roman times, about 1 m/year on average. Shipwrecks on the wave-cut terrace are broken up rapidly; this is additional evidence of the force of the waves.

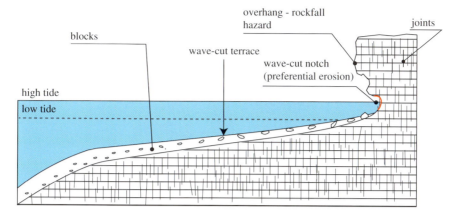

Fig. 9.6 Coastal erosion of cliffs in massive rocks (from [262]).

The wind often does not blow perpendicular to the coast. Each impact of waves on the coast causes a translation in water movement and the sediments carried by the water (Fig. 9.7). This causes a coastal current. The gradual movement of river mouths can be observed (Fig. 9.8).

In regions where this obliqueness is systematic, a broad-scale translation of sediments occurs. Sand bars occur next to and behind obstacles (Fig. 9.9, Fig. 9.10 and Fig. 9.11) where the current goes from laminar to turbulent, which causes the deposition of sediments.

A particular case of translation of material along coasts occurs where coasts are composed of rocks of variable hardness. The promontories and coves may have outcrops of the same rock, but the rock contains faults that weaken it. When a wave hits such a shore, the wave ray is refracted and it focuses on the promontories because of the shallows located at these capes (Fig. 9.12). When waves are slowed down in this way, they curve. However, the wave rays do not deviate enough to travel perpendicular to the coast. Thus, the debris from the concentrated erosion migrates toward the coves (Fig. 9.13). Here we have a mechanism that should theoretically cause a straight-line evolution of the coastal shape. In fact, it is often not efficient enough compared to the big difference in the hardness of certain rocks.

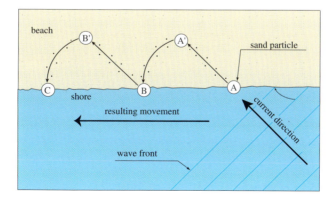

Fig. 9.7 Coastal current.

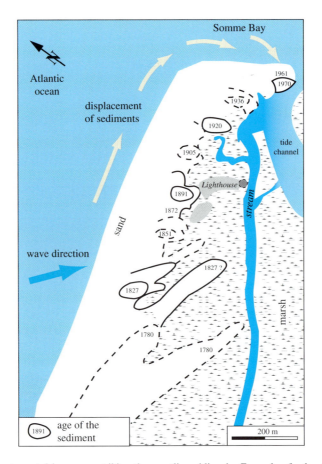

Fig. 9.8 Coastal current created by waves striking the coastline obliquely. Example of a barrier beach moving the mouth of the Somme River to the east, Picardie, France (from [168]).

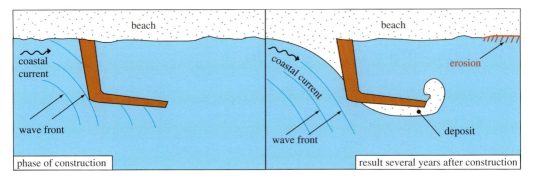

Fig. 9.9 Risk of silting up near coastal structures.

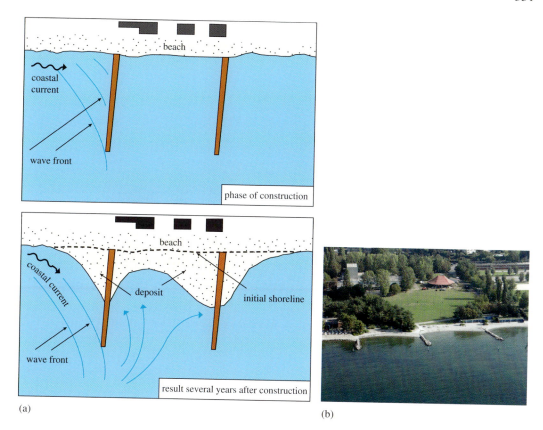

(a)   (b)

Fig. 9.10 (a) Beach protection using jetties perpendicular to the shore. (b) Example of the Bellerive beach, Lausanne. Photo. J.-M. Zellweger.

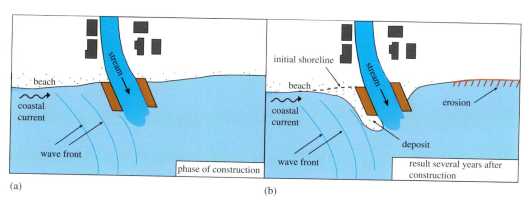

(a)   (b)

Fig. 9.11 Coastal currents in the process of obstructing the mouths of rivers along the shore. This phenomenon can be prevented by constructing barriers that transfer more sediments offshore from the river mouth, thus making it possible for the river to flow farther out to sea (a). Coastal currents gradually deposit their sediment along the margin of the barrier (b).

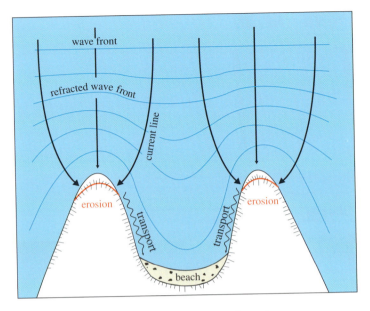

Fig. 9.12 Waves arriving at a non-linear shore.

Fig. 9.13 Sediments concentrated at the bottom of coves between the cliffs of Ireland. The size and rounding of the blocks show that the currents generating these deposits are very strong. GEOLEP Photo.

9.1.1.2 Tidal effects

We cannot describe littoral sedimentary processes without discussing tides, which cause significant sea level variations in certain regions of the oceanic coast.

By tides we mean the deformation of the Earth that results from its attraction to the surrounding celestial bodies. Each body exerts an effect proportional to its mass and inversely proportional to the square of the distance to the other body. Thus the Moon's influence is stronger (2.4 times) than the Sun's. Because of the Earth's rotation on its axis, a point on its surface is alternately subjected to strong and weak attraction. This creates deformations that are most visible in ocean waters. Rocks are also deformed but in a more tenuous manner.

How can we explain that a point located on the oceanic coast outside the polar regions experiences two rises and two drops in water level every day? How can we explain that the maximum and minimum levels change over time? To understand the phenomenon of ocean tides, we must consider not only the attractive forces but also the centrifugal forces induced by the revolution of the system. Several animations on the DVD show the movements involved and the forces mobilized to explain tides.

Centrifugal forces in the revolution of the Earth-Moon system

The Earth and the Moon can be considered as a system of two revolving bodies. The system turns around a center of mass C, also called a centroid or barycenter. Since the mass of the Earth and Moon are very different, the centroid is beneath the surface of the Earth. Its position is calculated by the equation for the center of mass for two bodies (see Problem 9.1). Each of the two bodies follows an elliptical trajectory, and one of the foci coincides with the centroid. In fact, because these ellipses have very low eccentricity, we can easily consider the movement to be circular around the point C. The system completes a revolution in 27.3296 days. It is important to remember that the Earth's rotation is left out of the following discussion because it has no effect on the swelling of water along the Moon-Earth axis.

Let us examine the Earth-Moon pair in its plane of revolution (Fig. 9.14). We can identify a point P on the Earth that lies along this plane. Let us consider what happens to the PT segment (segment connecting P to the center of the Earth) as the system performs its revolution. Moving from position 1 to position 2, the segment PT maintains the same orientation with respect to the universe because the Earth does not rotate during this movement. The center of the Earth has traveled through the arc of a circle of radius R = CT, or the distance from the center of the Earth to the centroid of the revolution. Because the orientation of the segment PT does not vary, point P describes the arc of a circle of the same radius as that of the center of the Earth. Moving from position 2 to position 3, these two circular trajectories are continued. Arriving again at position 1, the two circles are completed.

Let us attach a reference point to the moving Earth, an elementary mass placed at point P, for example. This reference point is subject to centrifugal force in the opposite direction of the center of the circle described by the point P. The same phenomenon occurs at the center of the Earth T and at all other points. This can be seen on figure 9.14. In position 1 of the system, the centrifugal force at the center of the Earth is along the Earth-Moon axis. At point P, it is identical to it in direction and intensity. The experiment can be repeated at all points on Earth; the centrifugal force is the same everywhere. In position 2, the centrifugal force at all points is oriented in the new direction of the Earth-Moon axis, and so on.

Attractive forces of the moon

In the case of the lunar tides, using a profile perpendicular to the plane of revolution of the Earth-Moon system, we can trace the vector of the force of attraction that the Moon exerts on

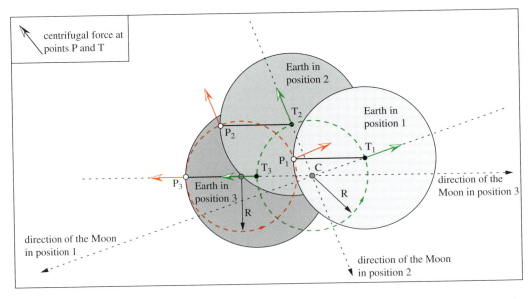

Fig. 9.14 Revolution of the Earth-Moon system around a center of mass C. Every point on Earth (for example, its center T or any point on the surface P) has a circular trajectory (here shown in three stages). Attention: the disk representing the Earth does not rotate; it keeps the same orientation in space. The centrifugal force exerted on matter at this location is a vector normal to the trajectory. Its intensity is equal to the product of the mass times the tangential velocity squared divided by the radius of the trajectory circle.

elementary masses inside the Earth and on its surface. These are slightly convergent vectors aimed at the center of the Moon (Fig. 9.15). Their intensity is not completely equal, although the Earth-Moon distance is much greater than the radius of the Earth.

Resultant force

We can now consider the resultant of the two types of forces on the plane perpendicular to the Moon's trajectory (Fig. 9.15). Let us recall that the plane of revolution of the Earth-Moon system is about 5° off the plane of the ecliptic. The axis of the Earth is inclined at a 28° angle to the line between the centers.

In the Earth-Moon system revolving around the center of mass C, the center of the Earth is the only point where the force of gravity exerted by the Moon is exactly equal and opposite the centrifugal force.

In near-polar regions, the centrifugal and gravitational forces are almost tangential. They are opposite and at the same time slightly oblique. The resultant is a small force that tends to attract matter toward the plane of revolution passing between the centers. The water level is thus depressed.

Near the equator on the side towards the Moon, the two forces are parallel but opposite. The Moon's force of attraction is greater. The resultant is a force directed toward the Moon. It has a tendency to raise the level of ocean waters, creating a first bulge.

In the area on the opposite side of the Earth, the centrifugal force exceeds the moon's force of attraction. The resultant this time faces away from the Moon, and it tends to raise the water level in this area. This is a second bulge.

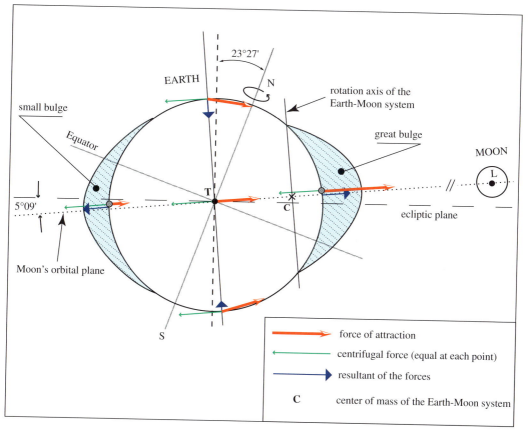

Fig. 9.15 View of forces due to the lunar tide in a plane perpendicular to the plane of revolution of the Earth-Moon system and passing through the centers of the two bodies.

In the equatorial regions in front of and behind the profile, the resultant of the two forces (the centrifugal force and the attractive force of the Moon) has a component directed toward the interior of the Earth, which creates a "hollow" in the sea level.

If we now let the Earth rotate on its axis, we see that a point near the equator passes through two tidal highs and two tidal lows every day. In reality the Moon moves slightly during this rotation. The period of this cycle is 12 hours and 25 minutes.

How can we explain the water movements that create these large volume differences over very large distances? If the water molecules had to move to the areas of the two humps, it would have to travel at a velocity of about 10,000 km (a quarter of the Earth's circumference) in 6 hours and then leave the area at the same velocity to cause the low tide. Such currents are not observed. In fact, the force that creates these humps on the Earth-Moon axis and raises the sea level behaves as if these water masses are subject to a smaller attractive force of the Earth, and thus less pressure. Outside the Earth-Moon axis, this attractive force is greater and it exercises a pressure wave that displaces the fluid toward the low-pressure area. This wave is transmitted gradually from water molecule to water molecule, therefore does not require large movements. The periodic variations of pressure are rapidly propagated and are the only ones capable of responding without delay to the variation of the tidal forces.

However, in very flat coastal areas the water does move, and as it does so it creates sig-nificant currents. As the tide rises, the current is called the *high tide* or *flood tide*. In the other direction, it is called *low tide* or *ebb tide*.

PROBLEM 9.1

Calculate the forces that govern the effect of lunar tides on an elementary mass located on the Earth-Moon axis on the Moon side and at the point opposite to it.

Look at the solution on the DVD

The height of the tides varies with time and geographic position. With time, the amplitude increases when the solar tide is added to the lunar tide, that is, when the bodies are at *syzygy*, which occurs during the full Moon (opposition) and new Moon (conjunction). In fact, both the solar tide and the lunar tide create two humps, but the amplitude of the solar tide is less (Fig. 9.16).

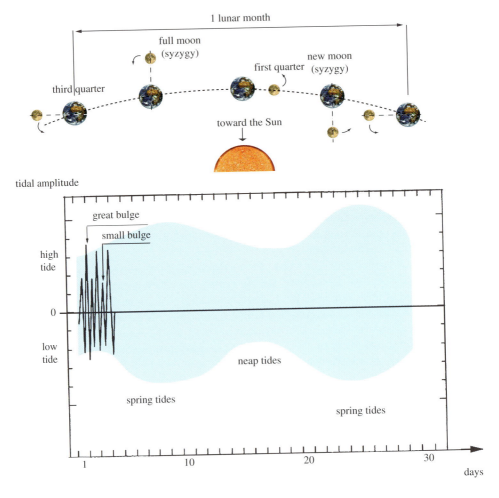

Fig. 9.16 Cumulative effect of lunar and solar tides.

The lunar and solar humps are added at opposition and at conjunction; these are the *spring tides*. The tides are minimal when the Sun is in quadrature with the Earth-Moon line; the lunar humps and solar humps are oriented at right angles and compete in raising the sea level; these are weak tides called *neap tides*.

Spatially, tides vary widely in amplitude. As a general rule, they measure about a half-meter in the deep open sea. The amplitude increases as the water column decreases. The shape of the coast also plays a role. Amplitudes increase in narrow sounds. Examples of this are the English Channel and its estuary zones, and some Canadian estuaries. Violent and dynamic currents can flow up into the river mouth areas and create phenomenal tidal heights. In such estuary zones, extreme amplitudes of twenty meters have been measured. In these cases, the tide rises after a significant delay. The so-called *"establishment"* of a port is the delay of the high tide related to the Moon passing the meridian at this location. The delay is negligible in ports in contact with the open sea. It can be as much as several hours for ports on shallow sea channels.

Tides can modify the hydrological conditions of coastal areas very rapidly. The sediments will follow water movements to some extent, especially in areas where the water is shallow. Tides play a fundamental role in estuaries (see § 9.1.1.3).

[277]

9.1.1.3   River mouth areas

There are two very different types of river mouths in coastal areas: deltas in regions where there is a notable hydraulic gradient and estuaries of rivers with a low or time-variable hydraulic gradient.

Marine deltas

When a river with a significant slope flows into the sea in an area of weak tides, it forms a delta similar to a lacustrine delta. If the water does not carry much suspended load, the river water flows at the surface of the sea because of the density difference due to salinity (Fig. 8.43). The planet's large rivers create considerable sediment masses that advance the continental slope. Their deltas can be seen on aerial photos as continental extensions shaped like a capital Greek delta Δ. The river flows seaward on these sediments and splits itself up into several meandering arms (Fig. 9.17).

Estuaries

In flat areas, rivers have more difficulty carrying their sediments to the sea, and this is all the more difficult if the coastal area has powerful tides. This configuration often occurs where fluvial valleys have been dug by the last Quaternary glaciation at a time when sea level was about one hundred meters lower than today. When the base level rose, seawater occupied the lower part of these valleys. This scenario describes the mouths of the Loire and the Seine, for example (Fig. 3.27).

An estuary (Fig. 9.18) is a progressive widening of the riverbed far upstream of the mouth. At the point of contact with the sea, the flow of the river encounters various shoreline sedimentary bodies created by the collision of previous river currents and coastal currents. At low tide, the fresh water and its suspended material flow slowly into the sea. At high tide the saline seawater penetrates into the mouth of the estuary and mixes with the fresh water. The result is water of moderate salinity, called *brackish water*. Fine particles are deposited, causing the depression to silt up. The brackish environment is an unusual ecosystem that contains unique fauna and flora.

Fig. 9.17 Marine delta of the Lena River in Siberia, one of the great rivers of the world. The delta is 250 km long. False color image made up of a composition of the seven frequencies analyzed by instruments. Landsat 7, 17/07/2000 Image courtesy of NASA Landsat Project Science Office and USGS National Center for EROS.

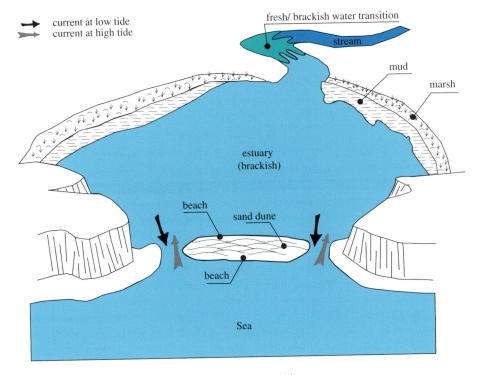

Fig. 9.18 Estuary morphology.

The tidal power plant at Lyvet was constructed several decades ago on the Rance estuary in the bay of St-Malo in Brittany (Fig. 9.19). It is an interesting example of energy diversification. However, it has created considerable silting up behind the dam that requires significant dredging on a regular basis.

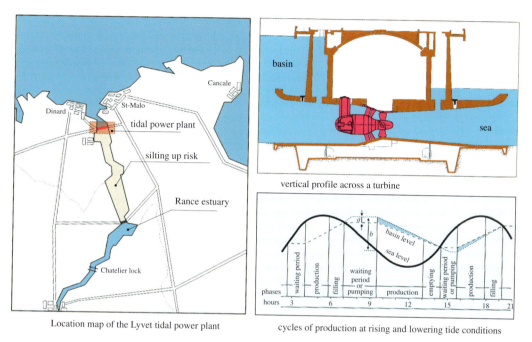

Location map of the Lyvet tidal power plant

cycles of production at rising and lowering tide conditions

Fig. 9.19 The tidal power plant at Lyvet on the Rance estuary, France. On the location map, the beige part shows the status of silting-up upstream of the dam. These silts, which are rich in calcium carbonate, are used as agricultural amendments (from [140]).

9.1.1.4 Continental slope

The continental shelf is bordered by a continental slope that has a gradient of about 5° that transitions to the ocean depths. The water column ranges from about 200 to 2000 meters.

Fine detrital particles transported by rivers are deposited on the shelf and the slope. The shelf also extends offshore through a process called progradation. The sediments are very loose and their pores are filled with water, so they are in a particularly precarious position in spite of the low angle of the slope. This situation leads to the slumping of the upper part of the slope. The sediments can flow down the slope as suspended matter, creating what are called *turbidity currents* (Fig. 9.20). These currents flow very fast, as shown by a major one that occurred in November 1929 in the estuary of St-Laurent (Canada). This turbidity current destroyed several telegraph cables. Because the time of the cable ruptures was known, it was possible to determine the velocity of the current: 80 km/h in high angle zones, 25 km/h at the end of the flow. This example shows the difficulties that can affect offshore structures.

Once the current reaches the foot of the slope, its velocity diminishes and this facilitates a second stage of sedimentation at the beginning of the continental rise (Sect. 9.2). The particles

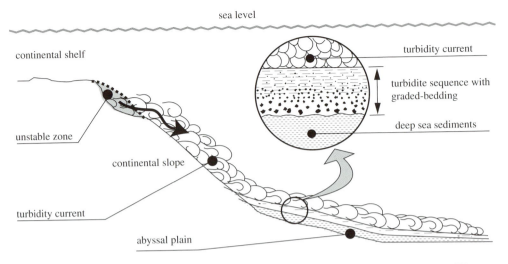

Fig. 9.20 Conditions leading to the formation of flysch by turbidity currents in deep sea fans (from [168]).

are deposited on the basis of grain size, with sands at the base followed by silts and finally clays. This sequence is a *turbidite*. As these instability phenomena are repeated, cyclic beds accumulate one on top of another, finally forming a type of abyssal delta called a deep sea fan.

Deep sea fans are complexes of great thicknesses of fine detrital sediments (several hundred meters to several kilometers thick). Many were formed in the Tethys trenches during Alpine orogenesis from the late Cretaceous to the beginning of the Tertiary. As a result of diagenesis, these turbiditic sedimentary complexes become a suite of rocks called flysch if they are included in an orogenic context (Chapt. 10).

9.1.2   Biogenic sedimentation

Biogenic continental margin sediments are the coastal facies of warm regions where detrital sedimentation is minor. These sediments form as a result of the growth of organisms in the near-shore area where the sea is shallow. This shoreline zone extends offshore along a very shallow slope that encounters the continental rise. The debris of organisms from the coastal zone accumulates in this location.

Under the term biogenic, we group sediments whose origin depends on organisms, either directly or indirectly. These are:

* Marine organisms, either plant or animal, that construct their skeletons of calcium carbonate that comes from carbonates dissolved in the water, (for example: coral reefs and encrusting algae);
* Debris of organisms that accumulate on the continental slope and are transported by currents (*biodetrital* sediments).

The formation of biogenic sediments on continental margins is closely linked to the ecology of these zones. Among the determining factors and their effects are the following:

* Water temperature and chlorophyllic activity (calcium carbonate equilibrium),
* Water depth (penetration of solar radiation),
* The nature of the bottom, rocky or loose sediments (ability of organisms to fasten to the bottom),

- Protection from waves (mechanical stability of the environment),
- Presence of a river mouth (turbidity, salinity, temperature, biologic factors).

Calcium carbonate equilibrium controls the solubility and precipitation of carbonates, in particular that of calcium carbonate or limestone. It is usually expressed as a series of reversible reactions that concern the water molecule, the solubility of carbon dioxide in water, the dissociation of carbonic acid, of hydrogencarbonate and calcite according to the following equilibrium equations:

$$H_2O \rightleftharpoons H^+ + OH^- \tag{9.1}$$

$$CO_2 + H_2O \rightleftharpoons H_2CO_3 \tag{9.2}$$

$$H_2CO_3 \rightleftharpoons H^+ + HCO_3^- \tag{9.3}$$

$$HCO_3^- \rightleftharpoons H^+ + CO_3^{2-} \tag{9.4}$$

$$CaCO_3 \rightleftharpoons Ca^{2+} + CO_3^{2-} \tag{9.5}$$

Each equation has an equilibrium constant that depends primarily on water temperature. It is a complex system. Generally and in a simplified manner, we can say:

$$CaCO_3 + CO_2 + H_2O \rightleftharpoons Ca^{2+} + 2(HCO_3^-) \tag{9.6}$$

If the equilibrium shifts toward the right by a contribution of CO_2, the carbonate (a solid on the left side) goes into solution as dissociated calcium and hydrogencarbonate ions. The carbonate is thus dissolved, just as in the case of karst erosion (Chapt. 7). If the equilibrium shifts to the left, the calcium hydrogencarbonates precipitate as calcite or aragonite, the basic minerals of limestone.

In high temperature areas, carbon dioxide is released from the water because of its low solubility. The equation that goes from right to left is favored and thus equilibrium moves to the left also. This tendency is reinforced when phytoplankton activity is significant (see § 9.2.2).

The formation of biogenic sediments is easily observed offshore of seacoasts in warm countries. We will look at present-day data to understand the conditions of sedimentary rock formation on the continental margin.

Figure 9.21 shows the typical situation of reef coastlines such as those of the Pacific islands. Proceeding offshore from land, there are generally three sedimentary zones.

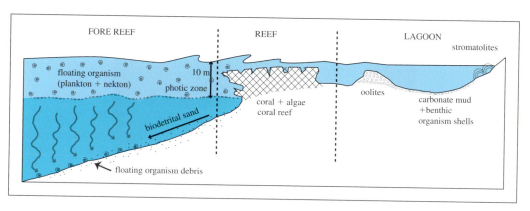

Fig. 9.21 Carbonate sediments along continental margins.

Lagoon

A lagoon is a shallow warm sea that is generally protected from large waves. Its seaward margin is a reef. There is a varied fauna living on the bottom (the **benthos**). The debris of this fauna forms biodetrital sand. The water has a strong tendency to precipitate carbonates in solution, as shown by **oolites** (limestone deposited in concentric envelopes around a tiny piece of organic debris or grain of sand). They are found near the reef or, in areas where waves are significant, as particles in saltation (Fig. 9.22).

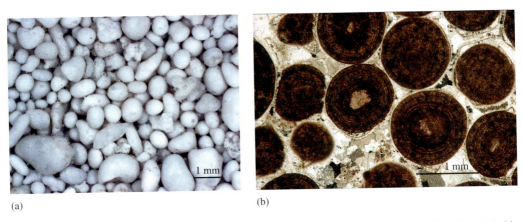

(a)
(b)

Fig. 9.22 Oolites: (a) present day oolites (Bahamas); (b) internal structure of an oolitic limestone (Upper Jurassic), thin section photo. Photos E. Davaud, University of Geneva.

In the internal part of the lagoon, in a calm environment, blue algae, which were important precursors of life on Earth, still form today as stromatolites. (Fig. 3.19 and 9.23).

Fig. 9.23 Laminated structure of a stromatolite. GEOLEP Photo, G. Franciosi.

Reef

The reef is a mechanical barrier to open ocean currents. The width of the reef can range from tens of meters to several kilometers. It can be emergent or it can grow completely under a shallow water column. The framework of the reef is composed of organisms attached the substrate, particularly corals, in symbiosis with various algae. The living portion of the reef grows on the dead part; it is limited to the photic zone (zone that is exposed to sufficient sunlight for photosynthesis to occur). Because of this restriction, the reef tends to grow laterally (Fig. 9.21). In the Pacific islands, the thickness of the reef is often significant because the island have subsided over time (Fig. 9.24).

Fig. 9.24 Coral reef, Strait of Tiran, Egypt. Photo E. Davaud, University of Geneva.

The sedimentary record of a reef is not an unconsolidated deposit as in the case of other biogenic deposits, but instead is a rocky foundation called a coralline limestone. Its erosion by waves and storms provides material to the lagoon and to the reef slope. It is permeated by intervals that facilitate easy water exchanges with the lagoon. Reefs are obstacles that pose major problems to navigation along coastlines. They can seriously limit the possibilities of port facilities.

Fore-reef

This is a transition zone between the reef and the great ocean bottoms. In the higher part, the fore-reef receives coarse reef debris transported by wave erosion. The size of this debris decreases toward the open sea. There is a gradual change of living species: bottom dwelling organisms (***benthic***) give way to floating and swimming organisms (***pelagic***). The latter produce biogenic sediments in the open ocean as we will discuss in section 9.2.2.

9.1.3 Evaporite deposits

The chemical composition of the sea should logically reflect the average chemical composition of rocks that make up the continents. Table 9.1 shows that this is far from being the case. Two substances illustrate this paradox:

- Silicon, the most abundant element in the Earth's crust, is present in minor quantities in sea water;
- Chlorine, an ion that is very concentrated in seawater, is, by contrast, present in rocks at very low concentrations.

Table 9.1 Comparison of the geochemistry of the continental crust and seawater, by weight percentage.

Hydrosphere Sea water [%]		Continental crust [%]	
H	10.6	O	41.2
O	86.5	Si	28
Cl	1.89	Al	14.3
Na	1.056	Fe	4.7
Mg	0.1272	Ca	3.9
S	0.08	K	2.3
Ca	0.04	Na	2.2
K	0.03	Mg	1.9
C	0.003	Ti	0.4
Br	0.065	C	0.3
Si	0.0001	H	0.2
		Mn	0.07
		Cl	<0.1

The origin of the salinity of seawater is explained by the considerable age of the first oceans and the processes that govern the solubility of mineral phases in water (§ 3.3.4).

Let us examine the case of silicon. Since rivers have contributed significant amounts of dissolved silicon to the oceans over the years, we might expect that the silicon concentration has continuously increased. But in fact, this is not the case. Only spatial variations in its concentration have been observed. To understand the exchange between dissolved silicon and solid silicon in the oceans, we must consider the solubility of the different solid forms of silica in ambient-temperature water. Quartz, which is the most stable form of silica, has a very low solubility of about 10 mg/l as orthosilicic acid H_4SiO_4 (the principal form of dissolved silica). The solubility of chalcedony is three times higher. The solubility of amorphous silica can exceed 100 mg/l H_4SiO_4. The silica content of the water is closely linked to depth. At the surface, the concentration is about 1 mg/l, in other words, very under-saturated with respect to all the forms of solid silica. This can be explained by the role of abundant planktonic organisms in the photic zone. These organisms ingest seawater and extract silicon from it to form their skeletons (for example, the shells of certain foraminifera) as amorphous silica. This biological intervention makes it possible to export silicon from the aqueous environment and import it into the solid environment at a concentration that is well below the chemical solubility limit. Once the organism has died and biological processes have stopped, the silica in the shell goes back into solution due to the under-saturation of the water. This tendency is reinforced as the organism sinks to deeper water since solubility increases sharply with pressure. During this path to the abyss, a significant portion of the silica returns to the liquid phase. This explains why its concentration exceeds 10 mg/l at a depth of 2000 m and it increases even more at greater depths. There remains a solid residue of siliceous oceanic sediments, which is finally exported from the aquatic environment. In summary, the supply of dissolved silicon in seawater does not increase in spite of the massive input, because of biological exporters that return it to the solid phase. The concentration of silicon in seawater at the surface is limited and controlled by the presence of planktonic organisms in the photic zone.

Sodium chloride, in contrast, is so soluble that the limit is reached only in closed seas in arid areas, such as the Dead Sea (Fig. 9.25). In seawater, its concentration is about 10% of its

Fig. 9.25 The quantity of salt in the Dead Sea exceeds its solubility limit. The excess NaCl makes white columns that outcrop at the water surface. West Bank of the Dead Sea. GEOLEP Photo, A. Parriaux.

solubility limit (360 g/l NaCl). This is why few salt-bearing rocks form in the ocean environment. The chlorine content in rocks is very low: the major part is present in pyroxenes and biotite which have concentrations of several hundred mg/kg. Chloride is present in seawater at high concentrations because of the effect of the long time periods involved. Chloride is also contributed to seawater from hydrothermal fluids.

Let us imagine a littoral zone containing a sound or lagoon that is well isolated from the open sea by a shallow or a sandbar (Fig. 9.26).

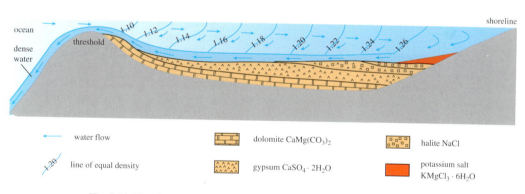

Fig. 9.26 Site of evaporitic sediment formation. Densities are expressed in t/m^3.

At high temperatures, water evaporates easily. The quantity of mineral salts in solution does not change, and the concentration of salts increases the water density. The water has increasing difficulty crossing a barrier to the open ocean. After some time, the solubility of a salt is exceeded, and it precipitates to maintain equilibrium. The different salts appear gradually as a function of their increasing solubility. The least soluble accumulate rapidly at the bottom, and the most soluble at the end of the sedimentation process. The density of the supernatant water increases as evaporation progresses. Thus we have precipitation of different salts on the basis of well-determined density thresholds:

- density $> 1.05 \cdot 10^3 \, kg/m^3$: beginning of carbonate precipitation, particularly primary dolomite $CaMg(CO_3)_2$ (this dolomite is called primary because magnesium is present in the mineral from the beginning, as opposed to secondary dolomite, which incorporates the magnesium later, during diagenesis, Chapt. 10);
- density $> 1.12 \cdot 10^3 \, kg/m^3$: beginning of gypsum precipitation $CaSO_4 \cdot 2H_2O$;
- density $> 1.21 \cdot 10^3 \, kg/m^3$: beginning of halite or rock salt precipitation $NaCl$;
- density $> 1.26 \cdot 10^3 \, kg/m^3$: beginning of precipitation of potassium salts, principally car-nalite $KMgCl_3 \cdot 6H_2O$ (the principal potassium mineral).

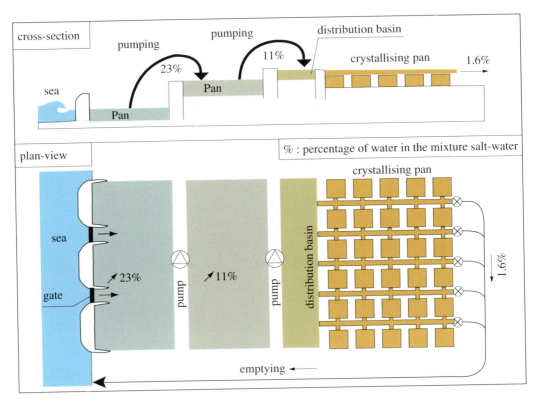

Fig. 9.27 The saltern is divided into two stages. The outer chamber is a lower pan that receives a certain amount of seawater. There the water undergoes primary evaporation and calcium carbonate is deposited. The inner chamber is located at a higher level. Water from the outer chamber is pumped there, evaporation continues, and calcium sulfate is precipitated. Then the water flows into the salt range, where it is ready to precipitate sodium chloride; it is pumped into antechambers where brine distributors direct it to "crystallising pans," multiple pans where there is a well-monitored soil covered by filamentous algae. There the salt crystallizes into twinned hopper-shaped crystals. When only 1.6% liquid remains, the salt is scraped up and the residual water containing magnesium and potassium chlorides is discarded. The salt is left in stacks so that the undesirable water drains away. These piles are grouped into a sort of pyramid that the salt-makers cover with Earth in order to protect it from the rain. Finally, the salt is crushed, iodized to prevent thyroid diseases, and packaged.

If the process goes to completion, halite will predominate (almost 80%) in the deposit. Often sedimentation is interrupted by the water removal as a result of storms or by the deepening of the sea. In this case, all that remains in the basin is dolomite or dolomite and gypsum, such as in the principal evaporitic rocks of the Tethys (Fig. 12.49). In the rare cases where the process continues, salt is formed, as in the Triassic in the region of the Bex mines in the Prealps or at depth in the Jura (Schweizerhalle mines).

Physically these deposits are solid crusts rather than loose sediments. They will become more consolidated during diagenesis.

The natural phenomenon of evaporite formation is imitated by man in numerous exploitations of salterns that occur along seas and oceans (Fig. 9.27).

9.2   The continental rise and the abyssal plains

The continental rise begins at a depth of around 2000 m and is an extension of the continental slope. This is the place where turbidity currents from the overlying slope are deposited. Offshore of the continental rise the topography of the marine floor becomes almost flat; this is the abyssal plain. The only topography that disrupts this monotony is the mid-oceanic ridges (Chapt. 6) and the trenches in subduction zones. The Marianas trench offshore from Japan is the deepest on Earth, more than 11,000 meters deep.

On the abyssal plain, the sedimentation rate is much lower than in coastal zones. For example, an E-W cross section across the Atlantic shows deposits 1000 m thick that were deposited over a period of 200 million years off the coast of Africa, an average thickness of 5 microns per year. Approaching the mid-oceanic ridge, the thickness decreases to zero. On the other side of the ridge, there is an almost symmetrical configuration of sediments.

Sediments accumulating on the abyssal plain can be grouped in the three classes used for the continental margin environment: detrital, biogenic, and evaporitic sediments:

9.2.1   Detrital sedimentation

Because of the distance from the coasts, sediments produced by continental erosion are limited to the finest particles: clays transported by rivers, eolian dust in the atmosphere and glacio-marine deposits (solid residues from the melting of icebergs). There is also ash and microscopic debris from continental volcanoes (sedimentary volcanic deposits).

9.2.2   Biogenic sedimentation

Organisms that live in the oceans can be divided into two sub-groups:

- Swimming animals (***nekton***) essentially fish, non-anchored mollusks, marine mammals;
- Floating microorganisms (***plankton***), less than a few millimeters in size.

The first group provides isolated fossils (external or internal skeletons, Chapt. 3) but they are only a small fraction of the global mass of sediments.

It is primarily the plankton that contributes to the biogenic sediments of the deep ocean. They decompose differently, depending on whether the organisms are animals (***zooplankton***) or plants (***phytoplankton***). The latter are responsible for the production of the majority of the oxygen in the Earth's atmosphere. By consuming carbon dioxide, they cause the water to be supersaturated in carbonates, according to reaction (9.6).

Many planktonic organisms have a mineral skeleton whose geometry is specific to the species, which makes it possible to identify them. The organism takes its skeletal material from the stock of mineral salts dissolved in the water. These organisms are either calcareous (aragonite or calcite) or siliceous (opal). Figure 9.28 gives an example of each of the four types. The calcareous organisms dominate in warm water, and the others are more abundant in cold regions.

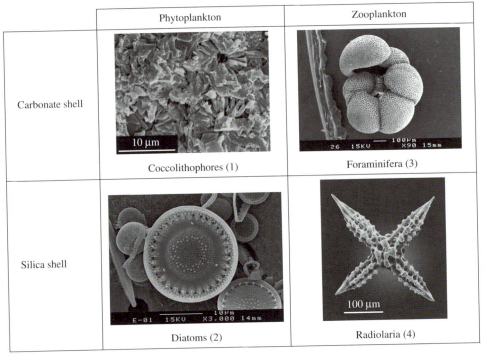

Fig. 9.28 Four present-day planktonic organisms that show the type of organism and the nature of their skeletons [27] Photos (2), (3) R. Martini, University of Geneva; photos (1), (4) P.O. Baumgartner, University of Lausanne. An animation on the DVD shows other planktonic organisms.

Similar to macrofossils, the remainders of dead organisms sink to the deeps by the force of gravity. During this long trajectory from the photic zone to the ocean bottom, they undergo a transformation process. In an oxidizing environment, the organic matter disappears due to mineralization. In a reducing environment, as is often the case in the deep ocean basins, organic particles are transformed into muds that may become petroleum and natural gas after diagenesis (Chapt. 10).

During its descent to the ocean deep, the skeleton dissolves according to very different processes depending on whether the skeleton is siliceous or carbonate. We have seen the mechanism of partial dissolution of silica shells of organisms during their migration to the deep (see § 9.2.2). The increasing pressure at depth causes an increase of solubility and thus favors dissolution. Although the majority of siliceous skeletons are dissolved in this way, there is a small percentage that persists, even at great depth. They form siliceous muds. Through diagenesis they become very hard rocks: for example, radiolarian muds become radiolarites (Chapt. 10).

At the surface, calcium carbonate is itself in supersaturation because of photosynthesis. At depth, carbon dioxide becomes more soluble with increasing pressure; water becomes more aggressive and the carbonate of the skeleton tends to go into solution as hydrogencarbonate and calcium ions, according to equation (9.6). As a result, there is a maximum limit beyond which solid calcium carbonate can no longer exist: this is the **CCD, *Calcite Compensation Depth***. Its position varies spatially as a function of the supply of carbon dioxide in deep water, which differs from ocean to ocean. The CCD is at about 5000 m in the Atlantic (Fig. 9.29). Its depth decreases to about 4300 meters in the Pacific, whose waters are richer in carbon dioxide. It can even vary within the same ocean. Caution should be used in applying present-day rules to ancient sediments; we know that the compensation maximum can decrease sharply if the productivity of the plankton decreases drastically. In addition, climatic and biologic factors (distribution of nutrients) play a significant role in the distribution of these sediments.

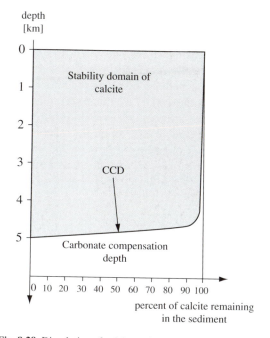

Fig. 9.29 Dissolution of calcite at depth. Case of Atlantic Ocean.

At depths above this chemical boundary, planktonic carbonate muds can accumulate. They produce fine, pasty limestones. Chalk is a particular type of calcareous sediments made up primarily of coccoliths (Fig. 9.28). It forms in shallow oceanic basins (less than 300 m) and undergoes only slight diagenesis.

Carbonate planktonic muds often have a rather high silica content. During diagenesis, the silica remains dispersed throughout the rock and forms *siliceous limestones*. Occasionally, the silica aggregates as *flint nodules* encased in chalk or limestone (Chapt. 10).

The distribution of carbonate and siliceous muds in present-day oceans thus depends on water temperature (latitude) and ocean depth. It can be presented schematically as follows:

*   siliceous muds near the poles and in very deep basins (Antarctic belt and North Pacific);
*   carbonate muds in equatorial zones and shallower basins (mid-oceanic ridges, Atlantic ocean).

This distribution of sediments can, to a certain degree, be applied to ancient marine basins. For example, Jurassic radiolarites in the internal zone of the Alps show a deep oceanic phase of the Tethys at this time.

Biogenic sediments are often associated with detrital sediments. One example is the mixture of calcareous and argillaceous muds. Diagenesis of these muds produces a rock that is very important for the engineer and is used for making cement: *marl* (Chapt. 10 and Problem 1.1).

9.2.3   Hydrochemical sedimentation

The abyssal deep produces neogenic clays (Chapt. 13) associated with submarine volcanism. These are deep sea *red clays* in the Pacific and in certain parts of the Indian Ocean. They belong essentially to the montmorillonite family and they are produced by the reaction of biogenic sediments and volcanic rocks with salts dissolved in seawater. Traces of iron and magnesium oxides give them a characteristic brown-red color.

In association with red clays from the deeps, we often find *polymetallic nodules* rich in manganese and iron. Other economically important metals such as nickel, copper, cobalt are present in smaller percentages. These nodules occur as rounded black bodies ranging in size from 1 to 10 cm. They form around a nucleus by chemical precipitation of metal oxides in concentric layers. The metals come from various sources: continental erosion, hydrothermal activity, and weathering of submerged volcanic rocks. Their precipitation is either purely chemical, as a result of oxidizing conditions, or biochemical through the intermediary of bacteria. The high concentrations of certain metals in these polymetallic nodules makes them an ore and their exploitation is under study; however, a dredging system that involves scraping the ocean floor raises ecological concerns.

[198, 222, 227, 258, 304]

PROBLEM 9.2

Figure 9.30 shows a simplified stratigraphic profile of the Western Swiss Plateau. It was constructed on the basis of various petroleum borings.

Describe the evolution of the sedimentary environment and the paleogeographic context by choosing several characteristic stages in this history. Consult chapter 10 for this problem.

Stratigraphic stages under consideration: see figure 3.11.

1. Carboniferous — coal
2. Permian — conglomerates and sandstones
3. Lower Triassic — anhydrite and salt
4. Upper Triassic — limestone, dolomite and anhydrite
5. Lias and Lower Dogger — claystones with ammonites
6. Middle and Upper Dogger — limestones with echinoderms
7. Lower Malm — marls with cephalopods
8. Middle and Upper Malm — coralline limestone
9. Purbeckian — lacustrine marls and sandstones
10. Valanginian — oolitic limestone
11. Lower Hauterivian — marls
12. Upper Hauterivian — sparry limestone
13. Eocene — karst filling by red ferruginous clays
14. Oligocene — lacustrine marls and sandstones
15. Miocene — marine sandstones and marls
16. Pliocene — absent
17. Pleistocene — till

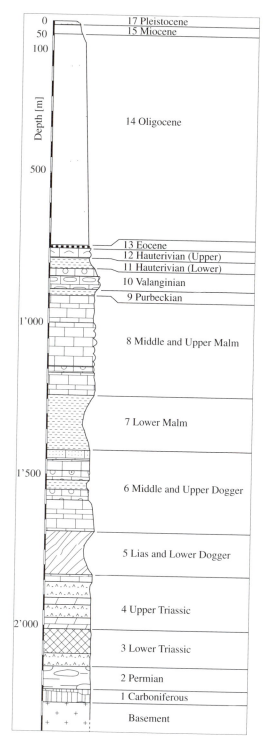

Fig. 9.30 Stratigraphic and lithologic (rock description) profile compiled from petroleum borings (Western Swiss Plateau).

Look at the solution on the DVD

Chapter review

Whereas the continental sedimentary environment produces predominantly detrital deposits, the marine environment involves more complex processes that include not only mechanical and hydraulic but also chemical and biologic factors. As a result there is a great variety of
sediments whose structures record the history of their formation. These sediments are often
much more extensive and continuous than continental deposits. Once transformed into rock by
diagenesis and brought to the surface by tectonic movements, these marine deposits make up
a large part of the rock masses that engineers encounter in their work. The generally stratified
nature of the sediments will give these rocks highly anisotropic properties.

10 Diagenesis and Properties of Sedimentary Rocks

In chapters 8 and 9, we studied the principal unconsolidated sediments formed in the Earth's continental and oceanic environments. We have discussed detrital sediments and their grain-size variations, the great diversity of biogenic sediments, and evaporite deposits formed by precipitation of salts from seawater. We have discussed the creation and composition of sediments and the nature of sedimentary structures. We will now see how sediments are transformed into rock. On land and in the sea, loose sediments are gradually solidified as they age and finally become sedimentary rocks.

We can describe how a layer of loose sand, such as present-day beach sand, turns into rock. Over millennia, new sediments bury the bed of sand, causing it to compact; the grains are gradually cemented by fragile bridges of mineral precipitates. As the pores fill, cementation proceeds. After several million years, the sediment becomes rock: sandstone.

Several general definitions will be useful:

- *Unconsolidated sediments* (or "soils" for geotechnical engineers): masses of particles that are not bound or are only loosely bound to each other.
- *Diagenesis*: the group of processes that transform unconsolidated sediment into rock. A synonymous term is *lithification*. In contrast to metamorphism (Chapt. 11), the mineralogical modifications that occur with diagenesis involve modest changes of temperature and pressure,
- *Sedimentary rock*: a mass of particles sufficiently bound together to behave like a single solid mass.

These definitions show that there is no clear-cut distinction between rock and unconsolidated sediments. Diagenesis is a process of continuous transformation; it would be arbitrary to set a boundary between the rock and sediment. Even so, it is obvious that unconsolidated sediments often belong to the present-day era (Quaternary), whereas older sediments have become soft rocks. This convention is practical for example in the Alps and their northern sedimentary basin, since there is a significant sedimentary gap between the most recent molasse (rocks that are about 10 Ma old) and deposits from the first Quaternary glaciations, which generally date from less than one million years ago. The molasse is thus classified as rock and the glacial moraines are considered unconsolidated sediments. The boundary is more fluid in large alluvial basins such as the Po Plain where sediments have accumulated continuously from the Tertiary until the present day.

Diagenesis is a physical process that leads to an increase in density and mechanical strength of the geological material. It is contrary to the process of *weathering*, a group of processes that progressively transform a rock into loose sediments (Chapt. 13). Referring to our sand and sandstone example, weathering of molasse buildings reduces the building stone to sand.

Diagenesis also acts on other loose materials that are not listed in the common classification of sediments, such as pyroclastic deposits and fault-filling materials. The lithification of these materials is covered in chapter 6 and 12, where their origins are also discussed.

After reviewing the principal processes that take place during diagenesis, we will describe the most important sedimentary rocks, beginning with the original sediments and the mechanisms that transform them (Fig. 10.1). We will then discuss their composition, their structure, and the principal technical properties relevant to the engineer's work: mechanical strength, foundation stability, suitability as a construction material, and hydrogeological properties.

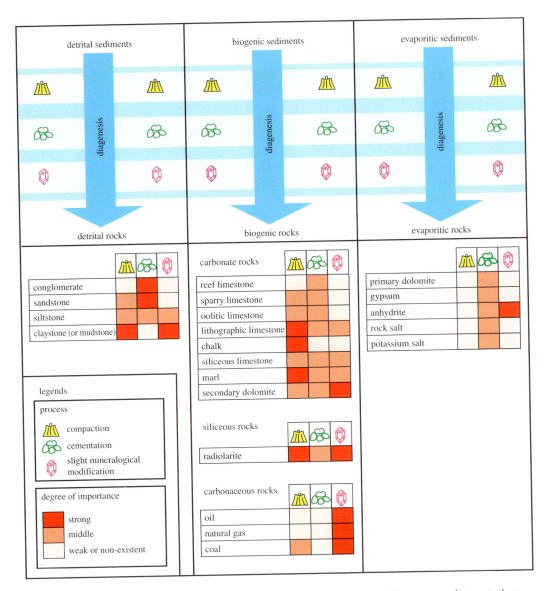

Fig. 10.1 From sediment to sedimentary rock. The principal sedimentary rocks and the processes that create them.

10.1   Diagenetic processes

Diagenesis includes complex physicochemical processes that cause the interaction of the solid sediment material and the water in pores. This original water that saturates the sedimentary rock during its formation, called ***connate water***, can be very rich in mineral salts. The solid–liquid reactions are often accompanied by biological factors. Our understanding of diagenetic processes is hampered by our inability to recreate long-term effects experimentally. We will discuss three principal diagenetic processes in this chapter.

10.1.1 Compaction

When a new layer covers a sedimentary bed, the overlying stratum exerts a mechanical load on the underlying one. The vertical stress on the contacts between the grains is equal to the total weight of the column of overlying sediments, if they are not in water, or to this weight less the Archimedes force if everything takes place in water (Problem 8.2). The underwater case is the most frequent.

We shall now define a fundamental concept in the compaction problem: ***porosity***. Total porosity is equal to the ratio of the pore volume to soil volume.

For saturated sediments with a porosity of 25%, the specific weight of the solid fraction is equal to $26.7 \, kN/m^3$. The grain-on-grain stress of a one-meter-thick cover is equal to:

$$(26.7 - 10) \cdot 0.75 = 12.5 \, kN/m^2 \qquad (10.1)$$

To clarify the order of magnitude of this pressure, the horizontal surface of a grain measuring $1 \, cm^2$ is subject to a weight of $12.5 \, kg$ for $100 \, m$ thickness of sediments. These are obviously considerable pressures.

This weight causes compaction of the granular structure. The grains shift to occupy a geometric space that takes up less space and expel interstitial water or air (Fig. 10.2).

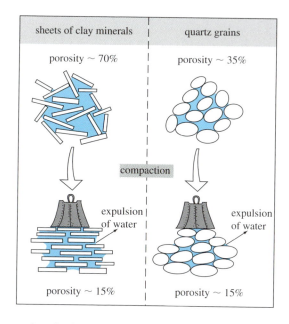

Fig. 10.2 Compaction of a clay skeleton and a granular skeleton (pseudo-spherical grains).

Spheroidal grains are not very deformable in themselves and sediments composed primarily of these grains generally do not compact to a significant degree. Clay particles have a laminar structure and can therefore undergo significant compaction. Clay mud with an original porosity of 70% can lithify to a mudstone with a porosity of about 10%. Figure 10.3 shows the relationship between porosity and burial depth for various grain sizes that was established by Selley, 1976.

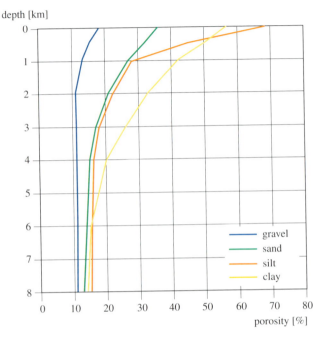

Fig. 10.3 Porosity reduction by compaction due to burial depth: average curves for various detrital classes (from [258]).

On the scale of geological time, the phenomenon of compaction responds immediately to an increase in pressure. It is different on the scale of engineering projects where the subsidence of impermeable ground under the load of a foundation may take place over several years. This time is necessary for the hydrostatic over-pressuring due to the overcharging to be resolved by the expulsion of pore water.

10.1.2 Cementation

In saturated sediments where the pore size is larger than a few tenths of a millimeter, the water is able to circulate. This may be connate water or groundwater of external origin. This fluid carries ions in solution that can precipitate when the conditions change from the initial equilibrium conditions. Crystals then gradually coat the walls of the grains and create solid links between grains (Fig. 10.4). Cementation causes the primary cohesion of these grains. If the process continues, cement can completely fill the pores.

The cement is predominantly calcite or amorphous silica that finally becomes quartz.

In some sediments, minerals do not grow homogeneously from ions in the water. Instead, they grow as nodules or concretions, creating a preferred location for precipitation (Fig. 10.5). Sometimes they surround organic debris present in the sediment to form fossiliferous nodules: the organic material is the nucleus for concentric precipitation (Fig. 10.6).

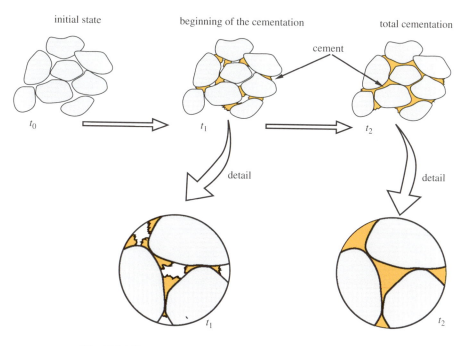

Fig. 10.4 Effect of cementation on a granular sediment (from [168]).

(a)   (b)

Fig. 10.5 (a) Calcite nodule produced by diagenesis of continental sands in the Sanga do Cabral Formation (Permian and Triassic) Mata region, Southern Brazil (diameter 5 cm). (b) Erosion causes nodules to accumulate at the foot of the slope, creating a surface that resembles a lawn-bowling court. (GEOLEP Photos, A. Parriaux).

A well-known type of nodule is a flint nodule (§ 10.3.2).

Cementation has implications for geological engineering. With cementation, material becomes less deformable and mechanically stronger. Porosity and matrix permeability are both greatly reduced.

Fig. 10.6 Cementation of nodules. A nodule containing a trilobite fossil (Paleozoic): interior and exterior of nodule.
GEOLEP Photo, L. Cortesi.

10.1.3 Mineralogical modifications

The mineralogical modifications of solid matter depends on pressure and temperature conditions and on the chemistry of the environment. If these factors change, the mineral may no longer be in its stability field. It is then possible for the mineral to change from its original form to a new one without magmatic melting. Metamorphism is an intense version of this solid recrystallization process, whereas diagenesis involves slight variations of thermodynamic conditions: the temperature remains low (T < 50°C) and the pressure is less than 100 MPa (1 kbar), which corresponds to a weight of 4000 m of sediments.

Some minerals, particularly amorphous ones or those in an initial stage of crystallization, are unstable as soon as the pressure or temperature rises. Time has the same effect. The principal transformations are the following (Chapt. 5):

- The aragonite (orthorhombic calcium carbonate) of mollusk shells is transformed into calcite (rhombohedral crystal symmetry);
- The amorphous to microcrystalline opal of siliceous organisms changes to chalcedony by dehydration, then to quartz (Fig. 10.22);
- Gypsum loses its crystallization water and is transformed into anhydrite at a depth of 300 to 700 m, with a great reduction in volume (Sect. 10.4);
- Detrital argillaceous minerals react with sea water to produce other types of clays (K is replaced by Mg);
- Calcareous muds may react with magnesium in sea water to form secondary dolomite (§ 10.3.1.2);
- Organic matter is gradually transformed into hydrocarbons (§ 10.3.3).

[160, 220]   We can now discuss the principal groups of sedimentary rocks.

10.2   Detrital rocks

Sediments in the detrital series range from blocks to clays, and as a result, they react differently to diagenetic processes.

Coarse sediments (boulders, cobbles, gravels, and sands) are not very compressible, and thus they do not compact very much (Fig. 10.1). Because of their large pores, considerable quantities of groundwater can travel through them. If the water is slightly oversaturated in mineral salts, these minerals can rapidly coat the grains and cement the rock. The grains are generally composed of already-formed rocks that are not very sensitive to slight recrystallization. Sands contain predominantly quartz, which is extremely stable chemically and mechanically.

Fine sediments (silts and clays) contain clay minerals and small quartz crystals. Aggregates of sheet-like clay minerals are highly compressible (Fig. 10.3) and cementation is more difficult because of the very low permeability. That is why argillaceous rocks remain slightly plastic. Mineralogical transformations of clay minerals are very common. For example, detrital clays are primarily composed of illite in temperate climates. In seawater, the potassium in the illite is replaced by magnesium to form swelling montmorillonite clays.

The principal detrital rocks are conglomerates, sandstones, siltstones, and claystone (or mudstone), in decreasing grain size.

10.2.1   Conglomerates

Conglomerates are the product of the cementation of boulders, cobbles, and gravels. If the grains are rounded, the conglomerate is called a ***puddingstone***. Angular grains form a rock called a sedimentary ***breccia***, which generally forms as a result of the collapse of a cliff (Fig. 10.7).

When conglomerates are well cemented, they can form massive rocks made up of layers that are several meters thick. A siliceous cement, such as that of the Permo-Carboniferous conglomerates of the Valaisian Alps, produces a very hard rock that can be found as erratic blocks far from their place of origin.

(a)   (b)

Fig. 10.7 Conglomerates: (a) Puddingstone from Mont-Pélerin (Oligocene), Swiss Plateau; this rock comes from the diagenesis of the first delta that formed at the foot of the Alpine chain at the beginning of its uplift; (b) Breccia with siliceous cement (Permo-Carboniferous, Valaisian Alps). GEOLEP Photos, G. Grosjean.

10.2.2 Sandstones

Sandstones are the diagenetic product of sands (Fig. 10.8 and Fig. 10.9). They are often composed of quartz grains. This explains why drilling through sandstones causes rapid abrasion of drilling tools. The nature of the cement is usually calcareous, siliceous, or argillaceous.

Fig. 10.8 Navajo Sandstone, Antelope Canyon, Arizona, USA. Detail of stratification in an erosional channel of the river. View from below. GEOLEP Photo, N. Schaffter.

Fig. 10.9 Navajo Sandstone, Page, Lake Powell, USA. The sediments show intensely crossbedded stratification. GEOLEP Photo, N. Schaffter.

Two *sandstones with calcareous cement* merit additional information for their importance in the Alpine geological environment: molasse and flysch.

In sedimentological terms, a *molasse* is defined as a shallow detrital deposit, either marine or lacustrine, that comes from the erosion of a mountain chain and is deposited at the foot of the mountains. Molasse includes several different types of rocks: sandstones, siltstones, marls, mudstones, conglomerates, lacustrine limestones and even coal. North of the Alps, the majority of the rocks of the plateau between Chambery and Vienna is composed of molasse.

Petrologically, the molasse is the most abundant rock in the molasse basin: it is a sand-stone that is rich in feldspar grains, with calcareous cement. It often contains green minerals, which give the rock a gray-green color. It occurs in meter-thick layers, with upwardly fining grain size. For many centuries, the molasse was the only building stone available on the Swiss Plateau. This stone gives the cities of Fribourg and Bern their typical architectural style. The molasse was the closest material to the work sites and it was easy to work. Since the development of mechanized transport, it is not used much any more because its high porosity makes it sensitive to freezing and its calcareous cement is not resistant to the acids created by atmospheric pollution.

Hydrogeologically the sandstone molasse can be a low-capacity aquifer when the calcium carbonate has been leached out. Many small springs occur throughout the area. Their large number makes up for their small individual discharge, and globally, this resource plays an important role in the drinking water supply of cities in the molasse plateau.

Flyschs (Fig. 10.10) are detrital rocks formed in the ocean from deposits that were originally deep sea fans caused by orogenic movements (Chapt. 9). In the Alps, flyschs have been tectonically transported by thrust sheets (Chapt. 12). The term flysch includes sandstones and

Fig. 10.10 Flysch of Fayaux sur Blonay, Lake Geneva Basin. The competent layers are sandstones, the schisty layers are claystones. GEOLEP Photo, J.-H. Gabus.

siltstones alternating with claystones, rocks that are characteristic of a turbidite sedimentary cycle. The sandstones are not very porous and occur in benches that are decimeters thick. They are good for construction and road pavement. The intercalated claystones promote landslides on slopes.

Sandstones with siliceous cement are very common on our planet. In Europe, the Permian red sandstone series is an old molasse that resulted from the erosion of the Hercynian mountains. These sandstones were used historically as building stone, as shown by numerous monuments such as the Cathedral of Strasbourg.

Arkose is a particular type of sandstone, generally with siliceous cement, produced by the diagenesis of granitic sand (Chapt. 13). It is composed of granitic minerals that have been transported only short distances, thus preserving their crystalline appearance. Some very massive arkoses can strongly resemble granites, except that the arkoses can be stratified. Other types of cement can occur depending on the chemistry of the water that percolated through the sediment (Fig. 10.11).

Fig. 10.11 An arkose dome of Kata Tjuta, a preserved relic of an old alluvial fan, thirty kilometers from Uluru (Ayers Rock), Central Australia. This arkose has an intense red color due to oxidation of iron present in the cement. GEOLEP Photo, L. Cortesi.

Quartzite is a siliceous rock that belongs to both sedimentary and metamorphic rocks. It will be discussed in Section 11.3.2.

10.2.3 Siltstone

Siltstones are cemented silts (Fig. 10.12). They are found in abundance in sandstone series, for example as thin beds in the molasse series and in flyschs. They often contain a bit of clay and as a result they are sensitive to variations of water content and to freezing.

10.2.4 Claystones

We use the term claystones (or mudstones) for argillaceous rocks, thus reserving the name clay for recent loose sediments (other definitions exist in the literature). Claystones (Fig. 10.13) are detrital rocks composed primarily of clay particles; accessory minerals are quartz (silty claystones) and calcite (transition to marls). In outcrop, claystones often have a layered appearance caused by the parallel organization of clay minerals during sedimentation and compaction

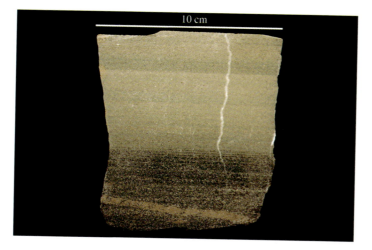

Fig. 10.12 Sandstone and siltstone with horizontal stratification showing sorting by grain size. The base of the layer contains green ferromagnesian minerals. Sawed surface. GEOLEP Photo, G. Grosjean.

Fig. 10.13 Triassic claystones (Valais, Switzerland). The schistose shape of the rock is clearly apparent. GEOLEP Photo, G. Grosjean.

(thus equivalent to the term "*shale*"). These rocks generally indicate sedimentation relatively far from the mouths of rivers.

North of the Alps, claystones are found in the molasse basin, in flyschs, and in the marine series of the Jura and the Alps.

Because of their weak cementation and their plastic behavior (Chapt. 12), claystones areas are prone to landslides. In underground projects, the presence of claystones can cause the swelling of walls if the rock contains clays such as montmorillonite.

Claystones are raw material for the terra cotta industry (tiles and bricks) and for the manufacture of cement in mixtures with limestone.

In terms of hydrogeology, claystones often function as an impermeable layer at the base of aquifers because of their lower permeability.

10.3 Biogenic rocks

We will discuss biogenic rocks in terms of their mineralogy rather than their depositional environment. We will describe carbonate, siliceous, and carbonaceous rocks in turn.

10.3.1 Carbonate rocks

Limestone, dolomite and marls are carbonate rocks. They represent different depositional environments that share some common characteristics.

10.3.1.1 Limestone

Regardless of whether the source of the calcium carbonate sediments is continental margins or the oceanic floor, they all become limestone through diagenesis. They share the properties of calcite:

- Low hardness. The hardness of calcite (3, see § 5.1.3.1) makes it much softer than silicate rocks such as granites or basalts. This difference results in much less abrasion of drill bits and rotary cutters of tunnel boring machines. Paradoxically, these soft rocks do not cause problems during excavation. The low hardness is compensated by the massive nature of many limestones, which makes them good foundation and construction rock in many cases.
- Solubility in acidic waters. We saw in chapter 7 that limestone often contains karst groundwater because water has been able to dissolve the rock and create large voids. Technically speaking, this karstification can locally reduce the mechanical strength of the subsurface. Collapses of these cavities have destroyed buildings (Fig. 10.14). Limestone aquifers play an important role in the water supply of vast regions of the planet.
- Alkalinity. Calcareous rocks provide an effective chemical buffer to soils, allowing them to react to acidification caused primarily by acid rain. One of the consequences of this effect is the retention of heavy metals in the soil. Infiltrating acidic water is quickly neutralized when it comes in contact with carbonates in the subsurface. This has the effect of immobilizing a majority of these toxic substances and preventing them from flowing into spring waters. Limestone soils are often so rich in carbonate solutions that calcite precipitates on organic matter and prevents its growth; the result is a decrease in fertility. In contrast, limestone or chalk amendments are added to acidic soils used to neutralize their acidity (Chapt. 14).

Limestone from various sedimentary environments can be classified according to their origin and the nature of the components that form the rock matrix, such as the presence of fossils, etc. We can distinguish limestones produced by precipitation of calcite on an immobile substrate and those made by sedimentation of calcareous particles. These particularities justify an additional note about the word "limestone." Here are some examples.

Reef limestone (or coralline limestone)
Through diagenesis, ancient reefs can become massive limestone layers, almost without stratification at the outcrop scale. The thickness of the formation and its structure may vary laterally over less than one kilometer. This type of limestone forms the majority of the cliffs in many mountainous areas, where overhanging limestone beds appear in several places. The

Fig. 10.14 Hazards of building on karstified limestones. A house that has subsided into a sinkhole in Florida, USA. Photo, Earth Sciences Photographic Archive, US Geological Survey.

stability of massive limestone cliffs is generally controlled by a more easily weathered layer at the base, which causes the limestone bench to overhang (Fig. 8.12). The matrix of the reef limestone causes coral fossils (Fig. 10.15), mollusks, and encrusting algae to stand out.

2 cm

Fig. 10.15 Reef limestone in the Jura Mountains, Mesozoic. GEOLEP Photo, G. Grosjean.

Sparry limestone (or **limestone with echinoderms**)

These are biodetrital limestones made up of echinoderm debris (Fig. 10.16). Echinoderm skeletons are made of very well-formed calcite crystals (the word "spar," of Germanic origin, denotes crystals with well-developed faces). When the organism dies, these crystals, which measure several millimeters, break up in the water current and are dispersed in the calcareous mud. When the rock is broken, cleavages of calcite crystals appear as multiple mirrors that reflect light.

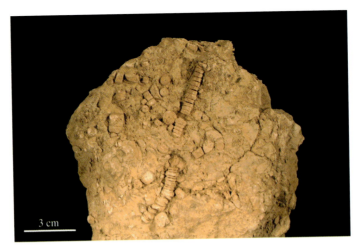

Fig. 10.16 Crinoid stem limestone (a crinoid is an echinoderm that is fastened to the substrate). GEOLEP Photo, G. Grosjean.

Oolitic limestone

This is a pure limestone that contains predominantly fine granules that have precipitated concentrically (Fig. 9.22). Oolites are calcareous precipitates that form around a nucleus of organic or mineral origin. For this reason these rocks are not purely biogenic. Oolitic limestone (like sparry calcite) is less massive than reef limestone; the thickness of the beds ranges from a decimeter to a millimeter, and they are often cross stratified. The cementation around the oolites may be incomplete. In this case, the limestone has high porosity. When hydrocarbons are present, oolitic limestone is a highly prized reservoir rock.

Lithographic limestone

The calcareous muds of the ocean floor can be transformed into a limestone with a very fine and regular matrix (Fig. 10.17). The beds are thin and well stratified. They sometimes alternate with layers of argillaceous limestone or marls.

Chalk

When a coccolith mud undergoes slight diagenesis, it becomes a soft, stratified rock that is friable and porous: chalk (Fig. 10.18). Chalk is found in large sedimentary basins such as the Paris and London basins, and it has given its name to the Cretaceous period. Like other limestones, chalk is subject to karstification and may thus play a significant role as an aquifer in these regions. However, the tunnel boring machines that dug almost entirely through chalk to excavate the tunnel under the English Channel encountered no significant water problems as they drilled rapidly through this rock. Aside from large inflows of seawater near England, the chalk proved to be relatively impermeable, thanks to interstratifications of more argillaceous layers that limited the vertical extent of karstification at depth.

Siliceous limestone

A siliceous fraction present in the calcareous mud (from sponge spicules, for example) can be transformed into quartz even as it remains homogeneously distributed throughout the matrix, unlike flint-bearing limestone (§ 10.3.2). The silica is not apparent to the naked eye, but it makes the rock hard enough to scratch a glass plate. This rock is prized for asphalt road

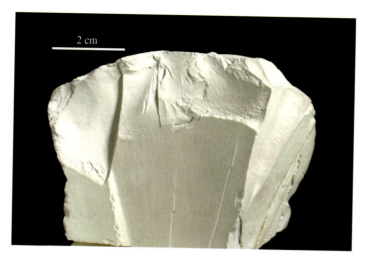

Fig. 10.17 Lithographic limestone. The fine texture and the regularity of the matrix are apparent. GEOLEP Photo, G. Grosjean.

Fig. 10.18 Chalk. The fine-grained matrix contains a fossil lamellibranch. GEOLEP Photo, G. Grosjean.

coatings and as railroad ballast, where a high degree of intergranular friction is required for a long period.

10.3.1.2 Other carbonate rocks

Two other rocks composed principally of carbonate are discussed here because of their importance to civil engineers.

Marls

If a calcareous mud contains a large component (about half) of detrital clay, diagenesis will create a rock that is between a true limestone and a true claystone: marl

(Fig. 10.19). Marl is generally harder than claystone, but keeps some plasticity. It is less soluble in water than pure limestone. When it dissolves, it leaves an argillaceous residue that leads to the development of a soil that is well suited to plant growth. These residues plug up the pores of the rock, which makes the marl impermeable. Marl is used in the production of Portland cement. In nature there is a continuous series between limestone and claystone, ranging from *argillaceous limestone* (5 to 35% clay), to marl (35 to 65% clay) and to *calcareous clay*.

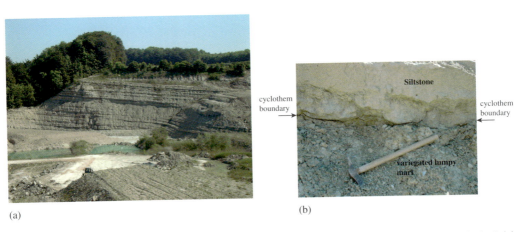

(a)   (b)

Fig. 10.19 Marl pit at Bois-Genoud in the Aquitanian molasse (west of Lausanne). Exploitation of marls for brick making. (a) General view. (b) Detail view. GEOLEP Photo, A. Parriaux.

Secondary dolomite

The reaction between calcareous mud and the magnesium in seawater creates secondary dolomite. This dolomite has a granular texture because it is an aggregation of small dolomite crystals; it is called saccharoidal dolomite (Fig. 10.20). It is porous and weak. Because it is a carbonate, it can be dissolved by water, but it is less affected by karstification than limestone. When digging a tunnel, if a high-pressure water-bearing saccharoidal dolomite unit is

Fig. 10.20 Saccharoidal dolomite from Binntal, Valais, Switzerland. The dark beds are pyrite crystals. GEOLEP Photo, G. Grosjean.

encountered, the rock is fluidized and begins to flow like a paste. This is what happened to the dolomite of the Piora syncline during the reconnaissance work for the Gotthard tunnel through the Alps (Fig. 10.21).

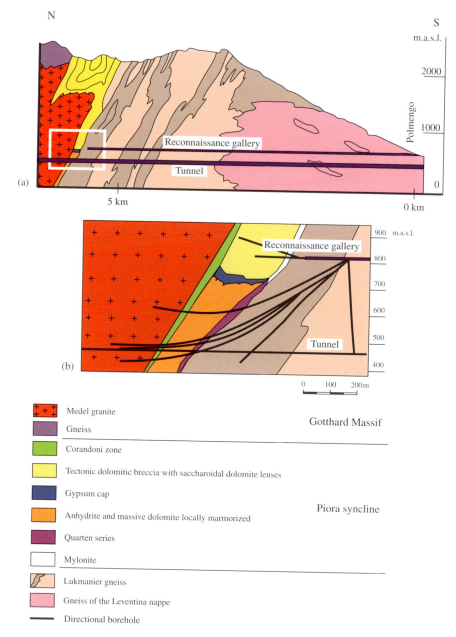

Fig. 10.21 Reconnaissance for the construction of the Alpine St. Gotthard tunnel: Pelmengo Gallery (Ticino). (a) General profile showing the Piora syncline squeezed between two crystalline massifs. (b) Detail of the deviated boreholes drilled from the end of the reconnaissance gallery to reach the sedimentary zone at a level of 300 m below, where the future tunnel will cross the syncline. By chance, the saccharoidal dolomite, still present at the level of the gallery, grades into anhydrite and massive dolomite along the axis of the project. Thus, the drilling of this zone with the TBM in fall 2008 was not difficult. Alptransit-Gotthard-AG document.

10.3.2   Siliceous rocks

Although less abundant than their carbonate counterparts, siliceous biogenic rocks are important to the engineer. Diagenesis of siliceous muds involves minor mineralogical transformations
(Fig. 10.22). The majority of the solid parts of siliceous organisms are made of opal, SiO_2
that is practically amorphous and hydrated. During the first ten million years or so, opal
undergoes a first crystallization that changes it to chalcedony, a microcrystalline form
of hydrated quartz. The chalcedony gradually loses its water and becomes quartz, a stable form of
SiO_2, after about a hundred million years (§ 5.2.1.1). The traces of the original organisms are
mostly wiped out by this transformation.

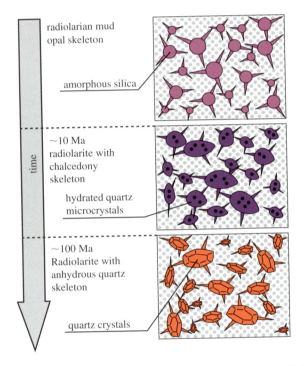

Fig. 10.22 Slight mineralogical transformations. Example of silica in radiolarites.

Radiolarites

The principal rock that comes from siliceous muds is radiolarite (Fig. 10.23). The matrix is
predominantly composed of fine quartz, giving the rock a flat vitreous appearance like polished
glass; traces of iron give it a strong red or green color. It is very hard and causes serious abrasion problems for tools used to drill through it. It occurs in beds that are tens of centimeters
thick, which sometimes alternate with lithographic limestones, as in the Alps.

Flint rocks

While not a dominant component, quartz can play a secondary role in carbonate biogenic
rocks. We saw this above with the example of siliceous limestone. Another case is that of

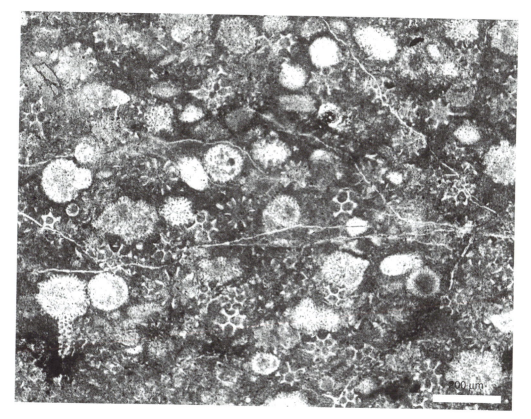

Fig. 10.23 Radiolarite. Thin section showing ghosts of radiolaria, later replaced by quartz. Pelitic and chloritic cement containing carbonaceous debris (Lower Carboniferous) Hersbach (Lower Rhine). P. Baumgartner Photo, University of Lausanne.

flint rocks (Fig. 10.24). During diagenesis chalcedony sometimes migrates through sediments toward points of concentration to form quartz nodules (Fig. 10.25). These are bizarrely shaped nodules that resemble branches or bones.

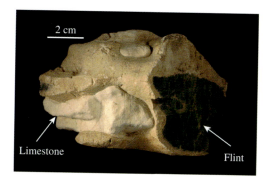

Fig. 10.24 Flint-bearing limestone. GEOLEP Photo, G. Grosjean.

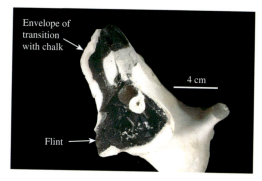

Fig. 10.25 Flint nodule present in chalk from the Paris Basin. GEOLEP Photo, G. Grosjean.

These rocks were the first "quarries" our ancient Paleolithic ancestors (75,000 to 10,000 years BC) used for materials to make tools and weapons because of the sharp way flint breaks. Flint is intensely imbricated in the limestone matrix in the Cretaceous of the Prealps or detached as in the chalk of the Paris Basin.

10.3.3   Fossil fuels

Scientific knowledge of geology in general would not have achieved the present stage of development if man had not had to exploit fossil fuels. Petroleum and coal geology are complex branches of the science that involve a particular type of diagenesis due to the organic component of the raw material. Coal geology and petroleum geology must be discussed separately because of the origin and the mobility of the organic matter.

10.3.3.1   Oil and natural gas

Liquid and gaseous hydrocarbons result from the transformation of primarily marine organic matter produced by the sedimentation of plankton. The original matter is composed of proteins, lipids and carbohydrates. Through the action of anaerobic bacteria in the absence of oxygen, this material is transformed into hydrocarbons via various mechanisms. The long organic chains undergo natural cracking and produce simpler molecules, called **kerogen**. Temperature and pressure conditions control these transformations.

A phase of categenesis (loss of CO_2, H_2O and N_2) facilitates the change from kerogen to hydrocarbons. Conditions favorable to petroleum formation (Fig. 10.26) occur at burial depths between 1500 and 3000 m. This is called the **oil window**.

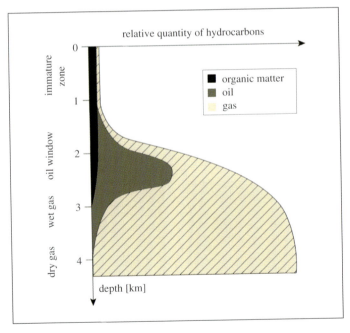

Fig. 10.26 Diagenetic conditions for the formation of petroleum and natural gas (from [288]).

Below a depth of 3000 m, a new stage of mineralization takes over and produces even simpler molecules that are known as natural gas, composed primarily of methane.

A very specific characteristic of hydrocarbons is their great mobility through geologic strata. During or after diagenesis, petroleum and natural gas perform a double migration (Fig. 10.27):

- Primary migration: because of their low specific weight, hydrocarbons have a tendency to move toward shallower rocks. They leave their source rocks and accumulate in a porous rock trapped under an impermeable layer. The host rock is called a ***reservoir rock***.
- Secondary migration: within the reservoir rock, oil and natural gas move toward high points in the stratum. They move toward tectonic structures where these high points occur: anticlines or folds on the margins of salt domes, for example. They are stable in such sites as long as the upper layer acts as a seal. A density-based stratification of the fluids is established: gas is at the top of the formation, followed by petroleum, then groundwater. These points are targets for hydrocarbon prospectors. Explorationists begin by locating favorable tectonic structures in marine sedimentary basins that are likely to produce oil, then they conduct geologic and seismic surveys (see § 4.1.7), and finally they drill into the reservoir layer. If the indications show that petroleum is present, they begin to exploit the deposit by drilling several wells and by decompressing and pumping the hydrocarbons.

[212, 288]

Oil sands are massively exploited by open-pit mining in Alberta, Canada (Lower Cretaceous McMurray Formation). They contain important reserves of very heavy crude oil.

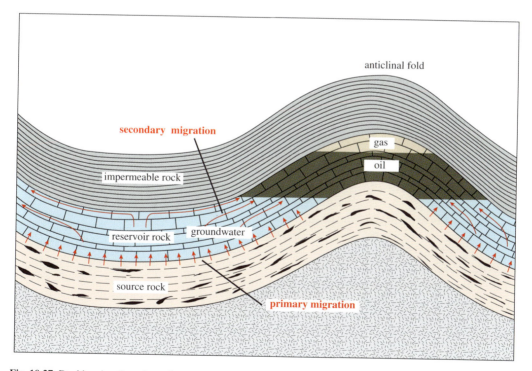

Fig. 10.27 Double migration of petroleum and natural gas toward deposits. See also the animation on the DVD.

Bituminous shales are old clays that are impregnated with petroleum. Their low permeability does not permit the extraction of large quantities of hydrocarbons by pumping. Nevertheless, since these are very thick formations, large hydrocarbon volumes can be stored in the rock. These deposits can be exploited in several ways:

- Mechanical extraction: the extracted bitumen is heated to 500°C; the most volatile hydrocarbons are released as a gas that is sent to a refining facility.
- In-situ treatment: hydraulic fracturing is used to fracture the formation. This is a process of making the rock crack around pressurized drill holes. Water vapor is injected into the fissured formation and then the vapor and hydrocarbons are extracted through boreholes.

When the cap rock of a deposit disappears, some of the petroleum can rise to the surface. In the process, volatile components are released and the deposit is enriched in highly viscous heavy hydrocarbons, creating an *asphalt* deposit. For example, in Val-de-Travers (near Neuchâtel, Switzerland), such deposits in the Cretaceous of the Jura were exploited from 1712 until 1986. These mines are now a tourist attraction.

World reserves of petroleum and natural gas continue to be significant because of offshore exploration and production techniques on the continental shelf (see also Chapt. 14).

10.3.3.2   Coal

In swampy basins, whether marine, brackish or lacustrine, organic matter comes primarily from plants and is very rich in lignin. Plant material decomposes more easily than the marine organisms that create petroleum. With burial and time, anaerobic microbial activity causes the loss of the volatile components in the plant material and a chemical reduction of molecules containing carbon and sulfur. This process is called *carbonization* (transformation of organic matter into coal). The result is an enrichment in carbon that may lead to the formation of graphite crystals.

With increasing pressure and temperature during burial, the following types of coal are produced:

- *Peat* contains about 55% carbon. This is the first stage of the series. Although peat is not coal, strictly speaking, it has long been used as a fuel in swampy areas, particularly in the Jura (§ 8.5.4). It is still used in some countries (Ireland and Poland, for example).
- *Lignite* contains 70 to 75% carbon. It is a poorly lithified coal with low caloric value. It is found in old Quaternary deposits (glacial interstades) and in the Tertiary. This type of coal is present in the alpine molasses. Its thin layers have been exploited for many centuries in the Belmont-sur-Lausanne and Oron regions of western Switzerland (Fig. 10.28). Some mines in this basin were being exploited as recently as World War II (Fig. 10.29).
- *Bituminous coal* has an average carbon content of 85%. This is the most common type of coal. It can be either matte or shiny. We differentiate it from anthracite by the fact that it leaves carbon stains on fingers. For many years, bituminous coal was distilled to supply gas to kitchens in cities. Coke is the residue from burning this fuel. This industry has practically disappeared with the development of natural gas.
- *Anthracite* contains between 92 and 95% carbon. This coal has undergone more of the lithification process, but has not yet reached the stage of metamorphic carbon, or graphite. Anthracite is found along with bituminous coal in the large subsidence basins of the Hercynian orogeny. It is the source of the name of the Carboniferous Period.

Fig. 10.28 Coal layer in the Chattien molasse, Oron (Canton of Vaud, Switzerland). The coal is interstratified in light-colored lacustrine limestones. GEOLEP Photo, G. Grosjean.

Fig. 10.29 Coal miner (end of 19th century). Photo from the Collection of the Cantonal Geological Museum of Lausanne, A. Bersier.

In Europe, carboniferous coals have been exploited on a large scale, but they are not used much today, except in Poland. They were abandoned because of their high sulfur content and the resulting atmospheric pollution (acid rain, respiratory illness), and because of high production costs. However, new less-polluting technologies have been developed to exploit this resource, including low-temperature combustion using limestone powder to neutralize the sulfuric acid, for example. The re-opening of some deposits has been considered. They would be exploited by in-situ combustion under partial oxygenation conditions to extract combustible gases.

10.4   Evaporite rocks

The diagenesis of the evaporite rocks is somewhat different. In some cases, recent sediments are already slightly indurated because of being encrusted on the floor of the basin (§ 9.1.3).

10.4.1   Primary dolomite

Mud composed of dolomite crystals becomes primary dolomite through the process of diagenesis. Primary dolomite occurs in decimeter-thick layers, with a fine matrix that resembles lithographic limestone (Fig. 10.17), although it is less soluble than limestone in acid. In outcrop, it is distinguished from limestone by its calcitic microfissures. These microfissures are dissolved more easily than the rock and thus occur in bas-relief. Primary dolomite is thus very different from secondary dolomite, which has a granular texture. Primary dolomite is also more stable underground. For the construction of hydraulic reservoirs in areas composed primarily of carbonate rocks, dolomite is preferred to limestones because it is less permeable.

10.4.2   Gypsum and anhydrite

In a sedimentary basin, hydrated calcium sulfate crystals (gypsum) form a white, slightly laminated crust. With diagenesis, gypsum loses large amounts of water and becomes anhydrite. The rock becomes very compact and massive, with millimeter-size crystals (Fig. 10.30).

Fig. 10.30 Massive Anhydrite. Underground hydroelectric plant at Bieudron (Valais, Switzerland). The crystalline nature is apparent. GEOLEP Photo, G. Grosjean.

When an anhydrite mass occurs near the surface as a result of tectonics, it reverts to gypsum (Fig. 10.31a). This is accompanied by a swelling of about 60% (Chapt. 13). The hydration can occur over depths of tens of meters. The same transformation can occur around a tunnel where anhydrite is present. This situation creates grave difficulties for an engineer. Swelling is maximum where the anhydrite alternates with beds of clay.

Gypsum and anhydrite are very soluble in water (Chapt. 13). However, dissolution pores occur only in gypsum. In fact, when water enters anhydrite through a fissure, the anhydrite will swell and seal the crack. Anhydrite is thus a tight formation, and is considered a good site for the storage of hazardous waste.

In contrast, a fissure in gypsum grows and forms large cavities. This is karstification (Fig. 10.31b), the same process that occurs in limestone. Gypsum thus makes a very permeable aquifer, but the water it contains is highly mineralized in $CaSO_4$. For this reason, this water cannot be used in public distribution systems. It can be exploited as mineral water if it is free from bacteria (Fig. 10.32).

(a)   (b)

Fig. 10.31 Gypsum habit. (a) Banded gypsum, made of pure (white) beds and beds containing a mixture of gypsum and clay (dark). GEOLEP Photo, J.B. Gabus. (b) A crater (doline) created by the dissolution of gypsum. Valley of Réchy, (Valais, Switzerland). GEOLEP Photo, A. Parriaux.

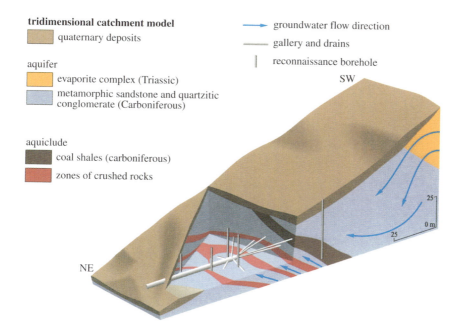

Fig. 10.32 The water collection system at Bouillets, Nendaz, Valais, Switzerland. Aproz mineral water (from [67]).

These sulfate-rich waters cause concrete to undergo alteration. The water reacts with ordinary cements, turns them into a paste, and causes swelling. The SO_4^{2-} anion attacks the hydrated silicates formed as the cement sets up. Aluminum cements should be used for concrete projects in regions where this type of water is present, because they are better able to resist this chemical reaction.

Construction in gypsiferous regions requires extra care. The engineer must carefully verify the continuity of the rock under the proposed building. Building collapses have occurred in these areas. The design and construction of water reservoirs is extremely problematic because of probable underground water losses. Let us also recall that gypsum is exploited for the manufacture of plaster (Chapt. 5).

10.4.3   Halite rocks

When accumulations of halite crystals undergo diagenesis the result is a more or less pure salt-bearing rock. Because of the low density of salt ($\rho = 2.1 \cdot 10^3 \, kg/m^3$) and its ductile behavior, salt rarely remains in beds that are parallel to the surrounding sedimentary rocks. Instead, salt forms ***domes*** or ***diapirs***. The original stratification of the beds is thus highly deformed into unfractured folds. The plasticity of salt causes slow deformation that generally does not cause deformation problems in underground cavities (Fig. 10.33). The plasticity and low hardness of salt make it easy to exploit salt mines. Salt is very soluble (about 360 g/l). For this reason salt is never found in outcrop in a temperate climate. The corollary is that the presence of salt at depth implies the absence of groundwater flow in the rock mass. Old salt mines can be used as geological storage sites for highly toxic waste, like the old mines in Gorleben, Germany.

Fig. 10.33 Chapel carved in salt, the Wieliczka mine, Poland, mined during the 13th century. The columns show no sign of creep. Photo A. Robin.

In many mines, particularly those with complicated geology, salt is extracted by circulating water through boreholes arranged in a radiating pattern around a center. The brine is then extracted and evaporated. This is the case of the Bex mines in the Prealps, which can be visited by tourists.

10.4.4 Potassium rocks

The last stage of the precipitation of seawater is rarely found in sedimentary rocks. Most of the time, the cycle ends before the process goes to completion or the last salts are re-dissolved by a marine invasion. One exception is the potassium salts of Alsace, which are part of an important evaporite series in the Tertiary Rhine area. They were mined until the beginning of this century, primarily for the manufacture of fertilizers. This industry has caused serious chloride pollution of the Rhine.

10.4.5 Cargneules

Cargneules (or *cellular dolomite*) are breccias composed of evaporite grains (dolomites, gypsum) and calcareous cement (Fig. 10.34). In the Alps, these rocks, which are generally located near the thrust plane of large thrust sheets, occur in the Triassic period. At shallow depths, the dissolution of the grains produces a rock with a very characteristic vesicular appearance. The presence of these voids associated with a brecciated structure gives the rock significant porosity that may make them good aquifers. The water is often highly mineralized by the dissolution of the evaporite grains. Cargneules pose significant stability problems for underground projects because of their fragility, their high water content, and the risk of liquefaction that they pose. When a tunnel traverses this type of rock, it is necessary to drain it in advance through boreholes drilled from the working face. The dissolution effect decreases with depth; the rock becomes less porous and has greater mechanical strength.

[1, 94, 251, 292, 293]

Fig. 10.34 Polygenetic cargneule with dolomitic constituents, Nax (Valais, Switzerland). The brecciated texture (angular grains) is readily visible. The gypsum constituents have disappeared, leaving large holes. GEOLEP Photo, G. Grosjean.

PROBLEM 10.1

Propose locations on the planisphere (Fig. 10.35) where the following sediments form today (one location per sediment). Which rock will they form once diagenesis is complete?

1. Carbonate-rich planktonic mud
2. Silica-rich planktonic mud
3. Oolithic sand
4. Quartz sand
5. Gypsum
6. Coral reef
7. Gravel
8. Salt
9. Deep sea red clays
10. Turbidites
11. Marine mud rich in organic matter
12. Pyroclastic deposit
13. Argillaceous mud containing calcium carbonate

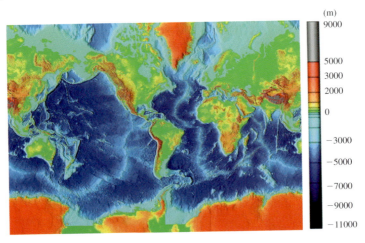

Fig. 10.35 Physical planisphere of continents and oceans. Photo NOAA.

Look the solution on the DVD

PROBLEM 10.2

A hydrocarbon research drilling at the front of a great delta gathered the following information:

Depth (m)		Geological description
From	To	
0	500	Quaternary: gravel and sands
500	1200	Pliocene: sands
1200	4100	Miocene: sands and silts
4100	5400	Oligocene: silts
5400	7900	Eocene and Paleocene: silts and clays
7900	8000	Upper Cretaceous: limestones

Questions

1. Why does the grain size of the sediments decrease with depth?
2. Given the depth at which these sediments are found today and their grain size, determine the average total porosity of each layer (use a figure from the book).
3. For the Oligocene (duration 11 Ma), calculate the average annual rate of sedimentation (thickness of fresh sediments per year) when the sediment was formed.
4. If we assume that the sediment from the Oligocene is composed entirely of quartz grains, what mass is deposed each year by the river, if for the sake of simplicity we suppose that the layer extends over an area of $300 \times 300\,km$ and its thickness is constant?
5. The current average flow rate of the river is $6000\,m^3/s$. If we assume that this flow was the same in the Oligocene and that transport was entirely in suspension, calculate the concentration in particles of the river's water.
6. The base of Miocene is considered as a layer to trap hydrocarbons; in what form are these hydrocarbons found?

Look at the solution on the DVD

Chapter review

Diagenesis transforms diverse sediments into a variety of sedimentary rocks. The three processes (compaction, cementation and slight mineralogical modification) strengthen the original geological material to varying degrees. Certain rocks become very strong. Others remain relatively weak, either because their constituent minerals are weak or because the period of diagenesis has been too brief. As a result, they are soft rocks. The consequences of diagenetic variability are significant for engineering applications. It is important to remember that all rocks formed by diagenesis can revert to the status of sediments through the reverse mechanism: weathering (Chapt. 13).

11 Metamorphism

In the previous chapter we showed how an unconsolidated sediment becomes an indurated rock. The principal agent in this transformation is the weight of younger sediments that have accumulated on top of the sediment. This burial is limited to a depth of a few kilometers, since diagenesis involves only slight mineralogical changes. In chapter 6, we saw that at a depth of several tens of kilometers in the Earth's crust, geological material undergoes partial melting and is thus transformed into magma. What occurs between these two extremes?

Metamorphism is a group of processes that affects a rock in the solid state; it does not involve a transformation to a magma. As a result of intense physical and chemical modifications, the mineralogical composition of the rock, the *texture* (crystal arrangement) and the *structure* (geometry at the outcrop scale) are profoundly affected. The principal factors that control metamorphism are pressure, temperature and fluid activity; they are interdependent. The modification processes are more effective if they occur over a long period of time.

What geologic circumstances cause metamorphism? Orogenesis is the primary cause. Let us examine how pressure and temperature conditions of an elementary volume of rock in the continental crust can evolve during a collision of tectonic plates (Fig. 11.1).

The first phase of orogenesis buries some parts of the lithosphere to a depth of several tens of kilometers, causing a pressure and temperature increase. This is called *prograde*

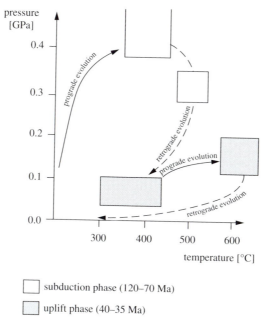

☐ subduction phase (120–70 Ma)

▨ uplift phase (40–35 Ma)

Fig. 11.1 Evolution of metamorphic conditions in the course of a cycle of orogenesis and erosion. Example of the Alps between Visp and Locarno (from [50]).

metamorphism. Thus it is easy to understand why metamorphism is closely related to tectonics (Chapt. 12).

What happens in the opposite case, as in the erosion of a mountain chain, for example? In discussing the theory of isostasy (see § 4.2.5.2), we noted that the wearing down of relief leads to a rebound of deep rocks toward the surface. As a result, our elementary volume of rock may experience a decrease of pressure and temperature: this is *retrograde metamorphism*.

Metamorphism creates a wide variety of rocks. Pressure and temperature conditions and geochemical fluxes are highly variable depending on their geologic settings and all types of rocks can be metamorphosed. We will concentrate here on the most common metamorphic conditions, that is, prograde metamorphism.

11.1   Transformation processes

Like diagenesis, metamorphism is the result of several mechanisms that act in parallel, but which may not all be active at the same time. We will discuss the three principal processes.

11.1.1   Mineralogical modifications

As a result of changes in physical conditions in the environment, such as *geostatic pressure* (equal to the weight of a column of sediments that covers the rock) and temperature, a mineral can become thermodynamically unstable; it can then be changed into a new mineral that is stable under the new conditions. In this process, it can also react with other mineral present in the rock or with fluids in the rock. For example, clay minerals are transformed into micas; a quartz-clay mixture will produce feldspars (Fig. 11.2); a clay-calcite mixture forms amphiboles or pyroxenes.

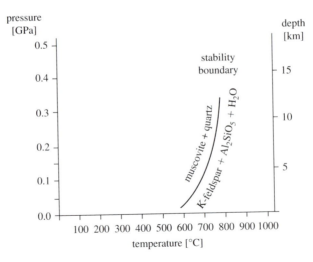

Fig. 11.2 Mineralogical transformation of a mixture of mica and quartz (from [45]). With increasing pressure and temperature, it becomes a potassium feldspar and an aluminum silicate. For the crystalline form of the aluminum silicate, see figure 11.9.

In addition, small crystals (detrital quartz, biogenic calcite) are replaced by new, larger ones. The internal structures of sedimentary rocks, such as fossils or sedimentary structures, are partially or completely obliterated by recrystallization (Fig. 11.3).

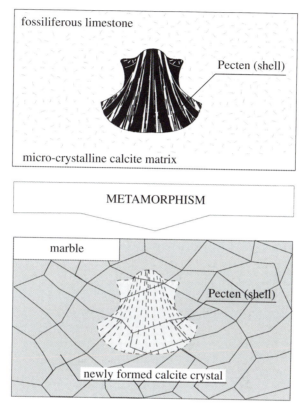

Fig. 11.3 Recrystallization obliterates the internal structure of a rock. The outline in the marble shows where the fossil was. See also figure 11.20.

We will see later how these diverse mineralogical changes produce new rocks whose composition depends on their material of origin.

On the scale of constituents within the rock, crystals can be re-oriented by pressure solution. Let us examine a layered mineral, for example, a mica, which is oriented parallel to the dominant pressure. The material is slowly redistributed: the atoms of a crystal that face the maximal pressure have a tendency to migrate into the pore fluid of the rock and recrystallize on the sides of the mineral, in the direction of the least pressure (Fig. 11.4). In this way, granite, which contains micas oriented in all directions, changes over time to a gneiss in which all the micas are parallel (Fig. 11.5). This process should not be confused with the re-orientation of minerals by a mechanical lamination of the rock (§ 11.1.3).

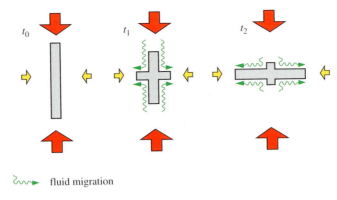

fluid migration

Fig. 11.4 Recrystallization of a mineral subjected to an anisotropic stress field. Example of a mica originally oriented in the direction of major stress (red arrows). Minor stress in yellow.

Fig. 11.5 Difference in texture between a granite (top) and a gneiss (bottom), shown by the reorientation of micas and feldspars. GEOLEP Photo, G. Grosjean.

11.1.2 Chemical modifications

The mineralogical modifications discussed above assume a chemically closed environment (the transformations are called "*isochemical*"). However, this is not always the case: the chemical balance of the environment can be changed by contributions from a neighboring environment or by the loss of chemicals. In this case, we talk of ***metasomatism***. This involves volatile substances such as H_2O, CO_2, F, Cl, and B. For example, the metamorphism of carbonates in the presence of silicates produces a reaction that leads to the crystallization of calcium silicates (marble with minerals) accompanied by the release of carbon dioxide, which can rise to the surface and create carbon-dioxide-rich spring water, as at Lower Engadin (eastern Swiss Alps) (Fig. 11.6).

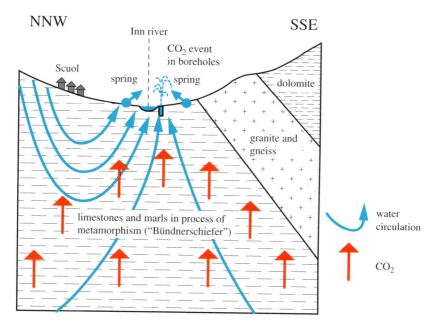

Fig. 11.6 Surface effects of deep metamorphism of carbonate rocks that contain impurities (for example clays or quartz-rich silt): Carbon-dioxide rich water of Lower Engadin (Eastern Swiss Alps) Graubünden. The carbon dioxide from the reaction with silicates is dissolved in the groundwater. The water becomes acidic and easily dissolves the surrounding rocks. The springs are thus highly mineralized and saturated with CO_2. Water spurted to a height of about ten meters from exploration boreholes due to the violent release that squirted the groundwater as a gas-liquid emulsion (the air-lift phenomenon). These boreholes were plugged naturally by the crystallization of calcite (from [28]).

11.1.3   Mechanical modifications

Mechanical stresses in the subsurface are often anisotropic due to their large-scale tectonic setting (see § 12.1.2). On a smaller scale, the core of a concentric fold is a location where a horizontal stress that exceeds a vertical stress. On the external part of this fold, it is just the opposite, and the horizontal stresses may even be tensile (see § 12.4.2).

When one stress dominates the others, for one of the reasons discussed above, there is a very slow lamination effect of the rock matrix, which causes dislocations of grains or minerals; this occurs at the center of zones of preferential deformation (Fig. 11.7). The planar or elongate minerals have a tendency to orient themselves parallel to each other in the direction of the extension. Inner transformations of the rock produce an anisotropic texture. However, it is also possible to create massive rocks such as gneiss and marble.

Various terms are used to describe the anisotropic nature of metamorphic rocks, but their definitions are rather ambiguous. The term *foliation* is used to characterize intensely metamorphosed rocks that have an alternation of layers of different mineralogy (quartzo-feldspathic layers and micaceous layers in gneisses). When the original rock is sedimentary, the resulting foliation is generally not parallel to the stratification; it is often perpendicular to the dominant stress in the rock mass. The term *schistosity* has several meanings. It can be used to describe the layered appearance of rocks of lower metamorphic grade such as slates and chlorite schists. The rock matrix here is finer and more homogeneous than in foliated rocks. In this case the schistosity may be parallel or oblique to the stratification in the original rock. The term

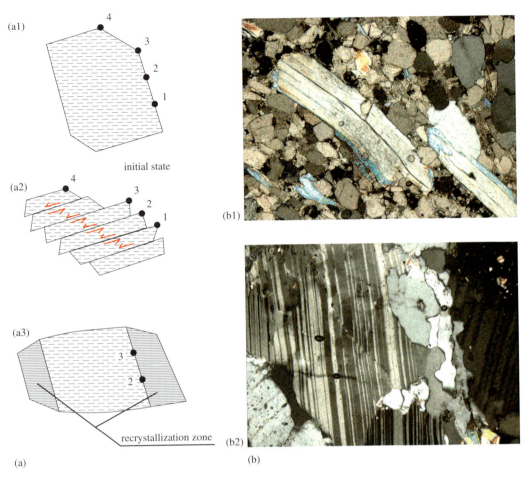

Fig. 11.7 Internal transformation of rocks due to metamorphic processes. (a) Comparison between mechanical deformation and recrystallization: (a1) Mineral prior to deformation. (a2) The mineral is deformed along shear planes; the four points are simply shifted. (a3) The material along the top and the bottom of the mineral is recrystallized on its edges; only the central points remain visible. (b) Thin sections showing the internal deformation of a crystal (width of the photos: 1.14 mm): (b1) Deformation of a mica crystal, bending of the crystal (dolomite, Val Piora, Ticino, Switzerland). (b2) Internal deformation of plagioclase. The twins are shifted with respect to each other (garnet gneiss from the Tremola series, Val Piora, Ticino, Switzerland). S. Schmidt Photos, University of Geneva.

schistosity is also used to describe the reorientation by recrystallization characteristic of rocks in parallel folds (See § 12.4.3). The ambiguities in these definitions can be explained by the close relationship between the different processes that cause recrystallization. In fact, metamorphism is generally accompanied by tectonic deformation at depth; the two phenomena act together to make a rock anisotropic.

It should be noted that the term *schist* is also used sometimes in argillaceous sedimentary rocks to emphasize a laminated structure resulting from compaction (for example, the schist-sandstone flyschs). This usage should be avoided.

11.2 Types of metamorphism

Here we will emphasize the physical variables of pressure and temperature. Using an X-Y diagram, we can define three principal categories of metamorphism (Fig. 11.8): *regional metamorphism* (thermodynamic), *contact metamorphism* (thermal) and *cataclastic metamorphism* (dynamic metamorphism).

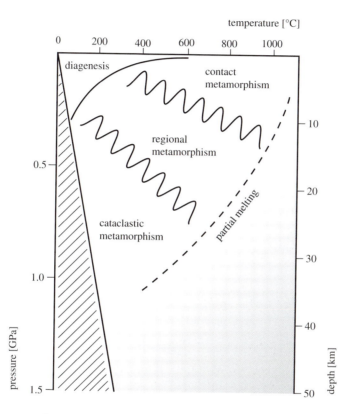

Fig. 11.8 The three types of metamorphism. The hachured area represents conditions that do not occur on Earth (from [273]).

Research shows that one initial material can produce different minerals depending on environmental conditions. An excellent example is the aluminum silicate Al_2SiO_5 (Fig. 11.9). Depending on the relative values of pressure and temperature, there are three different mineralogical forms: andalusite (orthorhombic), sillimanite (orthorhombic) and kyanite (triclinic). The presence of one of these minerals in a rock is used to determine three degrees of metamorphism, called "*zones*" (Fig. 11.10). Other zones can be defined on the basis of minerals that reflect environmental conditions.

In addition, metamorphic conditions are determined by the assemblage of minerals present in rocks after transformation; these assemblages define different *facies*. For example, the "greenschist" facies is characterized by the presence of chlorite, albite, epidote, and actinolite.

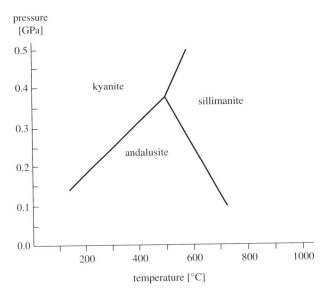

Fig. 11.9 Three different mineral forms of the same substance: aluminum silicate (Al_2SiO_5) as a function of metamorphic conditions (from [129]).

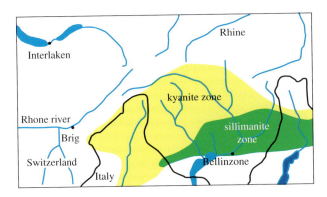

Fig. 11.10 Zones of Alpine metamorphism determined by the mineral form of aluminum silicates found in the rocks (Fig. 11.9) (from [102]).

11.2.1   Regional metamorphism

Regional or general metamorphism is the principal type. It is the area of increasing pressure and temperature shown on figure 11.8. Metamorphic intensity increases from diagenesis (near the origin of the axes) to partial melting and magmatism.

This type of metamorphism affects very large areas, larger than areas of magmatic intrusion or tectonic fractures. It occurs in subduction zones and in orogenesis (Fig 11.11). The vast majority of metamorphic rocks are created in regional metamorphism.

11.2.2   Contact metamorphism

Figure 11.8 shows contact metamorphism near the temperature axes. This means that it is essentially thermal metamorphism. Such conditions are found at depth in contact with magma chambers, or closer to the surface near volcanic chimneys or dikes (Fig. 11.11); it is a local phenomenon (up to several kilometers). This metamorphism creates a contact aureole of rocks that are increasingly metamorphosed closer to the intrusive rock (Fig. 11.12).

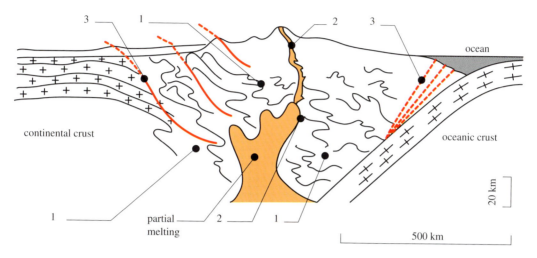

1   regional metamorphism, zone subjected to orogenesis and subduction zones
2   contact metamorphism, aureole around a magmatic body due to heat flow
3   cataclastic metamorphism, fault and overthrust zone

Fig. 11.11 The three types of metamorphism in a subduction zone.

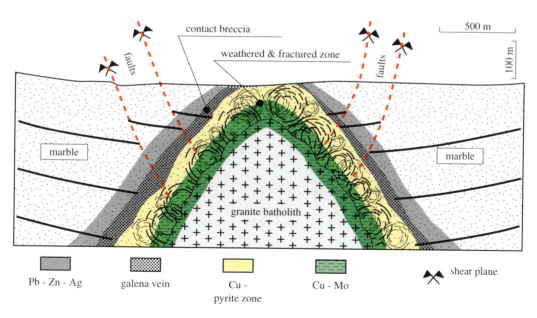

Fig. 11.12 Metalliferous deposit created by contact metamorphism. The hot granitic batholith metamorphosed the surrounding rock, here a limestone that becomes marble. The various isotherms around the intrusion are identified by a mineral zonation. Copper occurs near the intrusion, other metals on the periphery. Schematic diagram of a Bingham-type (Utah) porphyry copper.

High-temperature magmatic fluids cause metasomatism as they infiltrate into the surrounding rocks. A typical case is a granitic intrusion into a Mg-bearing limestone rock. This creates a *skarn*, a marble that contains silicate minerals. Fluids from the magma enrich the surrounding rock in major elements (Si, Al and Fe) and minor elements (Cl, F, and B). This process may form rare minerals (garnet, pyroxene, and calcic amphibole), and chemical reactions may produce exploitable *polymetallic deposits*.

11.2.3   Cataclastic metamorphism

Intense mechanical pressures can exist within tectonic fractures. They can deform the rock and transform it, even at low temperature, that is, at a relatively shallow depth (dynamic metamorphism, Fig. 11.8). These pressures can cause crushing, or *cataclasis*, of the original rock, and reduce it to a very fine crystal matrix (Fig. 11.13). The resulting material is a poorly aggregated debris (for example, *kakirites*) or it can become a new indurated rock through recrystallization (*mylonite*).

Fig. 11.13 Cataclastic texture, tectonic micro-breccia. Coarser elements are caught in a very fine matrix with gneissic structure. Cataclastic metagranodiorite of the Arolla series, Dent Blanche Nappe, Valais, Switzerland. Photo from the J. Bertrand Collection, University of Geneva.

Cataclastic rocks of the kakirite type have to be handled delicately when they are encountered in underground engineering projects (Fig. 11.14): It is not uncommon to suddenly penetrate a crushed zone when digging a tunnel within a well-mannered rock massif. This situation can cause collapses of the face or the roof, an earthflow, or a blockage of the worksite that can last several months.

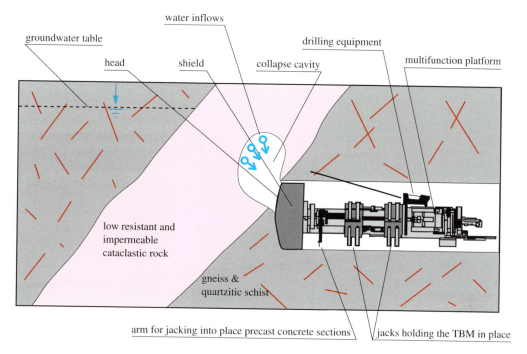

groundwater table

water inflows

drilling equipment

multifunction platform

head   shield   collapse cavity

low resistant and
impermeable
cataclastic rock

gneiss &
quartzitic schist

arm for jacking into place precast concrete sections   jacks holding the TBM in place

Fig. 11.14 A tunnel boring machine (TBM) excavating through cataclastic rocks; case of the Cleuson-Dixence gallery, Valais Alps, Switzerland. The low strength of these rocks, coupled with groundwater caused a collapse that blocked the machine (from [37]). An animation on the DVD shows videos of a TBM in its different operation phases.

PROBLEM 11.1

The history of alpine metamorphism along a profile through the heart of the chain (Visp-Lugano) was reconstituted by Colombi (1988). The author identifies two metamorphic cycles described in figure 11.1.

Questions

1. Transfer the six stages of these cycles onto the general metamorphism pressure-temperature diagram to conclude which types of metamorphism are concerned by this alpine history.
2. Calculate the geothermic gradient for each stage. What can you conclude from your calculations?
3. How can we explain the important difference between metamorphism conditions of the subduction and uplift phases?
4. If the series undergoing the metamorphism in question contains silica and aluminum, what mineral may form during the paroxysm stage of the subduction phase?

Look at the solution on the DVD

11.3   Principal metamorphic rocks and their properties

The variety and complexity of metamorphic processes and the diverse origins of initial rocks combine to generate a significant number of metamorphic rocks. The classification and nomenclature of these rocks are also complex.

Here we present the principal rocks that result from regional metamorphism. This requires us to define a series of sequences showing the gradual transformation of original material into rock as a result of increasing metamorphism (with the exception of the ultramafic sequence). We will consider six sequences in this book.

The different rocks created from these sequences are summarized in figure 11.15.

metamorphic sequences	initial rock	metamorphism intensity		
		weak	medium	strong
PELITIC	MUDSTONE	slate	chlorite schist	micaschist	paragneiss
QUARTZO-FELDSPATHIC	SILICEOUS SANDSTONE	quartzite		
	GRANITE	orthogneiss		
CARBONATE	LIMESTONE DOLOMITE	marble		
CALCAREOUS-PELITIC	MARLS	calc-schist		calcic amphibolite
MAFIC	GABBRO	aluminous amphibolite		
ULTRAMAFIC	PERIDOTITE	serpentinite*		

* retrograde metamorphism

Fig. 11.15 The principal metamorphic rocks in the six sequences of regional metamorphism.

11.3.1 Pelitic sequence

This sequence is treated first because it is particularly sensitive to metamorphic intensity. The starting material is a mixture of clay and quartz particles (for example, a silty mudstone, or shale). We thus have a chemical stock composed of Si, Al and various cations. A series of increasing pressure and temperature stages will produce rocks that grade into gneiss at the end of the sequence.

Slate

This mineralogical transformation is confined to clays and causes the formation of small embryonic white micas called *sericite*. Mica gives the rock a lustrous appearance. The rock does not appear to be crystalline. It breaks up into thin plates, is often black in color (slate), and is perfectly suitable as roofing material (Fig. 11.16).

In fact, the clays are sufficiently transformed so that schist resists freezing and water content variations for a period of several centuries. Over longer periods of time, the rock breaks up and weathers; this is why it is rarely found in outcrop.

Chlorite schist

The transformation of clays causes chlorite, a green phyllosilicate, to appear. The texture of the rock is very layered, although it does not have continuous quartz layers. This rock is sensitive to surface weathering and tends to break up into small pieces of foliated debris.

Fig. 11.16 Penrhyn Slate Company quarry, Middle Granville, NY, 1890. Illustration of the schistosity of roofing slate. Photo from Earth Sciences Photographic Archive, U.S. Geological Survey.

Mica schist

After the chlorite stage, the transformation of phyllosilicates leads to the formation of micas (muscovite and biotite). Quartz begins to recrystallize to form discontinuous layers. The rock structure remains finely schistose, but it has a slightly coarser grain size (minerals can be identified by the naked eye) and becomes stronger due to the quartz. Mica schists (Fig. 11.17) are slightly more resistant than chlorite schists, but they are still easily broken up during weathering.

2 cm

Fig. 11.17 Mica schist. Note: Chlorite schist has a similar appearance, with chlorite replacing the black or white mica. GEOLEP Photo, G. Grosjean.

Gneiss

In gneisses, the recrystallization is not limited to quartz, and feldspars begin to occur. With quartz, they form continuous beds of light-colored minerals separated from layers of phyllosilicates (foliation). These rocks are still very anisotropic, but they are strong and hard due to the quartzo-feldspathic framework. Gneiss is a good construction rock when it is massive. The gneiss in this sequence is called ***paragneiss***, to distinguish it from the gneiss produced from granite (orthogneiss). This distinction is often difficult to make.

11.3.2   Quartzo-feldspathic sequence

This sequence includes various silica-rich original rocks of any origin. In fact, there are two large families:

- detrital non-argillaceous and non-carbonate rocks: sandstones and conglomerates with siliceous cement; through metamorphosis they become quartzites;
- felsic magmatic rocks that become gneiss.

Quartzite

Quartzites are produced by the metamorphism of sandstones that contain quartz grains and siliceous cement. Even at low pressures and temperatures, the recrystallization of quartz grains and cement erases their different origins and their granular structure. The recrystallization process forms a monomineral homogeneous rock.

With increasing metamorphic intensity, quartzite ceases to evolve except that quartz grains continue to grow in size and ordinary quartz can be replaced by high temperature varieties. It then becomes difficult to distinguish between quartzite and a magmatic vein of quartz.

Quartzite is white, beige or green in color due to the presence of phyllosilicates. In the Alps, the Permo-Triassic quartzites occur in thick (tens of meters) beds that often contain sericite (small muscovite crystals) and abundant pyrite. In outcrop, the oxidation of pyrite causes a brown *patina* (rock surface exposed to weathering) to form on the rock surface. These rocks resist erosion during detrital transport (Fig. 11.18) because of the hardness of quartz and their low degree of physico-chemical weathering. For this reason, alluvium is highly enriched in quartzite as compared to its proportion in the watershed. The hardness of quartz causes the abrasion of drilling tools (boreholes, tunnel boring machines, etc.). In areas of tectonic crushing, quartzite beds can be weakened by intense micro-fracturing. Under these circumstances and in the presence of groundwater, quartzite can become a flowing rock that is very unstable in galleries.

Fig. 11.18 Quartzite. The pebble has a perfectly elliptical shape. A hammer blow shows that it has a monomineral crystalline texture. GEOLEP Photo, G. Grosjean.

Gneiss

We have seen that gneiss can be produced by the intense metamorphism of the pelitic sequence. It can also be produced from felsic magmatic rocks in the quartzo-feldspathic sequence. The constituent minerals of the rock undergo reorientation, particularly by the recrystallization of micas. Foliation develops, marked by a succession of quartzo-feldspathic layers and mica layers (biotite or muscovite). This type of gneiss (called *orthogneiss*) is some-what similar to the gneiss obtained from sedimentary rocks. Orthogneiss differs from rocks of magmatic origin because of the anisotropy of its properties; for example, it is less resistant to a shearing stress that is parallel to its foliation than to one that is perpendicular to its foliation. It is thus potentially less stable in walls. However, orthogneiss usually has good technical proper-ties. It is an excellent building stone (Fig. 11.19).

Fig. 11.19 When gneiss is finely schistose, it can be used as roofing material. However, the slabs are considerably heavier than slate. Example of the hamlet of Curogna (Cugnasco) Tessin, Switzerland. GEOLEP Photo, L. Cortesi.

11.3.3   Carbonate sequence

Limestone and dolomite are the two types of original rocks in this sequence. Depending on the purity of the carbonates, slightly different metamorphic rocks are produced.

Marble

At low to medium metamorphic intensities, limestone is transformed into marble by recrys-tallization of the calcite microcrystals. If the original limestone is pure, the result is a homoge-neous massive white crystalline rock that can be used for sculptures (Fig 11.20a). If the original rock contains iron-rich horizons, the result is a color-banded marble. The degree of oxidation of the iron determines the color of the rock: it may be green due to reduced iron, red from oxi-dized iron, and ochre and beige from iron combined with hydroxides. When the magnesium content of the carbonates is high, *dolomitic marble* is formed. If the limestone contains argilla-ceous or quartz impurities (within the matrix or in the sedimentary series) and if metamorphism is intense, there can be a mineralogical modification: CO_2 is released and calcic silicate miner-als are created. This type of marble is called *calc-silicate marble*. This marble occurs in layers of variable thickness that behave in a ductile manner at high pressure. Their internal structure is often finely folded.

When marble occurs in massive layers, they can be exploited as ornamental stone, as in the Carrara quarries in Italy, for example (Fig. 11.20b and 11.21).

(a)   (b)

Fig. 11.20 Marble. a) sample of a very pure marble. GEOLEP Photo, D. Marques. b) Marble quarry, Carrara, Italy. Open pit quarry, cutting and removal of blocks. GEOLEP Photo, G. Franciosi.

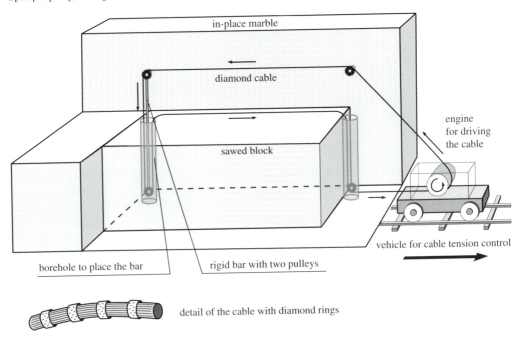

Fig. 11.21 Method of cutting marble blocks using diamond cables.

Marble's hardness is the same as calcite (3, see § 5.1.3.1) which makes it easy to work. Like limestone, marble is soluble in acidic water. It is thus subject to karstification and it can be an aquifer. The term marble, like granite, is also used in a very different sense by the construction industry, where it denotes a decorative polished rock, often without regard to its composition. Some artificial aggregates of colored rock are also called "marble."

11.3.4   Calcareous-pelitic sequence

Marls, which are a mixture of limestone and clay, are the point of departure for this sequence.

Calc-schist

These are finely-layered and schistose rocks that are slightly metamorphosed (Fig. 11.22). Calcite has recrystallized as in marble, creating competent layers. These layers are separated from schist layers, which come from the abundant clay component. Overall, this is a weak rock, very anisotropic and slightly permeable.

Fig. 11.22 The Col d'Olen calc-schist (Italian slope of Mont-Rose). GEOLEP Photo, G. Franciosi.

Amphibolite

As metamorphism intensifies, the rock loses its carbon as CO_2 is released. Feldspars and long prisms of calcic amphiboles crystallize. Amphibolite is composed of a fine alternation of amphibole layers and feldspar layers. It generally has a schistose structure.

11.3.5   Mafic sequence

As in the case of granite, mafic magmatic rocks are affected by prograde metamorphism. This process recrystallizes and orients the minerals without noticeably changing the mineralogy of the original matrix. The resulting rocks thus have a preferential orientation of ferromagnesian minerals, which are elongated like amphibole prisms. The structure of the rock also is more anisotropic.

Amphibolite

When subjected to medium to intense metamorphism, mafic rocks have a banded texture, made up of an alternation of plagioclase layers that are rich in calcium and layers of aluminous green amphibole (Fig. 11.23). This rock is more massive than the amphibole of the calcareous-pelitic sequence.

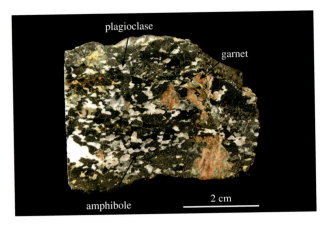

Fig. 11.23 Amphibolite, Swiss Alps. GEOLEP Photo, G. Grosjean.

It should be noted that mafic rocks are susceptible to retrograde metamorphism. Folded in mountain chains, they are transformed into metamorphic rocks that are rich in albite (white spots in the rock, a few millimeters in diameter), epidote, and chlorite (green matrix). They are called *prasinites* (Fig. 11.24).

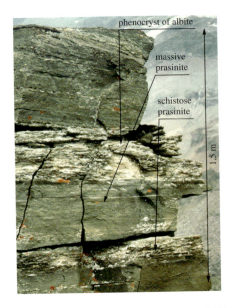

Fig. 11.24 Prasinite, Corno Rosso, Italian slope of Mont-Rose (GEOLEP Photo, G. Franciosi).

These rocks are common in the central part of the Alps. They are slightly anisotropic because of the occurrence of phyllosilicates. They are susceptible to surface alteration.

11.3.6 Ultramafic sequence

Ultramafic rocks come from the mantle, and as a result they are in great disequilibrium when they are folded into the upper parts of mountain chains. In that environment, they are subject to retrograde metamorphism.

Serpentinites

Rocks of deep origin composed predominantly of ferromagnesian silicates (olivine, in particular) undergo an almost complete transformation of their components as they move upward into the upper crust. These ferromagnesians silicates are very unstable at low pressure and temperature, and thus they are transformed by hydration to serpentine (§ 5.2.1.3). The resulting rock is a deep green, almost black in the dry state (Fig. 11.25) with a randomly-oriented texture. In underground construction, serpentines often have low strength due to the presence of microsurfaces polished by tectonic shearing that weaken the rock mass. In the Alps, they are found in thick series in tectonic thrust sheets within the mountain chain. They are the product of the metamorphism of ultramafic and mafic rocks from the oceanic ridges of the Jurassic Tethys.

Fig. 11.25 Highly tectonized serpentinite, Monte Gazzo, north of Genoa, Italy. GEOLEP Photo, G. Grosjean.

[13, 175, 188, 195, 264, 315, 317]

11.4   Identification of magmatic, sedimentary and metamorphic rocks

After defining the three large classes of rocks, it is necessary to discuss how to identify them. Geologists use a wide variety of methods and techniques that are beyond the scope of this book. In spite of that, attentive observation and easy-to-apply field techniques make it possible to identify most rocks when they exhibit their habitual facies (Fig. 11.26 and 11.27).

Equipment for petrologic identification consists of a steel scraper and a glass plate to determine the hardness of the minerals in the rock (extended to rocks by analogy), dilute HCl to characterize carbonate rocks, and a hand lens to identify minerals.

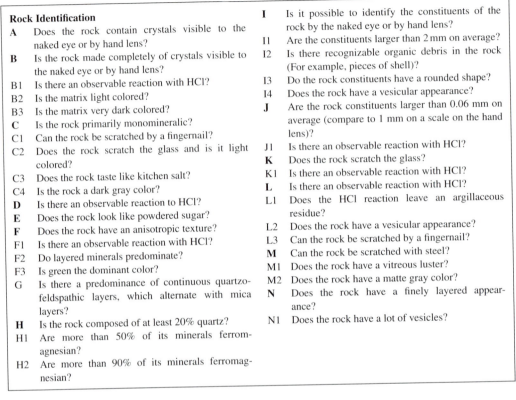

Rock Identification

A Does the rock contain crystals visible to the naked eye or by hand lens?

B Is the rock made completely of crystals visible to the naked eye or by hand lens?

B1 Is there an observable reaction with HCl?

B2 Is the matrix light colored?

B3 Is the matrix very dark colored?

C Is the rock primarily monomineralic?

C1 Can the rock be scratched by a fingernail?

C2 Does the rock scratch the glass and is it light colored?

C3 Does the rock taste like kitchen salt?

C4 Is the rock a dark gray color?

D Is there an observable reaction to HCl?

E Does the rock look like powdered sugar?

F Does the rock have an anisotropic texture?

F1 Is there an observable reaction with HCl?

F2 Do layered minerals predominate?

F3 Is green the dominant color?

G Is there a predominance of continuous quartzo-feldspathic layers, which alternate with mica layers?

H Is the rock composed of at least 20% quartz?

H1 Are more than 50% of its minerals ferromagnesian?

H2 Are more than 90% of its minerals ferromagnesian?

I Is it possible to identify the constituents of the rock by the naked eye or by hand lens?

I1 Are the constituents larger than 2 mm on average?

I2 Is there recognizable organic debris in the rock (For example, pieces of shell)?

I3 Do the rock constituents have a rounded shape?

I4 Does the rock have a vesicular appearance?

J Are the rock constituents larger than 0.06 mm on average (compare to 1 mm on a scale on the hand lens)?

J1 Is there an observable reaction with HCl?

K Does the rock scratch the glass?

K1 Is there an observable reaction with HCl?

L Is there an observable reaction with HCl?

L1 Does the HCl reaction leave an argillaceous residue?

L2 Does the rock have a vesicular appearance?

L3 Can the rock be scratched by a fingernail?

M Can the rock be scratched with steel?

M1 Does the rock have a vitreous luster?

M2 Does the rock have a matte gray color?

N Does the rock have a finely layered appearance?

N1 Does the rock have a lot of vesicles?

Fig. 11.26 List of questions related to the flowchart of figure 11.27.

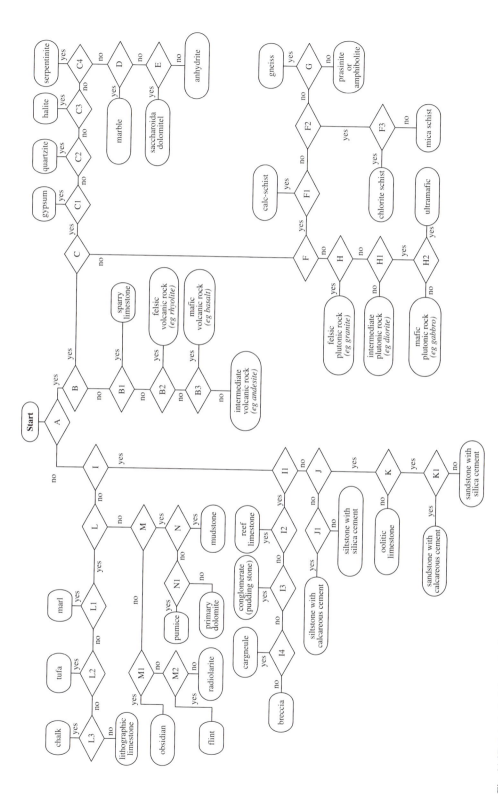

Fig. 11.27 Petrological determination flowchart. By systematically following this flowchart, the reader will find a rock or rock category. At this stage, the reader is advised to verify the likelihood of the result, and if it fails the test, he should try other possible branches of the chart. NB: this simplified determination method is useful for common rocks showing their typical features.

PROBLEM 11.2

Examine the photos of six rocks (Fig. 11.28a and f) and determine what they are with the help of the identification chart of figures 11.26 and 11.27.

Fig. 11.28 Photos of rocks from problem 11.2. GEOLEP collection, photos D. Marques

You also know the following:

Rock a: Scratches glass and reacts with HCI
Rock b: Is not scratched by a fingernail, does not scratch glass, does not taste like salt
Rock c: Does not scratch glass.
Rock d: Does not react to HCI, is not scratched by steel.
Rock f: Does not react to HCl, is scratched by steel.

Look at the solution on the DVD

Chapter review

Metamorphic rocks end our journey through the various categories of rocks that the engineer may encounter in his work. Given the richness of this group of rocks, the following should be kept in mind:

- It is useful to try to identify the original rock to appreciate the exact nature of the metamorphic rock and its structure in the rock massif,
- Although two rocks have the same name, they can be different in terms of their composition and their properties; this may be due to their history over several stages,
- Metamorphic rocks are often anisotropic. An analysis of their stability for an engineering project must consider the rock structure with respect to the geometry of the engineering project,
- The choice of laboratory tests and their interpretation should also take into account this anisotropy.

12 Tectonics

We have repeatedly observed that the Earth is subject to mechanical stresses that deform it in many ways: earthquakes, plate movement and collisions, etc. It is now time to analyze in more detail the stresses that cause these deformations and the ways in which rocks react to these stresses. We will see how we can use a deformed structure to deduce the stresses that created it.

12.1 Mechanical stresses in the subsurface

12.1.1 Physical definition

To understand the physical idea of **stress** within a rock mass, we can consider a solid at equilibrium (Fig. 12.1). All the forces acting on it have a null resultant. If we now cut this solid along any planar area a that passes through point A and if we remove this part, the equilibrium of the forces is disrupted because the removed mass no longer supports the face that remains in place. For the solid to regain its equilibrium, we must exert on the rupture surface a force F that is equivalent to the overall effect of the removed mass.

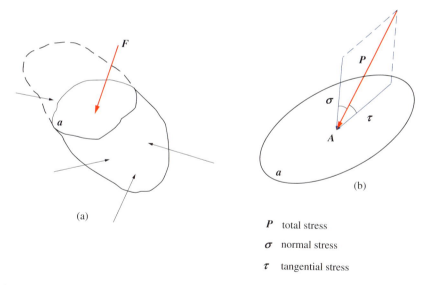

(a)

(b)

P total stress

σ normal stress

τ tangential stress

Fig. 12.1 Force and stress balancing a truncated solid. (a) Force F on the rupture area a. (b) Expressed as stresses.

This force F can also be expressed as the stress that is brought to bear on a surface unit da on which it operates and becomes the stress P at point A on plane a:

$$P = dF/da. \qquad (12.1)$$

On the contrary to pressure, which is a scalar, the mechanical stresses are expressed as vectors whose intensity is given in N/m² (or Pascals). Given the order of magnitude of these stresses, normally the kilopascal (kPa) or the megapascal (MPa) is used. They are equivalent to historically-used units as shown in the following expression:

$$1\,\text{bar} = 1\,\text{kgf/cm}^2 = 100000\,\text{N/m}^2 = 0.1\,\text{MPa} \tag{12.2}$$

To come back to our solid, the stress that is exerted on the surface and restores equilibrium is equivalent to the stress state that existed in the solid before it was cut. It is interesting to identify the stresses acting on surface a that was chosen randomly in the solid. To do this we decompose stress P into a stress normal to the surface (**normal stress** σ) and a tangential stress parallel to the same surface (**tangential stress** τ) (Fig. 12.1). This makes it possible to separate out the compressive and shear stresses. In the case where stress is perpendicular to the cut surface, the tangential stress is null.

In the same way that we have cut our solid, let us now extract a small cube of material, similar to a mini cavern, from an equilibrium point in the subsurface. For convenience we orient this cube in the primary directions of the Earth: x (N-S), y (E-W), and z (altitude). On each face of the cube we exert a stress to maintain the equilibrium of the forces and to keep the mini cavern open. Each face of the cube exerts a normal stress and a tangential stress on the rock body, and conversely. The components normal to the faces will be parallel to the x, y, and z axes. The tangential components are within the planes of the three dominant faces of the cube.

Finally, the total stress can be broken down as follows:

- three normal stresses σ_x, σ_y, σ_z,
- three tangential stresses τ_{xy}, τ_{yz}, τ_{zx},

If we modify the orientation of the cube at this point, while maintaining the same stress state in the rock body, the normal and tangential stresses will change. By drawing the stress vectors for all possible orientations of the cube, we obtain a vector envelope in the shape of an ellipsoid (Fig. 12.2). The three axes of symmetry of this ellipsoid are the directions that bear the **principal stresses**. This means that in planes normal to these principal stresses, the tangential stresses are null. The three principal stresses suffice to determine the ellipsoid and thus to determine the stress status of the rock mass at this point and the stress on any cut face that passes through the point.

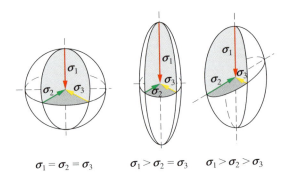

$$\sigma_1 = \sigma_2 = \sigma_3 \qquad \sigma_1 > \sigma_2 = \sigma_3 \qquad \sigma_1 > \sigma_2 > \sigma_3$$

Fig. 12.2 Principal stresses represented by the stress ellipsoid. The **maximum stress** is conventionally designated as σ_1 and the **minimum stress** is σ_3. The **intermediate stress** is σ_2.

In the general case, the ellipsoid can have many shapes, meaning that the stress field is entirely anisotropic. When two of the three stresses are equal, we obtain an ellipsoid of revolution. When the field is isotropic, the ellipsoid is a sphere.

12.1.2 Stress state in geological environments

The mechanical stress state in the subsurface is on a grand scale the resultant of forces linked to the solar system: the attraction of the Sun and the Moon and the centrifugal forces related to the Earth's movements. To them, we must add the stresses produced by the large thermal convection currents that drive plate tectonics and all the phenomena related to them.

On regional and local scales, stresses are caused by:

- The weight of rocks that overlie the reference volume;
- Tectonic forces in tectonically active regions;
- Local erosional history;
- Hillslope effects that move material toward valleys;
- The effects of temporary loads (glaciations, for example);
- The effects of surface and underground construction.

We can identify a series of representative geological conditions that show the distribution of natural stresses (Fig. 12.3). The state of the latter is very important for determining the stresses on the walls of underground projects. Deep tunnels and tectonically active zones are of primary concern (Fig. 12.4).

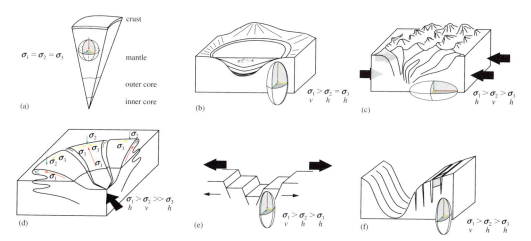

Fig. 12.3 Relationship between the three principal stresses due to geological situations (h = horizontal; v = vertical). (a) At depth (10 km), rocks become plastic. The stress becomes isotropic similar to that of a liquid. This stress is called lithostatic or geostatic. It is equal to the weight of the rock between that depth and the surface. (b) Sedimentary basin: σ_1 is vertical; σ_2 and σ_3 are horizontal and equal. This is the most common case in areas that are tectonically inactive and that have low relief. (c) Simple collision of continental plates: because of the tectonic pressure, the maximum stress is horizontal. (d) Punctual collision creating curved ranges like the Alps and the Himalayas: the horizontal stress normal to the major stress becomes very weak, causing fan-shaped tension ruptures. (e) Fault graben: the maximum stress is vertical; the horizontal stress perpendicular to the trench is the weakest. (f) Release of the sides of an erosional valley: the horizontal stress perpendicular to the valley is weaker than that parallel to it, because of the release.

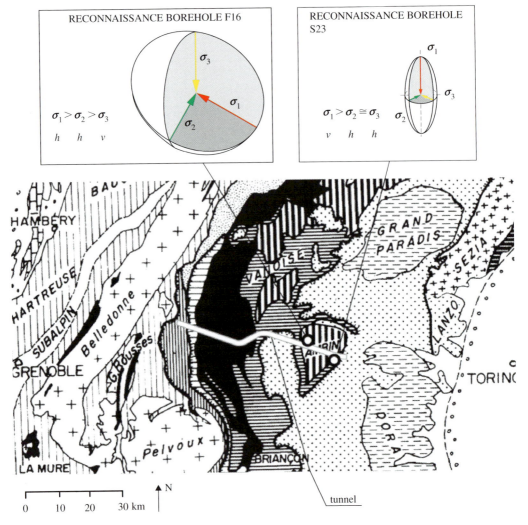

Fig. 12.4 Measurement of natural stresses along the trace of the future trans-Alpine tunnel between France and Italy. In the area of the Ambin massif (France), tests conducted in boreholes show major differences between tectonic units: The principal stress measured in borehole F16 (extreme NW part) of the rock massif is horizontal and typically shows the Alpine pressure; the stresses measured in borehole S23 show that the tectonic pressure had relaxed due to faulting and resembles those of a sedimentary basin (from [177]).

12.2   Stress-deformation relationship

Given a stress state, rocks undergo deformations whose geometry is related to this state. For years, geology has used deformation to determine the present and past tectonic stresses. This analysis is done on various scales: by observing fold geometry and mapping faults in the field, and by studying the deformation of the elements that make up the rocks (stretching of pebbles, deformation of ammonites to oval shapes, etc.) in the laboratory. Rock mechanics, with its development of laboratory tests and associated theories, offers a complementary approach to the stress-deformation relationship.

12.2.1 Laboratory tests

We know from physics that solid material can react in different ways to a mechanical stress. To visualize this relationship, we can use one of the most common tests in soil and rock mechanics: the compression test (first uniaxial, then triaxial test).

12.2.1.1 Uniaxial compression test

The test consists of placing a cylinder of the solid to be studied in a press that applies pressure on the top of the sample (Fig. 12.5). The sample might be a core of rock from a borehole. Comparators make it possible to measure the geometrical variations of the cylinder: the axial shortening and the radial extension. The **modulus of elasticity** [MPa] (or **Young's modulus**) and **Poisson's coefficient** [-] are calculated from these measurements.

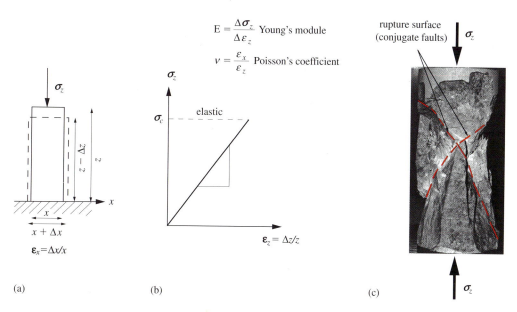

$$E = \frac{\Delta \sigma_z}{\Delta \varepsilon_z} \quad \text{Young's module}$$

$$v = \frac{\varepsilon_x}{\varepsilon_z} \quad \text{Poisson's coefficient}$$

$\varepsilon_x = \Delta x/x$

$\varepsilon_z = \Delta z/z$

(a)   (b)   (c)

Fig. 12.5 Uniaxial compression test. (a) Diagram of the test and measurements. (b) The modulus of elasticity is usually determined by the tangent to the curve $\sigma_z = f(\varepsilon_z)$ at one half of the rupture stress σ_c. (c) Example of a rock core after rupture along conjugate faults.

In the first scenario, the stress is applied in stages of increasing load that make it possible to chart an "increasing" strain-stress relationship (Fig. 12.6). When the stress is released, it can be observed whether the "decreasing" path is identical to the "increasing" one (absence of **hysteresis**) and whether the deformation disappears or not once the stress is totally released. A second test scenario places the solid under a constant stress over a long period, in order to measure any possible variation in deformation over time. These two tests are used to define three broad categories of theoretical solids in stress-deformation-time space (Fig. 12.6):

- **Elastic solid:** deformation totally disappears once the stress is released. The relationship can be linear (the slope is equal to the modulus of elasticity) or non-linear.
- **Plastic solid**: deformation is permanent after the stress is released.

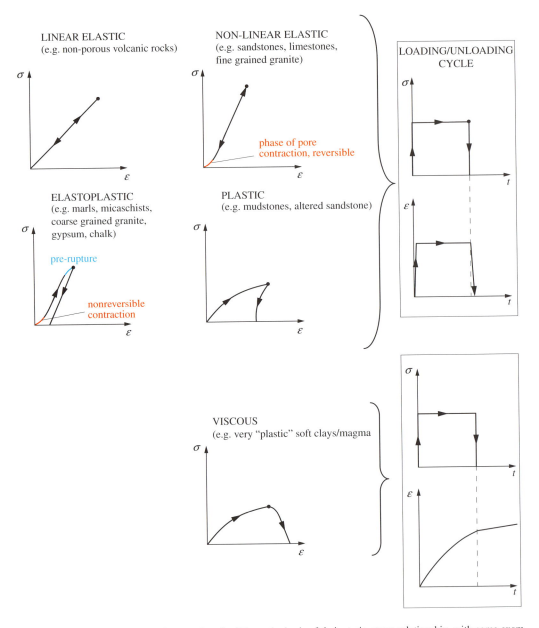

Fig. 12.6 The different theoretical categories of solids on the basis of their strain-stress relationship, with some examples of rocks in each category.

- **Viscous solid**: deformation changes depending on the period of time that the stress is applied, and it does not stabilize; here we speak of the phenomenon of **flow**.

Many rocks belong to a transition category called **elasto-plastic**, often involving an elastic section at the center of the function, surrounded by two plastic extremities.

We must pay attention to the many different meanings of the term "plastic." The meaning used here is the one used in solid mechanics. Geologists use the term to describe a material that deforms by flowing (for example, a plastic clay), which is equivalent to the viscous solids of soil mechanics. In this text, we will use the word plastic with both of these meanings, depending on the context.

If we advance the deformation to the point of rupturing the solid, we can distinguish two types of behavior (Fig. 12.7):

- ***Brittle behavior***: the solid cannot deform significantly without a clear and sudden fracture.
- ***Ductile behavior:*** the solid deforms throughout and in a striking way; rupture is poorly determined in space and time.

The maximum stress a rock can bear is equivalent to the ***uniaxial compressive strength***.

If we wish to analyze the conditions that govern rock rupture in more detail, it is useful to conduct a slightly more sophisticated compression test.

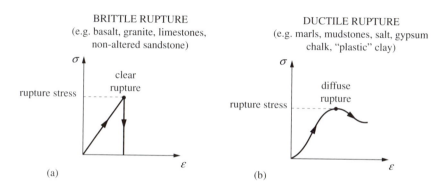

Fig. 12.7 Rupture behavior of rock samples. (a) Brittle rupture. (b) Ductile rupture. An animation on the DVD shows two tests performed in laboratory with the two types of behavior.

12.2.1.2 Triaxial compression test

This test is similar to the first one except that the sample is supported in a radial direction by a non-null stress to approximate subsurface conditions (Fig. 12.8). The major principal stress σ_1 is that of the press. The minor principal stress σ_3 is orthogonal to it. The medium principal stress σ_2 is equal to σ_3 in this test. These two stresses are applied to the sample by a hydrostatic container that surrounds the core. This test represents the particular case of an ellipsoid of revolution with a vertical axis (Fig 12.2, case 2 and Fig. 12.3b).

Let us consider a homogeneous and isotropic rock core. The difference between the axial and lateral stress creates shear effects in the rock mass. Let us imagine a potential oblique rupture plane that makes an angle θ with the axial stress. We know that if σ_1 is greater than σ_3, the stress tends to make the upper part of the core go down, as compared to the lower part. In fact, this compression test is actually a test of shear strength. The stress may cause the core to rupture if the difference of the two principal stresses is too high. But will the rupture really occur along this plane? To respond to this question, it is necessary to analyze how the shear stress is distributed in the core along different possible rupture surfaces.

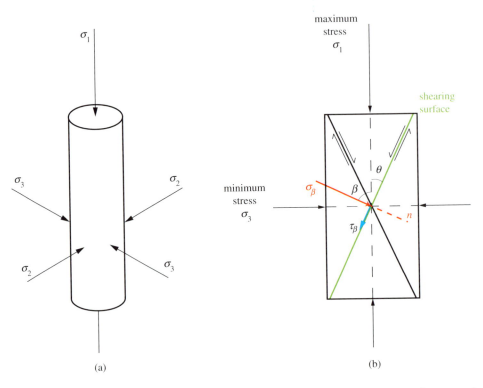

(a)   (b)

Fig. 12.8 Diagram of a triaxial test. Stress σ_2 is equal to σ_3. The deformation measurements are the same as in the uniaxial test. (b) Potential rupture surface by shearing of the core sample with definition of its obliqueness θ and the normal and tangential stresses exerted on this surface.

First of all, to understand the factors that control the shear strength of a joint in a rock (§ 12.3.1), let us observe what happens in a direct shear test when an opposing tensile stress is exerted on the two pieces of rock on either side of the joint (Fig. 12.9). The Navier-Coulomb criteria expresses *shear strength* τ_f of the surface very simply as a linear relationship with the stress σ that affects the two pieces of rock:

$$\tau_f = c + \sigma \cdot \mathrm{tg}\varphi \tag{12.3}$$

where
σ = stress normal to the shear plane
φ = angle of internal friction of the joint, which depends on surface conditions (aggregation, roughness, etc.)
c = cohesion (equivalent to the shear strength when the normal stress is null).
c and φ are intrinsic parameters of the rock.

Note that in soil mechanics the Navier-Coulomb criteria is also valid to characterize shear strength of cohesive soils. If now we return to the triaxial compression test, we will recall the assumption of the homogeneous and isotropic conditions, that is, the rock has the same values of c and of φ throughout the entire core sample and it is the same in all possible shear directions. There is thus a unique variable σ that itself depends on the principal stresses and the obliqueness θ (the angle of the maximal stress on the surface).

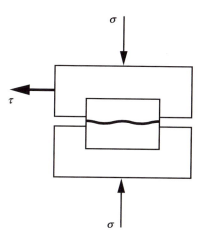

Fig. 12.9 Direct shear test on a rock discontinuity.

To determine which potential rupture plane will effectively cause the rupture of the sample for a given stress state, it is necessary to transpose the principal stresses into the axial system of the potential rupture surface under consideration. In other terms, it is necessary to project the tangential component of the two principal stresses onto the rupture surface and also project the normal components onto the normal to the surface. (Fig. 12.10a).

Stresses σ_1 and σ_3 each exert a shear effect on the potential rupture surface. To determine their combined effect, we cannot add them up in terms of stress; instead we must convert them into forces. To do this, we define an area A, a unit surface on the rupture surface (Fig. 12.10b). Let us designate $\boldsymbol{F}_1$ as the force that acts on area A corresponding to stress σ_1 and $\boldsymbol{F}_3$ as the equivalent of σ_3. For convenience, the potential rupture surface is defined by its pole, or normal. This normal makes an angle β with the maximum stress σ_1. Thus with respect to the obliqueness θ defined above, we have the ratio:

$$\beta = \frac{\pi}{2} - \theta \tag{12.4}$$

Force $\boldsymbol{F}_1$ is thus equal to stress σ_1 multiplied by the projection of area A on the plane normal to σ_1, or $\cos\beta \cdot A$:

$$\boldsymbol{F}_1 = \sigma_1 \cdot \cos\beta \cdot A \tag{12.5}$$

Following the same reasoning, Force $\boldsymbol{F}_3$ is equivalent to:

$$\boldsymbol{F}_3 = \sigma_3 \cdot \sin\beta \cdot A \tag{12.6}$$

The two forces each exert a normal force $\boldsymbol{N}_\beta$ (Fig. 12.10c) and a tangential force $\boldsymbol{T}_\beta$ (Fig. 12.10d) on the rupture surface. The angle β again plays a role, to project these forces onto the referential of the rupture surface. The two components of $\boldsymbol{N}_\beta$ are in the same direction and are additive:

$$\boldsymbol{N}_\beta = \boldsymbol{F}_1 \cdot \cos\beta + \boldsymbol{F}_3 \cdot \sin\beta \tag{12.7}$$

The two components of T_β are opposite and are subtracted:

$$T_\beta = F_1 \cdot \sin \beta - F_3 \cdot \cos \beta \qquad (12.8)$$

To return to the stresses (Fig. 12.10e) we use the expression of the forces, mentioning that this equality is written with a unit area $A = 1$. These equations become:

$$\sigma_\beta = F_1 \cdot \cos \beta + F_3 \cdot \sin \beta \quad \text{(per surface unit)} \qquad (12.9)$$

$$\tau_\beta = F_1 \cdot \sin \beta - F_3 \cdot \cos \beta \quad \text{(per surface unit)} \qquad (12.10)$$

Writing these expressions with the relationship to principal stresses (12.5) and (12.6), we obtain:

$$\sigma_\beta = \sigma_1 \cdot \cos \beta \cdot \cos \beta + \sigma_3 \cdot \sin \beta \cdot \sin \beta = \sigma_1 \cdot \cos^2\beta + \sigma_3 \cdot \sin^2\beta \qquad (12.11)$$

$$\tau_\beta = \sigma_1 \cdot \cos \beta \cdot \sin \beta - \sigma_3 \cdot \sin \beta \cdot \cos \beta \qquad (12.12)$$

From the rule on the duplication of arcs, we obtain the following:

$$\cos 2\beta = \cos^2\beta - \sin^2\beta = 1 - 2 \sin^2\beta = 2 \cos^2\beta - 1 \qquad (12.13)$$

$$\sin 2\beta = 2 \sin \beta \cdot \cos \beta \qquad (12.4)$$

From which we infer:

$$\sin^2 \beta = \frac{1 - \cos 2\beta}{2} \qquad (12.15)$$

$$\cos^2 \beta = \frac{\cos 2\beta + 1}{2} \qquad (12.16)$$

$$\sin \beta \cdot \cos \beta = \frac{\sin 2\beta}{2} \qquad (12.17)$$

Replacing these expressions in Equations (12.11) and (12.12), we obtain:

$$\sigma_\beta = \sigma_1 \cdot \frac{\cos 2\beta + 1}{2} + \sigma_3 \cdot \frac{1 - \cos 2\beta}{2} = \frac{\sigma_1 + \sigma_3}{2} + \frac{\sigma_1 - \sigma_3}{2} \cdot \cos 2\beta \qquad (12.18)$$

$$\tau_\beta = \frac{\sigma_1 - \sigma_3}{2} \cdot \sin 2\beta \qquad (12.19)$$

If we plot in a τ-σ axis system the geometric locations of points for an infinity of values of β, we obtain a circle centered at the abscissa $p = (\sigma_1 + \sigma_3)/2$ with a radius $q = (\sigma_1 - \sigma_3)/2$ (Fig. 12.10f). This circle is called a **Mohr circle**. The abscissa p of the center corresponds to the medium stress (medium between the two extreme principal stresses); the radius q, which describes the half-distance between the major stress and the minor stress, is called the **stress deviator**. These two definitions are unique to a space reduced to two principal stresses.

On the Mohr circle, we can easily determine the values of the normal stress and the tangential stress on a plane whose pole (normal) makes an angle β to the maximum stress by drawing an arc equal to 2β to the σ_1 axis. The point on the circle that is obtained in this way has the coordinates of the values that are being sought.

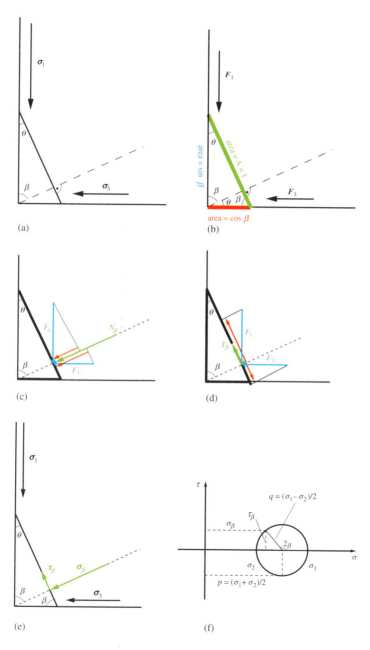

Fig. 12.10 Decomposition of the principal stresses σ_1 and σ_3 onto an oblique surface and the forces related to these stresses. (a) Representation of stresses. (b) Representation of forces. (c) Addition of the normal components of the forces. (d) Addition of the tangential components of the forces. (e) Return to the stress space. (f) Construction of the Mohr circle.

A triaxial test that has caused a sample to rupture and its corresponding Mohr circle describe the maximum strength of the sample under the particular conditions of the test stresses. However, it does not provide the rupture criteria of the rock under other stress states. To obtain this information, it is necessary to determine what rupture criteria best characterize the maximum strength of the rock under other states. The triaxial test makes it possible to construct this function experimentally; to do this, several tests must be done on samples that are as similar as possible, and different values of confining pressures σ_3 must be used during the tests. Through these tests, we obtain several rupture circles that allow an envelope called an **intrinsic curve** to be drawn, because it is characteristic of the material and is not dependent on the geometry of the rupture surface (Fig. 12.11). The interior of this envelope determines the field of non-rupture deformation, the envelope determines the transition to rupture. The shape of the envelope is often a parabolic curve. Its extension to the left of the X-Y axis indicates negative compressions, thus the **tensile strength R_t**. The equation of the parabola is the following:

$$\tau^2 - \frac{c^2}{R_t} \cdot \sigma - c^2 = 0 \qquad (12.20)$$

where
  c = cohesion (the y-intercept)

Once this rupture criterion is established experimentally, it is possible to simulate an anisotropic stress state in the shape of a Mohr circle. If it is within the field, there is no rupture of the rock. By increasing the contrast between the stresses, we reach a state where the circle becomes tangent, at P, to the rupture line (**limit equilibrium**). This point defines the angle β or the obliqueness θ of the surface that will have the minimum shear strength and will be the cause of rupture. The angle of internal friction of the rock under these stress states is the slope of the tangent to the circle at point P. We can also determine the rupture stresses for a surface of a given obliqueness θ' by tracing a straight line from the origin of axes and making an angle θ' with y-axis; the coordinates of point P' provide the stress state necessary to produce rupture in this direction.

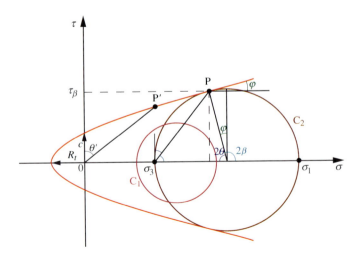

Fig. 12.11 Rupture envelope determined by several triaxial tests with increasing confining stress. Equilibrium states: overabundant (circle C_1), limit equilibrium (circle C_2). Determination of the obliqueness of the rupture surface.

We see in figure 12.11 that the angle of internal friction also located inside the Mohr circle between the radius of the circle leading to Point P and a vertical line. It is complementary to the angle 2θ. We can draw from this that the inclination of the fracture θ and the angle of internal friction φ are related by the relationships:

$$2\theta + \varphi = 90°$$ (12.21)

$$\theta = 45° - \frac{\varphi}{2}$$ (12.22)

In the case of an ideal homogeneous and isotropic rock, two symmetrical fracture planes around the maximum stress have the same probability of occurrence and often appear almost together. They are called ***conjugate faults***. It should be noted that rupture surfaces identified in this way are parallel to the direction of intermediate stress σ_2.

In fact, shear ruptures along conjugate faults in a rock that has not been previously ruptured are the result of the progressive appearance of en-echelon tensile microfissures along the shear zone (Fig. 12.12). If we enlarge the zone of the future fault, we see small open fractures oriented obliquely to the shear. Logically, their openings are almost perpendicular to the minimal stress σ_3. They result from rotational stresses induced by shearing at this scale. As the movement due to shearing increases, the microfissures take on a sigmoidal shape then end up by breaking, leading to the final rupture by shearing.

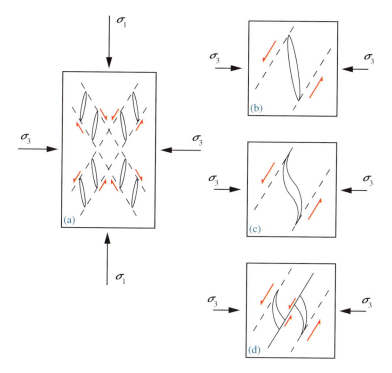

Fig. 12.12 En-echelon tensile microfractures caused by shearing. (a) Distribution along future conjugate faults with openings approximately normal to the minimum stress. (b) Microfracture at the beginning of shear displacement. (c) Sigmoidal microfissure for larger movements. (d) Final shear rupture.

Let us now see the effect of two important factors on the rupture. The first is a property of the rock itself and the second is related to the test conditions.

The effect of anisotropy

Rocks with constituents that have a preferred direction in the rock matrix are common, particularly among sedimentary and metamorphic rocks. These anisotropy directions are planes of mechanical weakness. It is evident that the orientation of a sample during a compression test will play a determining role in the strength measurement. This is illustrated in figure 12.13 where a schistose rock is placed at several possible orientations with respect to the principal maximum stress. The structural position of the rocks in the rock massif is thus a criterion of prime importance. The geologist provides this information to the engineer.

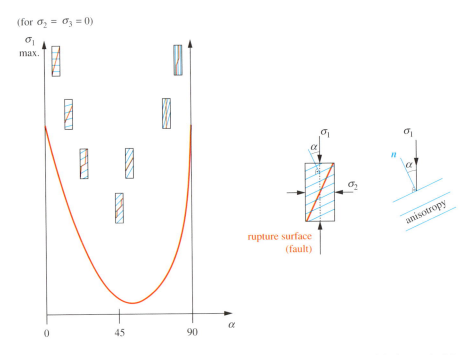

Fig. 12.13 The influence of the orientation of an anisotropic rock on the strength measured during a uniaxial compression test. The strength varies greatly between two extremes: the weakest is the coincidence of the critical obliqueness $\theta = \pi/4 - \varphi/2$ with the direction of anisotropy, approximately 45°. The maximum strengths are obtained when the direction of anisotropy is perpendicular to the maximum stress.

Effect of the confining stress on rupture behavior

Tests in which the confining stress increases can reproduce tectonic observations, aside from the effect of time. With a low supporting stress, ruptures are brittle and occur early, without deformation of the entire rock body (Fig. 12.14). For highly confined tests, rupture becomes ductile, similar to what is observed at great depths in mountain belts.

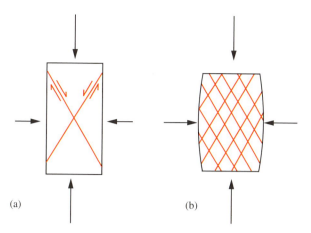

Fig. 12.14 Rupture shape as a function of the intensity of the confining stress in triaxial tests. (a) Brittle rupture in the case of low confining stresses. (b) Ductile rupture when the confining strength is high.

12.2.1.3   Case of previously ruptured rocks

The shear strength of preexisting rupture surfaces in rocks has been well studied. In fact, tests on rock cores are not generally representative of a rock massif that contains a whole series of discontinuities or joints: the strength properties in rock samples are often clearly higher than those of the entire rock body. This effect of scale is one of the major problems of engineering geology in many areas (mechanics, hydrogeology, etc.). It is thus necessary to characterize not only the strength of the intact rock but also that of the preexisting surfaces that it contains.

The shear strength of a smooth joint, without macroscopic asperity can be approximated by the Navier-Coulomb equation for non-cohesive soils (equation 12.3 with c = 0). It depends only on the normal stress and the angle of internal friction of the joint (Fig. 12.16, Function E).

When it comes to rough joint planes with indentations, this model no longer applies. In nature the roughness of joints occurs in many forms, sometimes random (for example, by dissolution of rock), and others well ordered (ripple marks, for example). For a mechanical calculation, we shall use a highly simplified geometry. Patton introduced an expression of shear strength for a joint whose roughness is compared to non-symmetrical saw-teeth that make an angle i with the shear direction (Fig. 12.15). If we move along first using a weak normal stress, we can avoid destroying the roughness. The movement implies a shearing on the edge of the teeth and the two pieces of rock move apart. It is obvious that the strength will be greater if the teeth are sharper, and is thus linked to the angle i. If the angle of internal friction of a smooth fissure in the rock is equal to φ, we can add angle i to φ in the expression for a smooth joint. Thus we have:

$$\tau = \sigma \cdot \mathrm{tg}(\varphi + i) \qquad (12.23)$$

This equation is represented by a first straight-line segment on the Mohr diagram, making an angle $\varphi + i$ with the axis of the normal stress and passing through the origin.

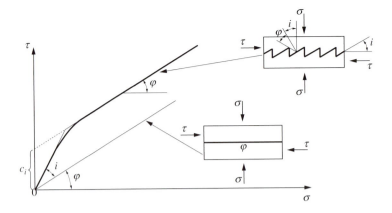

Fig. 12.15 Effect of roughness on shear strength of a rocky joint (from [211]).

When the normal stress is increased, the strength of the teeth is exceeded. The joint is then filled with the debris from the original roughness. On the Mohr diagram, we obtain a second straight-line segment whose slope is lower and that corresponds to the angle of residual friction, since it represents the post-rupture situation. As a first approximation, it is nearly equal to the angle of friction φ of the smooth joint. However, this straight-line intersects the y-axis at a non-null c_i origin, which corresponds to a cohesion.

12.2.1.4 Typical intrinsic curves and orders of magnitude

Figure 12.16 summarizes the usual shapes of intrinsic curves for different geologic materials

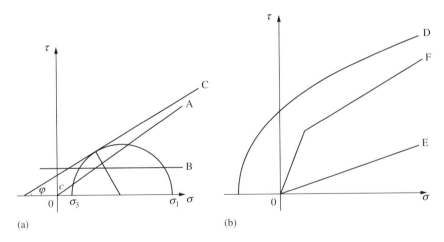

Fig. 12.16 Various shapes of intrinsic curves for geologic materials. (a) Unconsolidated materials with Navier-Coulomb straight lines. A: purely cohesionless soil; angle of internal friction is constant, null cohesion (example, sand). B: purely cohesive soil; elevated cohesion, angle of internal friction null (theoretical example of clay, but in reality the angle of internal friction is never null). C: soil composed of a mixture of cohesive and cohesionless particles (example of clay-rich silty sand). (b) Rocks. D. Non-previously ruptured rock; significant cohesion, non-constant angle of internal friction (parabolic curve). E. Rock previously ruptured along a smooth joint; null cohesion. F. Rock previously ruptured on a rough joint.

In addition to their wide variability in terms of mechanical behavior, composition, structure, degree of deformation or weathering, it is useful to be familiar with the orders of magnitude of the essential mechanical parameters of the principal rocks (Tab. 12.1).

[101, 179]

Table 12.1 Order of magnitude of some important mechanical properties of rocks: Uniaxial compressive strength of intact rock and angle of internal friction on pre-existing discontinuities (stratification or fracture joint). The cohesion on the discontinuity varies greatly from case to case, between 0 and several MPa. The modulus of elasticity ranges from 5000 MPa for soft sedimentary rocks to about 100,000 MPa for the hardest crystalline rocks.

Type of rock	Uniaxial compressive strength [MPa]	Internal friction angle of the joints [°]
Magmatic rocks (granites, basalts)	100–400	40–60
Massive metamorphic rocks (gneiss, marbles, quartzites)	50–350	35–50
Schistose metamorphic rocks (mica schists, chloritoschists)	10–100	20–40
Non-argillaceous sedimentary rocks (limestones, sandstones)	20–250	35–50
Argillaceous sedimentary rocks (mudstones, marls)	5–50	8–30

12.2.2   Rock deformation observed in the field

We have seen in laboratory testing that natural rocks can be divided into elastic solids with brittle behavior (geologists speak of ***competent rocks***) and more ductile rocks that may flow at ordinary temperatures if their water content is high enough (***incompetent rocks***). What do these look like in the field?

Moving from a laboratory sample to the field is a significant change in scale that introduces the component of heterogeneity and ***macro-anisotropy***, that is, a preferred orientation in the outcrop that is not visible in samples. In a sedimentary series, for example, there are variations in the competence of rocks from one layer to another (Fig. 12.17). When this series is subjected to compaction or stretching, brittle rupture of competent horizons will cause incompetent rocks to flow into open spaces in the fissures.

In addition, field observation often shows a paradox: rocks that appear brittle in the laboratory are folded in the field. This difference is due to several factors:

- Time (millions of years in nature compared to several days in the laboratory) makes ductile deformation possible through mineralogical and structural rearrangements, even if the rock is competent, thus preventing rupture;
- Rocks have ductile behavior at great depth because they are in an environment that is highly restricted in all directions;
- Temperatures are higher in the field, which gives the rock a significant viscosity.

These deep deformations occur predominantly in the course of metamorphism.

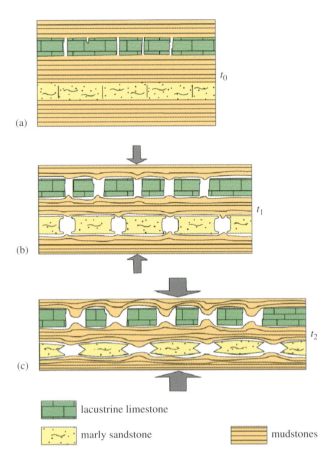

Fig. 12.17 Normal compression in a sedimentary series with layers of variable competencies. Example of the Chattian Molasse north of the Alps: in the lacustrine limestones, the fissures that naturally permeate the layer get wider, without deformation of the matrix; the marly sandstones move apart in segments, but deform slightly as boudins (this phenomenon is called "*boudinage*"); mudstones stretch plastically, without discontinuity, similar to a sheet of steel in a rolling mill, and replaces the open spaces in the fissures.

12.3   Brittle deformation

Deformation of competent rocks is resolved by a network of overall planar discontinuities that are grouped generally as breaks, fractures, or discontinuities. They can be distinguished on the basis of the movement of the two sides of the fracture.

12.3.1   Joints

When a competent rock series is subjected to a slight tectonic deformation, it causes the rock to break up along a multitude of small cracks, although the general structure of the mass is not modified and there is no visible movement along the discontinuities. Joints are generally classified into several sets (groups of joints with the same orientation). One common case is that of layered sedimentary rocks that have two sets of joints, one perpendicular to the strata and the other perpendicular to the first (Fig. 12.18).

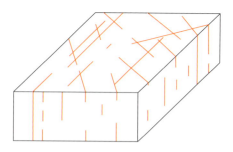
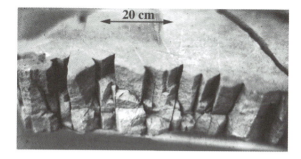

Fig. 12.18 Breakup of a rock mass due to jointing. Example of a jointed limestone layer. GEOLEP Photo, J.-H Gabus.

Joints are the most common fractures in rocks. They transform an originally monolithic mass into a juxtaposition of blocks. By studying excavations in a rock mass the geologist can determine fracture density, a parameter that depends primarily on joints.

With time, joints can be filled by recrystallization, which produces veins (Fig. 12.19).

Fig. 12.19 Open joints or faults are often filled by minerals that precipitate from groundwater. These veins are generally composed of calcite or quartz, as here in the granites at the Park of the Summer Palace in Beijing. GEOLEP Photo, A. Parriaux.

12.3.2 Faults

Faults are discontinuities along which shearing has occurred. Fault displacement is defined by a vector in space that connects two points that were in contact before the movement: the slip (Fig. 12.20). Faults are classified into three types depending on the direction of the displacement with respect to the rupture surface: normal faults, reverse faults (including thrust faults and thrust nappes) and strike-slip faults. These three types are also distinguished by the shape and orientation of the ellipsoid of the principal stresses that caused the faults.

12.3.2.1   Normal faults

Normal faults are fractures caused by a tensile stress on a rock series. An oblique rupture occurs; one of the blocks moves downward by sliding down the surface. This deformation causes a lengthening of the rock unit (Fig. 12.20a and 12.21). It represents the situation where the maximum principal stress is vertical, the minimum stress is in the direction of the lengthening and the intermediate stress is parallel to the fault.

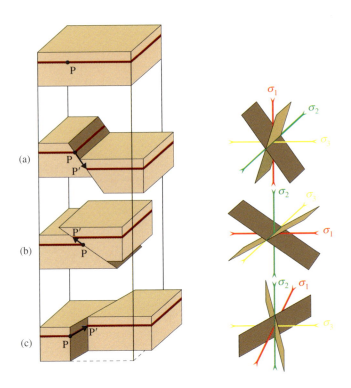

Fig. 12.20 Block diagram of different types of faults in relation to the orientation of the corresponding principal stresses. The ruptures correspond to conjugate faults. (a) Normal faults (b) Reverse faults (c) Strike-slip faults. PP' is the slip vector.

Normal faults occur in rift zones near spreading plates, or on a smaller scale, on the outside part of concentric folds.

12.3.2.2   Reverse faults

Reverse faults (Fig. 12.20b and 12.22) are the result of a horizontal compression of a rock series. The rock breaks along an oblique surface and one of the blocks moves by thrusting over the other. The result is a shortening of the rock unit. The maximum principal stress is horizontal in this case, in the direction of shortening.

Reverse faults occur in areas where two tectonic units collide.

Fig. 12.21 Example of normal faults in the Burdigalian Molasse of Prahins, Switzerland. GEOLEP Photo, A. Parriaux.

(a)   (b)

Fig. 12.22 Examples of reverse faults. (a) In Algeria, at Mostaganem. GEOLEP Photo, J.-H. Gabus. (b) Tectonic zone showing the uplift of the Tibetan Plateau (left side of photo) south of Golmud. GEOLEP Photo, A. Parriaux.

12.3.3 Thrusting

Some reverse faults can extend over a long distance and cut through large rock masses. These faults may have dips of less than 30° and can be almost horizontal. The rocks that dominate the discontinuity "thrust" over the lower series, which gives this type of faulting its name. Figure 12.48 shows these slightly inclined rupture surfaces in the Jura folds, as a result of Alpine thrusting.

12.3.2.4   Thrust nappes

If the thrusting extends over several kilometers, the thrust block is called a ***thrust nappe*** (Fig. 12.23), as compared to a fold nappe (§ 12.4.4).

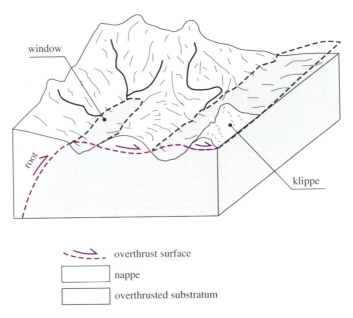

Fig. 12.23 Thrust nappe. When erosion reveals the rocks beneath the thrust nappe, the feature is called a "***window***." When a part of the thrust nappe is disconnected from its roots, it is called a "***klippe***."

Thrust nappes play a fundamental role in the architecture of large mountain belts. The extent of thrusting can be significant; in western Switzerland, the Préalpes Médianes thrust nappe moved a distance of 65 km over the region that became the High Calcareous Alps (Fig. 12.48).

12.3.2.5   Strike-slip faults

When a rock mass is subject to horizontal shearing, a sub-vertical rupture occurs and the shear is resolved by displacement of the fault blocks. In this case, the maximum and minimum principal stresses are horizontal (Fig. 12.20c and 12.24).

A famous example of strike-slip faulting is the San Andreas Fault of California (Fig. 4.1). Strike-slip faults also occur on much smaller scales.

12.3.4   Characterization of discontinuities in a rock massif

The characterization of discontinuities in a rock massif is a complex operation that requires field identification of geologic factors that affect their shear strength and, as a result, the stability of the rock mass. These factors are:

- Orientation of discontinuities in space: the geologist uses statistical measurements to define different sets of joints.

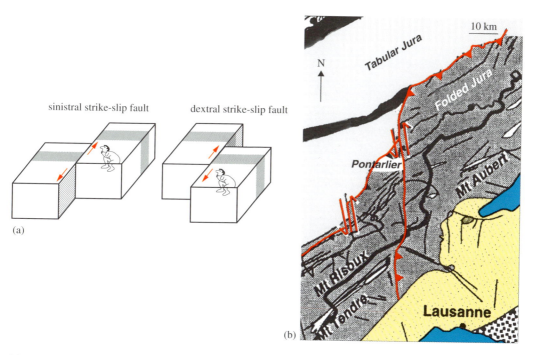

sinistral strike-slip fault   dextral strike-slip fault

(a)

(b)

Fig. 12.24 Strike-slip faults. (a) Displacement: If an observer located on one of the blocks looking at the fault sees that the other block moves to the right, the strike slip is dextral; in the opposite case, it is sinistral. (b) Example of sinistral strike slips that cut across the Jurassian arc; here the Pontarlier strike-slip fault offsets faults on both sides of the shear plane by several kilometers. The fragmentation of the rocks creates areas of lower relief that are more easily eroded, which makes them useful as roads through the mountains.

- Spacing: the statistical distribution of distances between discontinuities is measured for each set; this makes it possible to calculate the average size of blocks in the rock massif and the fracture density.
- Persistence: for the different sets, the geologist wants to know if the discontinuities have a significant spatial extent (such as stratification joints and fault) or a minor extent (jointing). Ruptures are more likely on the former since it is not necessary to break intact rock bridges during shear motion.
- Roughness: the calculation discussed above shows the importance of this factor; in the field, the geologist characterizes the roughness of the joints according to the size of the asperities and their shapes. In the case of roughness that is elongated lengthwise, such as ripple marks, the relative orientation of the feature and the shear movement must be determined because it has a great influence on the shear strength (Fig. 12.25).
- Opening: to be able to determine accurately the effective role of the roughness of the walls surrounding a fissure, it is necessary to measure the space between these walls. In the case of large openings, small-scale roughness has practically no effect.
- Filling: open joints can be filled with air, water or loose material. In the latter case, the material must be carefully described because it will determine the shear strength of the joint if the rocky asperities of the walls are not in contact (clay filling should be screened as a priority).

ORIENTATION OF THE RIPPLE-MARK CREST

RUPTURE STRENGTH τ_f ACCORDING
TO ORIENTATION OF THE SHEARING
MOVEMENT

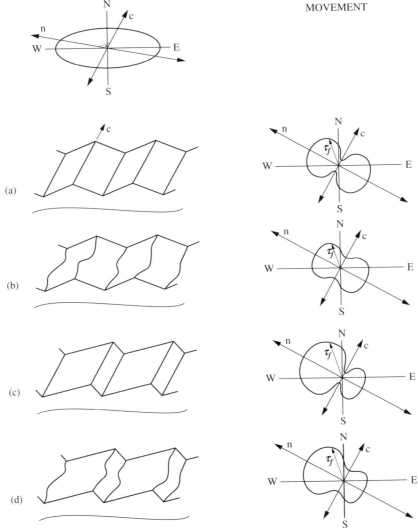

Fig. 12.25 The effect of interstratal roughness on shear strength when the roughness is due to directional sedimentary structures (anisotropy on the interstratal plane). Example of ripple marks in layer-on-layer movement. Direct shear tests were done in different orientations with respect to the orientation of the ripple marks (c = crest, n = normal to the crest). The maximum strength values τ_f, (shear strength) transferred onto a polar diagram make it possible to determine the appearance of the strength-direction of movement function. The curved line is the geometric place of all extremities of strength vectors in the various movement directions. (a) Ripple marks with symmetrical sides and a linear crest: there is maximum contrast between strength parallel to the crest, as compared to other directions. The strengths are influenced only by direction and not by the sense of movement (symmetrical lobes of the function). (b) Ripple marks with symmetrical sides but undulating crests: same general shape but less contrast between strength along the crest axis as compared to other orientations. (c) Asymmetrical ripple marks with linear crests: high contrast between strength in the direction of the crests and other directions. The sense of movement (not only the direction) influences the strength (asymmetrical lobes of the function). (d) Asymmetrical ripple marks with undulating crests: similar to case (c) but with less contrast.

- Hydrogeological behavior of joints: it is essential to determine if groundwater occupies the discontinuities permanently or temporarily.
- Weathering condition of the rock body: in the case of contact between the walls, the strength of the rock is to some extent influenced by the weathering state of the rocks in general and the rock near the joints in particular. It should be emphasized that discontinuities play a significant role in the propagation of weathering throughout a rock mass. It is possible that the rock is perfectly sound 10 cm away from a joint but is completely weathered in contact with the joint.

The characterization of these factors in the field is done at the outcrop, in pre-existing underground cavities, and in boreholes. One of the major difficulties is estimating the bias that affects measurements taken at the outcrop (for example, the opening is exaggerated by decompression of the rock mass, surface movement, weathering, and other factors) when we want to estimate those characteristics within the rock massif. Observations in underground cavities are also subject to these biases: plastic deformation near the walls, fracturing due to mining, etc. All these phenomena must be taken into account in the study of joints in rock massifs.

12.4   Ductile deformation

The deformation of incompetent rocks causes pervasive transformations of the rock structure. The principal non-brittle structures are caused by compressive stresses that create folds.

12.4.1   Folds

Folds are tectonic structures that allow pervasive deformation of a rock almost without rupture.

In the simplest and most common case of horizontal compression, a series of planes similar to sedimentary layers deforms into roughly sinusoidal cylindrical shapes similar to ripples as a tablecloth is moved across a table. A convex cylinder (bump) is called an *anticline* (Fig. 12.26) and the concave cylinder (creating a hollow) is called a *syncline* (Fig. 12.27). The *dip* provides the orientation and the inclination of the beds in the different parts of the fold.

When the resultant forces of compression have an application point at the same altitude on both sides of the strata pile, a straight symmetrical fold is created. The plane of symmetry is called the *axial plane*. The *axis* is the intersection of this plane with the surface of a rock layer (Fig. 12.28). The part with the tightest curve is called the *hinge* of the fold. It separates the two *flanks* or *limbs*.

If the resultant of forces has an application point at different altitudes, the axial plane is inclined (Fig. 12.29). When the axial plane is almost horizontal, it is called a *recumbent fold*.

The axis can be horizontal or inclined (Fig. 12.30). In the latter case, it is a *plunging fold* and there is an angle between the axis and the horizontal.

Erosion of a fold can cause the outline of the layers to appear in the topography: these are elongated bands on both sides of the axial trace (Fig. 12.30). If the axis is horizontal, the bands are parallel to each other. If it is inclined, the outlines join at the hinge of the fold, in the shape of a parabolic curve. In the case of an anticline, the old rock layers are at the center. In a syncline, the recent rocks occupy this position.

The geologist, mapping the rock outcrops on a topographic map and using the age of the rock layers, is able to establish the tectonic structure at depth. The study of fold geometry and

Fig. 12.26 Example of an anticlinal fold on the island of St. Honorat (France). GEOLEP Photo, J.-H. Gabus.

Fig. 12.27 Synclinal fold in the colored formations in the center of the Himalayas. Zaskar. GEOLEP Photo, A. Parriaux.

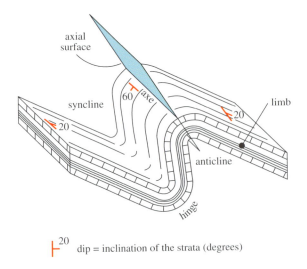

$\vdash^{20}$ dip = inclination of the strata (degrees)

Fig. 12.28 General characteristics of folds.

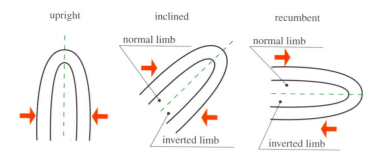

Fig. 12.29 Various positions of the axial plane.

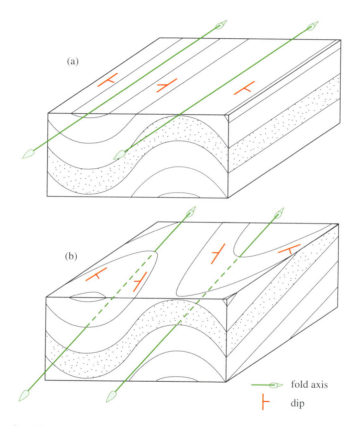

Fig. 12.30 Outline of rock layers after erosion of folds. (a) Folds with a horizontal axis cut by a horizontal surface. (b) Folds with a plunging axis cut by a horizontal surface.

small-scale deformation features of rocks allow to recreate the tectonic stresses that created them. For example, studies of deformation of fossils and pebbles have shown the stress orientation and how the rocky matrix has incorporated these movements (Fig. 12.31). These studies show various processes in the genesis of folds, making it possible to classify them into two large families: concentric folds and similar folds.

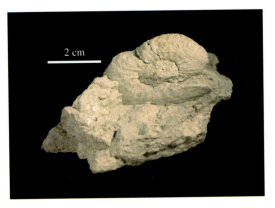

Fig. 12.31 Deep deformation of the rock matrix shown by the deformation of an ammonite. GEOLEP Photo, G. Grosjean.

Problem 12.1

You are to construct a bridge linking two sides of a valley. In order to do this, you must cross a mountain stream. At the outcrop, the geologic information indicates the presence of three formations: argillaceous sandstones, massive dolomite and schistose marls (Fig. 12.32). The project calls for the installations of three bridge piers at the locations indicated on the map.

Questions

- In what rock formation do you install the foundations of the bridge piers?
- How deep should the piers be?
- What type of pier do you choose (Fig. 1.1)?

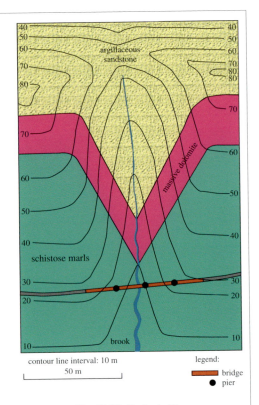

Fig. 12.32 Geologic Map

Look at the solution on the DVD

12.4.2   Concentric folds

Under this heading we group folds that have rounded contours, both small and large-scale. The boundaries between the layers are similar to arcs of concentric circles (Fig. 12.33). The original thicknesses of the layers, measured perpendicularly to the contacts between them, are almost unchanged. This type of fold is caused by compression at shallow depth (less than 10 km); it leads to a shortening of the rock.

The construction of folds as circular portions poses no problem for the outer rock layers. At the center of the fold, the innermost bed is completely folded over on itself. It is no longer possible for it to participate in the folding of deeper layers. Thus there is a disharmony caused by the detachment of the folded series from the substrate in the case of an anticline (Fig. 12.33). This detachment occurs in ductile rocks such as salt, anhydrite and claystones.

In the case of concentric folds, the transformation from a series of planar layers to a cylindrical shape causes deformation. Overall, the layers at the center are intensely compressed, those at the periphery are under tension. Two deformation mechanisms can occur: innermost deformation of the rock material or shearing at the contacts between layers.

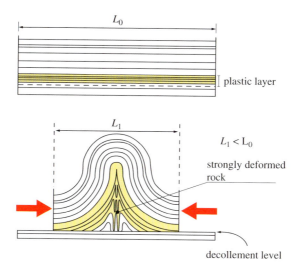

Fig. 12.33 Diagram of an anticlinal concentric fold.

12.4.2.1   Deformation by flexure folding

The rock series is treated as a single thick unit. The deformation is entirely resolved within the unit. The inner part of the unit is compressed and the outer part is under tension (Fig. 12.34).

This type of fold occurs in massive rocks. At the center, compression creates microfolds in the rock matrix with small and numerous reverse faults. On the outside, tension causes normal faults with downdropping of the arch on anticlines. Within a layer there is a surface where the rock is not deformed.

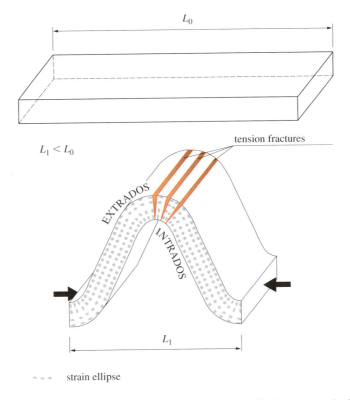

Fig. 12.34 Distribution of deformation in concentric folds caused by flexure (massive layers).

12.4.2.2 Deformation by flexural slip folding

In the case of finely stratified layers, deformation is resolved by shearing at the surface of the contact between layers, similar to the pages of a book as it is bent (Fig. 12.35). Deformation within layers is minor.

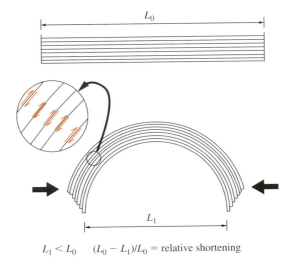

$L_1 < L_0$ $(L_0 - L_1)/L_0$ = relative shortening

Fig. 12.35 Distribution of deformation in concentric folds caused by flexural sliding (fine layers).

In nature, layer-on-layer sliding is facilitated by argillaceous beds that are often present between more competent layers. A good example is the case of rocks in the Jura Range (North of the Alps), which alternate between limestone and marls, with the exception of the thick limestone series of the upper Jurassic, which deforms by flexure (Fig. 12.48).

12.4.3   Similar folds

At great depth, folds evolve into more acute and complex shapes that deeply deform the rock matrix. The deformation is resolved by a very tight network of ruptures parallel to the axial surface of the fold; each of these surfaces undergoes shearing, slightly offsetting the stratigraphic contact (Fig. 12.36).

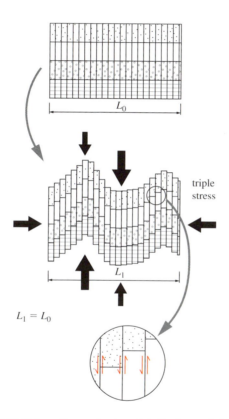

Fig. 12.36 Deformation mechanism of similar (or parallel) folds.

This fold is called a similar or parallel fold because it can be propagated at depth to infinity by fitting similar shapes into the fold. This type of folding does not cause shortening of the original rocks.

The deep shearing of the rock matrix leads to a reorientation of some minerals. Rocks that contain phyllosilicates, for example, show a mechanical reorientation of the layers in the shear planes. This phenomenon gives the rock an anisotropic texture called *schistosity* or *cleavage*. The intersection of the schistosity with the surface of the layers is the *cleavage–bedding intersection lineation* (Fig. 12.37).

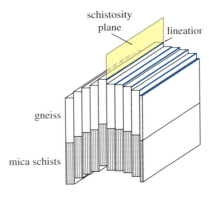

Fig. 12.37 Definition of schistosity and intersection lineation (in blue).

Because these folds are typical of high mechanical pressure conditions, they are often related to metamorphism. The metamorphism also contributes to a reorientation of minerals along schistosity planes. The layered structure of the original rock may completely disappear.

12.4.4 Fold nappes

When a recumbent fold advances several kilometers, it is called a fold nappe. The overturned limb continues to exist, at least partially. It undergoes lamination and even a slight metamorphism due to significant shear stresses. The western part of Switzerland is well-known to geologists throughout the world for the Morcles nappe that perfectly illustrates this definition (Fig. 12.38). Let us recall that thrust nappes have no overturned limb (fig. 12.23).

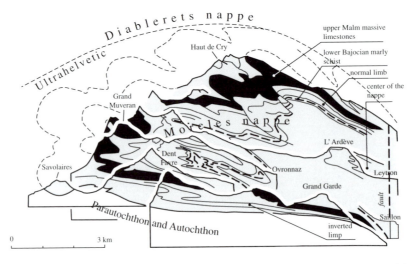

Fig. 12.38 Type example of a fold nappe: the Morcles nappe, (Vaud and Valais Alps). The thickness of the overturned limb is reduced near the root: the Urgonian-Aptian limestones, which are a hundred meters thick in the normal limb, are only a few meters thick. The lamination and a slight metamorphism have transformed them into a banded marble. This marble has been exploited at Saillon for several centuries (from [169]).

12.5   Geometric representation and treatment of structural elements

Many tectonic, engineering geology and rock mechanics problems require a representation of the structural elements of the subsurface in space in order to identify the intersections between them or with engineering projects. In practice the geologist first records all local structural information collected in the field on geologic maps and profiles, and then uses methods that deal with spatial relationships between these elements.

12.5.1   Mapping of structural elements

Geometrically, the principal structural data (orientation of geologic layers, schistosity, lineations, fold axes, etc.) can be reduced to straight lines and planes. The orientation and their inclination are first measured in the field with respect to cardinal points. Then these elements are recorded on the map at the location where the observation was made (Fig. 12.39).

Straight lines
The orientation of a straight line is determined by two angles: its azimuth (or strike) α (horizontal angle with respect to north, measured in a clockwise manner, $0° \leqslant \alpha \leqslant 360°$) and its dip β (angle with respect to the horizontal of a vertical plane containing the straight line, positive in the downward direction, $0° \leqslant \beta \leqslant 90°$). It is conventionally noted by a symbol identifying the type of line (fold axis, lineation, direction of a landslide) followed by α/β (three numbers for α and two numbers for β to avoid confusion). Straight lines can thus represent several structural elements and also topographic elements such as rivers and parts of engineering works such as the axis of a borehole or a tunnel. In a stereographic projection (§ 12.5.3), a straight line is represented by a point.

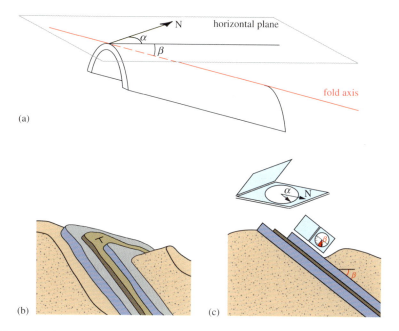

Fig. 12.39 Conventional representation of the orientation of straight lines and planes. The unit that is commonly used is the degree. (a) Case of a straight line (for example, fold axis); α = azimuth of the horizontal projection of the straight line; β = dip of the line. (b) Case of a plane (surface of a layer); three-dimensional view. (c) Vertical profile of azimuth α containing the straight line with maximum dip β.

Planes

The notion of dip is used to represent the inclination of geologic units. In practice we identify the straight line indicating maximum dip of the plane under consideration, then we give its orientation with the two angles α/β as described above for lines. By analogy, we note all planes in the same way: joint planes, schistosity or foliation planes, valley wall, road embankment, etc. In a stereographic projection, a plane is represented by a great circle (§ 12.5.3).

12.5.2   Geometric tools

All methods of descriptive geometry can be used. As a result of the development of computer technology, infographic representation tools are now available to handle a large number of two and three-dimensional problems (Fig. 12.40).

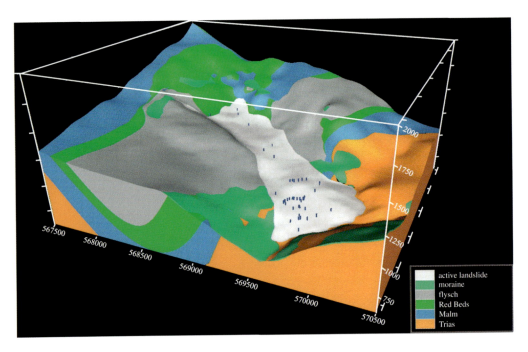

Fig. 12.40 Example of a computer model. The La Frasse Landslide (Prealps, Vaud, Switzerland) with the tectonic structure of the subsurface. GEOSHAPE Software, L. Tacher, GEOLEP.

However, many common problems can be resolved using simple methods that can be used in the field. Questions about the intersection of planes and straight lines can often be reduced to purely angular relationships where the notions of distances are not necessary. For such a case, a stereographic projection is a convenient and powerful tool.

12.5.3   Stereographic projections

We shall first describe a stereographic projection and its theory. Then we will discuss an example to provide a detailed illustration of a practical case typically encountered.

12.5.3.1 Theory

The principle of the projection is to translate all the geometric elements of a particular problem by attaching them to the center O of a sphere (Fig. 12.41). We ignore the distances between the objects under consideration. Each element (straight line or plane) cuts the sphere and leaves a trace. The trace of a straight line is two opposite points, whereas the trace of a plane is a great circle centered on the center of the sphere. In some problems it may be necessary to represent cones (geometric place of straight lines, envelope of planes, angle of friction around an axis, etc.); the trace of a cone of revolution consists of two small circles on opposite sides of the sphere. For each type of object, the representation can be simplified by considering only the lower hemisphere. These two half-traces perfectly identify the geometry of the elements and their orientation in space. In order to analyze the relationships between these traces, they are projected on an equatorial plane by a polar projection using the top of the sphere as the projection point. The intersection of the equatorial plane and the sphere identifies the circular work field whose center O coincides with the center of the sphere. The trace of a straight line is still a point, that of the plane is an arc of a circle, but it is no longer centered on O, and the trace of a cone of revolution is a small circle.

The Wulff stereographic net makes the treatment of projected elements in the circular work field easier (Fig. 12.41). It contains the projection of the traces of meridian planes dipping from 0 to 90°; these are the planes that cut the hemisphere into quarters and that have a diameter line in common with the stereo net (Diameter 1 Fig. 12.41), similar to terrestrial meridians but here they are on a horizontal axis; this cutting makes it possible to trace the great circles related to the planes. Their dip is graduated along a diameter line perpendicular to the preceding one (Diameter 2 Fig. 12.41). The projection of lines is done in the same way using the meridian plane that corresponds to their dip. The stereo net also allows the projection of cones of revolution of horizontal axis for opening angles from 0 to 90°. Their opening angle is graduated along

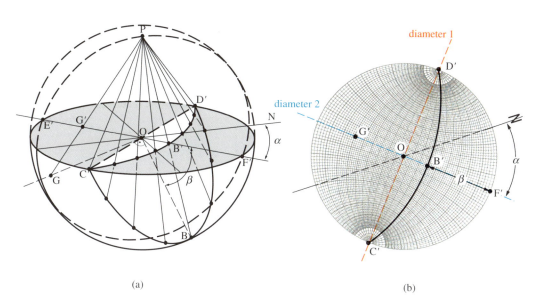

(a)   (b)

Fig. 12.41 (a) Diagram of the theory of the stereographic projection on the lower hemisphere. (b) Wulff stereo net with the projection of plane C'BD' and its pole G.

Diameter 1 on the stereo net. The gradation of the angles is identical on both diameters. The rotation of the stereo net around the center makes it possible to construct planes, straight lines, and cones according to their own compass directions.

When it is necessary to construct a large quantity of planes, the projection becomes very busy with numerous circles. This difficulty can be resolved by replacing planes by their normals. The trace of the normal in a hemisphere is called a *pole*. Using clouds of poles, it is possible to identify the number and average orientation of joint families in a rock massif.

12.5.3.2   Example of an application

A very common problem in engineering geology is to determine the stability of a rock mass that threatens to slide down a slope. Here is a realistic example of such a situation.

A new highway is going to cross a mound that is about fifteen meters high (Fig. 12.42). The axis of the road across the mound is NE-SW. The NW side is significantly higher than the SE side. The preliminary geologic survey indicates that the rock is probably metamorphic but it is masked by glacial deposits. A reconnaissance borehole shows that rock is present under a meter of moraine. It is a massive gneiss. In cores, the foliation dip has been measured at 50°. Given the massive nature of the rock it has been decided to remove the moraine and cross the mound, using a vertical-walled trench reinforced locally with anchors if necessary.

At the beginning of the trench excavation, the geologist measured the spatial orientation of the foliation in the gneiss: 066/50. As the excavation proceeded, evidence was collected showing that a large rock mass of the NW wall was threatening to slide. Its dihedral-shaped base was identified on one side by a bed of highly weathered mica schist, about 2 cm thick between two massive benches of gneiss, a thickness that had been lost in coring because of its low strength. On the other side, the wedge followed a fault oriented 195/60. The volume of the potentially unstable wedge was estimated at about 200 m^3. Taking the specific weight to be 26.5 kN/m^3, the estimated weight was about 5300 kN (equal to the weight of a mass of about 530 tons).

In order to determine the mechanical strength of these two discontinuities, the geologist took samples for direct shear tests in the laboratory. The foliation of the mica schist has a weak strength with an internal angle of friction of 18°. The strength of the fault is hardly better because its surface was very smooth due to tectonic movement. The angle of friction is about 27° on average. The weak cohesions measured in the laboratory were neglected for conservatism in the calculation of the stability of the wedge.

Before continuing the excavation, it was necessary to evaluate quickly the stability of the wedge and to determine if a reinforcement should be installed. In summary, we have the following data to work with:

- Plane p: orientation of the NW wall 135/90.
- Plane 1: orientation of foliation 066/50 (or $\alpha_1 = 66°$ and $\beta_1 = 50°$)
- Plane 2: orientation of the fault 195/60 (or $\alpha_2 = 195°$ and $\beta_2 = 60°$)
- Angle of internal friction of the foliation $\varphi_1 = 18°$.
- Angle of internal friction of the fault $\varphi_2 = 27°$.
- Weight of the wedge W $= 5300$ kN.

A stereographic projection will be used to solve this problem. But, let us first give some general recommendations on the technique for using the Wulff net.

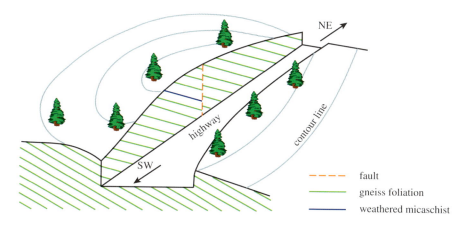

Fig. 12.42 Geometric diagram of the problem.

General instructions for the use of the stereo net

In order to construct the stereographic projection, we place a sheet of tracing paper on the Wulff stereo net and attach it with a thumb tack at the center O which will serve as an axis of rotation. On the tracing paper we draw the outline of the Wulff stereo net and the position of the north. Normally north is at the upper end of Diameter 1 (Fig 12.41). It is possible to trace the great circle related to any plane that contains a NS axis and plunges to the east (azimuth 090) or the west (azimuth 270). For example, a plane inclined 10° to the east (Plane 090/10) is a great circle near the edge of the stereo net. In contrast, a plane inclined 80° (plane 090/80) is near Diameter 1 on the stereo net. Its projection is the arc of a circle with a larger radius. The planes on the other side of the hemisphere are constructed in the same way. Two special planes are the horizontal plane that corresponds to the circle equivalent to the edge of the stereo net and the vertical plane represented by Diameter 1.

Planes that do not dip east or west but toward another azimuth α are constructed as described below after rotating the stereo net. First place a mark corresponding to the azimuth α of the maximum dip on the edge of the outer circle and then the rotate the stereo net to axis 2 (this causes a rotation of the stereo net from angle $\gamma = 90 - \alpha$ for a plane plunging to the east and $\gamma = 270 - \alpha$ for a plane plunging to the west). From here, the angle β is transferred (between 0 and 90°) by counting from the periphery to the center O of the stereo net. Thus the gradations of Diameter 2 make it possible to trace any great circle related to a given orientation.

Projection of data from the wedge problem

On the stereographic projection, with the aid of the Wulff stereo net, we construct the following (Fig. 12.43):

1. Plane 1 (foliation of weathered mica schist). We turn the tracing paper of an angle $\gamma = 90 - 66 = 24$ and we trace the great circle with dip $\beta_1 = 50$. We trace its pole (line N_1) on the vertical plane of the maximum dip of the plane and at 90° from the great circle, measured on Diameter 2.

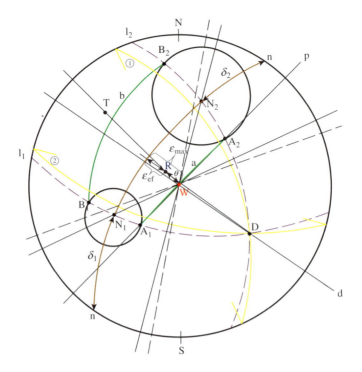

Fig. 12.43 Graphic resolution of the wedge problem by stereographic projection.

2. Plane 2 (fault). We turn the tracing paper of an angle $\gamma = 270 - 195 = 75$ and we trace the great circle that goes south from 60°. Its pole (line N_2) is constructed as in the previous example. Planes 1 and 2 are in yellow on the drawing.
3. Plane p of the vertical wall of the road, in black in figure 12.43. It is a diameter of the projection circle because its angle is 90°. We can immediately identify the straight line common to Planes 1 and 2 (intersection line). These two planes and their common straight line define a dihedron whose edge corresponds to the most probable line of displacement of the block. This straight line corresponds to point D of the projection, at the intersection of the two great circles of the two planes. We can read its orientation in space by making the vertical plane containing D (plane d) coincide with Diameter 2. Its azimuth, measured on the edge of the stereo net, is equivalent to $\alpha_D = 125°$. Its dip, read on Diameter 2, is $\beta_D = 31°$. The line of intersection is thus 125/31.
4. Construction of plane n (in brown on the figure) normal to the edge of the dihedron D. This great circle is at 90° to the projection of D and makes a dip angle complementary to that of D, or 59°. Its azimuth is that of D + 180°, or 305°. We thus obtain 305/59. It is obvious that it passes logically through normals N_1 and N_2 to Planes 1 and 2.

Finally we can measure two interesting angles, which are the apparent dips δ_1 and δ_2 of Planes 1 and 2 in the plane perpendicular to the edge. Let us recall that the normal plane n cuts Planes 1 and 2 along two lines, which are nothing more than normals to these planes. Their dip is measured on the great circle of plane n by using the angle gradation on Diameter 1. We read $\delta_1 = 48°$ and $\delta_2 = 36°$ (Fig. 12.44).

We must introduce yet another important piece of data into the model. This data, which has not been used so far, is the shear strengths of the two planes. As we neglected the cohesion

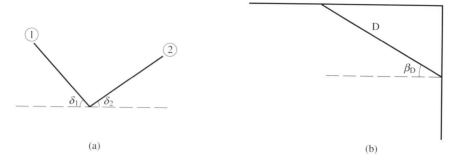

Fig. 12.44 Profiles. (a) Profile normal to line D of the dihedron, in plane n. (b) Vertical profile containing the edge of the dihedron.

of joints, the stability domain of a plane inclined by angle β, free on all sides, is particularly simple:

- If $\beta < \varphi$: the block is stable.
- If $\beta = \varphi$: the block is at equilibrium.
- If $\beta > \varphi$: sliding has occurred.

More generally, considering the possible movements in all directions in a plane by the application of a force R resulting from the various forces that act on the block (for example, the resultant of gravity and the anchoring stress), the boundary of the stability field corresponds geometrically to a cone that makes an angle φ around the normal to the plane. This cone is called the friction cone. In other terms, if the straight line bearing the resultant R of the forces is within the cone, the friction is sufficient to take up the stresses without exceeding the shear strength. The stability state is maintained.

We can now construct friction cones for Planes 1 and 2 around normals N_1 and N_2. In stereographic projection, the cones trace out small circles. To construct them, we use the gradations on Diameter 2 to trace the values of angles φ_1 and φ_2 from N_1 and N_2 respectively. It is necessary to trace these two circles by compass since neither N_1 nor N_2 is at the center of the circle (Fig. 12.43).

The geometric problem is now well determined within the static state. What about the movements theoretically possible for this rock wedge? As we envision whether it is necessary to consolidate it using an anchor, it is necessary to examine all possible movements. The only condition is that the block rests on at least one of its sides. Three types of slipping are theoretically possible:

- Case 1: Sliding along the edge with friction on the foliation and on the fault (friction on Plane 1 and Plane 2); the direction of movement in this case is unique and corresponds to edge D.
- Case 2: Sliding on the foliation only after detachment of the two sides of the fault (friction on Plane 1); different directions of movement are thus possible depending on the resultant of the forces because the edge plays no role.
- Case 3: Sliding on the fault only after detachment of the gneiss layers (friction on Plane 1); same remark as for Plane 1.

Is it possible to construct the boundaries between these three types of movement if the stability fields are determined? A resultant force whose azimuth would be near that of the edge D of the dihedron would obviously cause a stress on Planes 1 and 2. If such a force caused a movement in the SW direction, we can imagine that Plane 2 would be detached. To simplify, let us examine the problem in two dimensions by placing it in the plane normal to the edge of the dihedron (Fig. 12.45); Let us consider only Plane 1 and its normal N_1, so the possible movements correspond to case 1 and case 2. Let us first take a force $\boldsymbol{R}_b$ with a dip higher than that of the normal N_1. This force has a component $\boldsymbol{R}_b'$ tangential to Plane 1 which would be directed downward. The movement would take place with a support of the dihedron against Plane 2 (case 1). A force $\boldsymbol{R}_h$ with a dip less than the normal, in contrast, has a component $\boldsymbol{R}_h'$ tangential oriented upwards, which would open discontinuity 2 (case 2). We could repeat the same reasoning for movements along Plane 2 (case 1 or case 3). The limit position is thus the dip of the normal to the plane under consideration.

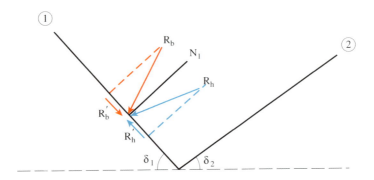

Fig. 12.45 Boundary between slipping along two planes and slipping on Plane 1 only.

In three dimensions, this limit corresponds to the plane containing the normal to the sliding plane under consideration and passing through the straight line of the dihedron. In this case, movement towards the SW, the limit is the plane passing through N_1 and the line D constructed with the great circles on the stereo net. This plane l_1 (in purple on figure 12.43) is oriented 191/55. Plane l_2, which marks the beginning of the detachment from Plane 1 toward the NE, is constructed by N_2; it is oriented 071/46.

Each limit plane cuts the friction cone of the plane under consideration along two lines A and B. They mark the stability boundary in the transition from movement on one sliding plane to movements on two planes. Lines A and B are constructed for planes l_1 and l_2 thus identifying the points A_1 B_1 and A_2 B_2 (Fig. 12.43). The stability limit of the sliding field on two planes (case 1) corresponds to the plane passing through lines A (plane a) and to the plane passing through lines B on the other side (plane b). These planes a and b (in green on the figure) are constructed using the great circles on the Wulff stereo net. Their orientation is the following: a: 314/88, b: 285/30. We thus obtain a stereographic illustration of the complete stability field of the problem, composed of two cones linked by an angular space between two planes. This field is found on both sides of plane n containing the normals to Planes 1 and 2. The plane shows the stability limit for descending movements of the dihedron along the edge. It will play the predominant role in our problem. Plane b marks the limit of ascending movements, useful for the case where we are consolidating the wedge with an anchor.

Let us examine the stability of the wedge under different conditions.

Stability without reinforcement

The only cause for instability is the weight of the wedge $\boldsymbol{W}$ (in red on Fig. 12.43). It is directed vertically. The line that supports the weight vector corresponds to the stereographic projections at the center O. We can see that it is practically on plane a, at the boundary of the stability field. If we are located on the plane passing through the force $\boldsymbol{W}$ and the side of the dihedron (Line D showing the direction of movement), we can read the angle between plane n of the normals and the maximum stability plane a on the gradation of meridians. It is the maximum angle of friction ε_{max} that can be activated in this plane. The angle between plane n and the resultant of the forces (in this case W) is the angle of friction ε_{eff} that is effectively activated. If the effective angle exceeds the maximum angle, the wedge is unstable. We can introduce the notion of the safety factor SF which is the ratio of the resisting forces over the driving forces. It is expressed directly by the ratio:

$$SF = \frac{tg\varepsilon_{max}}{tg\varepsilon_{eff}} \qquad (12.24)$$

If the $SF < 1$, the wedge is unstable. In this case, we note that the angles ε_{max} and ε_{eff} are practically equal (31°), with W on plane a. The safety factor SF is thus equal to 1. Equilibrium exists but is particularly precarious.

We can check this graphic result by making the same assumptions and solving the problem analytically. Using the angular ratios in planes n and d and trigonometry (especially the apparent dips δ_1 and δ_2 on plane n perpendicular to the edge), the safety factor SF can be expressed as follows, not demonstrated here:

$$SF = \frac{tg\varphi^*}{tg\beta_D} \qquad (12.25)$$

with

$$tg\varphi^* = \frac{\sin\delta_2 \cdot tg\varphi_1 + \sin\delta_1 \cdot tg\varphi_2}{\sin(\delta_1 + \delta_2)} \qquad (12.26)$$

$$tg\varphi^* = \frac{\sin 36 \cdot tg18 + \sin 48 \cdot tg27}{\sin 84} = 0.58 \qquad (12.27)$$

$$SF = \frac{0.58}{tg31} = 0.97 \qquad (12.28)$$

The graphic result is thus verified, within the uncertainties of the drawing. We note in passing that in this case, the equilibrium does not depend on the intensity of the weight vector because the cohesions on Planes 1 and 2 are set at zero.

It is evident that this result does not allow us to draw a conclusion on the stability of the wedge. We have not taken into account the uncertainties in the measurement of these parameters, the uncertainties on how well they are represented in space, or the phenomena that might threaten equilibrium (such as water pressure in joints 1 and 2). A reinforcement, such as an anchor will thus be necessary.

Case of stabilization by anchoring

We would like to study the effect of an anchor with an azimuth perpendicular to the wall, plunging 20° and having a tension $T = 2000\,kN$ (a force corresponding to the approximate weight of a 200 t mass). Will this anchoring significantly improve the stability of the wedge?

Outside the projection, we can trace a polygon of forces in the plane determined by the weight vector W and the tension T of the anchor (Fig. 12.46).

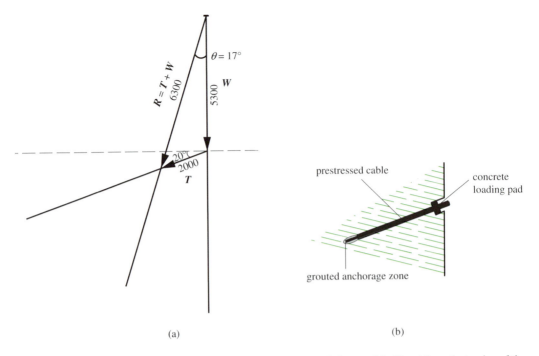

(a)   (b)

Fig. 12.46 Stabilization by an anchor. (a) Construction of the resultant **R** from weight **W** and from the tension of the anchor **T**. (b) Principle of anchor construction.

With **W** = 5300 kN and **T** = 2000 kN, the resultant **R** equals 6300 kN. Vector **R** makes an angle θ of almost 17° with the vertical. This angle can now be transferred to the plane that contains **W** and **T** to obtain the projection of the line containing the resultant **R** in the stereographic projections (in blue on figure 12.43). Obviously, the resultant of the forces closely approaches the center of the zone of equilibrium. We then trace the plane containing the resultant **R** and the side of the dihedron D. We measure the angles of friction on this plane. The angle of friction ε_{max} is essentially the same as for the case of weight only (31°). In contrast, the effective friction angle ε_{eff} is reduced to about 13°. The safety factor increases significantly:

$$SF = \frac{tg\varepsilon_{max}}{tg\varepsilon_{eff}} = \frac{tg31}{tg13} = \frac{0.6}{0.23} = 2.61 \qquad (12.29)$$

We can thus verify that the tension of the anchor can be reduced to a lower value. The case of a tension of 1000 kN, not shown on figure 12.43, would give an angle ε_{eff} of 20°, which brings the *SF* down to 1.65, a more reasonable number.

This example shows in detail how to use the stereographic projection to resolve angular geometry problems that are moderately complex and how the stereographic projection is applicable in engineering geology and rock mechanics.

12.6 The Alps: a tectonic model

We have seen in chapters 3 and 6 that the Alps are the result of the collision between two lithospheric plates: the European plate subducting under the African plate. Numerous deformations that resulted from this collision demonstrate the various responses of rocks to these stresses. Generally, deformation is brittle when it occurs at shallow depths and becomes ductile when pressure and temperature increase (Fig. 12.47).

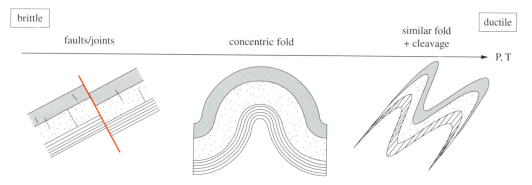

Fig. 12.47 General rule on the type of deformation observed as a function of the depth where the deformation occurred.

The profile across the Alps in western Switzerland (Fig. 12.48) makes it possible to identify several fields of different tectonic styles, from the European plate to the African plate, NNW to SSE.

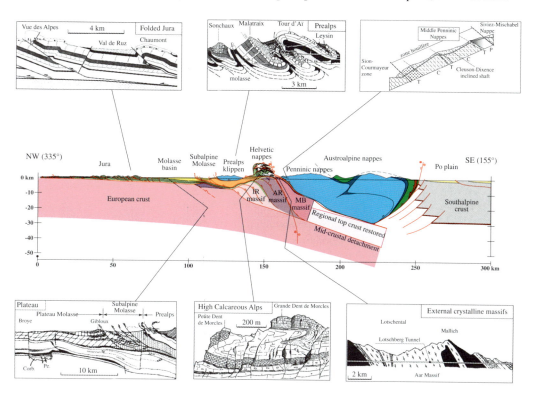

Fig. 12.48 General tectonic profile across the Alps, in western Switzerland. The close-ups show the details of the tectonic style of the different units: folded Jura, Swiss Plateau, Prealps, calcareous High Calcareous Alps, External Crystalline Massifs and internal thrust sheets. According to Burkhard, M., Sommaruga, A. Evolution of the western Swiss Molasse basin: Structural relations with the Alps and the Jura belt. Geological Society Special Publications, 134, 1998.

It is useful to compare this deformed structure to the paleogeographic conditions that existed before folding (Fig. 12.49). These are the locations of the large tectonic units prior to the collision, placed in their original position in the Tethys Sea.

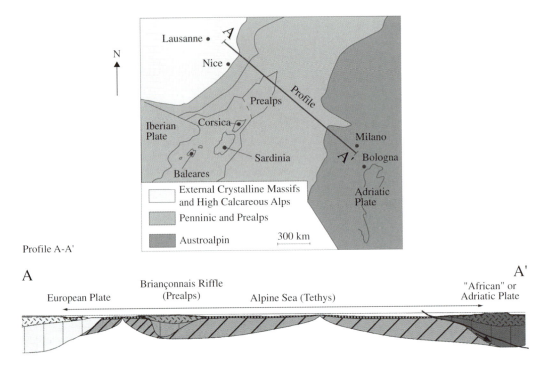

Profile A-A'

Fig. 12.49 Reconstruction of the sedimentary basin of the Alps at the beginning of the folding phase (early Cretaceous). This paleogeography shows the approximate location of present day features (cities, shapes of present-day coasts).

12.6.1 Jura

The rocks that make up the Jura were deposited along the NW margin of the Tethys (Chapt. 3), at the foot of the ancient Hercynian rock mass of the Vosges (Fig. 12.49). This belt has the most external deformations of the Alpine collision (Fig. 12.48). Franche-Comté corresponds to the tabular Jura, vast unfolded limestone plateaus that are separated by faults. The Bresse rift valley was thrust several kilometers. The internal part of the mountain belt, the folded Jura, is a succession of rather simple anticlinal and synclinal folds similar to parallel folds, with square shapes ("box" folds). The anticlines are slightly thrust over the synclines. These folds have detached from the primary substratum along a layer of Triassic evaporites. In summary, the Jura belt is predominantly shaped by brittle tectonics (faults, thrusting, strike-slip movements) and by parallel folds with near-vertical axial planes.

12.6.2 Molasse plateau

The rocks of the Swiss Plateau were deposited slightly more offshore than those of the Jura (Fig. 12.49). The Swiss Plateau is divided into two areas, the exterior and the interior (Fig. 12.48), as follows:

- In the ***Molasse Plateau*** area, the rock series is tabular, crossed by sub-vertical faults in various directions.
- In the ***sub-Alpine Molasse*** area, the rocks are pinched into several tectonic units under pre-Alpine thrust sheets; faults separating these units are inclined toward the Alps, in the direction of the bedding.

12.6.3 Prealps

This is a more complicated tectonic area. On the sub-Alpine Molasse, there are several intensely interpenetrated nappes that have a sedimentary origin much farther to the south (Fig. 12.49). The connection between these roots is totally eroded today; the lost connection can be reconstructed "in the air".

Among these nappes, let us talk about the Préalpes Médianes nappe that have two sub-units with very different tectonic styles. In the plastic Préalpes Médianes nappe, these are large folds with parallel axes, intensely folded and stretched, with axial planes inclined toward the Alps (Fig. 12.48). The rigid Préalpes Médianes nappe, on the contrary, is deformed particularly along the faults breaking up the thick series of limestones and dolomites (for example, the region of Gumfluh south of Château-d'Oex). The roots of the Préalpes Médianes nappes are found above Zermatt.

12.6.4 High Calcareous Alps

The rocks of the High Calcareous Alps, mountains with a majestic profile that border the Rhone Valley upstream of Lake Geneva, were deposited more offshore than the Swiss Plateau and form a cover for the External Crystalline Massifs (Fig. 12.49).

They belong to the Helvetic thrust sheet, of which the Morcles nappe is also a part (Fig. 12.38). These thrusts contain large folds with parallel axes, often with a recumbent axial plane, and the development of clear axial schistosity (Fig. 12.48 and 12.50). Faults often cut the continuity of the folds.

Fig. 12.50 View of the large folds in the Diablerets Nappe, Sex Rouge. GEOLEP Photo, G. Grosjean.

12.6.5 External crystalline massifs

These are massifs of material of predominantly Hercynian origin. The units on the exterior of the Alps (massifs of Aiguilles Rouges, Belledonne and Pelvoux) have preserved structures from that period almost intact; these are parallel folds of great depths. Massifs closer to the zones of high Alpine metamorphism (massifs of Mont-Blanc, Aar, Gotthard and Tavetsch) show a replacement of old structures: the schistosity of the rocks is often near vertical, elongated in the direction of the Alpine folding (Fig. 12.48).

12.6.6 Internal thrust nappes

The southern side of the Rhone in Valais is a complex stack of thrust nappes (Fig. 12.48). This is the location of the most intense Alpine deformation in the mountain belt: parallel folds, schistosities, lineations. These nappes are subdivided into two paleogeographic areas: the Penninic area and the Austroalpine area, with the latter belonging to the African plate (Fig. 12.49).

The *Penninic* area includes the lower thrusts of the Alpine complex. They are composed mainly of rocks of the continental crust, remains of oceanic crust (for example, the Aiguilles Rouges of Arolla, Valais) and associated sediments.

The *Austroalpine* area includes upper thrust nappes (for example, the gneisses of Cervan (the Matterhorn) and the Dent-Blanche), which show the intense internal deformation of Hercynian origin. In the SSE, these thrusts have their roots in a very upright position in the principal scar of the collision. In this location there is an outcrop of ultramafic rocks from the African mantle.

12.6.7 Present-day deformation

The comparison between old and present-day elevation measurements, reprocessed by modern geodetic methods, shows vertical movements in the Alpine area, particularly in Switzerland. If we choose an area that we assume to be invariable (region of Aarburg on the border between the Molasse Plateau and the Jura) we see that the Alps are systematically rising compared to this point, with average velocities on the order of 1 mm/a. The central region of the Alps shows maximum velocities of 1.5 mm/a. The Jura, on the contrary, is slightly subsiding with velocities of several tenths of a millimeter per year (Fig. 12.51). This observation can be compared to some active Alpine faults with displacements on the same order of magnitude. According to the analysis of earthquakes it seems that these are for the most part strike-slip movements and normal faults that are active today. We can thus conclude that the Alps are still in the process of deforming. The science that studies these present-day movements is called *neotectonics* (Fig. 12.52). Geologists and engineers who study sites for nuclear waste storage pay a lot of attention to these deformations, which may have an impact on the integrity of a repository over its long lifespan.

In the near future, geodetic scientists will be able to measure minute horizontal movements such as altitude, particularly due to the rapid technical progress of satellites (Global Positioning System). It will then be possible to show areas of compression or extension and variations in the velocity of movement. This will make it possible to interpret the movements of rock masses and to attribute the cause to one or another phenomenon at depth.

[60, 102, 108, 181, 211, 219, 223, 224, 225]

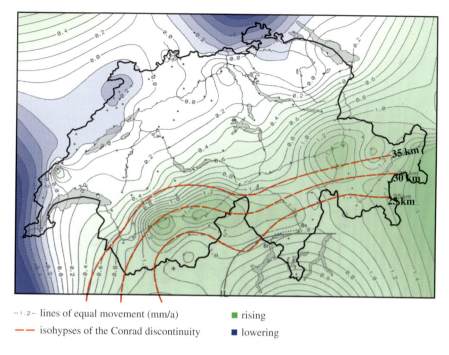

─¹·²─ lines of equal movement (mm/a)   ■ rising

── ── isohypses of the Conrad discontinuity   ■ lowering

Fig. 12.51 Map of recent vertical movements in Switzerland (from [13]).

Fig. 12.52 Neotectonics is revealed by faults that are still moving today, whether gradually or during earthquakes. By digging large research trenches, Chinese geologists have identified displacements in Quaternary sediments NE of the Himalayas that show significant movement in very recent times. Yanking-Huailai Basin, NW of Beijing. GEOLEP Photo, A. Parriaux.

PROBLEM 12.2

In a poor region with a tropical climate, a village constructed on a plateau must deal with a chronic shortage of drinking water. The only water source is a spring at the base of a large mountain that dominates the village. The women go to this spring for water and carry it back in buckets. This spring does not always provide enough water. Sometimes it even goes dry. A humanitarian organization has sent you there to evaluate a new source of potable water.

You begin by conducting a geologic survey of the region and drawing a cross section (Fig. 12.53). You observe that the area is made up of sedimentary formations (see the series of the legend of figure 12.53).

These layers have been deformed by tectonic movements and they have the morphology shown on the geologic cross section.

Questions:

- On the cross section, identify:
  - a syncline, draw its axial plane, and comment on its position
  - an anticline, and draw its axial plane
  - a normal fault
  - a reverse fault
  - a thrust fault and indicate the approximate amount of displacement

- What influence does the tectonic structure have on the presence of groundwater? Considering the nature of the rocks, draw the principal zones where there should be an accumulation of groundwater.

- Where would you locate a new well for the village and what techniques would you use to produce the water from this aquifer?

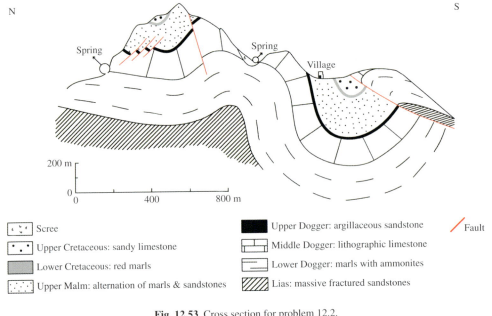

Fig. 12.53 Cross section for problem 12.2.

Look at the solution on the DVD

Chapter review

Rocks did not await the arrival of engineers and their construction projects to deform. The deformation status of a rock mass is a result of its long tectonic history. The natural stress state results from the latter phases of this history. Over these long periods, rocks in nature have behaved differently than rocks in laboratories: elastic rocks have been folded like ductile rocks due to factors that are difficult to reproduce experimentally, the effect of time, in particular. In planning projects, engineers must take into account the different scales of tectonics, in order to predict the shape of the rock bodies as precisely as possible and to integrate the role of deformation in the structure of the rock massif. As in sedimentology, an understanding of the causes of deformation will be a great help in reducing uncertainties in the planning and construction of engineering projects.

13 Weathering

Weathering is an important process in the transformation of geologic materials, as important as diagenesis (Fig. 13.1). In contrast to diagenesis, which changes loose sediments into a consolidated material called rock, weathering turns solid rock into loose material.

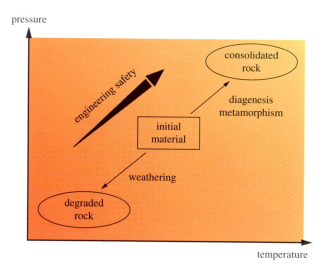

Fig. 13.1 Weathering, the reverse process of diagenesis and metamorphism.

Weathering includes a group of processes that transforms rocks into mechanically weak, easily eroded material. Weathering affects mineralogy and the physical structure of a rock. *Surface weathering* is controlled by climate and shallow groundwater flow; it extends to a depth of several meters. It is the cause of soil genesis (*pedology* = soil science). *Deep weathering* alters the mineralogical composition of rocks at depth in areas of circulating hydrothermal fluids (§ 6.4.1).

We will focus on surface weathering because it is directly involved in many shallow engineering projects. Surface weathering leads to erosion and thus has practical consequences for the engineer, both in terms of hydrogeology and the mechanical strength of the ground. The parameters used to characterize solid rock no longer apply to weathered rocks. What modifications must the engineer incorporate into the project to account for weathering?

The answer to this question is complex. It involves a consideration of many different situations in which both weathering and the geological material play an important role. In this text, we will examine only the principal factors of weathering.

13.1   Weathering processes

Here we distinguish between the agent (rainwater, for example) and the process (dissolution). In the following classification, we give priority to the processes (Fig. 13.2).

For clarity we will discuss weathering processes individually.

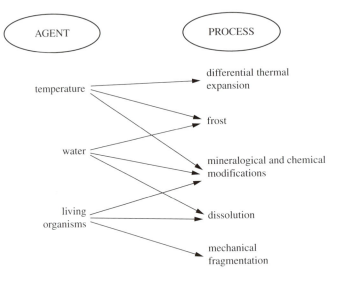

Fig. 13.2 The most common weathering processes.

13.1.1 Thermal processes

The ground surface and the upper part of the zone of seasonal temperature variation are subject to significant and repeated temperature changes that cause the physical weathering of geological materials. Here we will distinguish between processes that involve a change of the physical form of water and those that do not.

13.1.1.1 Temperature variations

Minerals have different thermal expansion coefficients (Tab. 13.1).

Table 13.1 Volumetric Thermal Expansion Coefficients of some Important Minerals.

Mineral	Volumetric thermal expansion γ [K^{-1}]
quartz	$34 \cdot 10^{-6}$
feldspar	$15 \cdot 10^{-6}$
calcite	$24 \cdot 10^{-6}$ (parallel to the optical axis)

Drastic temperature variations cause shearing effects at the contacts between minerals, causing a breakup of the rock structure. For example, granite turns into ***granitic sand*** (sand made up of quartz and feldspar from the granite) (Fig. 13.3 and Fig.13.4). This physical weathering is generally accompanied by chemical weathering.

Such a process is particularly active in regions where the climate exhibits stark contrasts, as in deserts, for example: in the Sahara, diurnal-nocturnal variations are often several tens of degrees.

In order for alteration caused by differential expansion to be really effective, it must be accompanied by other processes. The most effective of these are variations in water content.

Fig. 13.3 Weathering of porphyritic granite in a humid tropical climate. Durban, South Africa. The minerals are still in place in the rock but they are no longer firmly connected. Once eroded and transported by water, they will form a granitic sand layer. GEOLEP Photo, A. Parriaux.

Fig. 13.4 The process of granular disintegration in a granite massif in the Vosges, France. Several balls of altered rock are all that remain in this mass, most of which has been turned into sand. GEOLEP Photo, A. Parriaux.

13.1.1.2   Freezing

We will not discuss the deep freezing that occurs in the Arctic countries, which was covered in chapter 8. Here we focus our attention on the temporary surface freezing that occurs in temperate latitudes.

When the winter temperature goes below the 0°C barrier, interstitial water causes the surface soil to freeze. Because water expands almost 10% when it freezes, it damages the rock. It is useful to compare the effects of freezing on loose material and on hard rocks.

Freezing of unconsolidated material

First let us consider materials that have large pores, such as gravels or coarse sands. Near the surface, they are generally not water saturated. Surface water drains downward to ground-water-bearing layers. These sediments contain very little capillary water. As they freeze, some air is expelled, but there is no deformation of the sediment framework. In the rare cases where these near-surface sediments are saturated, they may be the location of springs that release heat and prevent the ground from freezing. These spring zones can be easily located in winter by the fact that the snow around them melts quickly.

In contrast, fine-grained loose materials (silts and clays) are very sensitive to freezing. They are often saturated near the surface because of the presence of perched groundwater. When this water freezes, the expanding ice pushes the grains upward causing a slight upheaval of the soil. When these fine grounds are not saturated, a second phenomenon causes the material to disintegrate: *cryosuction*. (Fig. 13.5a). When an ice crystal forms in a pore of finely granular soil that contains water and air, it exerts a suction potential on the water in neighboring pores (just as a dry soil does) due to the reduction of the liquid water content; this water migrates toward the ice crystal as an active film that covers the grains and freezes at the surface of the crystal. This phenomenon leads to the accretion of ice as horizontal lenses at certain places in the soil. The ice pushes the mineral material out of the soil. All of this leads to significant swelling. When it thaws the soil does not re-establish itself homogeneously: the part of the sediment occupied by the ice lenses subsides deeply. Silty materials are more sensitive than clays to cryosuction because clays are not permeable enough to allow significant migration of water toward the ice lenses.

This phenomenon is well known to highway geotechnical engineers, who recognize the necessity to install thaw barriers to prevent the deterioration of the road coatings during thaw periods (Fig. 13.5b). On main roads, frost sensitive soils are replaced by gravel to the freezing depth. This depth is calculated on the basis the local climatic conditions. North of the Alps, for example, it occurs at a depth several decimeters to a meter.

Freezing is an issue that must be considered in the design of caps over highly toxic waste repositories. The waste is covered with a multi-layer cover that must prevent the penetration of meteoric water into the landfill to the maximum possible extent. One of the principal components of this multi-layer cover is a clay layer (Fig. 13.6) that must be extremely watertight. But the sealing capability of clay is greatly decreased if it is subject to freeze-thaw cycles because these cycles cause discontinuities that are more permeable than the clay matrix. For this reason, this layer must be placed at a depth where it will not be affected by the freezing front.

Freezing of rocks

Fine porous rocks such as marls or fine-grained sandstones behave similarly to the fine unconsolidated sediments described above. Ice forms in fine lenses, parallel to stratification, causing the rock to disintegrate into sheets (Fig. 13.7).

Other rocks are sensitive to freezing if they have preexisting fissures. A fissure that is open to the surface and whose base is impermeable will tend to collect water. In the beginning, the fissure is plugged up by ice that confines the water to the fissure. As the freezing continues, the water in the fissure will freeze. The expansion exerts a stress on the sides of the fissure, causing it to widen. This phenomenon continues until one side of the fracture breaks loose, which may cause a rock fall. Alpine climbers and quarry operators recognize the risks of rock falls during alternating periods of freezing and thawing (Fig. 13.8).

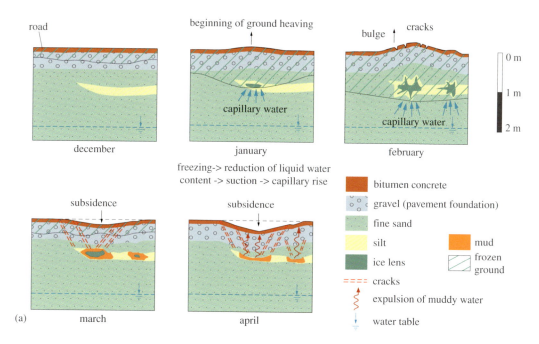

Fig. 13.5 (a) Freezing of road foundations. (a) Diagram of the freezing process. An animation on the DVD shows the different phases of the frost-thaw cycle. (b) Examples of roads in Ulan-Bator that are subject to annual temperature variations from −40 to +30°C. When they thaw, the bearing capacity is reduced and cracks appear in the asphalt coating. GEOLEP photo, A. Parriaux.

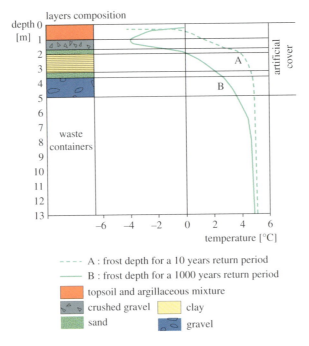

A : frost depth for a 10 years return period
B : frost depth for a 1000 years return period
topsoil and argillaceous mixture
crushed gravel   clay
sand   gravel

Fig. 13.6 Cap studied for the radioactive waste landfill at Aube (northeastern part of France). Simulation of water seepage and freezing in the cap with various atmospheric scenarios. Research on air temperatures over a period of several decades allowed researchers to make a statistical estimate of the annual intensity of freezing (using Gumbel's law). By extrapolation, they obtained the probable freezing conditions for a period of 1000 years. These forecasts were compared to historical records; For example, Garnier (1967) reports on the archives for the 1570–1571 winter: "The winter was so cold from the end of November until the end of February that for three months the rivers were frozen enough to support all carriages." On the graph are reported simulations of a 10-year freeze on dry sediments (Curve A: 16 days of freezing at −10°C) and a 1000-year freeze (Curve B: 86 days of a freeze at −10°C). It can be seen that the clay layer resists the maximum of the 1000-year freeze (from [111]). Climate change must be taken into account in the scenarios.

Fig. 13.7 Freezing has caused the breakup of an argillaceous rock into sheets, Petit Muveran, Switzerland. GEOLEP Photo, A. Parriaux.

Fig. 13.8 The effect of freezing on a fissured rock. Exploded block at Stock Kangri, 5200 m, Ladakh, Himalaya. GEOLEP Photo, A. Parriaux.

To determine if a rock can be used as an exterior building stone, it is subjected to freezing strength tests in the laboratory (Fig. 13.9). Samples are placed in a thermal container where repeated simulations of freeze-thaw cycles are conducted. The samples are placed in contact with water surfaces that ensure a capillary water supply to the rock. The volumetric expansion of the rock is measured during each freeze-thaw cycle. These values are reported graphically and are used to construct a swelling curve.

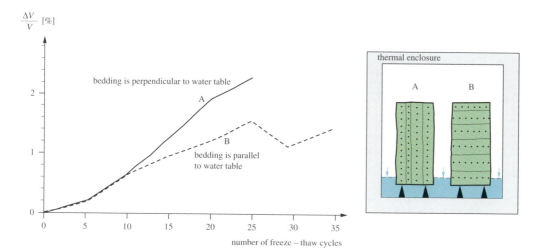

Fig. 13.9 Test of freezing strength of rocks: examples of molasse sandstone for the repair of monuments constructed of molasse in the historic center of Lausanne. V = sample volume (from [89]).

13.1.2   Physicochemical processes

Physicochemical weathering processes are reactions between solids and water. The water can be surface water or groundwater. Biology also plays an important role in some reactions. We distinguish here between dissolution, which causes a solid to go into solution, and mineralogical modifications involving hydration, which occur in the solid state.

13.1.2.1   Dissolution

For many minerals, dissolution in pure water is less efficient than in slightly acidic water. This is especially true for carbonates. Various acids can be found in water that percolates through the subsurface:

- Acids from rain (HCl, HNO_3, H_2SO_4, H_2CO_3) partially of anthropogenic origin;
- Acids from plants and soils (H_2CO_3, humic acids).

High in the mountains, plants are replaced by lichens, which also contribute to the dissolution of their rocky substrate. Some of them are specific to the composition of the rock; for example, the famous yellow-green lichen *Rhizocarpum geographicum* grows only on siliceous rocks (Fig. 13.10).

Fig. 13.10 The lichen *Rhizocarpum geographicum* is specific to siliceous rocks. (a) Lichen encrusted on granite. GEOLEP Photo, A. Pantet. (b) Thin section showing the penetration of its filaments into microfractures in previously fragmented granite. A. Puiz Photo, Natural History Museum of Geneva City.

The direct role of bacteria in the weathering of minerals is still poorly understood. We know that in the case of sulfur, sulfur-oxidizing bacteria such as *Thiobacillus* act as catalysts in weathering reactions. They are capable of making sulfuric acid and calcium sulfates. This type of weathering causes the corrosion of limestone rocks and calcareous sandstones. This is called degradation by blistering or patchy degradation.

A fundamental distinction is made between complete dissolution (or congruent), which as its name implies, causes a solid phase to go completely into solution, and incongruent dissolution, which generates new solid phases.

Congruent dissolution

The majority of minerals that are salts and some oxides are completely soluble in water. The solid goes into solution as ions. Ionic solubility is given by the solubility product K_s:

$$[C]^n \cdot [A]^m = K_s$$

where C = cation; A = anion; n = number of cations; m = number of anions.

The solubility product depends on pressure and temperature. Solubility in water can also be described by a concentration at saturation. For the principal minerals, the solubility in pure cold water is as follows, in decreasing order:

- Halite (NaCl)   360 g/l
- Anhydrite ($CaSO_4$)   3 g/l
- Gypsum ($CaSO_4 \cdot 2H_2O$)   2 g/l
- Calcite ($CaCO_3$)   0.014 g/l
- Quartz (SiO_2)   0.012 g/l
- Dolomite ($CaMg(CO_3)_2$)   0.0003 g/l

The solubility of calcite and dolomite in acidic water increases by several orders of magnitude (calcite dissolution reactions (§ 9.1.2).

Dissolution leads to the enlargement of fissures in the rock, then to the creation of significant cavities: this is *karstification* (Fig. 13.11).

(a)   (b)

Fig. 13.11 Karst morphology. (a) A slab of limestone begins to be dissolved along flutes parallel to the slope direction (lapiaz), which correspond to the direction of rainfall runoff and to the water flow at the top of the bedrock. Pierre du Moëllé, Prealps, Vaud, Switzerland. (b) As the flutes deepen, eventually all that is left are limestone blades and pillars with impressive shapes. Rock Forest of Yunan in southern China. GEOLEP Photos, A. Parriaux.

Conversely, groundwater that contains bicarbonates loses a portion of its dissolved carbon dioxide at a spring. Equation 9.6 moves to the left and limestone precipitates on every type of substrate it encounters at a spring: The result is a porous crust called *tufa* (Fig. 13.12). Tufa has been used for many years as a building stone because it is easy to cut and it is light and strong. However, it has the disadvantage of collecting soot from urban pollution in its pores. When tufa is stratified, it is called *travertine*. It is still exploited today and is used as an ornamental rock (Fig. 13.13).

Fig. 13.12 Tufa outcrop. Gorges de Covatannaz, Jura, Vaud, Switzerland. GEOLEP Photo, G. Franciosi.

5 cm

Fig. 13.13 Slab showing travertine's stratification. GEOLEP Photo, G. Grosjean.

Karstification also occurs in gypsum. Dissolution can pose significant problems for engineering project foundations and can cause reservoirs to leak (Fig. 13.14).

Cargneules are in a category by themselves. Here, dissolution affects the gypsum elements in particular (§ 10.4.5), which makes the rock highly porous.

Fig. 13.14 A sinkhole in a soccer field in the city of Bex, Prealps, Vaud, Switzerland. A doline is present in the gypsum under the sediments of an alluvial fan in Avançon. Nothing suggested its presence at this site. GEOLEP Photo, P. Turberg.

Incongruent dissolution

Most of the rocks in the crust are silicate in nature. They are not very soluble but the dissolution that does take place produces new silicate minerals, often in the clay family. One relatively simple example is that of orthoclase (Fig. 13.15). Dissolution can be broken down into a series of stages. These slow alteration processes are as complex as the minerals they affect. The description of these processes is beyond the scope of this book. As a general rule, we can say that silicates that crystallize from magma at high temperatures are more susceptible to weathering than those that crystallize at lower temperatures. (§ 6.3.2.1). This is due to the fact that the thermodynamic stability field of high temperature minerals is very different from the pressure and temperature conditions at the ground surface.

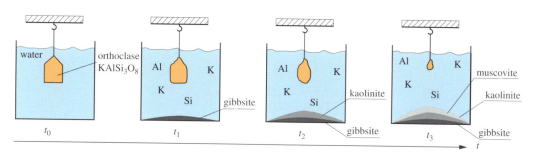

Fig. 13.15 Stages in the incongruent dissolution of orthoclase. At t_0 an orthoclase crystal is placed in contact with water. At t_1, the weathering of orthoclase leads to Si, Al, and K ions going into solution; simultaneously, gibbsite $Al(OH)_3$ precipitates. Weathering continues (t_2) and kaolinite precipitates. In the final stage (t_3), muscovite forms.

In practical terms, these reactions are much slower than salt reactions.

The neogenesis of clay minerals must be monitored because it can cause rocks to become more sensitive to weathering and more ductile. Incongruent dissolution may play a major role in transforming rock into soil.

13.1.2.2 Mineralogical modifications

In this paragraph we will discuss the hydration of minerals. The incorporation of H_2O groups in the crystalline skeleton leads to swelling. Rock can come into contact with water naturally as a result of groundwater circulation in the subsurface or artificially as a result of engineering projects such as tunnels. Tunnels can modify the hydration status of the ground because they bring water or humid air into massive rocks that had previously been dry. Mechanical decompression of rock around the projects facilitates this process. Two rocks in particular undergo this hydration:

- Anhydride ($CaSO_4$) changes into gypsum ($CaSO_4 \cdot 2\ H_2O$) and theoretically swells 60%; in reality, the maximum swelling occurs when anhydrite alternates with clay beds (Fig. 13.16 and Fig. 13.17);
- Clay rocks have a tendency to incorporate hydrated cations between the clay mineral sheets, and this controls their spacing and the swelling of the rock. Not all clays react in the same way to variations in water contact. Mineralogical analysis and tests must be conducted to determine engineering risks.

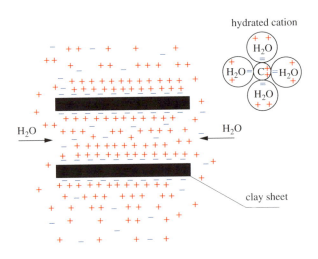

Fig. 13.16 Process of swelling by hydration. Introduction of hydrated ions into the spaces between clay sheets.

These phenomena take place rapidly enough that they must be taken into account in underground projects. As an example, two tunnels dug through anticlines in the northern Jura of Switzerland caused spectacular swelling due to Triassic anhydrite at one location and Opalinus clays at another.

- In the Belchen highway tunnel between Oensingen and Basel (Switzerland) the highway foundation swelled 90 cm in the anhydrite zone; repeated repairs were necessary to keep the road in operation.

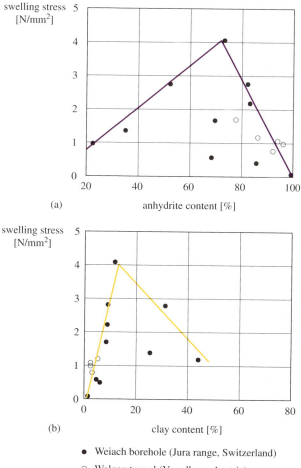

(a) anhydrite content [%]

(b) clay content [%]

● Weiach borehole (Jura range, Switzerland)
○ Walgau tunnel (Vorarlberg, Austria)

Fig. 13.17 Anhydrite's susceptibility to swelling, depending on the clay content. The case of Triassic anhydrites of the western Jura (Weiach boring) and the northern calcareous Alps (Walgau Tunnel). (a) Swelling as a function of anhydrite content. (b) Swelling as a function of clay content. Note: The same samples were used for both graphs (from [173]).

- The Bötzberg railroad tunnel, between the Aar Valley and Fricktal, has swollen 45 cm in 31 years (Fig. 13.18).

In such a case, a tunnel with a cylindrical section is recommended to resist the forces of swelling. In fact, a circular section has maximum strength, like a whole arch, without any zone of weakness.

Swelling can also affect clay-rich unconsolidated sediments by increasing their water content during excavation (mechanical decompression and presence of water in the excavation). This can cause significant deformation of building foundations.

Conversely, these sediments are prone to dessication shrinking when the water content decreases. Significant subsidence and polygonal cracking (Fig. 13.19) may accompany this shrinking. Thus extreme droughts can cause significant damage to buildings with shallow foundations in areas with clay-rich subsoils. For example, an analysis of more than five hundred cases of building cracks in France following the 1989 drought showed the vulnerability of constructions that lack deep foundations (Fig. 13.20).

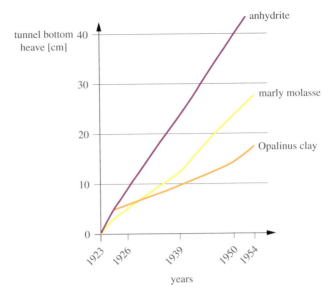

Fig. 13.18 Example of the effects of swelling on the foundation of the Bötzberg tunnel (eastern Jura, Switzerland) on various formations that are prone to swelling.

Fig. 13.19 Shrinkage cracks in clays. These clay muds were produced by gravel washing. GEOLEP Photo, G. de los Cobos.

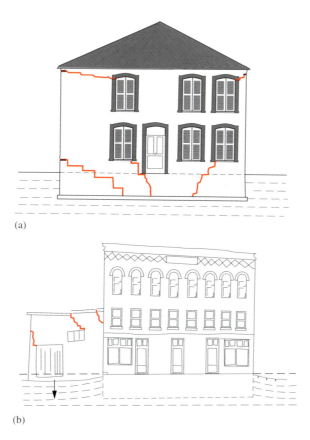

Fig. 13.20 Subsidence of buildings as a result of dessication of clay soils. (a) Typical configuration of cracks in a fragile structure on a shallow foundation and clay soil during a dry period. (b) Cracks in an annexe on a shallow foundation built next to a main building.

Mineralogical modifications that cause swelling also occur during the reaction between gypsum and hydraulic binders. This occurs especially during the treatment of fine-matrix, water-rich loose sediments, where the mixing with cement is supposed to increase the strength of the foundation. If the sediments contain gypsum, there is a reaction between the cement hydrates and the sulfates, which creates a calcium sulfo-aluminate called ***ettringite***, a mineral that has exceptional swelling capability (Fig. 13.21). Such a phenomenon can also occur with sulfide minerals such as pyrite, in particular when the aggregates are excavation material from tunnels or mining waste. At first the sulfur oxidizes to sulfates, then the ettringite appears. The gypsum reaction during stabilization to lime produces another swelling mineral: ***thaumasite***.

13.2   Catalog of weathering-prone materials

To summarize the various processes discussed above, we can draw up a general list of unconsolidated sediments and rocks that are vulnerable to one or more of these processes (Tab. 13.2). The engineer needs to be aware of the cases mentioned and study these environments carefully before construction.

Fig. 13.21 Swelling of a road before it was put into service. The foundation was constructed by stabilization with cement of tunnel excavation material containing gypsum marl. The interstitial water in the fill contained locally more than 2000 mg/l sulfate in solution. The company had failed to verify the water geochemistry of the fill material and used ordinary cement. The road had to be entirely demolished and rebuilt with a conventional foundation. Yverdon-les-Bains. GEOLEP photo, A. Parriaux.

Table 13.2 List of the main types of terrain that are susceptible to freezing, dissolution, or chemical and mineralogical modifications.

		frost	dissolution	mineralogical and chemical modifications
unconsolidated material	organic soil	high	weak	medium
	lacustrine silt	high	very weak	weak
	till	weak	very weak	very weak
sedimentary rocks	cargneule	weak	weak	weak
	calcareous sandstone	weak	very weak	very weak
	marls	high	very weak	medium
	limestone	very weak	medium	very weak
	anhydrite	very weak	high	high
	gypsum	very weak	high	medium
	saccharoidal dolomite	medium	very weak	medium
Magmatic and metamorphic rocks	granite	very weak	very weak	weak
	mica schists	weak	very weak	very weak
	calcschist	weak	weak	very weak

very weak   weak   ****** medium   high

13.3   Extent of weathering at depth

Weathering is essentially a surface process. In homogeneous terrain, weathering decreases with depth. Pedology generally confirms this concept because it shows the following superposition of layers that range in thickness from several decimeters to several meters (Fig. 13.22):

- Organic layer,
- Horizon of geological material that is physically and chemically disintegrated,
- Horizon of geological material that is primarily physically disintegrated (the rock is decompressed and fragmented),
- Non-decomposed bedrock.

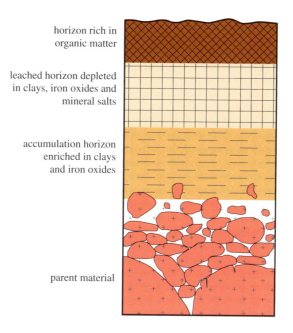

horizon rich in
organic matter

leached horizon depleted
in clays, iron oxides and
mineral salts

accumulation horizon
enriched in clays
and iron oxides

parent material

Fig. 13.22 Typical alteration profile of a soil.

In real geological environments, homogeneity rarely exists. This explains why it is possible to find unaltered sediment near the surface, resting on less well-preserved layers. This case is typical of formations made up of alternating beds of resistant rocks and easily eroded rocks, such as flysch, for example: the very dense and well-cemented sandstones are much stronger than the intercalated mudstones.

Another important factor should be taken into account: circulating groundwater. It can often be aggressive and cause weathering deeper than the levels where it circulates in aquifer rocks (Fig. 13.23). In the case of cold groundwater, these depths may be several tens of meters or even hundreds of meters in fissured or karstic aquifers. Deeper water may be thermal and may have unusual chemistry. This water can create weathered zones that are sometimes encountered in tunnels.

In humid tropical regions, the combined effect of rainfall and high temperatures causes a very advanced dissolution of silicates. Silica and alumina are leached out with the other cations at depth. All that remains at the surface are insoluble iron oxides that form a hard red crust that can be several meters thick, called a ***laterite*** (Fig. 13.24). In some areas, aluminum is concentrated in horizons that constitute the primary deposits of this metal: ***bauxite***.

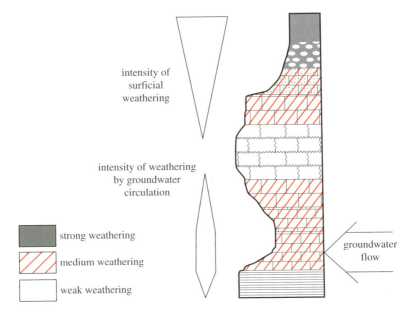

Fig. 13.23 Distribution of surface alteration in a natural profile. The phenomenon occurs particularly at the level of the soil, but also more deeply, as a result of preferential pathways of groundwater.

Fig. 13.24 Lateritic soil produced by intense, high-temperature leaching of a geologic substrate. South Africa. GEOLEP Photo, A. Parriaux.

13.4   Engineering concerns

Weathering is a natural process that takes place over time. Each process acts on the geologic environment according to its own kinetics. Degradation can occur in several hours in some cases (for example, the swelling of clays) or over a period of several thousand years (mineralogical weathering of magmatic rocks and karstification of carbonates, for example). This variable reactivity is a particular concern for the engineer, who must consider geological risks in terms of the lifespan of a structure or a development.

However, the kinetics of some weathering processes is strongly influenced by human activities, particularly by civil engineering projects. Rocks that swell in tunnels are one example. The natural weathering regime can be disrupted, with potentially detrimental consequences for the engineer.

Today's engineer must be concerned not only with the effects of a structure during its functioning life, but also with its consequences after it has been abandoned. For example, what are the effects of rock weathering around an abandoned mining tunnel that is no longer maintained? To simply wall up its entrance is not a solution that is compatible with a policy of sustainable development; in some cases collapses can be expressed on the surface and can cause modifications of groundwater flow. During the construction of a civil engineering project, it is crucial to determine the possible types of weathering and their consequences. The safety factor, SF is used to calculate the level of safety appropriate for a structure and it should be used also for the safety evolution after the end of its useful life (Fig. 13.25).

We have cited several practical effects of weathering, but we can group these consequences more systematically into two categories of effects: mechanical and hydrogeological.

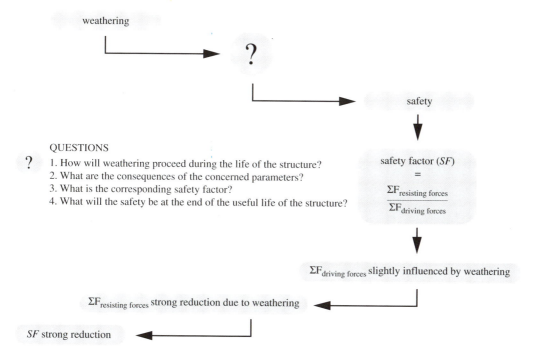

Fig. 13.25 Questions for the engineer on the relationship between construction and weathering. The safety factor SF describes the ratio between resisting force and driving force in terms of instability; when the ratio is less than 1, the structure becomes unstable.

13.4.1   Weathering effects on mechanical properties

A weathered material always has less mechanical strength than its original material. We can show this by a simple compression test conducted on a core of rock (Fig. 13.26). A weathered rock ruptures before an unweathered one. Also the weathered rock often behaves less elastically than the solid rock. This phenomenon is fundamental to slope instability (Fig. 13.27).

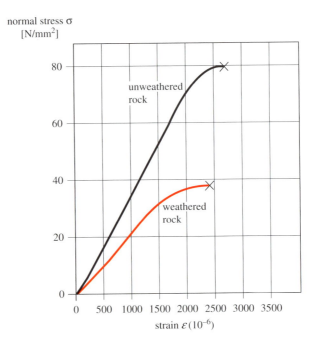

Fig. 13.26 Difference in stress-strain behavior between an intact rock (siliceous conglomerate from the El Juncal Quarry, Zamora, Spain) and a weathered rock (same rock quarried in the 18th century and altered by atmospheric agents). The cross means rupture.

Fig. 13.27 The large Hope Landslide in British Columbia (Canada) was due to the weathering of felsic and mafic rocks in discontinuities, among other factors. The discontinuities are zones of weakness filled with a fine alteration matrix composed predominantly of chlorite and fibrous amphibolite. The mass of 48 million cubic meters of material gave way abruptly in 1965, burying the road under 50 m of debris and killing four people. It was believed for a long time that this event was due to an earthquake, because of its perfect coincidence with shocks measured on a seismograph in the region; in fact, vibration spectrum studies showed that the shocks were the result of the abrupt slide, and not the cause. GEOLEP Photo, A. Parriaux.

Weathering also affects the suitability of rocks for building stone or ornamental stone (Fig. 13.28).

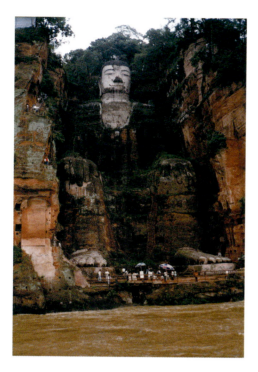

Fig. 13.28 Soft sandstones are well suited to sculpting for monuments, sometimes very large monuments. In the 8th century the Buddha of Leshan was carved into a red sandstone cliff along a tributary to the Yangtze River in China. It is 71 meters tall. Weathering of the rock has required numerous restorations over the years. GEOLEP Photo, A. Parriaux.

13.4.2 Weathering effects on hydrogeological properties

Weathering creates porosity in environments that were not porous originally. This is called secondary porosity, the same as in fissured rocks. There are three principal types:

- Dissolution of the cement of a detrital rock, such as the molasse sandstones, which are reduced to sand;
- Karst dissolution of limestone and gypsum;
- Breakup of crystals in magmatic rocks.

This transformation can thus turn an impermeable rock into an aquifer. A perfect demonstration is the molasse sandstones that make up a significant part of the Swiss Plateau. The surface is decalcified to a depth of about 5 to 10 meters, and below that the sandstone cement causes the rock to be almost impermeable. Thus a groundwater-bearing layer occupies the pores of the weathered sandstone, giving rise to a myriad of small springs that are often exploited as drinking water. A remarkable example is the Pierre-Ozaire gallery, where the city of Lausanne collects water from numerous springs at the contact between the weathered molasse and the unweathered molasse in the Jorat Massif (Lake Geneva Basin) (Fig. 13.29).

Fig. 13.29 Water-bearing nature of the molasse, as a result of the decalcification of the sandstone. GEOLEP Photo, S. Kilchmann.

PROBLEM 13.1

The Swiss electric company Romande Energie and the water utility of the city of Lausanne together exploit the Bornels spring, near the village of L'Etivaz, in the Vaud Prealps. The collected water is transported via a long underground supply line to the heights of Montreux where a part of it goes through a turbine. The other part is sent to Lausanne for drinking water.

Originally, the water flowed from a large spring into a pasture at the foot of limestone cliffs at the western end of the Gumfluh massif. The spring occurs at the contact between carbonate aquifer rocks and the schistose-sandstone flysch that forms an essentially water-tight base (Fig. 13.30).

In 1896 the spring water was captured by a original gallery that went up the principal groundwater outlet (Fig. 13.31); it went through more than 200 m of diverse carbonate and evaporite rocks, before reaching a water-bearing limestone massif. A complex system of galleries was dug into this zone to capture the maximum amount of water (northeastern extremity). This old gallery had suffered significant deformation that had been difficult to control (see reconnaissance done in the 1980s. Fig. 13.32).

In the 1960s the decision was made to dig a new gallery that would approach the limestone massif area along a straight line (Fig. 13.31). When the gallery was built in 1968, it was quickly invaded by several tens of cubic meters of mud (Fig. 13.33). This unstable zone was walled up and the gallery was deviated to avoid it.

In the 1980s, this same zone showed signs of instability: the concrete had turned to paste, and revealed the framework; the sidewalls and the vault had difficulty holding the rock. A drilling chamber was installed downstream of the site and three radiating boreholes were drilled to determine the nature of the rock in this sector. Figure 13.33 shows the geological map.

Questions

Analyze the source of the problems encountered in these two projects. Propose a plan for a section of a gallery to cross this zone under the best possible conditions. Draw the trace of the gallery on the geologic map and make a list of construction methods.

Fig. 13.30 Tectonic drawing showing the location of the Bornels gallery for Problem 13.1. From the Swiss national map, sheet 1265 at 1:25,000. Reproduced with authorization of Swisstopo (BA056911).

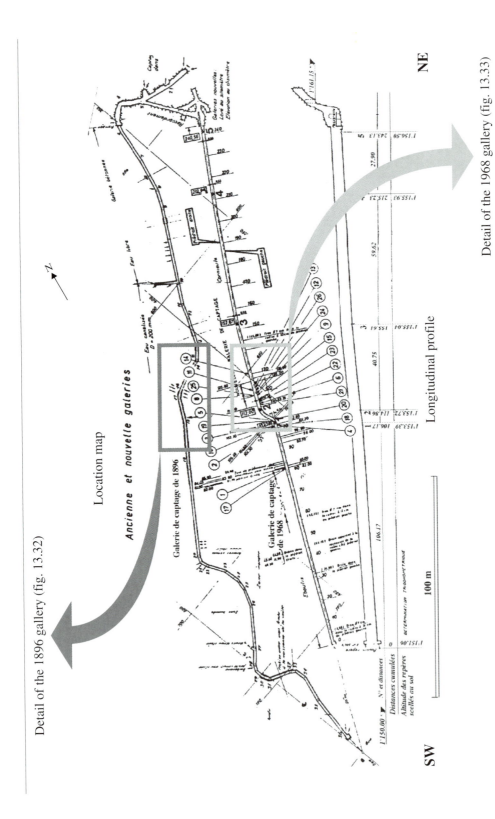

Fig. 13.31 Location map of the Bornels galleries of 1896 and 1968, with longitudinal profile of the gallery of 1968 (for Problem 13.1).

Detail of the 1896 gallery (fig. 13.32)

Location map

Detail of the 1968 gallery (fig. 13.33)

Ancienne et nouvelle galeries

Longitudinal profile

NE

SW

100 m

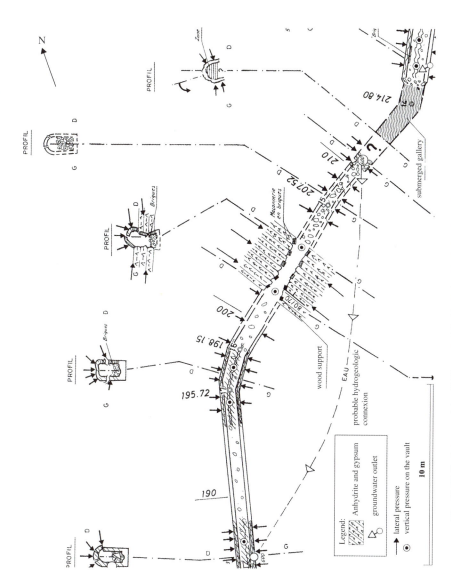

Fig. 13.32 Partial map of the 1980 status of the Bornels gallery constructed in 1896 (for Problem 13.1).

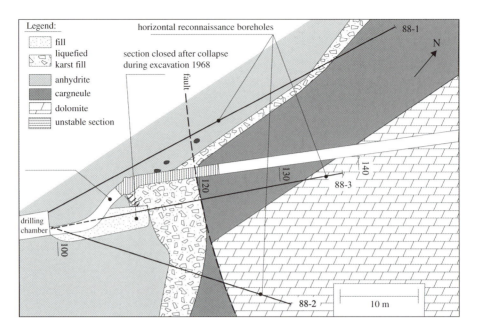

Fig. 13.33 Geologic map of the 1968 Bornels gallery (for Problem 13.1). Plan-view map at the gallery level.

Look at the solution on the DVD

13.5   Screening for weathering

Understanding the weathering of geological materials and their effects requires complex analysis. It involves a series of methods that consider the problem on different scales, from minerals to the entire rock mass. Let us cite some of them:

- Mineralogical analysis (polarizing microscope, X-ray diffraction, etc.) to identify minerals typical of alteration; (Fig. 13.34)
- Hardness tests on samples (for example, the ***Schmidt hammer*** can determine hardness by the way a hammer bounces off a rock).
- Geomechanical tests (compression tests, tensile strength, shear strength, etc.);
- Geological survey of the area (mapping of weathered zones);
- Geophysical reconnaissance to measure seismic velocity contrasts (low velocity for altered rocks) or electrical resistivity (low electrical resistivity of weathered rocks due to the water content in the pores); geophysics is a particularly elegant method of determining the weathering depth from the surface. For example, the ripability of rocks for civil engineering works can be estimated by seismic measurements (Fig. 13.35).

[19, 65, 81, 314]

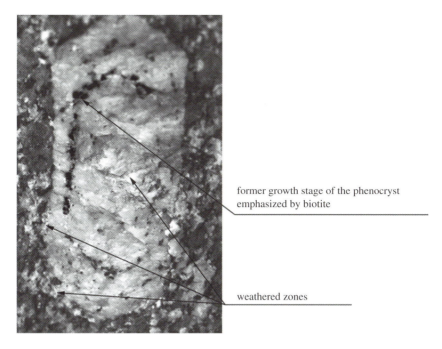

Fig. 13.34 Weathering of an orthoclase phenocryst. The cloudy and milky patches are weathered areas. The crystal system is being destroyed and is partially replaced by new microscopic minerals (kaolinite, illite, etc.) The feldspar gets it reddish tinge from iron oxides produced by the alteration of biotite into chlorite. GEOLEP Photo, J.-H. Gabus.

Fig. 13.35 Rippability. (a) Depending on the degree of weathering, rock masses can be excavated with a ripper, thus avoiding conventional mining techniques. The machine puts all its weight on the rock with the tooth out in front, thus breaking it. This technique is suitable for finely bedded rocks. It is often used for the excavation of road cuts, as here in metamorphic rocks in Mongolia. GEOLEP Photo, A. Parriaux.

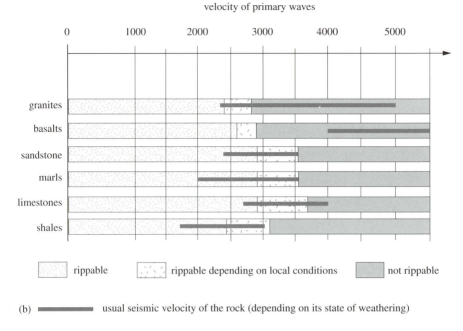

velocity of primary waves

Fig. 13.35 (b) The ripability for different types of rock and different types of machines can be determined on the basis of the seismic velocity of the rock mass (see § 4.1.7); indeed, the seismic velocity is strongly influenced by the state of weathering of the rock.

Chapter review

Weathering has a significant effect on the technical properties of rocks. There are widely varying repercussions in the different engineering fields of activity. The effects of weathering are generally negative in civil engineering projects due to the decrease in mechanical strength. They can be beneficial in the domain of environmental engineering because of the fact that the breakup of the rocky substratum promotes the growth of soils and the accumulation of ground-water resources.

14 Geology's Role in the Major Issues Facing Society

We would be remiss ending a book on geology without a discussion of the place of geology and its specializations with respect to the current and future concerns facing humanity. Although we have seen numerous examples of concrete contributions of geology in previous chapters, it is useful at this point to discuss several big challenges facing society and to show the fundamental role geology may play in resolving them. These challenges can be grouped into two large categories:

- Land development and natural resources;
- Environmental protection

14.1 Land development and natural resources

Land development and the management of natural resources are so strongly related that they can hardly be treated separately. Geology is closely related to land development because of regional planning and the management of subsurface resources. Geology is also essential in all phases of construction and infrastructure management (building construction, transport links, water supply, energy production, waste disposal, etc.). To describe these multiple contributions, let us examine the various resources necessary to human beings.

14.1.1 Food resources

Soil fertility depends primarily on geological and climatic factors. The great variety of soil substrates in the world shows the extent to which soil fertility varies, even within the same climate. One common example is soils that have formed from crystalline substrates and are thus poor in carbonates. Acids from rain and the mineralization of organic matter make them acidic, which limits their fertility. For example, the northern regions of Europe and America have very acidic soils, not for the climatic reasons that cause leaching in equatorial zones, but for purely geologic reasons: either the soil rests directly on crystalline rocks or on unconsolidated sediments that have formed from them.

Because of geologic or climatic reasons, a good part of the planet is characterized by highly acidic soils. In these areas, corrective measures are necessary to neutralize this natural defect and make the land more productive (Fig. 14.1). These corrective measures require geological resources: carbonate rocks. The first method is called liming: the addition of lime from calcinated carbonate rock reduces soil acidity. The second is amending the soil with crushed limestone (powdered calcite), but this neutralization method is slower. See Equation (9.6, § 9.1.2). For magnesium-poor soils, dolomite can be added since it provides this essential element to the soil. These operations are often limited because of large distances between quarries and the area to be treated but also because of the high cost and reduced financial capacity of the countries involved.

It is easier to use water to affect soils than to change the soil composition directly: the activity of the soil is strongly influenced by whether the water is flowing or is stagnant. Integrated management, including the monitoring of water inflows and drainage, plays an essential role in the long-term fertility of soil. Also, drainage is directly influenced by geological conditions

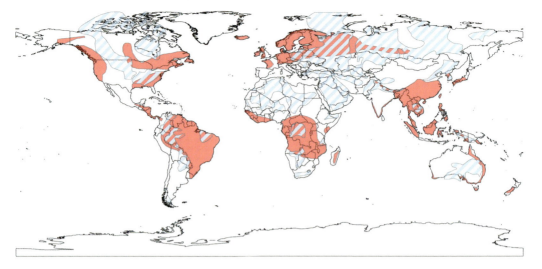

Fig. 14.1 The red area shows the extent of soils affected by excess acidity. Map based on the World map of soil resources edited by the FAO (1998) including podsols, acrisols, and ferralsols. The hachured blue area shows the regional extent of carbonate sediments that can be used as amendments to decrease soil acidity (map of geologic provinces "basins and platforms" of the USGS). The area hachured in red and blue represents acidic soils although carbonate rocks are present at depth; this situation can be caused either by a very intense lixiviation of the soil or by a cover of unconsolidated sediments between the soil and the rock, for example, the silica-rich moraines from the Scandinavian ice sheet in Poland.

present under the pedologic cover. We discussed the importance of water in chapter 7, not only for human consumption but also for food production (see Fig. 7.2). Aquifers are very large natural "reservoirs" that are effectively protected from human activities (see § 7.4.9). Two tendencies on the global scale reduce their potential:

– The overexploitation of aquifers, which is knowingly practiced in arid lands (Fig. 14.2)
– Water pollution from a wide variety of substances and activities (see § 14.2.5).

Fig. 14.2 The cultivation of desert regions using wells installed in non-replenishing aquifers cannot be considered sustainable management, either hydrogeologically or politically. Example of agriculture in the Negev Desert (Israel). Crops are irrigated with drip systems using deep groundwater at the foot of the Sinai massif. The water level in the wells decreases every year. GEOLEP Photo, A. Parriaux.

Geology (hydrogeology in particular) provides methods for discovering new aquifers, learning how they function, determining the age of the water, their rate of renewal, and their vulnerability. The following characteristics are indispensable for the sustainable development of resources:

- Long term management of the "reservoir" to prevent depletion of the resource
- Establishment of protection zones to protect aquifers from wastes due to human activities
- Systematic exploration to define potential resources for future availability, including poorly known deep aquifers
- Techniques to recharge aquifers artificially to improve the yield of the reservoir
- Rehabilitation of polluted aquifers, to make it possible to reuse abandoned reserves after they have been decontaminated (see § 14.2.5).

These measures have economic and political costs that unfortunately often make them difficult to apply, thus prolonging the short-sighted management of some supply systems.

A global policy of ending hunger will necessarily become a world policy on water resources.

A new source of stress is developing in the area of food production: biofuels. This agriculture consumes immense amounts of water, fertilizers, and pesticides. Again, the three fundamental resources, food, water and energy, are highly interdependent. Food and energy are more and more in competition in the use of the agricultural potential.

[61, 312]

14.1.2 Energy resources

The 21st century will be a pivotal period in the management of energy resources. The depletion of hydrocarbon deposits and the need to reduce greenhouse gases are responsible for these changes. Just as geology took the lead in the exploration and extraction of fossil fuels, it will be in the forefront during the transitional phase. Geology will discover new oil and natural gas deposits and develop new extraction technologies that have become more common since the sharp rise in crude oil prices began in 2004:

- Heavy oil to be extracted by solvents
- Oil condensates from some gas deposits
- Bituminous shales (see § 10.3.3.1)
- In-situ volatilization of coal (see § 10.3.3.2)

In reality, production from these non-conventional hydrocarbon deposits entails more serious obstacles than it would appear. The necessary investments are huge, as are energy and environmental costs. At the present time, production of these new resources does not exceed several percent of global hydrocarbon production. But the situation could change rapidly with the development of the geologic storage of carbon dioxide, which may be implemented synergistically with hydrocarbon production (§ 14.2.4.2). The spectacular lowering of oil price of 2008–2009 should not modify the long term evolution.

In terms of non-carbon energy, geology will be useful in determining the potential of radioactive ores. Although the future of nuclear energy is uncertain, it must be recognized that the fears concerning climate change have generated renewed interest in it. In the long term, nuclear fusion will also require exploration for new deposits.

In the area of combustion of fossil fuels and nuclear energy, geology is active in areas beyond the production of "fuels" themselves: since 1970, it has been involved in the elimination of undesirable by-products from the production of nuclear waste, and more recently in the removal of excess carbon dioxide (see § 14.2.4).

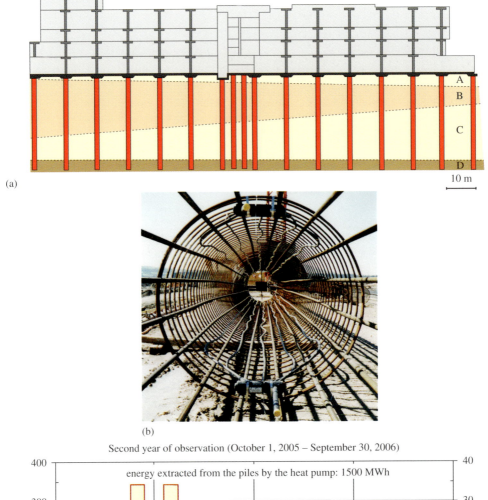

(a)

(b)

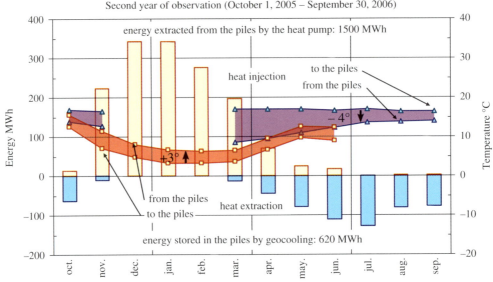

Second year of observation (October 1, 2005 – September 30, 2006)

(c)

Among the modern forms of energy that produce neither waste nor carbon dioxide, geothermal energy plays an exponentially expanding role as shown by the various techniques described in § 4.4.2. Geothermal probes that are totally installed within commercial circuits are beginning to cause serious problems in the management of underground urban space (see § 14.1.3). They are beginning to be used by builders, especially in foundations on piles. The Zurich airport recently built a terminal that is heated and cooled by energy piles (Fig. 14.3). An animation on the DVD shows how the system works to produce heat and cold.

The Vienna metro is pioneering the domain of energy geostructures with the construction of four stations that produce heat and cold from the subsurface through the use of exchangers in the walls and foundation slabs [298]. The Vienna (Austria) engineers are developing exchanger systems for the concrete linings of the tunnels.

298

Stimulated geothermal systems at great depth (see § 4.4.2.1) have some unresolved problems that prevent them from being totally operational. One major concern is earthquake activity that can occur in urbanized areas as a result of the hydraulic fracturing process, as happened in Basel (Switzerland) in 2006 (Fig. 4.65).

[128, 148, 199, 246, 289]

14.1.3 Underground resources

The subsurface is also a spatial resource for construction. Cities are growing and consuming larger amounts of surface area to such a point that one of the last remaining degrees of freedom is the third dimension. In the upward direction, this opportunity has already been exploited as skyscrapers that have shot up in all the major cities of the Planet. In contrast, downward development is less common, for several reasons:

– The geology of the urban subsurface is not well-known, and construction may reveal unpleasant surprises
– Old pollution may be discovered when the subsurface is excavated and this will require costly treatment
– The discovery of an archeological site may slow down a project for years
– Subsurface construction is still more costly than surface construction
– People are uncomfortable in many underground spaces.

In spite of these unfavorable elements, large cities can no longer ignore the possibility of this available space under our feet. Most cities have already installed underground infrastructures, for the most part transport systems and parking garages. But these works were built following a sectored approach (Fig. 14.4), that is, to satisfy a need expressed at a given

---

Fig. 14.3 (a) Terminal E of the Zurich airport (Switzerland) has several hundred piles because of the poor foundation conditions due to the lacustrine sediments. (A and C: silt and fine sand. B: argillaceous silt. D: moraine). A great part of the piles are equipped with heat exchangers. (b) Geothermal pile with exchanger tubes attached to the reinforcement to be incorporated into the concrete. (c) The subsurface is used to heat the building in winter and cool it in summer by geocooling, by drawing the frigories from the subsurface. The graph shows the energy balance for the second year of operation. The beige bars show the thermal energy extracted by the piles during the cold season; this heat is produced by removing 3°C from the coolant fluid in the piles (red graph). The blue bars are the cooling energy during the summer; this cold causes a warming of 4°C in the primary circuit (purple graph). The temperature in the piles is well above 0°C at the end of the winter, thus avoiding any risk of freezing the ground next to the piles. The efficiency of the system is excellent: the ratio between the energy produced (hot and cold) and the electrical energy for the heat pump and circulation pumps is 5. As a result, there is a payoff period of 6 years for the additional installation cost of the system, which is higher than a conventional one (from [200]).

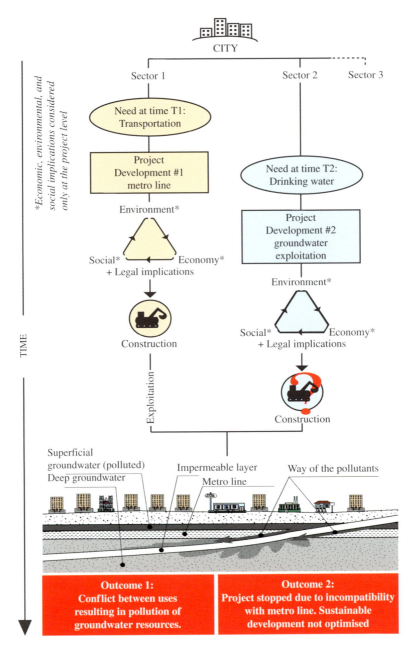

Fig. 14.4 Example of the use of urban subsurface by the sectored approach. At time t_1, the city needs a transport infrastructure. The city creates a metro system by optimizing the conditions specific to this need. At time t_2, the city needs to increase its potable water supply. An aquifer that is present at depth under the city can satisfy this need. However, it is likely that this water resource has been degraded by the construction of the metro. Tunnels can provide pathways between a polluted surface aquifer and deep resources, which is a typical case of negative interaction (diagram below the figure). The use of this natural reservoir becomes difficult, if not impossible. The use of the subsurface has in this case pre-empted other possible uses of this geologic space.

moment. The sectored approach does not include sustainable management of the overall potential of the available resources of the city's underground. This process has caused potential subsurface resources to be wasted.

The subsurface of a city is an environment that can be used for more than underground construction. In fact, it has four principal resources (Fig. 14.5):

- Space for infrastructure
- Groundwater
- Energy
- Geomaterials.

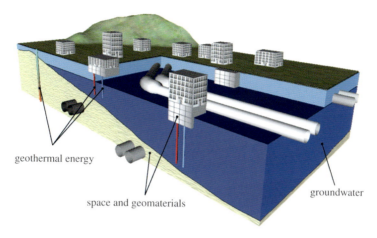

Fig. 14.5 The four principal resources of the urban underground, according to the DEEP CITY project.

For a research project called "DEEP CITY", we have introduced the idea of multiple uses of the urban subsurface (from [205]). It consists of considering the uses cited above to evaluate the potential resources of the geologic space by developing synergies between uses, (for example, space and geomaterials, space and geothermics) and avoiding harmful interferences, such as between space and groundwater (Fig. 14.6).

This multi-use approach calls for a new concept of planning: three-dimensional regional planning, which manages, preserves, and develops the subsurface in a sustainable manner (Fig. 14.7).

An example of a typical interaction among various underground resources involves the phenomenon of water level rebound. This phenomenon occurs in most cities where groundwater resources are used and where significant growth has occurred during their industrialization. For example, the rise in the groundwater level in the center of Paris after industry moved to the suburbs has made it necessary to pump a large amount of water, the equivalent of a water supply for 50,000 inhabitants, out of the metro system on a permanent basis, simply to keep the subway dry. Research into this kind of phenomena requires a systemic analysis (Fig. 14.8).

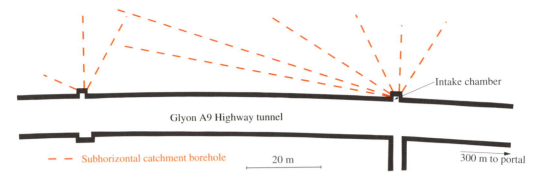

Glyon A9 Highway tunnel

– – Subhorizontal catchment borehole |—— 20 m ——| 300 m to portal

Intake chamber

Fig. 14.6 Interactions between space and groundwater resources are often negative, as shown in figure 14.4. There exist cases where synergy can be found, by collecting potable water from a tunnel designed for another use, for example. Case of the Vevey-Montreux water collection system in a highway tunnel where drainage boreholes collect potable water from a karst aquifer. Plan view.

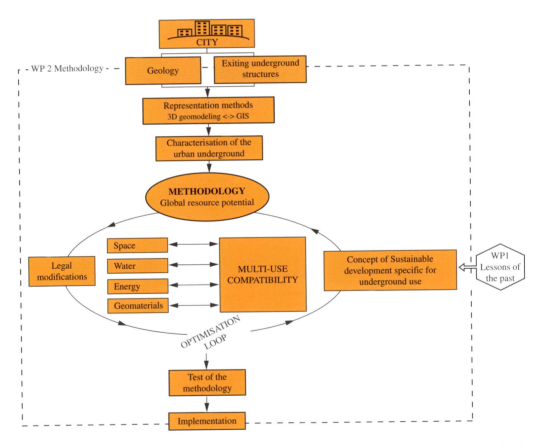

Fig. 14.7 The multi-use approach advocated by the DEEP CITY project. In contrast to the sectored approach, the multi-use approach establishes the total potential of the subsurface of the city, taking all four available resources into account. The first step is the construction of a three-dimensional geologic model of the subsurface, including the existing infrastructure. The next step is long-term planning for the underground space as a function of its overall potential, by optimizing the synergies between resources (multi-use construction – geothermics for example) and by establishing rules of incompatibility (for example, construction – drinking water) (from [205]).

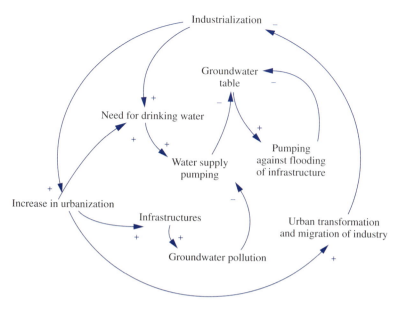

Fig. 14.8 Example of a systemic analysis of the interaction between the uses of urban underground resources: diagram showing causes of groundwater rebound. The causal diagram has been prepared on the basis of the understanding and interpretation of the phenomena governing the analyzed system. An arrow implies a causal relationship between two factors; the plus sign (+) indicates a variation of the two factors in the same direction, for example, more industrialization implies a larger urban growth; the minus sign (−) signifies a variation in the opposite direction. The causal diagram is a tool that helps in the analysis of a system. It makes it possible to identify feedback loops. For example, "industrialization," "urban growth," and "urban transformation and migration of industries" are a feedback loop: a variation of one of these factors (for example, an increase in industrialization) will have an effect on itself as a result of the overall variations in the system. These causes that affect themselves create what is called a non-linearity in the dynamic evolution of the system, which is particularly difficult to model (from [30]).

Geologists, civil engineers, urban planners, architects, economists and sociologists are the key-actors of a sustainable development of urban underground, itself a key-point of an harmonious future of the entire city.

6, 42, 137]

14.1.4 Mineral resources

The sorting of waste and the systematic recovery of industrial products have reduced the need for raw materials. This partially explains the slump in the mining sector at the end of the 20th century. Since the last century, general economic growth has boosted the consumption of ores. Strong development in countries that were formerly developing countries has clearly increased the demand, as shown by China and India, for example. This is true for the usual metals (iron, aluminum, manganese, copper, zinc) and the less common metals (lead, nickel, cadmium, tin, etc.). We must add industrial minerals such as diamonds, graphite, feldspars, talc, kaolin, bentonite (see also geomaterials § 14.1.5), phosphates (for fertilizers and cleaning products), halides from salt and potash mines. Rare metals have experienced an impressive increase in demand because of new technologies (Fig. 14.9). The consequence of this mineral boom has been a steep increase in prices and also the reactivation of the mining industry in the old producing countries that had planned to close their installations.

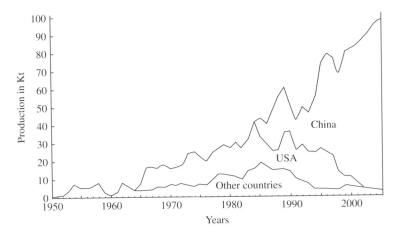

Fig. 14.9 The production of rare earth minerals has changed profoundly between 1950 and 2005 (the production decrease due to the 2008–2009 crisis is not included in the graph). Demand has exploded because these elements are used in new technologies (lasers, luminescent screens, etc.). In addition, China has become both a consumer and producer of these minerals because its Bayun Obo iron mine in Inner Mongolia produces bastnaesite (a rare-earth fluorocarbonate) as a by-product. American deposits that are now no longer in activity used to produce monazite (a rare-earth phosphate), which is also the principal source mineral for thorium.

According to the statistics of the British Geological Service (World Mineral Production 2001–2005) [118], the most spectacular increases between 2001 and 2005 have been for the following products:

– chromium and tungsten: >50%
– Iron ore and diamonds: between 40 and 50%
– cobalt, manganese, molybdenum and tin: between 30 and 40%
– bauxite: 26%.

The economic crisis since autumn 2008 has strongly reversed this tendency, but probably only temporarily. These spectacular variations show how difficult it is to manage natural resources exploitation and the involved investments.

For all these products, geology provides basic knowledge that can be used with numerous geophysical methods and geochemistry. The same can be said for mining production (spatial distribution of concentrations, three-dimensional structural model, research on the extent of deposits).

As in numerous other areas of engineering, the question of "how to produce most cheaply" has become less important compared to the question of "how to reduce environmental impact in the long term." This is particularly true for the mining sector. This can be shown by looking at the number of companies that have gone into bankruptcy protection to avoid paying remediation costs for long-term damage caused by their activities. There are three types of damage:

– Significant subsidence (Fig. 14.10) caused by the collapse of pillars left in the deposit to ensure short-term excavation safety; the subsidence occurs over very large regions and will last for several centuries. There are frequent accidents caused by old unmarked mine shafts. In Wallonia (Belgium), for example, more than one accident per week occurs in the old shallow coal mining areas ([213]). Today's mines use a technique called "cut and fill" which means that the miners fill the galleries with barren rock (Fig. 14.11). This has

Fig. 14.10 Surface subsidence at the Ridgeway Mine (Australia) (from [309]).

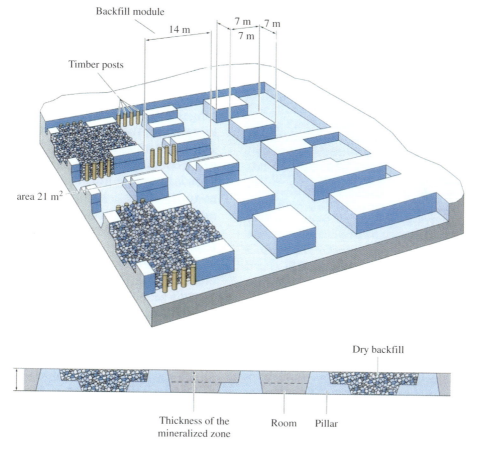

Fig. 14.11 The "cut and fill" mining method: the excavated space is filled with barren rock from mining, using various techniques of emplacement (solid rock debris or mud stabilized with cement). This principle has numerous advantages from the point of view of sustainable mining activity: it stabilizes the mine, prevents the deferred effects of subsidence and puts the barren rock back into the mine, thereby preventing the problem of the leaching of tailings. It is however more costly than the traditional technique. An example is the copper mine in Plokowice-Sieroszwice in the Polish Sudete that exploits a shallow horizontal deposit. The ceiling is supported over the long term by materials that have been put back into the chambers (from [9]).

the double advantage of preventing debris piles on the surface and their damage while at the same time stabilizing galleries for the long term. The tailings damage that affects the countryside (although numerous tailings heaps have been there for such a long time that they are almost part of the "natural" landscape) especially the acidity of soil and water caused by the oxidation of sulfides; the catastrophe of Aberfan (Fig. 14.12) reminds us that tailings stability can be a problem.

–   The water and the residues from the treatment of ores are often toxic (cyanide, arsenic, heavy metal residues, etc.).

Fig. 14.12 The principal "mountains" of Wales are artificial: they are mine tailings. One of these artificial mountains caused the dramatic landslide of Aberfan on October 21, 1966. The barren rock from the coal mine, a coarse and non-cohesive material, began to slide and liquefy. The flow traveled at about 30 km/h and was about ten meters thick when it entered the village. It destroyed a school, killing 147 people, 116 of them children. There were multiple causes of the accident: heavy rains, poor management of the tailings deposits, the slope angle of the fresh material had become excessive for the altered material, because of the argillaceous and graphitic interstrata. Picture Borough of Merthyr Tydfil.

It is clear that mining creates very complex environmental situations. Mining geology and environmental geology can together resolve these problems and in industrialized countries today, a mining project includes a post-mining management plan. These measures also have high costs. This is why this industry has for the most part moved to developing countries where the environmental requirements are often minimal or even non-existent. But these are also [9, 118] "ticking time bombs" for the Earth's environment.

14.1.5   Geomaterials resources

Geomaterials are products extracted from the geosphere that have undergone some degree of treatment in order to be used as construction material. There are various types of geomaterials:

–   Gravel and sand as aggregate for concrete, asphalt coatings, railroad ballast and under-layers for building foundations

- Clays for the terra cotta industry, special clays (kaolin, bentonite)
- Quartz sands for glass and glass wool manufacturing
- Siliceous rock for the manufacture of mineral wool
- Limestone and marl for cement manufacturing
- Gypsum for the manufacturing of plaster
- Hard rocks for crushed aggregate, rip-rap, cut stone, slabs for pavement and roofs
- Scoria for light aggregates
- Ornamental rock

These products that we use every day are produced from quarries that are often open pits, with the exception of ornamental rock quarries such as the Carrara marble facility. The exploitation of quarries involves geology in the exploration for deposits and the production methods to ensure good stability of the walls. Quarries also impact the environment, particularly groundwater (see § 7.4.11.2 and Fig. 7.80).

What is the future of the quarry industry?

In terms of the materials excavated from construction worksites, the principle of sustainable development will considerably modify current practices. In the urban environment, disposing of excavated soil in a landfill outside the city is becoming more of a problem. Importing costly, and particularly, heavy geomaterials (e.g. gravel) over long distances is also not sustainable. This transport causes significant difficulties. The trend now is to encourage builders to evaluate materials taken from urban excavations and use them as much as possible on site or nearby. This earthen material does not usually conform to geotechnical specifications and must be improved in order to be usable.

The types of materials excavated from the subsurface at worksites (not including demolition materials) can be classified into the six following categories (Fig. 14.13):

- Type I: Valuable materials that can be used with or without washing and screening in the geomaterials industry: alluvium and hard rock debris for aggregate, blocks for walls and erosion protection, pure gypsum for the manufacture of plaster, marl and limestone for the cement industry (for these two last applications the volume of materials present must be large enough).
- Type II: Materials that can be used without treatment for earthworks for transport infrastructures or for landscaping; unconsolidated sediments and rock debris with a low clay and water content.
- Type III: Materials that can be used at another worksite with on-site treatment: fill material that contains significant amounts of water can be treated with lime or cement prior to compaction.
- Type IV: Materials with high water content that require stabilization in a factory to be transformed into artificial aggregates: mineral slurries.
- Type V: Material that can cause pollution for geologic reasons: primarily evaporites (gypsum, anhydrite, rocks that contain chlorides) and rocks rich in sulfides. Depending on the concentration of soluble material, it can be used as the central body of embankments if it is confined by impermeable layers. If the content of soluble material is too high, it should be disposed in landfills.
- Type VI: Material that can pollute the environment for anthropogenic reasons. If the contaminants are minerals (e.g. heavy metals), they can be managed as type V. If the substances are organic, the material should be incinerated before use.

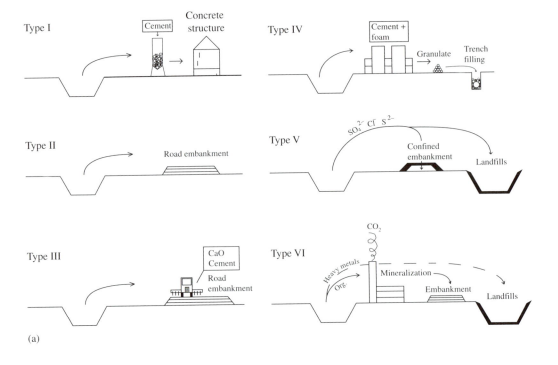

(a)

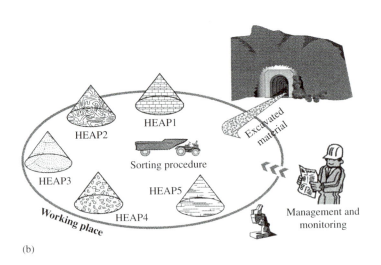

(b)

Fig. 14.13 (*Continued*)

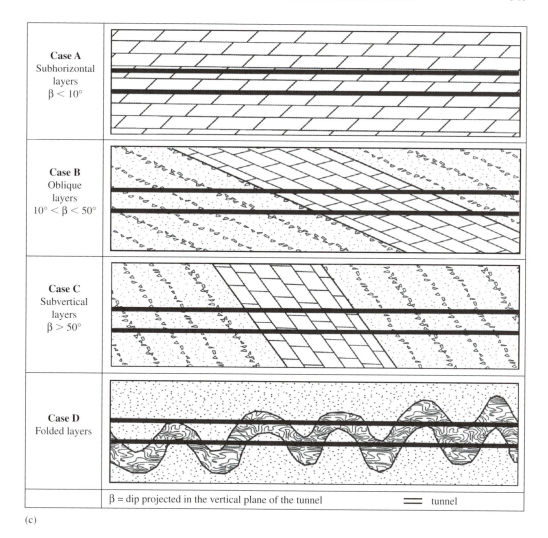

Case A Subhorizontal layers $\beta < 10°$	
Case B Oblique layers $10° < \beta < 50°$	
Case C Subvertical layers $\beta > 50°$	
Case D Folded layers	

β = dip projected in the vertical plane of the tunnel   ⬛ tunnel

(c)

Fig. 14.13 (a) Types of geologic materials extracted from worksites and possibilities for their use. (b) Principle of sorting materials as they come out of a tunnel under construction. (c) Possibility for sorting according to geologic structure (vertical profiles along the axis of the tunnel). Case A: thick layers parallel to the axis of the tunnel, ideal conditions. Case B: slightly oblique layers, long passage zones from one layer to another, not favorable to sorting. Case C: very upright layers, good conditions for sorting if the layers are thick enough. Case D: highly folded layers, not favorable for sorting (from [69]).

Thus the material produced from urban excavations must be included in the list of general underground resources. At the present time, most countries recycle demolition materials (concrete, asphalt coatings, etc.). At large tunnel worksites excavation material is systematically sorted to be used as much as possible (Fig. 14.13).

In contrast, very few countries make the best use of materials from urban excavation, except for those that have very little space for landfills. Japan is the pioneer in this area (Fig. 14.14). To make use of materials possible, Japan has the opportunity to dispose of cements coming partly from China at very low prices, with no comparison to prices in the West where such treatment is still prohibitively expensive.

Fig. 14.14 Use of materials from urban worksites: from theory to practice. An urban factory for making worksite slurries in downtown Tokyo. The slurries are dumped from trucks into a basin where they are homogenized. They are then mixed with an emulsifier and cement (about 200 kg/m³) to make a suspension that sets up to form a hard and very light material. It is used to make artificial aggregates to fill urban trenches and also light fill for roads (fill that is poured into place in 10-centimeter-thick layers) (from [185]).

This management policy, which seems laudable at first glance, requires deeper examination. The large quantities of cement required for treatment create a problem in terms of energy and global ecological balance. We know that cement manufacture consumes phenomenal quantities of energy and releases equally phenomenal amounts of carbon dioxide into the atmosphere (about 5% of the total global greenhouse gases). These consequences should be examined in light of the overall environmental condition of the Planet.

[12, 62, 232]

14.1.6   Resources of the space

Space should be considered a resource, even though it is used only for scientific research at the present. In chapter 2, we discussed numerous aspects of the planets and satellites of the solar system. We can see from these high-altitude and surface photographs that the scientific reconnaissance is primarily geologic. The first astronauts to set foot on the moon also conducted geologic observations. After the end of the manned Apollo missions, space exploration has been conducted by robot geologists (Fig. 14.15).

Reconnaissance missions to Mars (for example Phoenix, which landed successfully in 2008) will hopefully make it possible to add biological samples to these mineral analyses. What will we do with these future discoveries, in terms of practical applications?

It is a big question, and one that does not really require an answer, since at present the continuation of space research is fuelled by our simple scientific curiosity and our passion of geologist to explore time and space on a large scale.

[272]

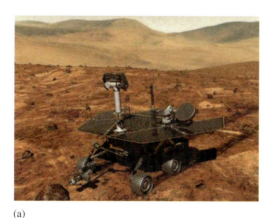

(a)

(b)

(c)

Fig. 14.15 Geologic exploration of the planet Mars in 2006. (a) The robot geologist: the Opportunity probe. (b) Panoramic false color view of the Victoria crater (diameter 740 m) by Opportunity; the crater is a trap for eolian sediments and has several rocky escarpments on the margin. (c) Detail of the layered geologic structure of one of the rocky cliffs. NASA document. An animation on the DVD shows Opportunity boring a sample of soil.

14.2   Environmental protection

We have just discussed several aspects of the environment, particularly natural resources. We must discuss several other ecological challenges where geology is involved and that society must seek to resolve.

14.2.1   Ecosystems and biodiversity

What does geology have to do with ecosystems and biodiversity, issues that seem at first glance to be purely biological? In fact, since geology is largely responsible for the formation of soils (see § 13.3) and their hydrogeologic conditions, it affects the living environment of animals and plants. It is in fact one of the limiting conditions of ecosystems. For example,

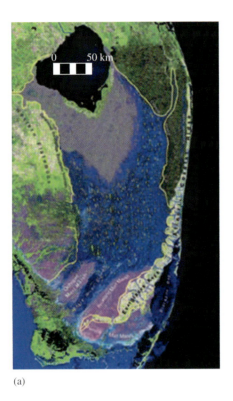
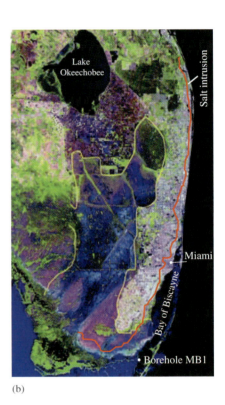

(a)   (b)

Fig. 14.16 The southeastern part of Florida is a well-documented example of the effects of geological conditions on the biodiversity of a coastal aquatic environment. The map from the beginning of the 20th century (a) compared to the situation at the end of the century (b) shows the effects of man's activities. The construction of channels in the Everglades swamps for agriculture and the overexploitation of the coastal aquifer beneath the city of Miami have modified the flow conditions in the subsurface. This hydrogeologic disturbance also has repercussions on the biodiversity along the shore. The Bay of Biscayne, south of Miami, now has a high-salinity marine ecosystem that extends almost to the shore. Borings installed at some distance from the shore, such as boring MB1 for example (c) show that conditions were different in the 19th century, before the intense human activity in the region. The sediments dating from that time contain ostracods, typical of a non-saline environment. During the 20th century, the biodiversity has changed. Benthic foraminifera have appeared (d), which is evidence of increasing salinity (in blue on the diagram). The reason for this change is the disruption of fresh-water underground flow from the aquifer into the sea, which ensured a transitional ecosystem (e). The overexploitation of the aquifer by the intense pumping done during the 20th century has not only put an end to the contribution of freshwater, but has also caused the saline water boundary to move inland (b). If the geological conditions had been different, this modification of the coastal ecosystem would not have taken place (from [136, 236]).

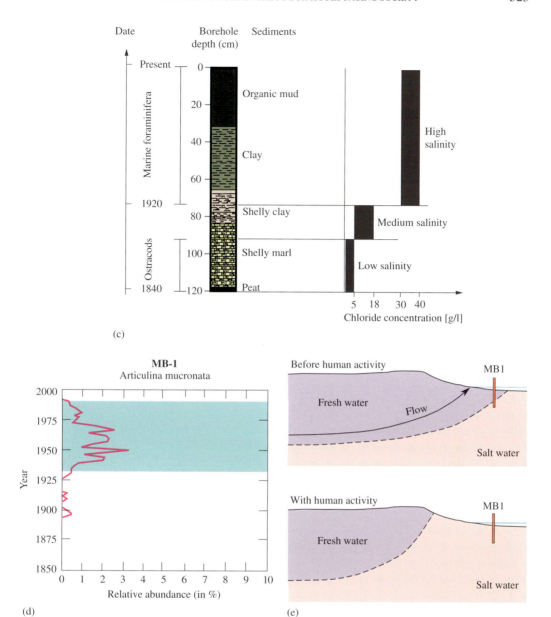

Fig. 14.16 (Continued)

variations in water content over space and time in wet ecosystems are often dictated by the geologic conditions of a site. Hydrogeology also causes salinity variations that directly affect the biodiversity of surface environments. (Fig. 14.16). Paleontology also highlights the influence of geological conditions on ecosystems and on the evolution of species throughout Earth history. An outstanding example is the large extinctions linked to the aggregation of the continents at the time of Pangea. In other terms, an understanding of present-day phenomena in the subsurface and their interactions with the surface is often necessary to understand the biosphere.

Rapid geodynamic phenomena, such as volcanic ash eruptions, floods, mudflows and landslides also affect ecosystems (see § 14.2.2).

[136, 194, 236]

14.2.2   Natural hazards

Geologic risk is an important component of the natural hazards that threaten society. Various types of geologic risk-causing phenomena have been described in the chapters of this book:

- Slope instability hazard (§ 8.2)
- Seismic hazard (§ 4.1.6)
- Water-related hazards (Chapt. 7)

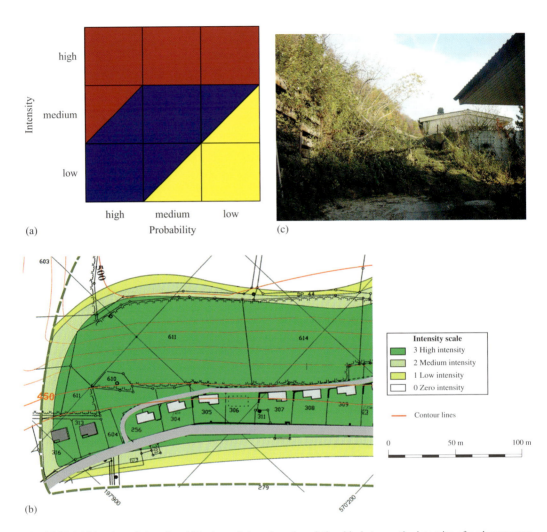

Fig. 14.17 (a) Mapping of slope instability hazards based on the relationship between the intensity of a phenomenon and its occurrence probability. Yellow indicates a minor hazard, blue indicates a medium hazard, and red indicates a major hazard. (b) Example of the hazard map of landslides and flows from the alteration of the marly molasse, Mont-Vully, western Switzerland. Map for events with a return period of $T = 100$ years. Houses that are too exposed to hazards (in light gray on the map) will be demolished after the hazard mapping is completed (ABA-GEOL document). (c) Slides and flows invading the houses. GEOLEP Photo. A. Parriaux.

Two types of activities are done to reduce these risks:

- Land development codes: This cartographic exercise begins by making a map to show the various types of phenomena, followed by a hazard map based on the relationship between the intensity of an event and its probability, as shown in figure 14.17. Regions classified as having higher hazard degree are designated as unsuitable for construction. A high-hazard designation may cause a significant conflict if the land parcel had previously been authorized for construction prior to the compilation of the hazard map.
- Engineering safety measures: When land development codes cannot provide effective measures, safety can be ensured by technical means. These measures either reduce instability (drainage, mechanical stabilization) or protect threatened structures (protective walls or dikes).

Experience in this area has shown that hydrogeologic action is often more effective than mechanical reinforcement in the stabilization of slopes that are prone to sliding. Depending on the condition of the subsurface and the size of the slide area, there are various techniques for using groundwater to promote slope stability (Fig. 14.18).

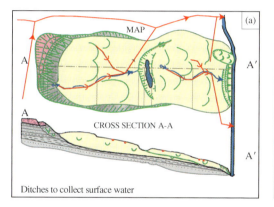
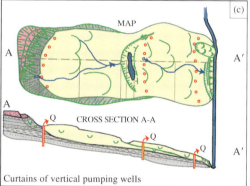
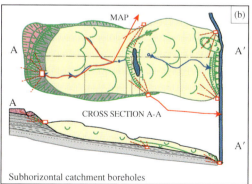
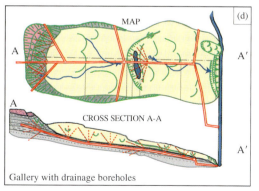

Fig. 14.18 Landslide stabilization techniques that use groundwater management: application to a hypothetical slide presented in map view and cross section. (a) Collection of surface water by ditches or drainage trenches and removal of water from the slide; a necessary but generally insufficient measure. (b) Drainage boreholes radiating from a drilling chamber in stable ground on the periphery of the landslide mass. (c) Well curtain to extract groundwater by pumping or siphons; the wells may be sheared if the slide movement is not rapidly stabilized. (d) Gallery with drainage wells beneath the surface of the slide; an effective but costly solution.

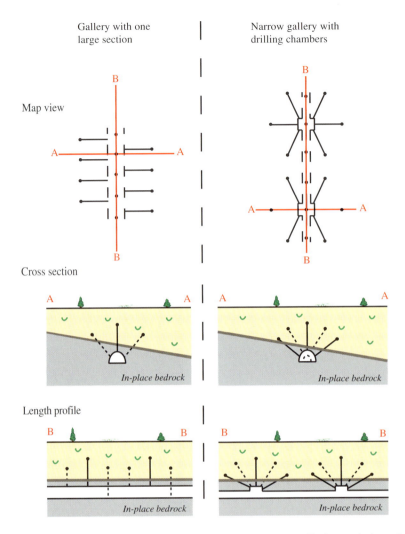

Fig. 14.19 Galleries combined with drainage boreholes are one of the most effective methods to draw down the groundwater table in and under a landslide. The gallery has either a large diameter to allow drilling at all points (left) or a smaller diameter (dotted line) to diminish the cost and the boreholes are done from enlarged segments of the gallery (right).

Large deep landslides often require substantial drainage using galleries that contain radiating drainage boreholes (Fig. 14.19).

The issue of slope instability lies largely within the domain of geology for the mapping of landslides, the description of the properties of unstable formations (see [201] for example), an estimate of the relationship between intensity and frequency, and the design of stabilization projects.

[56, 57, 119, 132, 143,183]

14.2.3 Climate change

The problem of climate change has become a major concern for society. Increasing temperatures have been observed planet-wide since the beginning of the industrial era (about 1850) and are highlighted by temperature statistics (Fig. 14.20) and the retreat of the glaciers,

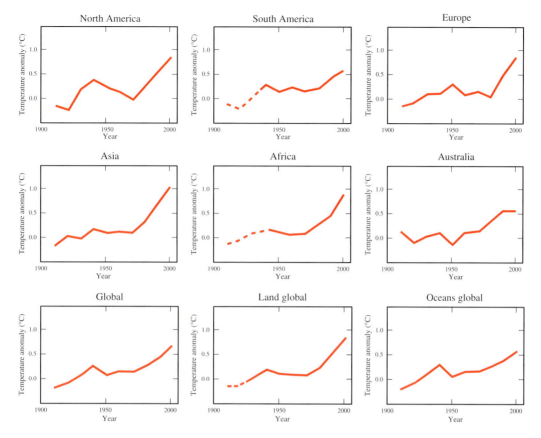

Fig. 14.20 Temperature increases in various regions of the world. The increase is not linear: there is an inflection in the middle of the 20th century and an intense acceleration recently. Warming is more noticeable on land than in the oceans, which explains a higher gradient in the northern hemisphere (from [23]).

including those in the Alps, since the end of the Little Ice Age (see § 7.3.3.1). This temperature increase has accelerated during recent decades.

In some places, even more significant changes have been recorded. For example, in Switzerland the change has been almost a half-degree per decade (Fig. 14.21), twice that of countries located near the oceanic coasts.

Climate change is a certainty. However, its cause is the object of lively controversy. The majority of scientists (see for example [23]) attribute the change to an anthropogenic increase of greenhouse gases (see § 7.2.2). A small minority thinks natural factors are dominant; they base their arguments on the significant climate changes that the Earth has experienced throughout its history, particularly during the Quaternary (for example [2]). It is true that significant temperature variations have occurred before man had any influence. However, the rate of the current change favors the thesis of an anthropogenic origin.

Many research groups are developing numeric climate models on both global and regional scales. Due to the complexity of the phenomena involved, their numerous feedback loops, and the role of inertia, which may be very significant, the models do not define a precise climatic evolution. In addition, the models must take into account scenarios of energy production and consumption that may vary greatly depending on future political decisions, particularly on the

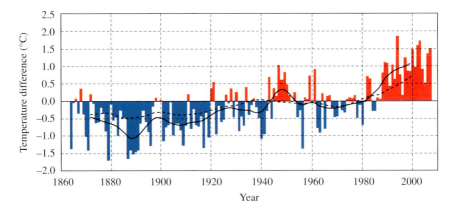

Fig. 14.21 Temperature history in Switzerland for the last century and a half. Compared to the reference period of 1961–1990, the number of colder years (in blue) or warmer years (in red) clearly shows the evolution toward warming. The weighted average for 20 years (continuous black line) confirms the acceleration of the warming, which is more noticeable in Switzerland than the average in the northern hemisphere (dashed black line). Source: Meteoswiss.

development of non-fossil-fuel energy. The future of nuclear energy will play a fundamental role in this balance. The uncertainties of the model results ([23]) predict a temperature increase of between 2.4 and 6.4°C by the end of the 21st century, depending on the assumptions selected.

Whatever the uncertainties, the engineer must take climate change into account because it will affect many of the geological processes that shape the Earth (from [204]), for example). Some of these are:

- A rise in sea level as a result of the combined action of the melting of the polar ice sheets (the dominant process) and the thermal expansion of the water column. Dikes will have to be built or built higher along coastal areas and also along rivers inland of their mouths. Depending on the actual extent of the sea level rise, which could range from 0.2 to 0.6 m by 2100 (according to [23]), populations will have to be moved and roads will have to be modified because they often follow coastlines and rivers.

- The melting of permafrost and a decrease in soil freezing. In Arctic areas, permafrost will disappear in some cases, and in other cases its upper surface will move downward, creating significant foundation problems for structures and their supporting infrastructure, which will require costly repair work. In mountainous regions, slope instability will occur (landslides, mudflows) and will be exacerbated by an increase in extreme hydro-meteorological events. However, great prudence is called for in predicting the effect of warming on the depth of freezing. The research we have conducted in the Alps shows that the relationships between temperature – precipitation – snow – frost is very complex (Fig. 14.22). Climate warming scenarios simulated in these regions paradoxically show an increase of frost depth at a moderate altitude (Fig. 14.23). The modification of the frost extent will also affect the distribution of runoff and groundwater recharge as well as the type of instability phenomena.

- Higher frequencies of extreme rainfall events. These will have consequences on the gravity-related phenomena that affect slopes: rock falls, landslides, mudflows, debris flows, erosion, and floods. Engineers will need to simulate numerous possible scenarios

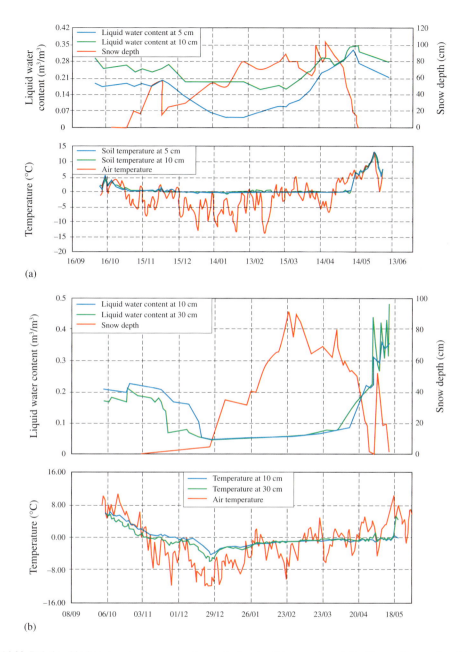

Fig. 14.22 Relationship between snow cover, water content and soil temperature at the Hannigalp experimental site (Valais Alps). (a) Winter 2000–2001, snowy, no soil freezing. (b) Winter 2001–2002, not much snow, significant soil freezing (from [20]). The hydrological balance during these two winter seasons is different as a result of the presence or absence of frozen soil. When the soil is frozen, its permeability is decreased, but it is not impermeable. This is because the soil is not totally saturated when it freezes. However, the reduction in permeability is tangible because it reduces the percolation flux into the subsurface by about 30%, particularly where a sheet of ice forms at the contact between the snow and the soil. This significant variation in the distribution of deep percolation and surface water has an effect on slope stability: during snowy winters without soil freezing, intense percolation into the subsurface during the snowmelt will cause increased risk of deep slides. In contrast, during winters with frozen soil, the melt water will concentrate in the upper part of the soil and cause an increased risk of superficial mudflows.

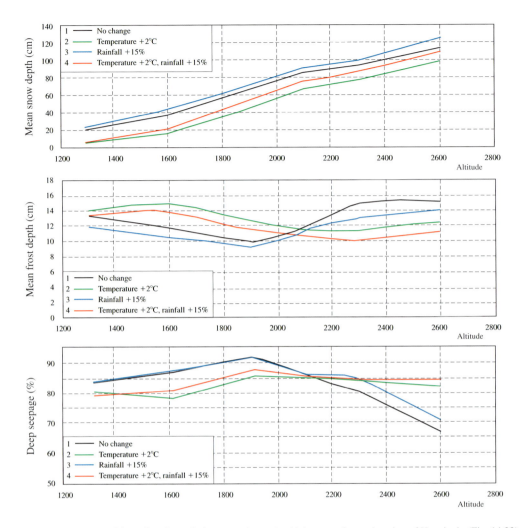

Fig. 14.23 Simulation of four climatic evolution scenarios at the Alpine experimental station of Hannigalp (Fig. 14.22) using the COUP model. Scenario 1: No change from the present climate. Scenario 2: Temperature increase only: 2°C. Scenario 3: Increase in rainfall: 15%. Scenario 4: Combination of scenarios 2 and 3 (from [20]). Calculation of the average thickness of the snow cover, the depth of frozen soil and the percolation flux toward deep aquifers. The result of these simulations is given as a function of altitude. It has been calculated that the scenarios with temperature increases (2 and 4) show a reversed behavior with altitude: at a medium altitude, the snow thickness is less than the present thickness because of warming; but at the same time, the absence of snow correlates with more soil freezing in spite of the temperature increase; at high altitudes the warming would not be sufficient to reduce the thickness of the snow cover; the permafrost should continue to disappear there, just as present-day observations indicate.

of climate change to design infrastructures to counteract the increased potential for these natural catastrophes.

- Changes in water flow in mountainous regions (see § 7.3.1.2). These systems may be substantially modified. Basins where rainfall-snow conditions are now prevalent may change into systems where rainfall is dominant. This means that water flow will be more continuous in rivers during the winter and spring floods will be moderate. The low flows of summer will be lower since they will be on a longer recession curve after the flood of the springtime. This retreat from extreme discharges should lead to positive effects on river erosion. Basins with snow-glacial conditions will evolve toward snow conditions due to the progressive disappearance of most small and medium-sized glacial features. As a result, high summer water will occur earlier in the year and will contribute to spring floods. As a result, a negative effect on erosion can be expected since amount of water that used to flow during the summer will now arrive simultaneously with the high spring precipitation. The management of hydroelectric reservoirs will change as a result.

As an engineer learns about the effects of climate change, he is likely to feel a degree of uncertainty. Fundamentally, this reaction is healthy because it leads him to ask questions and expand his intellectual modesty related to the sustainable design of engineering projects. To respond concretely to this new challenge he must avoid panic and maintain a cool head with regard to climate change. The engineer must adapt to this new situation without calling into question his entire professional experience. The mechanisms that affect the durability of a structure remain the same. Only the boundary conditions are changing. The hydrological and geologic phenomena are also the same, but the locations where they occur may change, just as the intensity–probability ratio established in the past has changed.

At this stage, given the uncertainties and the complexity of the system of "structures that interact with the environment", we propose some fundamental rules to incorporate into the process of the design and maintenance of infrastructures.

Design Phase

- Consider the project in a broader context (for example, a communication pathway is both a linear structure and a part of a three-dimensional context).
- Review the available reconnaissance on geological and hydrological hazards
- Emphasize serial risks
- Use additional prudence in the choice of transportation routes in order to adapt to the geologic and hydrologic environment. This will involve multi-disciplinary work (engineer – geologist – hydrologist – snow specialist – glaciologist)
- Increase safety factors in projects and test their stability based on several possible scenarios that may be unfavorable
- Plan projects that are easily adaptable to future climatic conditions.

Use and Maintenance Phase

- Establish a local observation network (meteorological, hydrological and geological) with identification of warning factors within the risk areas.
- Evaluate the data, identify tendencies of actual change, and if necessary, reinforce the safety of structures and their environment.
- Pay particular attention to the review of the operational status of drainage structures.

It will also be necessary to observe how the climate actually evolves and to plan as early as the design phase how to modify structures in order to make it possible to adapt them to actual conditions at the least cost.

[23, 71, 309]

14.2.4 Geologic waste disposal

The subsurface has been used historically for the disposal of solid waste (chemical or radioactive) and recent proposals call for it to be used for the disposal of a gaseous waste: carbon dioxide.

14.2.4.1 Nuclear and chemical wastes

The waste management industry processes radioactive and chemical materials to produce a final waste that must be permanently isolated from the biosphere. For years, these products have been placed in casks and disposed at the ocean bottom or buried somewhere beneath the land surface. Legislators have now introduced the notion of "final disposal," a term that is well understood in the political and legal realm. Technically speaking, disposal at sea obviously does not meet this criterion. The only viable solution is the geologic disposal of wastes. The geologist, with his knowledge of Earth history, has doubts about the notion of "final" disposal. Instead, he prefers to speak of long-term isolation. This period of time depends on the type of waste, its toxicity, the ability of the contaminants to leach into the environment, and the practical means of maintaining its isolation. Nuclear waste disintegrates according to well-known decay chains and time periods. For nuclear material, the long-term period necessary for geologic isolation can be defined (it is several hundred thousand years for highly radioactive waste). For non-fissionable products, such as heavy metals, an isolation time cannot be calculated.

The geologic storage of waste is based on the principle of multiple barriers (Fig. 14.24):

- *Artificial barriers* that immobilize the wastes in insoluble and watertight matrices such as glass, cement, or bitumen, within stainless steel containers. The artificial barriers have a lifespan limited to several thousand years.

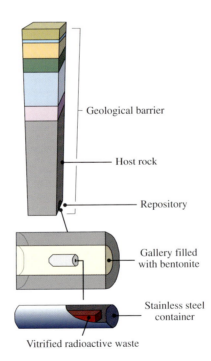

Fig. 14.24 The concept of multiple barriers used in geologic disposal of nuclear waste. These are primarily artificial barriers to retard leaching: vitrification of waste, stainless steel containers, and bentonite filling of the galleries. The geologic barrier is the host rock and the other formations that isolate it from the surface.

- *Natural* or *geological barriers* are rock units that separate the material from the surface. The most fundamental barrier for the safety of the repository is the rock unit in which the material is placed, which is called the "*host rock*."

To be considered as a possible host rock, a rock must have certain characteristics (Fig. 14.25).

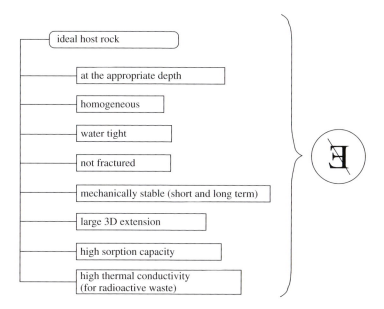

Fig. 14.25 Necessary qualities for a host rock for nuclear waste disposal. No such ideal rock exists in nature, but several rock types approach it.

The "geodiversity" of the Earth has produced such a wide variety of rocks that several can be considered candidates for host rocks since each of them have advantages and disadvantages (Tab. 14.1).

The potential candidates for hosting nuclear waste are essentially clay or marl formations (Fig. 14.26), anhydrite, salt, and crystalline rocks.

The long-term reliability of these nuclear waste repositories is a delicate problem for the geologist, the engineer, and society, because the isolation guarantee has to last for a very long period of time. If we consider the events that have occurred over the past hundreds of thousands of years, including several glaciations, we realize that it is very difficult to predict the future. This is why agencies that manage this question study and model multiple possible geologic scenarios to test the safety of repositories. This is another area where the geologist will play a key role in our future.

14.2.4.2   Carbon dioxide

Recently scientists have become concerned with various ways of disposing of carbon dioxide, a gaseous waste that in itself is not toxic but is responsible for climate change

Table 14.1 Some rocks and unconsolidated sediments that can be considered, at first glance, as possible host rocks for the geologic storage of waste.

Rock or rock family	Advantages	Disadvantages
Granite	High abundance High thickness Good mechanical behavior Joints often have low permeability	Low sorption capacity Difficult to predict the permeability of fractures Deep open fractures due to hydrothermal activity
Basalt	High abundance High thickness Very homogeneous Good mechanical behavior Sometimes low vertical permeability due to intercalations containing confined volcanic pores	Permeability sometimes high due to columnar jointing Lava tunnels
Volcanic tuff	High abundance High thickness in paleo-valleys Low permeability High sorption capacity	Heterogeneous medium Low thermal conductivity Poor mechanical strength
Claystone-mudstone	Very low permeability Sedimentological continuity Very rare open joints High sorption capacity	Poor mechanical behavior (swelling, creeping) Difficulty to maintain the accessibility of the repository over a long period => depth < 800 m Low thermal conductivity
Marly series	High abundance High thickness High sorption capacity	Poor mechanical behavior Carbonate intercalations that may contain deep karst conduits
Salt	Self healing of fissures due to creeping Very low permeability and water content High thermal conductivity	High permeability in sandstone channels with complex geometry Poor mechanical strength Sometimes corrosion due chloride waters
Anhydrite	Very low permeability Self healing of fissures High thermal conductivity Good mechanical behavior over the short term if the rock is massive	Often impure Self healing process not effective if groundwater is flowing (open system) Long term swelling if clay layers are present in the anhydrite Container corrosion
Unconsolidated material		
Fine lacustrine or glacio-lacustrine sediments or waterlain till	Very low permeability High thickness High homogeneity High sorption capacity	Higher horizontal permeability Poor mechanical behavior (subsidence, slide)
Lodgment till	Low permeability good mechanical strength	Heterogeneous medium with some permeable channels that are difficult to detect Rather low sorption capacity

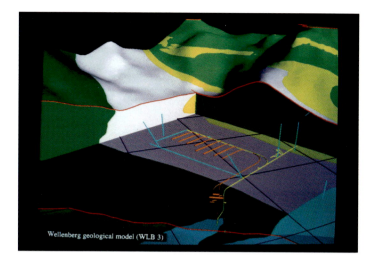

Fig. 14.26 Three-dimensional geologic model for a disposal site (middle radioactivity waste) planned for Wellenberg (Swiss Alps). This site, which provides a high degree of isolation, was temporarily abandoned because of opposition from the population of the canton. Model GEOLEP L. Tacher.

(see § 14.2.3). In the search for a place to dispose of carbon dioxide, a "carbon sink," geology and oceanography are collaborating to study the following:

- *Ocean storage*: One solution involves dissolving the CO_2 at depth. We saw in § 9.2.2 that with increased pressure it is possible to dissolve a larger quantity of carbon dioxide. The water becomes more acidic and attacks biological and geological carbonates. The CO_2 is thus immobilized as bicarbonate ions that cannot cause climate problems (see the calcium-carbon equilibrium equations in § 9.1.2). A second solution is storage in trenches at depths of about 4000 m. Under these conditions, the CO_2 is liquid and has a higher density than water (Fig. 14.27). The carbon dioxide forms a "lake" of heavy liquid, which is immobile at the base of the trench.

- *Geological storage*: Injection of CO_2 as a supercritical fluid in porous geologic layers isolated from the surface (Fig. 14.27). Geologic structures similar to those that contain oil and gas (see § 10.3.3.1) are suitable for this purpose. In fact, old deposits are ideal targets because they are known for their excellent natural isolation. In economic terms, abandoned wells can be profitably used for this purpose. Also, CO_2 injection into a deposit through wells other than the producing well can improve the extraction of liquid and gaseous hydrocarbons in a deposit that is still producing. It may also be possible to distill in-situ coal deposits by the injection of carbon dioxide.

A great deal of research is being done around the world to resolve the carbon dioxide problem. This research includes feasibility studies of the concept, verification of the capacity for long-term storage, and an evaluation of economic and environmental suitability. In terms of sustainability, research is focused on natural isolation using geologic traps and on technical issues of closure (for example, determining how to permanently plug an injection well once the waste is in place). A great concern in some oil fields is the state and the number of old abandoned wells. In-situ pilot studies are being conducted in various places, particularly in Norway (Sleipner), Canada (Weyburn), and Algeria (In Salah). The political pressure is so intense today that significant progress is anticipated in the coming years.

[35, 72, 122, 135, 193]

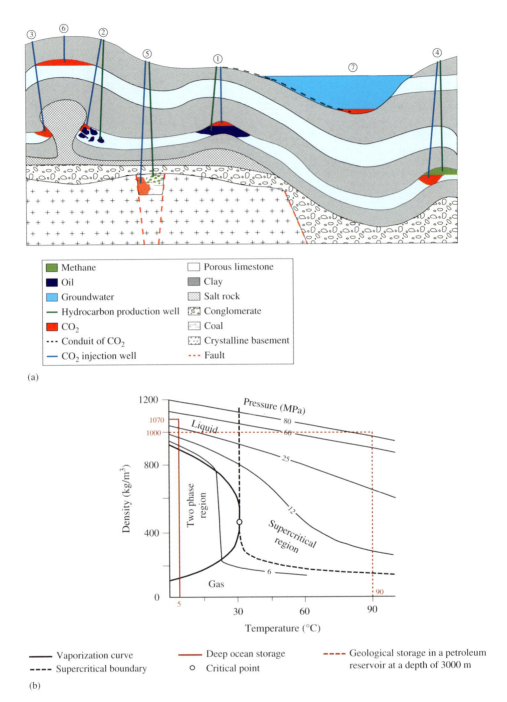

(a)

(b)

Fig. 14.27 (a) Principle of geological and oceanographic sequestration of carbon dioxide. Drawing not to scale. 1: Active petroleum deposit. 2: Petroleum recovery in a deposit while it is being exploited. 3: Depleted petroleum deposit. 4: Recovery of natural gas from a deposit undergoing exploitation. 5: Enhanced recovery of methane from coal. 6: Storage in deep groundwater. 7: Liquid carbon dioxide storage at the ocean bottom. (b) Diagram showing the physical state of the carbon dioxide. For storage in oceanic trenches, the CO_2 is a liquid with a density slightly higher than that of water (about $1.07 \cdot 10^3 \, kg/m^3$). To inject it into a petroleum deposit at a depth of 3000 m, the carbon dioxide is in a supercritical fluid state, with a density of about $0.95 \cdot 10^3 \, kg/m^3$ (assuming a geothermal gradient of 30°C/km).

14.2.5   Subsurface pollution

The subsurface is certainly the least known component of the environment. It has also been the most misused. These two "handicaps" have arisen through the application of the principle, "out of sight, out of mind." Throughout time, Man has used the subsurface as a place to put things he no longer wants and things that cause problems if left at the surface. The subsurface has been used as a convenient "trash can" since considerable volumes of waste can be put there, and often these wastes do not smell bad and do not contaminate the surface water. These short-term benefits are often a deficit in terms of sustainable development. But before discussing this topic, let us examine the sources of the waste.

14.2.5.1   A history of pollution

A historical analysis is interesting because it shows how the relationship between Man and the environment has evolved over centuries, and points out certain mechanisms of this complex relationship that touches on all areas of science and the social sciences.

Period I: Prior to the Industrial Revolution (Pre-19th century)

Before Man had machines, his actions were very limited in space and scope. Even so, pollution existed. Traditional crafts had developed significantly and they certainly caused pollution. For example, the tanning of hides required large quantities of mercury, which caused high mortality rates among persons who practiced this profession and also caused the contamination of these sites. We find traces of these sites even today. Traditional mining that produced metals were also sources of pollution. The first human populations in cities of this period caused serious microbiological pollution of the water. Sanitary conditions in cities were disastrous: sewers flowed directly into streets and houses often used water from wells located near the sewers. This biological pollution has left only historical evidence.

Period II: Uncontrolled Industrialization (19th and 20th centuries until about 1970)

For more than one hundred and fifty years, industrialization upset the natural equilibrium. It started slowly and later accelerated greatly. The force behind this development was solely socio-economic. Industrialization was "uncontrolled" in the sense that practically no one considered the environment in an industrial project.

The result was ecological degradation that reached its height between the end of World War II and the 1970s. This period is the most devastating in history because the economy grew at a maximum and machine power increased tenfold while the notion of environmental protection was still in its infancy. Most of the contaminated sites that we have inherited were created during this period: immense deposits of garbage and chemicals, industrial sites, and military sites.

Period III: Beginning of Environmental Protection (1970 to 1990)

At the end of the 1960s there was a veritable awakening about environmental problems in the developed countries. Politicians transmitted this social concern to governments. In the 1970s the first laws were passed to create the legal basis for protection of the environment. For example, in Switzerland, the 1971 federal law on water protection divided the country up into various protection zones depending on the vulnerability of water resources. The federal law on environmental protection followed about ten years later with the introduction of environmental impact studies.

At this time, the states took responsibility for the new task of environmental management of the country and its resources. Environmental departments were formed in all government administrations. The most important goal was to avoid the creation of new contaminated sites. No one dared to speak openly about cleaning up old sites.

Period IV: Birth of the idea of sustainable development (1990 to today)

At the end of the century, the idea of sustainable development gradually entered our minds. In 1987, the World Commission on the Environment and Development, called the Brundtland Commission after its president, gave a definition of sustainable development that is now known worldwide. It is development "that lets all populations now living on Earth satisfy their needs without compromising the possibilities of future generations." Future generations, like the present one, have the right to a healthy environment. It is important not to limit sustainable development to the environment. Economic prosperity and protection of the natural foundations of life are necessary to satisfy our material and spiritual needs. Only a united society will be capable of equitably distributing economic wealth, preserving the values of our societies, and making prudent use of natural resources. Sustainable development assumes equality in the treatment of its three components: environment, economy, and society.

This fourth period logically follows the ideas of the preceding period. It relates human activities to the notion of long-term sustainability. This way of thinking encourages us to approach the problem of remediating contaminated sites, which is one of the keys of sustainability. The correction of past errors is onerous in terms of economic and political implications and the technical difficulties that it confronts. But this "reasonable" period also calls our attention to the existence of a significant boundary between the civil and military spheres. The armed conflicts at the end of the 20th century generated considerable environmental impacts in just a few days, such as the hydrocarbon pollution of the Kuwait war. This permanent "military exception" should not be neglected in the global evaluation of environmental protection.

14.2.5.2   Classification of pollutants and their behavior in the subsurface

This classification of pollution is based on:

* Its distribution in space (point-source or non-point-source)
* Its distribution in time (continuous or accidental pollution)
* The socio-economic issues generating the pollution (Fig. 14.28).

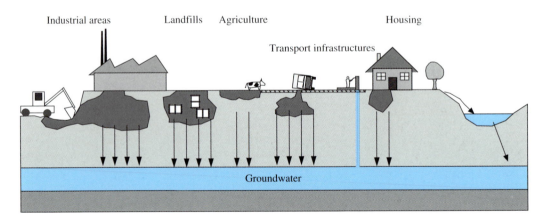

Fig. 14.28 The principal human activities that cause pollution of the subsurface (from [60]).

Table 14.2 Classification of causes of subsurface pollution: distribution in space, time and by socio-economic issues (from [60]).

Origin of the pollution	Spatial distribution			
	Diffuse (Non-Point Source)		Punctual (Point Source)	
	Time distribution			
	continuous	accidental (one-time occurrence)	continuous	accidental (one-time occurrence)
Agriculture, gardens	Manure, sludge from wastewater treatment plant, phytosanitary products		Cracked manure tanks	Emptying of manure tank Unauthorized release of chemical products
Housing	Diffuse leaks from wastewater conduits		Small losses from domestic oil tanks	Overflow during filling of oil tanks
Industry	Diffuse releases of pollutants in industrial zones	Nuclear accident	Continuous releases of pollutants from storage buildings	Accidental releases of pollutants. Unauthorized releases of chemical products
Transportation	Roads : salt de-icing, oil, gasoline residues, metal debris Rail : herbicides			Road and rail accidents
Polluted sites			Former industrial sites, old dumps, military areas, accident sites	

The intersection of these variables creates the classification summarized in table 14.2.

Contaminated sites result from either old releases or industrial sites that handled pollutants (for example, refineries and chemical industries) or the locations of accidents involving vehicles that were transporting toxic products.

There are multiple ways of classifying products that contaminate the subsurface, depending on whether a chemist, biologist, geologist, toxicologist, or legal specialist is doing the classification. We will describe the most up-to-date classification used in groundwater contamination. It focuses on the physical characterization of the contaminant and its relationship with water. This classification includes three principal families of products.

Particulate contaminants

These are mainly microbiological pollutants including germs and pathogenic viruses. Because they are present in the organisms of the living creatures they inhabit, they enter into the environment by the excretion of fecal matter. The most dangerous microbes for man are those of human origin. These microbes pollute groundwater when they leak out of defective sewer pipes (Fig. 14.29).

Germs from animals come from manure used in agriculture. The leaching of solid farm fertilizers and manure during rainy periods causes these microbes to penetrate into the subsurface. If the geologic substrate has good interstitial porosity, the bacteria can be easily retained

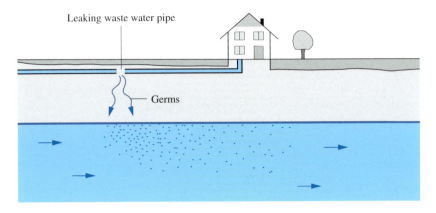

Fig. 14.29 One of the most common types of particulate pollution results from releases from the domestic wastewater network. In this case, the particles are germs and viruses, some of which may be pathogenic (from [60]).

in these pores. However, viruses generally pass through these natural filters because of their smaller size. Aquifers with fissure or karst porosity are not very effective in filtering these substances out of water.

The lifespan of these organisms in the subsurface is limited to some tens of days, so there is no long-term contamination generated by this type of pollution. Groundwater protection zones can be set up in these cases and they can be coupled with facility safety measures and restrictions on the spreading of manure.

Water-soluble contaminants

We include in this group components whose major flux is transported in solution in groundwater. After a period of time, these substances can spread out over the entire thickness of the saturated zone due to the hydrodynamic dispersion of water and molecular diffusion (Fig. 14.30).

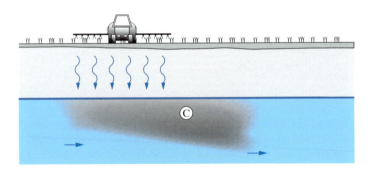

Fig. 14.30 Contamination of the subsurface by water-soluble pollutants. Agriculture, especially fertilizers and pesticides, is largely responsible for this pollution (from [60]).

Two families can be distinguished chemically:

- Mineral Products. These products include highly soluble mineral salts (for example, halide, sulfate, nitrate and borate salts). They can be released from landfills; chlorides are also released by the desalting of roads, nitrates from agriculture.
- Soluble Hydrocarbons. Non-point source pollution from roads (see § 7.4.11.2) is an illustration of the interaction between civil engineering and groundwater. Some of these hydrocarbons are soluble in water. These are mono-aromatic hydrocarbons in the BTEX series, which stands for benzene, toluene, ethylbenzene, and xylene (Fig. 14.31), which are made from single benzene rings. Because of their high volatility, they are not generally found in groundwater (from [210] and [216]). These pollutants are released into the environment either by dispersal from roadways or by leaking tanks or by accidents. MTBE (methyl tert butyl ether) is very different because it has low volatility and is highly persistent in the subsurface. The case of this oxidizing gasoline additive is very instructive for environmental history. It illustrates an error in political ecology in the USA caused by a partial vision of the environment. The Clean Air Act Amendments of 1992 required the petroleum industry to use MTBE in gasoline to promote combustion and thus decrease the production of carbon monoxide and ozone. This measure was favorable to air quality but did not take into account the fact that MTBE was very soluble in water (MTBE is 30 times more soluble in water than benzene), highly mobile in groundwater, and not very degradable. The consequence of this partial benefit has been widespread hydrogeological pollution around numerous service stations with leaky product tanks. MTBE plumes are much more extensive than BTEX plumes, for example. In the USA alone, thousands of public water supply wells have been contaminated by MTBE. California has banned its use in gasoline since 2003. In Europe, opinions on MTBE are mixed.

Non-soluble contaminants

We must distinguish two families of contaminants: insoluble solids and non-miscible liquids.

Insoluble solids: heavy metals

Heavy metals are dangerous because they bioaccumulate in organisms. They are generally only slightly soluble in water. The pH of the environment and the redox potential have a major effect on the mobility of these elements. We can summarize the mobility of the different metals in the soil and the subsurface in the following way, whether of geologic or anthropogenic origin:

- K, V, Rb, Cr, Si, B and Ba are leached slowly depending on the weathering processes in the soil, independent of the pH
- Cu and Pb are retained in soils
- Mn, Zn, Al, Fe are sensitive to the acidity of the soil. At low pH, they are leached.

A large part of the heavy metals thus precipitate at the base of the soil, at the decalcification front, due to the high pH there. The soil then behaves as an accumulator of metallic pollutants. There are two sources of accumulated metals: atmospheric and geologic. Aquifers beneath the soil are not much affected by heavy metals in the soils if the aquifers contain carbonates and have interstitial porosity. In contrast, siliceous and carbonate aquifers with fissure or karst porosity are more intensely affected by the introduction of heavy metals. It is the same for aquifers that are not covered by soils, particularly in mountains. Radioactive contaminants belong

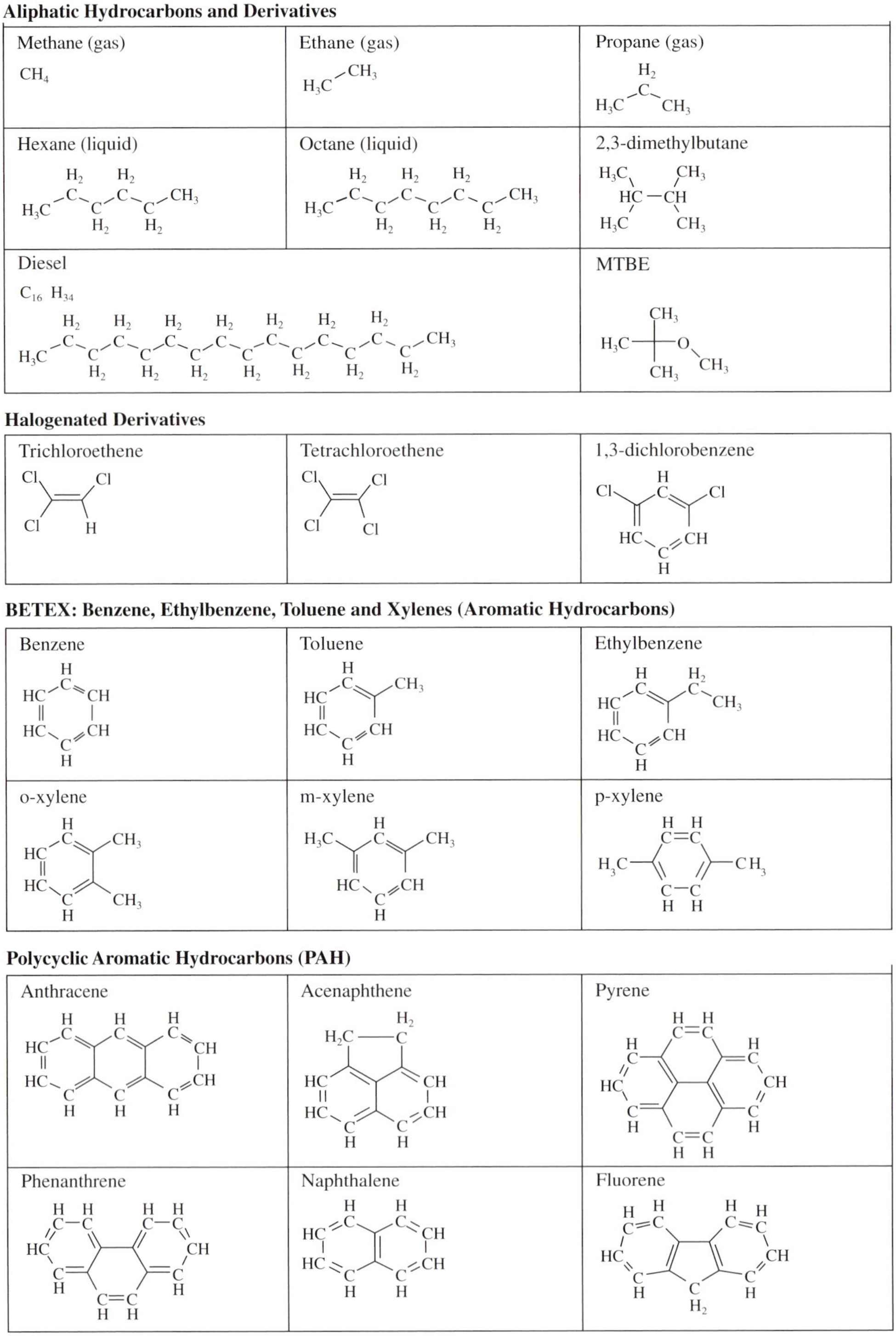

Aliphatic Hydrocarbons and Derivatives

Methane (gas)	Ethane (gas)	Propane (gas)
Hexane (liquid)	Octane (liquid)	2,3-dimethylbutane
Diesel		MTBE

Halogenated Derivatives

Trichloroethene	Tetrachloroethene	1,3-dichlorobenzene

BETEX: Benzene, Ethylbenzene, Toluene and Xylenes (Aromatic Hydrocarbons)

Benzene	Toluene	Ethylbenzene
o-xylene	m-xylene	p-xylene

Polycyclic Aromatic Hydrocarbons (PAH)

Anthracene	Acenaphthene	Pyrene
Phenanthrene	Naphthalene	Fluorene

Fig. 14.31 The principal hydrocarbons found in underground pollution, based on their chemical classification.

to the class of heavy metals (with the exception of tritium). Numerous industrial sites and land-fills are contaminated with such products (for example, radium), particularly as a result of the manufacture of watch faces and phosphorescent paints. Nuclear plant sites also have anomalies; for example Chernobyl (Ukraine) after the large accident in 1986.

Liquid non-miscible pollutants (NAPL)

These non-miscible substances are common in organic chemistry, particularly in the area of hydrocarbons. This family of pollutants is often called NAPL: Non Aqueous Phase Liquids. Contaminated formations may contain four phases (Fig. 14.32):

- The mineral framework, granular or rocky (+ solid residues of contaminants)
- Gases
- Groundwater
- The liquid contaminant.

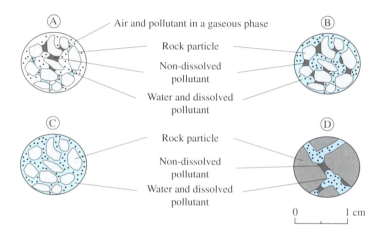

Fig. 14.32 Behavior of contaminants within pores. A: Unsaturated zone. B: Saturated zone with non-dissolved contaminant. C: Saturated zone with dissolved contaminants. D: Contaminants in a fissured aquifer (from [60]).

In the case of volatile contaminants, the air in the subsurface is also polluted. For this reason, scientists often analyze gas samples extracted from the unsaturated zone to detect the presence of deep pollution. This is particularly true for chlorinated hydrocarbons (Fig. 14.31).

Depending on whether a substance is lighter or heavier than water determines the distribution of these contaminants at depth.

Light non-miscible contaminants (LNAPLs)

Often referred to by their name "Light Non Aqueous Phase Liquids," these products, whose density is less than $1 \cdot 10^3 \, kg/m^3$, accumulate on the surface of the saturated zone of the aquifer

(Fig. 14.33). These are petroleum products, particularly fuels and lubricants. They are composed predominantly of low-toxicity aliphatic hydrocarbons (Fig. 14.31).

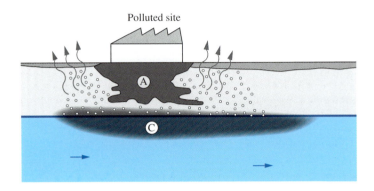

Fig. 14.33 Pollution caused by a light volatile organic compound (LNAPL). Letters same as figure 14.32 (from [60]).

Dense non-miscible contaminants (DNAPL)

[60]   These products are called "Dense Non-Aqueous Phase Liquids". They are particularly harmful because they penetrate to great depths because they are denser than water (Fig. 14.34).

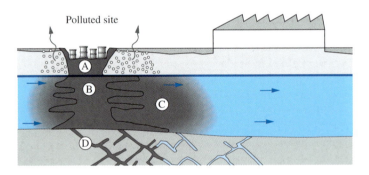

Fig. 14.34 Pollution caused by a dense volatile organic compound (DNAPL). Letters same as figure 14.32 (from [60]).

Most of them are chlorinated solvents, for example, trichloroethylene and tetrachloroethylene (Fig. 14.31). These products were first used by weapons factories during World War II and later by non-military industries as degreasers in the machine industry, in the semiconductor industry, in photography and as solvents at dry cleaners. The polycyclic aromatic hydrocarbons PAHs (Fig. 14.31) also belong to this family. They are present at high concentrations in tars. They are also indicators of highway pollution (see § 7.4.11.2).

14.2.5.3 Evolution of contaminants in the subsurface

Once a site is contaminated, the pollutants disperse throughout the underground environment as a result of the fluid movement of air and water. The mobility of the contaminants depends primarily on two main factors:

- Their physical and chemical state: solid, colloidal, dissolved (ionic or non-ionic), gaseous
- Their affinity for the geologic environment: the interactions between the contaminant and the geologic matrix are numerous and complex. Substances that are not retained on mineral particles travel with the transporting fluid. They are called *conservative tracers.* The others, which luckily are the majority of pollutants, are retained or retarded by interactions with the grains.

If we set aside volatilization, which is a return to the atmosphere through the non-saturated zone of the subsurface, the other processes essentially involve transport by groundwater in aquifers, particularly in the saturated zone. The transport of dissolved substances, which is generally the most rapid method of migration, is simulated using models that take into account five major processes, three of which involve the movement of the pollutant with the movement of water, one that involves retention of the substance, and one that decreases the mass of the moving substance:

- *Advection* (or convection). This is the movement of substances dissolved in water as a result of the movement of the water itself. Advection is generally calculated using Darcy's Law (see § 7.4.3).
- *Hydrodynamic dispersion* (Fig. 14.35): Water molecules actually travel very differently than indicated by Darcy's Law. Flow within pores with complicated geometries (interstitial, fissure, or karst porosity) means that molecules have trajectories of variable length and that they move at different velocities (longitudinal dispersion); in addition, their trajectories deviate in three dimensions from the Darcy flow line, creating cones of dispersion (transverse dispersion).
- *Molecular diffusion*: This is the transport of substances without the movement of water; that is, transport by ionic migration in proportion to the concentration gradient.
- *Sorption*: This term includes processes that retain a substance on the mineral grains (interaction between solid and solute). These processes are adsorption (attachment of a solute to a solid surface), absorption (attachment of a solute on a porous particle), ionic exchange (between the grain and the solution), chemical reactions with the grain, with or without precipitation. The reverse mechanism, which involves the movement of solutes from the solid into solution, is called desorption. We generally use the simplifying assumption that the quantity of solute that attaches to the grains is proportional to the quantity of solute in the liquid phase.
- *Degradation*: This is a group of processes that destroys the transported substance. It involves the radioactive decay of radionuclides (see § 3.1.1) and either the chemical or biological degradation of unstable substance in the underground environment (for example, the oxidation of organic molecules such as fuel oil or the death of bacteria). Biodegradability acts on the majority of natural organic products, and acts with more difficulty on certain artificial molecules such as PCBs and chlorinated hydrocarbons. The redox potential of the medium greatly affects the degradation of organic matter. In fact, mineralization consumes an important part of the available oxygen and turns the chemical environment into a reducing one, where anaerobic transformations are no longer effective in the evolution of organic matter.

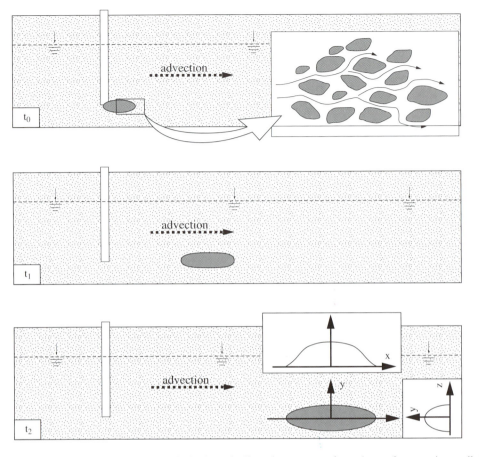

Fig. 14.35 Combined role of advection and hydrodynamic dispersion on a one-time release of a tracer in a well. By dispersion, the tracer expands to form an ellipsoid with its long axis parallel to advection (longitudinal dispersion). The two other axes are the horizontal and vertical transverse dispersions.

14.2.5.4   Evaluation strategy

With an understanding of the scope of subsurface pollution, society is developing a policy to systematically identify sites that may contain contaminants. The reconnaissance procedure for these sites involves a risk analysis based on three principal criteria (Fig. 14.36):

- *Pollution potential*. This is the mass of contaminating products at a contaminated site, taking into account the physico-chemical form of the products, their solubility, and their toxicity. Example: the subsurface of an old refinery contains 60 tons of gasoline with lead additives contained within the pores of the ballast of the tank foundations.
- *Migration potential*. This is the possibility that the contaminants may disseminate through the various transport mechanisms in the underground environment. Migration depends on artificial and natural barriers. The latter are mainly geological in nature. Depending on migration potential the pollution can be considered effective or latent.

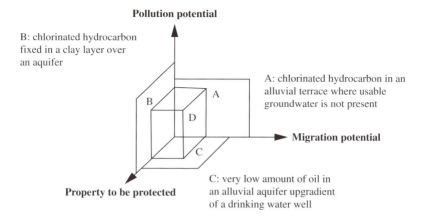

Fig. 14.36 Graphical representation of the three fundamental criteria in the risk analysis of a contaminated site. Several results are possible depending on whether the characterization factors are high or low on the various axes. The most dangerous cases are those that have high factors for all three criteria. A, B and C are examples of cases with 2 high criteria, D with 3.

Example: a site has a high migration potential if the waste rests directly on high-permeability alluvial gravels.

- **_Property protection._** If the resources are useful to man and nature then they must be protected from pollution. There are four such resources: air, soil, surface water and groundwater.

This type of analysis makes it possible to identify the most critical cases in terms of the environment. By considering the environment in a technical and economic feasibility study, it is possible to assign priorities to remedial activities. Because the migration potential plays a preponderant role in this analysis, we understand why geology is a key factor in the treatment of contaminated sites.

Conducting this type of analysis requires the data collection at the site. The procedure for evaluating suspect sites, and later declaring them contaminated sites, is gradual because of economic criteria. This is due to the large number of sites to be investigated in parallel at the beginning of a campaign, before focusing on sites that actually turn out to be hazardous to human health and the environment. This procedure has direct implications on reconnaissance methods. Preliminary studies must be low in cost. As the accumulated data points to a hazard, the methods become more sophisticated. In this sense, the exploration of contaminated sites is similar to all geological prospecting (ores, petroleum, water, etc.).

This classic approach is related to two political issues:

- The idea of a contaminated site is emotionally charged for the populations who live in the area and even outside of it.
- Significant economic consequences result from designating a piece of property as contaminated.

14.2.5.5   Pollution reduction methods

These activities include pollution prevention and remediation measures. The first type has been described for groundwater resources (see § 7.4.10). Regarding remediation, depending on the risk evaluation, a site will either be decontaminated or isolated:

- **Decontamination:** When a contaminated site is releasing significant pollutants into the environment, there is good reason to remediate it by decontamination. This involves the excavation and treatment of the contaminated material and putting the treated material back into the excavation, or the extraction of contaminants by wells, or the immobilization of toxic materials in place, or even the in-situ treatment of contaminants (Fig. 14.37). Contaminant extraction methods using wells are particularly suitable in the case of deep and extensive pollution. Many sites that were contaminated with light hydrocarbons as a result of accidents have been treated in this manner. In the case of heavy hydrocarbons (DNAPL), the methods are more complicated, especially when the products are volatile, which is often the case. Regarding in-situ decontamination, intense research in environmental biotechnology is being done to select micro-organisms capable of attacking these substances, most of them chlorinated hydrocarbons, which are not very degradable (Fig. 14.38).

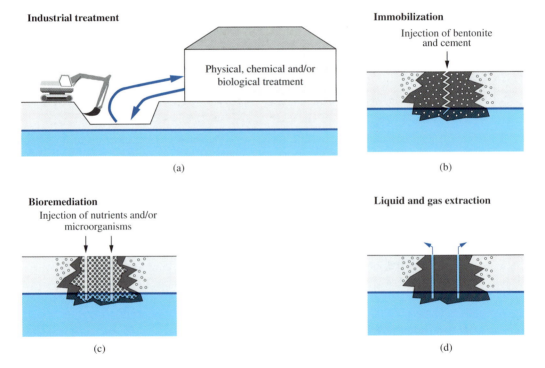

Fig. 14.37 Remediation techniques for contaminated sites. (a) Complete excavation of the contaminated zone with industrial treatment. (b) In-situ immobilization by fixing contaminants to injected binders (bentonite-cement). (c) Bioremediation with injection of specific micro-organisms and nutrients. (d) Extraction of gases and liquids by wells (from [60]).

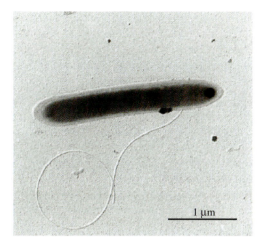

Fig. 14.38 Example of Dehalobacter bacteria which dechlorinates tetrachloroethene to cis-1,2 dichloroethene. Photo by Christof Holliger, EPFL.

- ***Isolation:*** When a site has latent pollution, it can be advisable, at least initially, to isolate the site. This involves various methods of creating a barrier around the site (Fig. 14.39). The barrier can be physical; for example, watertight panels installed around the site at the surface and at depth (encapsulation). Underground construction and grouting techniques have progressed greatly, which makes such operations possible. Barriers can also be purely hydraulic, especially along the sides of the contaminated site. Well curtains that draw down the water table around the site prevent the movement of contaminants out of the site. This system is economical to use but has the disadvantage of bringing large quantities of water to the surface. The great majority of this water is clean water but it is mixed with a small amount of contaminated water. This situation does not favor efficient and economical treatment. Sites that have been isolated will likely be candidates for decontamination once the more urgent sites are cleaned up.

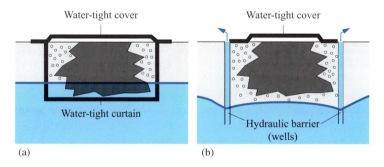

Fig. 14.39 Remediation by isolation. (a) Encapsulation. (b) Hydrogeologic isolation (from [60]).

14.2.5.6 Perspectives on the evolution of subsurface pollution

Like other environmental threats, subsurface pollution is a result of many different factors. Its specific characteristic is its inertia. It is often invisible and does not cause problems for long periods of time. When it is detected, it is often too late to save groundwater resources. Risk evaluations and decontamination projects are becoming increasingly complex. However, a community has measures to combat this threat to public health and the environment. Measures related to land development that take into account the hydrogeologic vulnerability of aquifers are the best prevention tool.

The improvement in laboratory techniques has revealed new substances that we are not prepared to combat. This is particularly the case of organic micro-contaminants (mainly medications, pesticides, and viruses) because they cause significant concerns to entities that use surface water to produce drinking water. Groundwater often contains agricultural micro-contaminants but is generally unaffected by medicinal products and viruses, except for karst aquifers and wells that are heavily recharged by rivers. Research is being done to determine the effective impact of these substances on health and the environment, to replace the most harmful and persistent molecules, and to develop new techniques to retain them in water treatment plants.

Given this subsurface pollution problem, geology will be a leading factor in determining the source of contamination, the contaminant's pathway from the source, its extent, the velocity of its movement, and the use of confinement techniques for toxic substances so that they do not reappear in the biosphere.

[55, 117, 131, 308]

14.3   Conclusion

Throughout this book, we have demonstrated that the Earth is a living and complex environment. It is the workplace of the engineer. The complex Earth is more difficult to understand than a human manufacturing plant, whose behavior is perfectly known. This is why for a long time some engineers have tried to minimize the importance of the Earth's complexity. The challenges to society that we have discussed in this chapter show that engineers can no longer follow this approach. On the contrary, the understanding of the "geologic system" is an essential key for meeting these challenges. No single person or single discipline has the know-how; it will require a team approach. This is the end of the gap between the natural sciences and the technical sciences; and both of these sciences are reaching out to the human sciences. Tomorrow even more than today, the general design phases for infrastructure and their integration into the spatio-temporal environment will require the integration of a number of factors related to the complexity of the land, particularly the "geologic factors". For example, instead of being purely linear projects, roads will be three-dimensional. They will be designed with respect to sustainability and will be more harmoniously integrated in the evolving territory.

This change in the engineer's profession is not easy to embrace. Once engineers get adjusted to this new worldview, they will find their mission enriched by the simple fact that they will work together with numerous other disciplines and thus they will be more involved in the fundamental concerns of society. It is imperative that the engineer succeeds in meeting this challenge. The geologist and geology will surely aid him in his quest.

Bibliography

[1]   ADAMS, A.E., MACKENZIE, W.S. et GUILFORD, C., *Atlas of Sedimentary Rocks Under the Microscope*, Longmann, 1984.

[2]   ALLEGRE, C., *Ma vérité sur la planète*, Edition Plon – Fayard, 2007.

[3]   AMBERGER, G., BERGIER, J.F., GÉROUDET, P., MONOD, R., PITTARD, J.-J., REVACLIER, R., SAUTER, M.-R., *Le Léman. Un lac à découvrir*, Office du Livre S.A., Fribourg, 1976.

[4]   ANDRÉ, J.-C. et ROYER, J.-F., Les fluctuations à court terme du climat et l'interprétation des observations récentes en terme d'effet de serre, *Comptes Rendus de l'Académie des Sciences* – Series IIA – Earth and Planetary Science, Volume 328, Issue 4, February 1999, pp. 261–272.

[5]   ARAGNO, M., GOBAT, J.M., MATTHEY, W., *Le sol vivant, bases de pédologie et biologie des sols*, Presses polytechniques et universitaires romandes, Lausanne, 2003.

[6]   ARIMOTO, R., Eolian Dust and Climate: Relationships to Sources, Tropospheric Chemistry, Transport and Deposition, *Earth-Science Reviews*, Volume 54, Issues 1–3, June 2001, pp. 29–42.

[7]   ARNOULD, M., *Altération bactérienne des pierres calcaires*, A.A. Balkema, Rotterdam, 1988.

[8]   ASTIER, J.L., *Géophysique appliquée à l'hydrogéologie*, Masson & Cie, Paris, 1971.

[9]   ATLAS COPCO ROCK DRILLS AB, *Mining Methods in Underground Mining*. Second Edition 2007.

[10] ATTEWELL, P.B., FARMER, I.W., *Principles of Engineering Geology*, Chapman and Hall Ltd, London, 1976.

[11] AUDOUZE, J., ISRAËL, G., *Le grand atlas de l'astronomie*, Encyclopaedia Universalis France S.A., Paris, 1988.

[12] AURIOL, J.C. et al., *Promoting optimal use of local materials*, PIARC, 2007.

[13] B.R.G.M., *Gisements français des Pb, Zn, W, Sn, Au, As, barytine, fluorine, talc*, Mémoire du B.R.G.M. n°112, Délégation générale à la recherche scientifique et technique, 1980.

[14] BABIN, CL., *Eléments de paléontologie*, Armand Colin, Paris, 1971.

[15] BACHMANN, H., *Erdbebensicherung von Bauwerken*, Birkhäuser Verlag, Bâle, 1995.

[16] BANNER, J. L., Radiogenic Isotopes: Systematics and Applications to Earth Surface Processes and Chemical Stratigraphy, *Earth-Science Reviews*, Volume 65, Issues 3–4, May 2004, pp. 141–194.

[17] BARD, P.-Y., MÉNEROUD, J.P., DURVILLE, J.L., et MOUROUX, P., Microzonage sismique. Application aux plans d'exposition aux risques (PER), *Bull. Liaison Laboratoires des Ponts et Chaussées*, Numéro Spécial «Risques Naturels», pp. 150–151, 130–139, 1987.

[18] BARETS, A., FERRAGNE, A., GAYET, J., GROS, J.CL., HABIB, P., LEVÊQUE, P.CH., MAURY, V., MORLIER, P., PANET, M., PRIOU, P., THIBAULT, CL., WIGNIOLLE, E., *Géologie appliquée au génie civil, au génie nucléaire et à l'environnement*. Tome 1, Technique & Documentation – Lavoisier, Paris, 1984.

[19] BAUER-PLAINDOUX, C., TESSIER, D., GHOREYCHI, M., Propriétés mécaniques des roches argileuses carbonatées: Importance de la relation calcite-argile. *Comptes Rendus de l'Académie des Sciences* – Series IIA – Earth and Planetary Science, Volume 326, Issue 4, February 1998, pp. 231–237.

[20] BAYARD, D., *The effect of seasonal soil frost on the alpine groundwater recharge including climate change aspects*. Thèse N°2709, EPF Lausanne, Suisse, 2003.

[21] BELL, F.G., *Engineering Geology*, Blackwell Scientific Publications, Oxford, 1993.

[22] BENNETT, M.R., GLASSER, N.F., *Glacial Geology; Ice Sheets and Landforms*, New York, John Wiley and Sons, Inc., 1996.

[23] BERNSTEIN, L., BOSCH, P., CANZIANI, O., et al., *Climate Change 2007 – Synthesis Report, Summary for Policy Makers*. Intergovernmental Panel on Climate Change, Cambridge University Press, Cambridge, UK and New York USA.

[24] BERRY, L.G., MASON, B., *Mineralogy: Concepts, Descriptions, Determinations*, W.H. Freeman and Company, San Francisco, 1959.

[25] BESNER, J., *The sustainable usage of the underground space in metropolitan area*. Conference of the Associated Research Centers for Urban Underground Space, Torino, 2002.

[26] BIANCHETTI, G., ROTH, P., VUATAZ, F.-D. & VERGAIN, J.: Deep Groundwater Circulation in the Alps: Relations between Water Infiltration, Induced Seismicity and Thermal Springs. The Case of Val d'Illiez, Wallis, Switzerland. *Eclogae geol. Helv.*, 85/2, 1992, pp. 291–305.

[27] BIGNOT, G., *Les microfossiles*, Bordas, Paris, 1982.

[28] BISSIG, P., Die CO_2-reichen Mineralquellen von Scuol-Tarasp (Unterengadin, Kt. GR). *Bull. angew. Geol.* 9/2: 39–47, 2004.

[29]   BLOSS, F.D., *Crystallography and Crystal Chemistry: an Introduction*, Holt, Rinehart and Winston Inc., New York, 1971.

[30]   BLUNIER, P., TACHER, L. et PARRIAUX, A., *Systematic approach of urban underground resources exploitation*. 11th ACUUS International Conference, Athens, Greece, 2007, pp. 43–48.

[31]   BONNEAU, M., SOUCHIER, B., *Pédologie, 2. Constituants et propriétés du sol*, Masson, Paris, 1979.

[32]   BOUËT, M., *Climat et météorologie de la Suisse romande*. Payot, Lausanne, 1972.

[33]   BRATER, E.F., KING, H.W., *Handbook of Hydraulics*, McGraw-Hill Inc., New York, 1976.

[34]   BROWN, G.C., MUSSETT, A.E., *The Inaccessible Earth: an Integrated View to its Structure and Composition*, Chapman & Hall, London, 1995.

[35]   BRYANT, S., *Geologic CO_2 Storage: Can the Oil and Gas Industry Help Save the Planet?*, Journal of Petroleum Technology, SPE 103474, 2007, pp. 98–105.

[36]   BURGER, A., RECORDON, E., BOVET, D., COTTON, L., SAUGY, B., *Thermique des nappes souterraines*, Presses polytechniques et universitaires romandes, Lausanne, 1985.

[37]   BÜRGI, C., 1999. *Cataclastic Fault Rocks in Underground Excavations – a Geological Characterisation*. Thèse de doctorat n°1975, EPFL, Lausanne.

[38]   BURGI, C., PARRIAUX, A., FRANCIOSI, G., REY, J.P., *Cataclastic rocks in underground structures – terminology and impact on the feasibility of projects (initial results)*, Engineering Geology, vol. 51, num. 3, 1999, p. 225–253.

[39]   CAILLÈRE, S., HÉNIN, S., RAUTUREAU, M., *Minéralogie des argiles, 2. Classification et nomenclature*, INRA et Masson, Paris, 1982.

[40]   CAILLEUX, A., CHAVAN, A., *Détermination pratique des roches*, SEDES, Paris, 1971.

[41]   CAMPY, M., MACAIRE, J.J., *Géologie des formations superficielles. Géodynamique – faciès – utilisation*, Masson, Paris, 1989.

[42]   CARMODY, J. and STERLING, R., *Underground Space Design: A Guide to Subsurface Utilization and Design for People in Underground Spaces*, Van Nostrand Reinhold, New York, 1993.

[43]   CASTANY, G., *Traité pratique des eaux souterraines*, Dunod, Paris, 1967.

[44]   CERNICA, J.N., *Geotechnical Engineering: Foundation Design*, John Wiley & Sons, New York, 1995.

[45]   CHATTERJEE, N.D. & JOHANNES, W., Thermal Stability and Standard Thermodynamic Properties of Synthetic 2M1- Muscovite $KAl_2(AlSi_3O_{10}(OH)_2)$. *Contributions to Mineralogy and Petrology*, 48, pp. 89–14, 1974.

[46]   CLARK, I., FRITZ, P., *Environmental Isotopes in Hydrogeology*, Lewis Publishers, New York, 1997.

[47]   CLUFF, L.S., Peru earthquake of May 31, 1970. Engineering Geology Observations. *Seismological Society of America Bulletin*, 61:3, 1971, pp. 511–521.

[48]   COCH, N.K., *Geohazards: Natural and Human*, Prentice Hall, New Jersey, 1995.

[49]   COLE, J.W., MILNER, D.M. and SPINKS, K.D., Calderas and caldera structures: a review. *Earth-Science Reviews*, Volume 69, Issues 1–2, February 2005, pp. 1–26.

[50]   COLOMBI, A., *Métamorphisme et géochimie des roches mafiques des Alpes ouest-centrales (géoprofil Viège-Domodossola-Locarno*. Thèse de minéralogie. Mémoires de géologie (Lausanne) n°4, 1988.

[51]   COULOMB, J., *La Constitution physique de la Terre*, coll. «La Science d'aujourd'hui», Albin Michel, Paris, 1952.

[52]   COVEY, C., BARRON, E., The role of ocean heat transport in climatic change. *Earth-Science Reviews*, Volume 24, Issue 6, February 1988, pp. 429–445.

[53]   COVEY, C., THOMPSON, S.L., Testing the Effects of Ocean Heat Transport on Climate. *Global and Planetary Change*, Volume 1, Issue 4, December 1989, pp. 331–341.

[54]   COX, A., *Plate Tectonics and Geomagnetic Reversals*, W.H. Freeman and Company, San Francisco, 1973.

[55]   C.R.C AUSTRALIA, *Organic Micropollutants in water*. The Cooperative Research Centre (CRC) for Water Quality and Treatment, 2007.

[56]   CRUDEN, D.M. and VARNES, D.J., *Landslide types and processes*. In: Turner A.K.; Shuster R.L. (eds) Landslides: Investigation and Mitigation. Transp Res Board, Spec Rep 247, pp. 36–75, 1996.

[57]   CRUDEN, D.M., *The Multilingual Landslide Glossary*, Bitech Publishers, Richmond., British Columbia, for the UNESCO Working Party on World Landslide Inventory, 1993.

[58]   DANIEL, J.-Y., BRAHIC, A., HOFFERT, M., SCHAAF, A., TARDY, M., *Sciences de la Terre et de l'Univers*, Librairie Vuibert, Paris, 1999.

[59]   DAUMAS, M., *Histoire de la Science. Des origines au XXe siècle*, Collection Encyclopédie de la Pléiade (n°5), Gallimard, Paris, 1957.

[60]   DAVIS, G.H., *Structural Geology of Rocks and Regions*, John Wiley and Sons, New York, 1984.

[61]   DE CASTRO, J., *Biofuels – An Overview*, Final report, 2007.

[62]   DENIS, F., *Existing specific national regulations applied to material recycling*, SAMARIS, 2004.

[63]   DE QUERVAIN, F., *Die nutzbaren Gesteine der Schweiz*, Kümmerly et Frey, Berne, 1969.

[64]   DE QUERVAIN, F., *Steine schweizerischer Kunstdenkmäler*, Manesse Verlag, Zurich, 1979.

[65]   DE QUERVAIN, F., *Technische Gesteinskunde*, Birkhäuser Verlag, Bâle, 1967.

[66]   DELECLUSE, P., El Niño et sa prévision. *Comptes Rendus de l'Académie des Sciences* – Series IIA – Earth and Planetary Science, Volume 328, Issue 4, February 1999, pp. 281–288.

[67]   DEMATTEIS, A., HESSKE, S., PARRIAUX, A., TACHER, L., *Hydrologischer Atlas der Schweiz. Dreidimesionales Modell der Grundwasserfassung, Haupttypen der Grundwasserleiter*, Berne, 1997.

[68]   DERBYSHIRE, E., Geological hazards in loess terrain, with particular reference to the loess regions of China. *Earth-Science Reviews*, Volume 54, Issues 1–3 , June 2001, pp. 231–260.

[69]   DESCOEUDRES, F., DUMONT, A.-G., PARRIAUX, A., VULLIET, L., DYSLI, M., ROBYR, P., FONTANA, M., FRANCIOSI, G., *Utilisation des matériaux d'excavation de tunnels dans le domaine routier.* Mandat-ASTRA (1998/094), n° 1006, mars 2002.

[70]   DESVISMES, P., *Atlas photographique des minéraux d'alluvions*, Mémoire du Bureau de recherches géologiques et minières n°95, B.R.G.M., Nantes, 1978.

[71]   DIMENTO, J. and DOUGHMAN, P., *Climate Change – What It Means for Us, Our Children, and Our Grandchildren*, MIT Press, 2007.

[72]   DOOLEY, JJ. et al., *Carbone Dioxide Capture and Geologic Storage, A core element of a global energy technology strategy to address climate change*, The Global Energy Technology Strategy Program (GTSP), 2006.

[73]   DREWRY, D., *Glacial Geologic Processes*, Edward Arnold, London, 1986.

[74]   DRISCOLL, F.G., *Groundwater and Wells*, Johnson filtration Systems, St. Paul Minnesota, 1986.

[75]   DUCHAUFOUR, P., *Pédologie, 1. Pédogenèse et classification*, Masson, Paris, 1977.

[76]   DUCHAUFOUR, P., *Précis de Pédologie*, Masson & Cie, Paris, 1970.

[77]   DUFFAUT, P. and LABBÉ, M., *From underground road traffic to underground city planning.* Conference of the Associated research Centers for Urban Underground Space, Torino, 2002.

[78]   DUNCAN, N., *Engineering Geology and Rock Mechanics*, Volume 1 and 2, Leonard Hill, London, 1970.

[79]   DUPASQUIER, S., PARRIAUX, A., *Types de pollution des eaux souterraines.* Atlas Hydrologique de la Suisse, feuille n°7.5, 2002.

[80]   DÜRST, A., *Weltbild. Studiendokumentation*, Eigenverlag, Zurich, 1987.

[81]   DYSLI, M., *Le gel.* Complément au Traité de génie civil, Presses polytechniques et universitaires romandes, 1993.

[82]   EMBLETON, C., KING, C.A.M., *Glacial Geomorphology*, New York, John Wiley and Sons, 1975a.

[83]   EMBLETON, C., KING, C.A.M, *Periglacial Geomorphology*, New York, John Wiley and Sons, 1975b.

[84]   EVANS, D.M., *The Denver Area Earthquakes and the Rocky Mountain Arsenal Disposal Well. The Mountain Geologist*, vol. 3, 1966.

[85]   EYLES, N., *Glacial Geology: An Introduction for Engineers and Earth Scientists.* Pergamon Press, Oxford, 1983.

[86]   FAIRHURST, C. *et al.*, International Geomechanical Commission, 1999. Underground Nuclear Testing in French Polynesia: Stability and Hydrology Issues. Paris: La Documentation Française.

[87]   FARMER I.W. *Engineering Properties of Rocks.* Chapman and Hall, London, 1968.

[88]   FAURE, G., *Principles of Isotope Geology*, John Wiley & Sons, New York, 1987.

[89]   FÉLIX, C., *Essais de dilatation linéaire isotherme par absorption d'eau sur des grès.* Chroniques des matériaux de construction. Chantiers/Suisse, vol. 15, 12/84.

[90]   FER, I., LEMMIN, U. et THORPE, S. A., Observations of Mixing Near the Sides of a Deep Lake in Winter. *Limnol. Oceanogr.*, 47, 2002, pp. 535–544.

[91]   FER, I., LEMMIN, U., THORPE, S. A., Winter Cascading of Cold Water in Lake Geneva. *Journal of Geophysical Research*, 107(C6), 13.1–13.16, 2002.

[92]   FETTER, C.W., *Contaminant Hydrogeology*, Macmillan Publishing Company, New York, 1993.

[93]   FLINT, R.F., *Glacial and Quaternary Geology*, John Wiley and Sons, New York, 1971.

[94]   FLUGEL, E., *Microfacies of Carbonate Rocks.* Springer Verlag, Berlin, 2004.

[95]   FLUTEAU, F., La dynamique terrestre et les modifications climatiques. *Comptes Rendus Géosciences*, Volume 335, Issue 1, January 2003, pp. 157–174.

[96]   FORD, T.D., *The Science of Speleology*, Academic Press, London, 1976.

[97]   FOREL, F.A., *Le Léman:* Monographie limnologique, 3 vol., 1892–1904.

[98]   FRANCIS, P., *Volcanoes: A Planetary Perspective*, Oxford University Press, Hong-kong, 1995.

[99]   FRANÇOIS, L., FAURE, H., PROBST, J.L., The global carbon cycle and its changes over glacial–interglacial cycles. *Global and Planetary Change*, Volume 33, Issues 1–2, June 2002, pp. vii–viii.

[100] FREI, W., HEITZMANN, P. & LEHNER, P., Swiss NFP-20 Research Program of the Deep Structure of the Alps. In: Deep Structure of the Alps. *Mém. Soc. géol. Suisse* 1, 1990.

[101] FREY, F., *Analyse des structures et milieux continus: mécanique des solides*, Traité de génie civil, vol. 3, Presses polytechniques universitaires romandes, Lausanne, 1973.

[102] FREY, M., MÄLMANN FERREIRO, R., 1999. Alpine metamorphism of the Central Alps. *SMPM* (Bulletin suisse de Minéralogie et Pétrographie) vol. 79/1, 1999, pp. 135–154.

[103] GEISER, F., *Comportement mécanique d'un limon non saturé. Etude expérimentale et modélisation constitutive*. Thèse de doctorat EPFL n°1942, mars 1999.

[104] GIGNOUX, M., *Géologie stratigraphique*, Masson & Cie., Paris, 1960.

[105] GIRAULT, J., *Caractères optiques des minéraux transparents: tables de détermination*, Masson & Cie, Paris, 1980.

[106] GIROD, M., *Les roches volcaniques, pétrologie et cadre structural*, Doin Editeurs, Paris, 1978.

[107] GOGUEL, J., *Géologie de l'environnement*, Masson & Cie, Paris, 2002.

[108] GOGUEL, J., *Traité de tectonique*, Masson & Cie, Paris, 1965.

[109] GOGUEL, J., *Application de la géologie aux travaux de l'ingénieur*, Masson & Cie, Paris, 1993.

[110] GRÜNTHAL, G., (ed.), *European Macroseismic Scale 1998*. Cahiers du Centre Européen de Géodynamique et de Séismologie, volume 15, Luxembourg, 1998.

[111] GUINLE-THÉNEVIN, I., *Influence des valeurs extrêmes (millennales) des données physiques de l'environnement naturel sur le sol et le proche sous-sol – Application sur sites de stockage de déchets*. Thèse de doctorat, ENSMP, Paris, 1998.

[112] HÄBERLI, W. BÖSCH, H., SCHERLER, K., OSTREM, G., WALEN, C.C., *World glacier inventory status 1988*, IAHS (ICSI)-UNEP-UNESCO, Switzerland, 1988.

[113] HÄBERLI, W., Creep of Mountain Permafrost: Internal Structure and Flow of Alpine Rock Glaciers, Mitteilungen der Versuchsanstalt für Wasserbau, *Hydrologie und Glaziologie* Nr. 77, 1985.

[114] HAMBREY, M., ALEAN, J., *Glaciers*, 2nd ed., New York, Cambridge University Press, 2004.

[115] HAMBREY, M., *Glacial Environments*, Routledge, London, 2003.

[116] HANTKE, R., *Eiszeitalter1,2,3. Die jüngste Erdgeschichte der Schweiz und ihrer Nachbargebiete*. Band 1–3, Thoune – Ott Verlag, 1978, 1980, 1983.

[117] HARTER, T., *Groundwater Quality and Groundwater Pollution*. University of California, Division of Agriculture and Natural Resources, Publication 8084, 2003.

[118] HETHERINGTON, L., BROWN, T., BENHAM, A., LUSTY, P., IDOINE N., *World Mineral Production 2001–2005*, British Geological Survey, 2007.

[119] HIGHLAND, L., *Landslide Hazards*, USGS Fact Sheet FS-071-00, US Geological Survey, 2000.

[120] HATCH, F.H., WELLS, A.K., WELLS, M.K., *Petrology of the Igneous Rocks*, George Allen & Unwin, London, 1972.

[121] HAUBER, L., Der Südliche Rheingraben und seine geothermische Situation. *Bull. Ver. Schweiz*, 1993.

[122] HEDDLE, G., HERZOG, H. and KLETT, M., *The Economics of CO2 Storage*, Laboratory for energy and the environment, Publication No. LFEE 2003-003 RP, 2003.

[123] HEIM, D., *Tone und Tonminerale*, Ferdinand Enke Verlag, Stuttgart, 1990.

[124] HEITZMANN P., LEHNER P., MÜLLER St., PFIFFNER A., STECK A. (Hrsg.), *Deep Geological Structure of the Swiss Alps: Results of the National Research Programme (NRP)* 20, Birkhäuser Verlag AG, Basel, 1996.

[125] HESSKE, S., PARRIAUX, A., BENSIMON, M., Geochemistry of springwaters in Molasse aquifers: Typical mineral trace elements. *Eclogae geol. Helv*. 90/1, 1997, pp. 151–171.

[126] HJULSTRÖM, F., Transportation of Detritus by Moving Water, Trask P.D., *Am. Asso. Petrol. Geol.*, pp. 438–460, 1939.

[127] HOEK, E., BRAY, J., *Rock Slope Engineering*, Chapman & Hall, London, 1997.

[128] HOFINGER, H. and ADAM, D., *Tunnels as energy sources – Practical applications in Austria*. Proceedings of the 14th European Conference on Soil Mechanics and Geotechnical Engineering, Madrid, Spain, 2007.

[129] HOLDAWAY, M.J., Stability of Andalusite and the Aluminium Silicate Phase Diagrams. *American Journal of Science*, 271, 1971, pp. 97–131.

[130] HOLTZ, R.D., KOVACS, W.D., *Introduction à la géothechnique*, Editions de l'Ecole Polytechnique de Montréal, New Jersey, 1999.

[131] HOORMAN et al., *Agricultural Impacts on Lake and Stream Water Quality in Grand Lake St. Marys, Western Ohio*, Water, Air and Soil Pollution, 2008.

[132] HUNGR et al., *A review of the classification of landslides flow type*, Environmental and Engineering Geoscience, v. 7, n°3, 2001.

[133] HURLBUTT, C.S., KLEIN, C., *Manual of Mineralogy*, 21th edition, John Wiley and Sons Inc., New York, 1985.

[134] HURTIG, E., STILLER, H., *Erdbeben und Erdbebengefährdung*, Springer Verlag, Berlin, 1984.

[135] I.A.E.A, International Atomic Energy Agency, *A Long Term Storage of Radioactive Waste: Safety and Sustainability*, Vienna, 2003.

[136] ISHMAN, S., BREWSTER-WINGARD, L., WILLARD, D., *Preliminary Paleontologic Report on Core 37, from Pass Key, Everglades National Park, Florida Bay*, Open-file report 98-122. U.S. Geol. Surv., 1998.

[137] ITA WORKING GROUP 4. *Planning and mapping of underground space – an overview*. Tunnelling and Underground Space Technology, 15(3), 271–286, 2000.

[138] IVES, J.D., BARRY, R.G., *Arctic and Alpine Environments*, Methuen & Co Ltd, London, 1974.

[139] JÄGER, E., HUNZIKER, J.C., *Lectures in Isotope Geology*, Springer, Berlin, 1979.

[140] JIGOREL, A., PITOIS, F., *Gestion expérimentale des apports sédimentaires dans l'estuaire de la Rance, France*. Proceedings Eighth International Association for Engineering Geology and the Environment, Vancouver, Canada, 21–25 septembre 1998, pp. 3859–3865, 1998.

[141] JOHNSON, R.B., DeGRAFF, J.V., *Principles of Engineering Geology*, John Wiley & Sons, New York, 1988.

[142] JONES, B.W., *The Solar System*, Pergamon Press, Oxford, 1984.

[143] JONES, J.L., *Mapping a Flood...Before It Happens*, USGS Fact Sheet 2004-3060, US Geological Survey, 2004.

[144] JOUENNE, C.A., *Traité de Céramiques et Matériaux minéraux*, Editions Septima, Paris, 1975.

[145] JUNG, J., *Précis de Pétrographie: Roches sédimentaires métamorphiques et éruptives*, Masson & Cie, Paris, 1969.

[146] KAUFMANN, W.J., *Universe*, W.H. Freeman and Company, New York, 1991.

[147] KAY, M., COLBERT, E.H., *Stratigraphy and Life History*, John Wiley & Sons, New York, 1965.

[148] KENNETH, J.H., *Hydrologic cells for recovery of hydrocarbons or thermal energy from coal, oil-shale, tar-sands and oil-bearing formations*, 1999.

[149] KLEIN, C., HURLBUT, C.S., *Jr. Manual of mineralogy, after James D. Dana.*, 21st edition. Ed. Wiley & Sons, Inc. New York, 1999.

[150] KLEINHANS, M.G., Sorting in grain flows at the lee side of dunes. *Earth-Science Reviews*, Volume 65, Issues 1–2, March 2004, pp. 75–102.

[151] KÖNIG, M.A., *Geologische Katastrophen und ihre Auswirkungen auf die Umwelt. Vulkane, Erdbeben, Bergstürze*, Ott Verlag, Thoune, 1994.

[152] KRAMER STEVEN L., *Geotechnical Earthquake Engineering*, Prentice Hall, 1996.

[153] KUKLA, G., GAVIN, J., MILANKOVITCH, *Climate Reinforcements. Global and Planetary Change*, Volume 40, Issues 1–2, pp. 27–48. Global Climate Changes during the Late Quaterny, January 2004.

[154] KUMP, L.R., Palaeoclimate: Foreshadowing the Glacial Era. *Nature* 436, 333–334, 21 July 2005.

[155] KÜNDIG, R., MUMENTHALER, T., ECKARDT, P., KEUSEN, H.R., SCHINDLER, C., HOFMANN, F., VOGLER, R., GUNTLI, P., *Die mineralischen Rohstoffe der Schweiz*, Schweizerische Geotechnische Kommision, Zurich, 1997.

[156] LAMBE, T.W., WHITMAN, R.V., *Soil Mechanics*, John Wiley & Sons, New York, 1999.

[157] LANG, H.-J., HUDER, J., *Bodenmechanik und Grundbau. Das Verhalten von Böden und die wichtigsten grundbaulichen Konzepte*, Springer Verlag, Berlin, 1990.

[158] LANGGUTH, H.R., VOIGT, R., *Hydrogeologische Methoden*, Springer Verlag, Berlin, 1980.

[159] LAROUZIÈRE DE, F.D., *Dictionnaire des roches d'origine magmatique: Manuels et méthodes*, Editions du BRGM, Orléans Cedex, 1989.

[160] LARSEN, G., CHILINGAR G.V., *Developments in Sedimentology: Diagenesis in Sediments and Sedimentary Rocks*, Elsevier, Amsterdam, 1979.

[161] LASSALLE, F., *Die Philosophie Herakleitos des Dunklen von Ephesos*, F. Duncker, Berlin, 1858.

[162] LEEDER, M.R., *Sedimentology: Process and Product*, Chapman & Hall Ltd, London, 1995.

[163] LEET, L.D., JUDSON, S., KAUFFMAN, M.E., *Physical Geology*, Prentice-Hall Inc., New Jersey, 1978.

[164] LEGROS, J.P., *Cartographie des sols, De l'analyse spatiale à la gestion des territoires*, Presses polytechniques et universitaires romandes, Lausanne, 1996.

[165]   LEOPOLD, L.B., WOLMAN, M.G., MILLER, J. P., *Fluvial Processes in Geomorphology*, W.H. Freeman and Company, San Francisco, 1964.

[166]   LETOURNEUR, J., MICHEL, R., *Géologie du génie civil*, Armand Colin, Paris, 1971.

[167]   LOZET, J., MATHIEU, C., *Dictionnaire de Science du Sol*, Technique et Documentation – Lavoisier, Paris, 1990.

[168]   LUDMAN, A., COCH, N.K., *Physical Geology*, McGraw-Hill Book Company, New York, 1982.

[169]   LUGEON, M., ARGAND, E., *Atlas géologique de la Suisse, 1:25 000, feuille 485 Saxon-Morcles. Notice explicative*. Commission géologique de la soc. helv. des sciences naturelles, 1937.

[170]   LUGEON, M., *Barrages et Géologie. Méthodes de recherches. Terrassement et imperméabilisation*, Librairie de l'Université, Lausanne, 1979.

[171]   MABBUTT, J.A., *Desert Landforms. An Introduction to Systematic Geomorphology*, Volume 2, MIT Press, Cambridge, 1977.

[172]   MACKENZIE, W.S., GUILFORD, C., *Atlas of Rock Forming Minerals in Thin Section*, Longman Group Limited, London, 1980.

[173]   MADSEN, F.T., NÜESCH, R., Juradurchquerungen – aktuelle Tunnelprojekte im Jura. SIA-Dokumentation D 037, 1989.

[174]   MASON, B., *Principles of Geochemistry*, John Wiley & Sons, New York, 1966.

[175]   MASON, R., *Petrology of the Metamorphic Rocks*, George Allen et Unwin Ltd., London, 1978.

[176]   MATHER, A.S., CHAPMAN, K., *Environmental Resources*, Longman, Essex, 1999.

[177]   MAYEUR, B., FABRE, D. (1999-a) – Mesure et modélisation des contraintes naturelles. Application au projet de tunnel ferroviaire Maurienne-Ambin. *Bull. Eng. Geol. & Env.*, vol. 58, pp. 45–59.

[178]   MC BIRNEY, A.R., *Igneous Petrology*. Jones and Bartlett Publishers International. London, 1993.

[179]   MEANS, W.D., *Stress and Strain: Basic Concepts of Continuum Mechanics for Geologists*, Springer Verlag, New York, 1976.

[180]   MEARS, B. JR., *The Changing Earth. Introduction to Geology*, D. Van Nostrand Company, New York, 1978.

[181]   MEGHRAOUI, M., Failles actives et trace des séismes en surface: l'approche paléosismologique. *Comptes Rendus de l'Académie des Sciences* – Series IIA – Earth and Planetary Science, Volume 333, Issue 9 15 November 2001, pp. 495–511.

[182]   MEYER, O., WERLEN, M., PFAMMATTER, C., EYER, E., Die Hubbrücke «ªSaltina» in Brig-Glis Bericht aus Schweizer Ingenieur und Architekt Nr. 50, 11 Décembre 1997 (IAS).

[183]   MIDDELMANN, M.H. (Editor), *Natural Hazards in Australia*: *Identifying Risk Analysis Requirements*, Geoscience Australia, Canberra, 2007.

[184]   MIDDLEMOST, E.A.K., *Magmas and Magmatic Rocks: An introduction to igneous petrology*, Longman Group, New York, 1985.

[185]   MIKI, H., IWABUCHI, J. et CHIDA, S., *New Soil Treatment Methods in Japan*, 2005.

[186]   MILANOVIC, P.T., *Geological Engineering in Karst. Dams, Reservoirs, Grouting, Groundwater, Protection, Water Tapping, Tunneling*. Zebra Publishing, Ldt, Belgrade, Yugoslavia, 2000.

[187]   MINSTER, J.B., JORDAN, T.H., Present-Day Plate Motion, *Journal of Geophysical Research*, v. 83, pp. 5331–5354, 1978.

[188]   MIYASHIRO, A., *Metamorphism and Metamorphic Belts*, George Allen and Unwin Ltd, London, 1973.

[189]   MOORE D.P., RIPLEY, B.D., and GROVES, K.L., 1992. Evaluation of Mountainslope Movements at Wahleach; in *Geotechnique and Natural Hazards*; Vancouver, BiTech Publishers, pp. 99–107.

[190]   MORET, L., *Manuel de paléontologie animale*, Masson & Cie, Paris, 1966.

[191]   MOTTANA, A., CRESPI, R., LIBORIA, G., *Minéraux et roches*, Fernand Nathan et Cie, Paris, 1981.

[192]   MUSY, A., SOUTTER, M., *Physique du sol*. Presses polytechniques et universitaires romandes, Lausanne, 1991.

[193]   N.A.G.R.A, *Mont Terri rock laboratory: research on safe geological disposal of radioactive waste*, 2006.

[194]   NAEEM et al., *Biodiversity and Ecosystem Functioning: Maintaining Natural Life Support Processes*, Issues in Ecology Nº 4, Ecological Society of America, 1999.

[195]   NICOLINI, P., *Gîtologie et exploration minière*, Technique et documentation – Lavoisier, Paris, 1990.

[196]   NOËL, P., *Technologie de la pierre de taille. Dictionnaire des termes couramment employés dans l'extraction, l'emploi et la conservation de la pierre de taille*, Société de diffusion des techniques du bâtiment et des travaux publics, Paris, 1965.

[197]   NORDELL, B., Thermal pollution causes global warming. *Global and Planetary Change*, Volume 38, Issues 3–4, September 2003, pp. 305–312.

[198] OTTMANN, F., *Introduction à la géologie marine et littorale*, Masson & Cie Editeurs, Paris, 1965.

[199] PAHUD, D., FROMENTIN, A. and HUBBUCH, M., *Heat exchanger Pile System of the Dock Midfield at the Zürich Airport. Detailed Simulation and Optimisation of the installation. Final Report*. Swiss Federal Office of Energy, Switzerland, 1999.

[200] PAHUD, D., *Mesures et optimisation de l'installation de pieux énergétiques du Dock Midfield, Aéroport de Zurich*. Office fédéral de l'énergie, Berne, 2007.

[201] PANTET, A., PARRIAUX, A., THELIN, PH., *New method for in situ characterization of loose material for landslide mapping purpose*. Engineering Geology 94, pp. 166–179, 2007.

[202] PARRIAUX, A., Contribution à l'étude des ressources en eau du bassin de la Broye. Thèse Ecole polytechnique fédérale de Lausanne, n°393, 1981.

[203] PARRIAUX, A., *Problème de l'intégration des carrières dans une protection globale de l'environnement avec des exemples de la Suisse*. Communication au «2° convegno di geoingegneria» de Turin. Bolletino dell' Associazione Mineraria Subalpina, 1991, pp. 731–743.

[204] PARRIAUX, A., *Routes et changement climatique en montagne : un problème chaud à résoudre la tête froide*, Routes/Roads N° 338, 2008.

[205] PARRIAUX, A., BLUNIER, P., MAIRE, P., TACHER, L., *The DEEP CITY Project: A Global Concept for a Sustainable Urban Underground Management*. 11TH ACUUS International Conference, Underground Space: Expanding the Frontiers, 10–13 September 2007, Athens, Greece, pp. 255–260.

[206] PARRIAUX, A., MAYORAZ, R., Fragilité des ressources en eau du Jura ou le jeu des probabilités à la Vallée de Joux. *Bull. Soc. Neuch. Sc. Nat.*, 113, 1990.

[207] PARRIAUX, A., MAYORAZ, R., MANDIA, Y., *Impact Assessment of Deep Underground Works on a Mineral Water Resource in an Alpine Evaporitic Context*. Memoires of the XXIInd Congress of IAH, EPFL, Lausanne, Part 2, 1990, pp. 1249–1258.

[208] PARRIAUX, A., Problème de l'intégration des carrières dans une protection globale de l'environnement avec des exemples de la Suisse. Communication au «2° convegno di geoingegneria» de Turin. *Bolletino dell' Associazione Mineraria Subalpina*, pp. 731–743, Turin, 1991.

[209] PARRIAUX, A., TACHER, L., *La ville par en dessous: vers un aménagement tridimensionnel du territoire*. Actes de la Conférence CISBAT oct. 2001, Lausanne, 2001, pp. 43–48.

[210] PARRIAUX, A., TARRADELLAS, J., SPACK, L., BENSIMON, M., *Interaction entre les routes et l'environnement souterrain*. Bulletin n°462 de la VSS et Office fédéral des routes, Berne, janvier 1999.

[211] PATTON, F.D., *Multiple Modes of Shear Failure in Rock*. 1st Congress of Int. Soc. Rock Mech., Lisbon, vol. 1, 1966, pp. 509–513.

[212] PERRODON, A., *Géodynamique pétrolière: Genèse et répartition des gisements d'hydrocarbures*, Deuxième édition, Elf Aquitaine et Masson, Pau et Paris, 1985.

[213] PETIT, D., *La gestion de l'après mine : Exemples étrangers*. Annales des Mines, 2004.

[214] PETIT, J.R., JOUZEL, J. et al., *Climate and Atmospheric History of the Past 420 000 Years from the Vostok ice core in Antarctica*. Nature 399, 1999, pp. 429–436.

[215] PETIT, M., Incertitudes scientifiques et risques climatiques. *Comptes Rendus Géosciences*, Volume 337, Issue 4, March 2005, pp. 393–398.

[216] PIGUET, P., *Road runoff over the shoulder diffuse infiltration: real-scale experimentation and optimization*. Thèse N°3858, EPF Lausanne, Suisse, 2007.

[217] PLANTON, S., DÉQUÉ, M., DOUVILLE, H., SPAGNOLI, B., Impact du réchauffement climatique sur le cycle hydrologique. *Comptes Rendus Géosciences*, Volume 337, Issues 1–2, January–February 2005, pp. 193–202.

[218] POMEROL, CH., BABIN, CL., LANCELOT, Y., LE PICHON, X., RAT, P., RENARD, M., *Stratigraphie. Principes, Méthodes, Applications*, Doin Editeurs, Paris, 1987.

[219] PRICE, N.J., COSGROWE, J.W., *Analysis of Geological Structures*, Cambridge University Press, Cambridge, 1990.

[220] PURSER, B.H., *Sédimentation et diagenèse des carbonates néritiques récents*, Tome 1, Editions Technip, Paris, 1980.

[221] PUTZGER, F.W., *Historischer Atlas zur Welt- und Schweizer Geschichte*. Verlag Sauerländer Aarau, Berlin, 1981.

[222] RACKI, G., CORDEY, F., Radiolarian Palaeoecology and Radiolarites: is the Present the Key to the Past? *Earth-Science Reviews*, Volume 52, Issues 1–3, November 2000, pp. 83–120.

[223] RAGAN, D.M., *Structural Geology: An Introduction to Geometrical Techniques*, John Wiley and Sons, New York, 1985.

[224]   RAMSAY, J. G., HUBER, M.I., *The Techniques of Modern Structural Geology, Volume 1: Strain Analysis*, Academic Press Inc., London, 1983.

[225]   RAMSAY, J.G., HUBER, M.I., *The Techniques of Modern Structural Geology, Volume 2: Folds and Fractures*, Academic Press Inc., London, 1987.

[226]   RAYNAUD, D., LORIUS, C., Climat et atmosphère: la mémoire des glaces. *Comptes Rendus Géosciences*, Volume 336, Issues 7–8, June 2004, pp. 647–656.

[227]   READING, H.G., *Sedimentary Environment and Facies*, 2nd Edition, Blackwell Scientific Publication, Oxford, 1986.

[228]   RECORDON, E., *Dynamique des eaux souterraines*. Cours polycopié, Université de Neuchâtel, 1969.

[229]   REEVES, H., *Patience dans l'azur: l'évolution cosmique*. Editions du Seuil, Paris, 1988.

[230]   REEVES, H., *La plus belle histoire du monde*. Le Seuil, collection «Science ouverte», Paris, 1995.

[231]   REEVES, H., *La première seconde*. Le Seuil, collection «Science ouverte», Paris, 1995.

[232]   REID, J.M. et al., *Final Report: Alternative Materials ALT-MAT*, Project funded by the European Commission under the Transport RTD Programme.

[233]   REINECK, H.E., SINGH, I.B., *Depositional Sedimentary Environments*, 2nd Edition, Springer Verlag, Berlin, 1980.

[234]   REINSCH, D., *Natursteinkunde: Eine Einführung für Bauingenieure, Architekten, Denkmalpfleger und Steinmetze*, Enke Verlag, Stuttgart, 1991.

[235]   RÉMÉNIÉRAS, G., *L'hydrologie de l'ingénieur*, Eyrolles, Paris, 1972.

[236]   RENKEN, R.A., DIXON, J., KOEHMSTEDT, J., ISHMAN, S., LIETZ, A.C., MARELLA, R.L, TELIS, P., RODGERS, J. et MEMBERG, S., *Impact of Anthropogenic Development on Coastal Ground-Water Hydrology in Southeastern Florida 1900–2000*. Circular 1275, U.S. Geol. Surv., 2005.

[237]   RIKITAKE, T., *Earthquake Prediction. Developments in Solid Earth Geophysics 9*, Elsevier Scientific Publishing Company, Amsterdam-Oxford-New York 1976.

[238]   RINEHARD, J.S., *Geysers and Geothermal Energy*, Springer-Verlag, New York 1980.

[239]   ROBIN, A., *La terre: ses aspects, sa structure, son évolution*. Larousse, Paris, 1902.

[240]   ROCHE, M., *Hydrologie de surface*, Gauthier-Villars, Paris, 1963.

[241]   ROGER, J., *Paléontologie générale*, Masson & Cie, Paris, 1997.

[242]   ROMANOWICZ, B., 3D structure of the Earth's Lower Mantle. *Comptes Rendus Géosciences*, Volume 335, Issue 1, January 2003, pp. 23–35.

[243]   ROSALES, E., *Analyse a posteriori de la traversée du Jungfraukeil par le tunnel de base du Lötschberg*. Travail de diplôme du Cycle postgrade en géologie de l'ingénieur et de l'environnement EPFL, 2002.

[244]   ROTH, E., POTY, B., *Méthodes de datation par les phénomènes nucléaires naturels. Applications*, Masson & Cie, Paris, 1985.

[245]   ROUBAULT, M., *Détermination des minéraux des roches au microscope polarisant*, Ed. Lamarre-Poinat, Paris, 1991.

[246]   RYBACH, L., *Geothermal energy: sustainability and the environment*, Geothermics 32, pp. 463–470, 2003.

[247]   RYBACH, L., MUFFLER, L.J.P., *Geothermal Systems. Principles and Case Histories*, JohnWiley & Sons Ltd., NewYork, 1981.

[248]   SANDERS, J.E., *Principles of Physical Geology*, John Wiley & Sons, New York, 1981.

[249]   SCHNEEBELI, G., *Hydraulique souterraine*, Eyrolles, Paris, 1966.

[250]   SCHNEIDER, T.R., *Stauanlage Zeuzier – Geologisch-geotechnisch-hydrogeologische Aspekte des Mauerdeformationen*. Wasser Energie Luft, 7/8, 1980, pp. 193–200.

[251]   SCHOLE, P.A., *A Color Guide to the Petrography of Carbonate Rocks: Grains, Textures, Porosity and Diagenesis*. AAPG Memoir 77, 2003.

[252]   SCHWAB, H., MÜLLER, R., *Le passé du Seeland sous un jour nouveau, les niveaux des lacs du Jura*, Editions Universitaires de Fribourg, 1973.

[253]   SCHWARTZ, S.E., *Uncertainty Requirements in Radiative Forcing of Climate Change*. Air & Waste Manage. Assoc. 54, 2004, pp. 1351–1359.

[254]   SCHWARZACHER, W., Repetitions and Cycles in Stratigraphy. *Earth-Science Reviews*, Volume 50, Issues 1–2, May 2000, pp. 51–75

[255]   SCLATER, J.G., PARSONS, B., JAUPART, C., Oceans and Continents: Similarities and Differences in the Mechanism of Heat Loss, *Journal of Geophysical Research*, vol. 86, 1981.

[256]   SEGUIN, M.K., *La Géophysique et les propriétés physiques des roches*, Les Presses de l'Université Laval, Québec, 1971.

[257]   SELBY, M.J., *Hillslope Materials and Processes*, 2nd edition, Oxford University Press, New York, 1993.

[258]   SELLEY, R.C., *An Introduction to Sedimentology*, 2nd Edition, Academic Press Inc., London, 1982.

[259]   SERRUYA, C., *Problems of Sedimentation in the Lake of Geneva*, Verth. Internat. Verein. Limnol. Bd. 17, 1969.

[260]   SHAW, E.M., *Hydrology in Practice*, Van Nostrand Reinhold (UK), Wokingham, 1983.

[261]   SINNIGER, R., HAGER, W., *Constructions hydrauliques*, Traité de génie civil, vol. 15, Presses polytechniques et universitaires romandes, Lausanne, 1989.

[262]   SKINNER, B.J., PORTER, S.C., *The Dynamic Earth. An Introduction to Physical Geology*, John Wiley & Sons, New York, 1989.

[263]   SMALLEY, I.J., *Loess: Lithology and Genesis*, Dowden, Hutchinson & Ross Inc., Stroudsburg, 1975.

[264]   SMIRNOV, V., *Géologie des minéraux utiles*, Editions MIR, Moscou, 1988.

[265]   SMITH, B.J., WRIGHT, J.S., WHALLEY, W.B., Sources of Non-Glacial, Loess-Size Quartz Silt and the Origins of «Desert Loess». *Earth-Science Reviews*, Volume 59, Issues 1–4, November 2002, pp. 1–26.

[266]   Société Suisse pour la géothermie. *Géothermie des tunnels*. Energies renouvelables 1/2003, p. 28.

[267]   Société Suisse pour la géothermie. *La chaleur de la Terre: une énergie propre et durable pour tous / Sondes géothermiques et grands bâtiments*. Energies renouvelables 1/2004, p. 28.

[268]   Société Suisse pour la géothermie. *La Géothermie*. Energies renouvelables 6/2002, p. 28.

[269]   Société Suisse pour la géothermie. *Les pieux de fondation pour chauffer les bâtiments*. Energies renouvelables 2/2003, p. 28.

[270]   SOOD, M.K., *Modern Igneous Petrology*, John Wiley and Sons, New York, 1981.

[271]   SOUTTER, F., 1913. *Note sur la construction du raccourci Frasnes-Vallorbe. Entreprise du Tunnel du Mont-d'Or*. Extrait du Bulletin technique de la Suisse romande, numéros des 10 et 25 octobre, 25 novembre et 25 décembre 1913.

[272]   SPARROW, G., *The Planets*, Quercus Publishing, 2006.

[273]   SPEAR, F.S.,Metamorphic Phase Equilibria and Pressure-Temperature-Time Paths. *Min. Soc. Amer., Monograph Ser.*, 799 p., 1993.

[274]   SPREAFICO, M., WEINGARTNER, R., *Atlas hydrologique de la Suisse*. Office fédéral des eaux et de la géologie, 1992.

[275]   SPREAFICO, M., WEINGARTNER, R., *The Hydrology of Switzerland*. Rapports de l'OFEG, Série Eaux, n°7, Berne, 2005.

[276]   STEVENSON, D.J., Styles of Mantle Convection and their Influence on Planetary Evolution. *Comptes Rendus Géosciences*, Volume 335, Issue 1. January 2003, pp. 99–111.

[277]   STRAHLER, A.N., *The Earth Sciences*, 2nd Edition, Harper International Edition, Singapore, 1971.

[278]   STRASSER, A., WEIDMANN, HOCHULI, P.A., Sédimentation postglaciaire fluviatile et palustre près d'Avenches (Suisse): Implications climatiques, *Bull. Soc. Frib. Sc. Nat.*-Vol. 88, 1999, pp. 5–26.

[279]   TARDY, Y., *Le cycle de l'eau*, Masson, Paris, 1986.

[280]   TELFORD, W.M., GELDART, L.P., SHERIFF, R.E., *Applied Geophysics*, 2nd Edition, Cambridge University Press, New York, 1990.

[281]   TERASMAE, J., *Postglacial History of Canadian Muskeg. In Muskeg and the Norhern Environment in Canada*, University of Toronto Press, Toronto, 1997.

[282]   TERMIER, H., TERMIER, G., *Histoire de la Terre*, Presses Universitaires de France, Paris, 1992.

[283]   TERMIER, H., TERMIER, G., *L'évolution de la Lithosphère I: Pétrogenèse*, Masson & Cie, Paris, 1955.

[284]   TERZAGHI, K., PECK, R.B., MESRI, G., *Soil Mechanics in Engineering Practice*, John Wiley & Sons, New York, 1996.

[285]   TESTA, S.M., WINEGARDNER, D.L., *Restoration of Contaminated Aquifers. Petroleum Hydrocarbons and Organic Compounds*, Lewis Publishers, Florida, 2000.

[286]   THURNER, A., *Hydrogeologie*, Springer Verlag, Vienne, 1967.

[287]   TIMOSHENKO, S.P. ET GOODIER J.N. *Theory of Elasticity*. 2nd Ed. McGraw-Hill Book Co. Inc., NewYork, 1951.

[288]   TISSOT, B.P., WELTE, D.H., *Petroleum Formation and Occurrence*, Springer Verlag, Heidelberg, 1984.

[289]   TOTAL,*Extra-heavy oils and bitumen: Reserves for the future*. The know-how series, Exploration and Production, 2007.

[290]   TOUCHART. L., *Limnologie physique et dynamique, une géographie des lacs et des étangs*, Paris, L'Harmattan, 2002.

[291]   TRIPLET, J.P., ROCHE, G., *Météorologie générale*, Ecole nationale de la météorologie, Paris, 1971.

[292]   TUCKER, M.E., *Sedimentary Petrology. An Introduction*. Blackwell Scientific Publ., Oxford, 1987.

[293]   TUCKER, M.E., WRIGHT, P., *Carbonate Sedimentology*, Blackwell scientific Publ., Oxford, 1990.

[294]   TULLEN, P., *Méthodes d'analyse du fonctionnement hydrogéologique des versants instables*. Thèse de doctorat EPFL n°2622, août 2002, Lausanne.

[295]   TUREK, V., MAREK, J., BENES, J., *La grande encyclopédie des fossiles*, Gründ, Paris, 1993.

[296]   TURNER, J.S. et CAMPBELL, I.H., Convection and Mixing in Magma Chambers. *Earth-Science Reviews*, Volume 23, Issue 4, August 1986, pp. 255–352.

[297]   University of California La Jolla, TP Whorf Scripps, Mauna Loa Observatory, Hawaii, Institution of Oceanography (SIO), California, United States, 1999.

[298]   UNTERBERGER, W., HOFINGER, H., MARKIEWITZ, R. et DIETMAR, A., *Running hot and cold in Vienna*. Tunnel and tunnelling International, 2005.

[299]   VEDER, CH., *Landslides and their Stabilization*, Springer Verlag, New York, 1981.

[300]   VOGT, J., *Les tremblements de terre en France*, Mémoire du bureau de recherches géologiques et minières n°96, BRGM, Orléans Cedex, 1979.

[301]   VON MEISS, P., RADU, F., *Vingt mille lieux sous les terres. Espaces publics souterrains*, Presses polytechniques et universitaires romandes, Lausanne, 2004.

[302]   VULLIET, L., KOELBL, O., PARRIAUX, A. ET VÉDY, J.-C., 2003. *Gutachtenbericht über die Setzungen von St. German, im Auftrag der BLS Alptransit AG*. Rapport d'expertise non publié.

[303]   WALKER, C., WARD, D., *Les Fossiles*, Bordas, Paris, 2002.

[304]   WALTER, M.R., *Developments in Sedimentology: Stromatolites*, Elsevier, Amsterdam, 1976.

[305]   WALTHAM, A.C., *Foundations of Engineering Geology*, Chapman & Hall, London, 2003.

[306]   WASHBURN, A.L., *Periglacial Processes and Environments*. Edward Arnold (Publishers), London, 1973.

[307]   WASSON, J.T., *Meteorites. Classification and Properties*, Springer Verlag, Berlin, 1974.

[308]   WATER AND RIVERS COMMISSION, *Groundwater pollution*, Water facts 10, Australia, 1998.

[309]   WATSON, R.T., *Climate Change 2001: Synthesis Report*, Third Assessment Report of the Intergovernmental Panel on Climate Change, 2001.

[310]   WEBER, H.H., *Altlasten. Erkennen, Bewerten, Sanieren*, Springer Verlag, Berlin, 1993.

[311]   WEIDMANN, M., *Erdbeben in der Schweiz*, In Zusammenarbeit mit dem Schweizerischen Erdbebendienst, Verlag Desertina, Chur, Schweiz, 2002.

[312]   WHITNEY, A. and LAMOND, E., *Liming Acid Soils*, Kansas State University Agricultural Experiment Station and Cooperative Extension Service, Department of Agronomy, 1993.

[313]   WINDLEY, B.F., *The Evolving Continents*, John Wiley & Sons, London, 1995.

[314]   WINKLER, E.M., *Stone: Properties, Durability in Man's Environment*, Springer Verlag, Wien, 1994.

[315]   WINKLER, H.G.F., *Petrogenesis of Metamorphic Rocks*, 3rd Edition, Springer Verlag, New York, 1974.

[316]   YALIN, M.S., *Mechanics of Sediment Transport*, 2nd Edition, Pergamon Press, Oxford, 1977.

[317]   YARDLEY, B.W.D., *An Introduction to Metamorphic Petrology*, Longman Scientific and Technical, New York, 1989.

[318]   ZÁRUBA, Q., MENCL, V., *Landslides and their Control*, Elsevier, Amsterdam, 1969.

[319]   ZEHNDER, A. J. B., R. SCHERTENLEIB, et JAEGER, C. C., *Le défi de l'eau*. Gas Wasser Abwasser 2:3–8. [EAWAG-Nr. 02598] 1999, pp.131–136.

[320]   ZHIZHIN, M., BATTAGLIA, J., DUBOIS, J. et GVISHIANI, A., Syntactic Recognition of Magnetic Anomalies along the Mid Atlantic Ridge, *Comptes Rendus de l'Académie des Sciences* – Series IIA – Earth and Planetary Science, Volume 325, Issue 12, December 1997, pp. 983–990.

[321]   ZÖTL, J.G., *Karsthydrogeologie*, Springer Verlag, Wien, 1974.

Index